AF800743

75
Topics in Organometallic Chemistry

Series Editors

Matthias Beller, Leibniz-Institut für Katalyse e.V., Rostock, Germany

Pierre H. Dixneuf, Université de Rennes 1, Rennes CX, France

Jairton Dupont, UFRGS, Porto Alegre, Brazil

Alois Fürstner, Max-Planck-Institut fur Kohlenforschung, Mülheim, Germany

Frank Glorius, WWU Münster, Münster, Germany

Lukas J. Gooßen, Ruhr-Universität Bochum, Bochum, Germany

Steven P. Nolan, Ghent University, Ghent, Belgium

Jun Okuda, RWTH Aachen University, Aachen, Germany

Luis A. Oro, University of Zaragoza-CSIC, Zaragoza, Spain

Michael Willis, University of Oxford, Oxford, UK

Qi-Lin Zhou, Nankai University, Tianjin, China

Aims and Scope

The series *Topics in Organometallic Chemistry* presents critical overviews of research results in organometallic chemistry. As our understanding of organometallic structure, properties and mechanisms increases, new ways are opened for the design of organometallic compounds and reactions tailored to the needs of such diverse areas as organic synthesis, medical research, biology and materials science. Thus the scope of coverage includes a broad range of topics of pure and applied organometallic chemistry, where new breakthroughs are being achieved that are of significance to a larger scientific audience.

The individual volumes of *Topics in Organometallic Chemistry* are thematic. Review articles are generally invited by the volume editors. All chapters from Topics in Organometallic Chemistry are published Online First with an individual DOI. In references, *Topics in Organometallic Chemistry* is abbreviated as Top Organomet Chem and cited as a journal.

Luis M. Martínez-Prieto
Editor

Surface Functionalized Metal Catalysts

With contributions by

E. A. Baquero · R. Bofill · A. Bordet · B. Chaudret · M. D. Fernández-Martínez · L. Francàs · J. García-Antón · M. Gómez · C. Godard · J.-F Lahitte · P. Lara · Á. Lozano-Roche · G. Martí · J. Mazarío · K. Philippot · R. Pich · D. Pla · J.-C. Remigy · N. Romero · X. Sala · A. Suárez · O. Suárez-Riaño · V. Varela-Izquierdo · Y. Zhang

Springer

Editor
Luis M. Martínez-Prieto
IIQ
Instituto de Investigaciones Químicas
(CSIC-Universidad de Sevilla)
Seville, Spain

ISSN 1436-6002 ISSN 1616-8534 (electronic)
Topics in Organometallic Chemistry
ISBN 978-3-031-73840-1 ISBN 978-3-031-73841-8 (eBook)
https://doi.org/10.1007/978-3-031-73841-8

© The Editor(s) (if applicable) and The Author(s), under exclusive license to Springer Nature Switzerland AG 2024
This work is subject to copyright. All rights are solely and exclusively licensed by the Publisher, whether the whole or part of the material is concerned, specifically the rights of translation, reprinting, reuse of illustrations, recitation, broadcasting, reproduction on microfilms or in any other physical way, and transmission or information storage and retrieval, electronic adaptation, computer software, or by similar or dissimilar methodology now known or hereafter developed.
The use of general descriptive names, registered names, trademarks, service marks, etc. in this publication does not imply, even in the absence of a specific statement, that such names are exempt from the relevant protective laws and regulations and therefore free for general use.
The publisher, the authors and the editors are safe to assume that the advice and information in this book are believed to be true and accurate at the date of publication. Neither the publisher nor the authors or the editors give a warranty, expressed or implied, with respect to the material contained herein or for any errors or omissions that may have been made. The publisher remains neutral with regard to jurisdictional claims in published maps and institutional affiliations.

This Springer imprint is published by the registered company Springer Nature Switzerland AG
The registered company address is: Gewerbestrasse 11, 6330 Cham, Switzerland

If disposing of this product, please recycle the paper.

Preface

This Volume of *Topics in Organometallic Chemistry* presents a collection of review chapters centered on surface functionalized metal catalysts. Catalysis with surface functionalized metallic systems is receiving significant attention due to the ability to precisely control the reactivity of surface active sites.

Metal catalysts can be functionalized in different manners depending on the nature of the nanocatalyst and its potential application. For example, metal nanoparticles (MNPs) can be generated/immobilized on molecularly modified surfaces (MMSs) or functionalized polymeric membranes as revised in chapters "Metal Nanoparticles on Molecularly Modified Surfaces and their Application in Catalysis" and "Metal Nanoparticles on Polymeric Membranes Applied in Catalytic Hydrogenations", respectively. The main advantage of creating such hybrid materials is to benefit from their potential properties compared to classical supported MNPs (i.e. better control over the growth of MNPs, improved nanoparticle stabilization, incorporation of new active sites, etc.).

On the other hand, MNPs can also be functionalized using the ligands that act as stabilizers in colloidal systems. The use of stabilizing ligands is an effective way to control the activity and selectivity of MNPs. Just like in organometallic complexes, these surface ligands can influence the electronic and steric properties of MNPs, thereby changing their catalytic properties, as discussed in chapters "Influence of Capping Ligands on Metal Nanoparticle-Driven Hydrogen Evolution and CO_2 Reduction Reactions", "N-Heterocyclic Carbene Modified Metal Nanocatalysts", "Metal Nanoparticles Functionalized with Hydrosoluble Ligands" and "Organometallic Nanoparticles for Magnetically Induced Catalysis in Solution". In addition, the interaction between substrates and surface ligands can influence catalyst activity. The nature of the coordinated ligand can either enhance or inhibit the reactivity of a specific functional group. Therefore, surface ligands offer the potential to control catalyst properties in MNP catalysis, which has been an approach mainly used in homogeneous catalysis.

Ligand-stabilized MNPs can be later immobilized onto solid supports to enhance the stability and recyclability of the colloidal systems. Similarly, ligands can be used to functionalize supported MNPs, thus altering their surface properties, as described in chapter "Ligand-Capped Heterogeneous Catalysts from Groups 8–10". This innovative approach aims to enhance the selectivity of heterogeneous catalysts, which are crucial for industrial applications. Thus, controlling the synthesis, characterization, and application of surface-functionalized metal catalysts opens new frontiers in classical heterogeneous catalysis.

I would like to express my gratitude to all the authors for their contributions to this volume, which will hopefully generate new ideas and lead to future advances in this field. I also want to acknowledge the editorial team for their invaluable support.

Seville, Spain
September 2024

Luis M. Martínez-Prieto

Contents

Top Organomet Chem (2024) 75: 1–46
https://doi.org/10.1007/3418_2024_121
© The Author(s), under exclusive license to Springer Nature Switzerland AG 2024
Published online: 7 October 2024

Metal Nanoparticles on Molecularly Modified Surfaces and Their Application in Catalysis

Yuyan Zhang and Alexis Bordet

Contents

Abstract The development and understanding of catalytic systems composed of metal nanoparticles (MNPs) on molecularly modified surfaces (MMSs) have attracted tremendous attention in the field of materials chemistry in recent years, especially for applications in catalysis where it opened a conceptually new strategy for the design of multifunctional catalysts (MNPs@MMSs) with tailor-made reactivity. MNPs@MMSs are hybrid materials assembled through judicious associations of supports, molecular modifiers, and MNPs, using a combination of molecular ("homogeneous") and material ("heterogeneous") approaches. These complex yet fascinating materials have shown great potential in catalysis, where the synergistic interactions between MNPs@MMSs components provide important benefits including for example enhanced catalytic performance (e.g., activity, selectivity) and improved life-time. Thus, MNPs@MMSs stimulate interdisciplinary research

Y. Zhang and A. Bordet (✉)
Max Planck Institute for Chemical Energy Conversion, Mülheim an der Ruhr, Germany
e-mail: alexis.bordet@cec.mpg.de

activities involving organometallic chemistry, material science, synthetic organic chemistry, and catalysis. This chapter presents a non-exhaustive overview of the various methods available to prepare and characterize MNPs@MMSs materials, and provides selected examples of the application of MNPs@MMSs to a wide range of catalytic transformations including reduction, oxidation, and coupling reactions.

Keywords Catalysis · Characterization · Metal Nanoparticles · Molecularly Modified Surfaces · Preparation

1 Introduction

The continuous development of catalytic technologies is of ever-increasing importance to cope with the advent of alternative renewable energy sources and chemical feedstocks. Catalysts are essential to control the activation and formation of chemical bonds, which are crucial steps to convert renewable carbon sources (e.g., CO_2 and biogenic substrates and intermediates) into value-added chemicals. Thus, considerable efforts in the field of chemistry are devoted to the development of multifunctional catalytic systems possessing exciting new properties and opening new synthetic strategies. In this context, the combination of inorganic and organic components within one system is attracting tremendous attention to produce hybrid materials with unique properties [1–5]. Hybrid or composite materials made of inorganic and organic species have been prepared and used since ancient times (e.g., in paints) and became particularly attractive to scientists in the past century thanks to major advances in preparation and characterization methods. Nowadays, hybrid materials find widespread application including for example in the fields of biomedicine, optics, electronics, energy, and catalysis [3]. In catalysis, the combination of inorganic and organic components is typically inspired by nature, and intimately linked to the long-lasting goal of approaching and mimicking the complexity of active sites found for example in enzymes [3, 4].

In this context, molecularly modified surfaces (MMSs) belong to the family of hybrid materials and consist of two components – a support material and an organic molecular structure – assembled on the molecular scale [6–8]. A major drive to develop such hybrid materials is to benefit from synergistic effects arising from the combined properties of molecular and solid components assembled in a single system. The immobilization of metal nanoparticles (MNPs) on MMSs generates complex multicomponent MNPs@MMS systems that possess several attractive features, especially for applications in catalysis [8, 9]. Potential beneficial properties as compared to more classical supported MNP catalysts include for example improved MNP stabilization [10–13], control over MNP growth [13–15], tunable wettability [7, 16, 17], mass transport [7, 9], and the incorporation of reactive functional groups [8, 18–20]. The synthesis of MNPs@MMS systems is highly versatile, allowing a myriad of variations in the assembly of support materials, surface molecular modifiers, and MNPs, and acting as a toolbox for the modular design of multifunctional catalytic systems with new and enhanced catalytic properties [4, 8].

In this chapter, the main strategies and techniques to prepare and characterize MMS and MNPs@MMS will be described. Principal applications of MNPs@MMS materials in catalysis will be summarized, particularly focusing on reduction, oxidation, and coupling reactions.

2 Synthesis and Characterization of MNPs@MMSs

2.1 Molecularly Modified Surfaces (MMSs)

2.1.1 Support Materials and Molecular Modifiers

A plethora of combinations of support materials and molecular structures can be used to prepare MMSs. For example, commonly used solid supports include SiO_2 [21–23], metal oxides (e.g., Al_2O_3, TiO_2, ZrO_2, Fe_3O_4) [24–26], zeolites [27–29], aluminum phosphates [30], clays [10], carbons (e.g., activated carbon, graphite, graphene, graphene oxide, carbon nanotubes) [31–33], metal-organic frameworks (MOFs) [34, 35], cellulose acetate [36–38], and polymers [39]. The choice of support materials to prepare MMSs is driven by a combination of factors including for example its thermal, mechanical, and chemical stabilities, textural properties (e.g., surface area, pore size, pore volume), crystallinity, conductivity, and surface functionalities.

The second component of MMSs is the surface molecular modifier, which can be any molecular structure ranging from small organic molecules [40] to ionic liquids [7, 8, 41], polymers [42, 43], and organometallic complexes [44, 45]. Molecular modifiers can also contain specific functionalities (e.g., acidic/basic functional groups, hydrophobic/hydrophilic side chains, etc.), and open the way to the preparation of multifunctional materials.

Two main approaches for the surface modification of support materials with molecular modifiers can be identified according to the nature of the interactions existing between the material's surface and the molecular modifier.

2.1.2 Preparation of MMSs by Physisorption

The simplest approach is called physisorption and involves the non-specific adsorption of molecular modifiers on the support surface, resulting in non-covalent interactions between the two components, for example through dispersion forces, hydrogen bonding, or weak electrostatic interactions (Fig. 1).

The non-specificity of the physisorption approach enables the use of almost any type of support material and molecular modifier, and provides good control over the loading of molecular modifier on the surface, although multiple layers of molecular

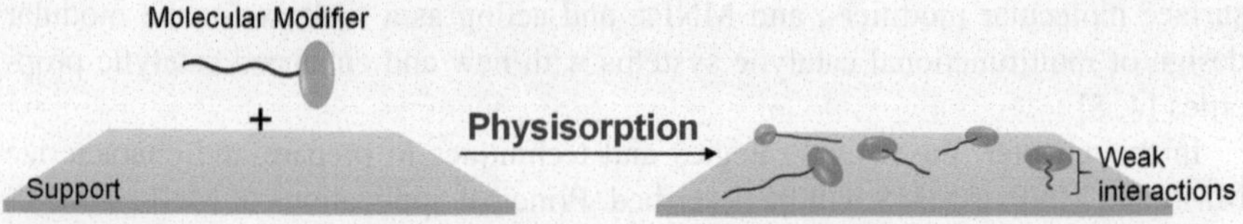

Fig. 1 MMSs prepared by physisorption, also sometimes referred to as "Class I" [4, 46] hybrid materials

modifier can be formed. This approach proved particularly suitable for the preparation of supported catalysts with ionic liquid layers (SCILLs) [7]. Physisorption is an exothermic process characterized by the release of heat and described thermodynamically by the equation $\Delta H = \Delta G + T\Delta S$. Physisorption results in a more ordered molecular state associated with a decreasing degree of freedom, and the enthalpy change (ΔS) is thus negative. Spontaneous adsorption is also associated with a negative change in Gibbs free energy ΔG. As a result, the enthalpy change ΔH is negative, the physisorption process is exothermic, and thus temperature-dependent. The preparation of MMSs by physisorption can be accomplished by employing various strategies, including in particular:

- *Impregnation* [47, 48]: MMSs can be prepared by impregnation through the addition of a solution of the molecular modifier to the support material, followed by solvent evaporation to obtain MMSs. If the amount of molecular modifier solution matches the total pore volume of the support, the approach is called "wetness incipient impregnation." Alternatively, "wet impregnation" implies the use of an excess of molecular modifier solution as compared to the total pore volume of the support.

- *Equilibrium adsorption* [49]: this approach consists in the adsorption of a molecular modifier from a solution to a support through weak interaction, e.g., electrostatic van der Waals interactions, until the establishment of an equilibrium between the amount of molecular modifiers on the support and in solution. In practice, the support material is stirred in a solution of the molecular modifier, and the resulting material is filtered and washed to eliminate any excess molecular modifier.

- *Ion exchange*: ions present in supports such as natural clays are exchanged with those of molecular modifiers (e.g., ionic liquids) in solution, and resulting materials are filtered and washed to remove any excess molecular modifier, similar to equilibrium adsorption.

As compared to simple impregnation, equilibrium adsorption and ion exchange methodologies typically provide stronger interactions between molecular modifiers and support materials, as well as a more homogeneous coverage of the support by the molecular modifiers.

More sophisticated approaches such as the vapor deposition of ionic liquids have also been proposed for the preparation of MMSs [50–52], although their application remains mainly limited to the investigation of model surfaces [51].

2.1.3 Preparation of MMSs by Chemisorption

While the preparation of MMSs via physisorption is simple and versatile, the surfaces created through this approach are prone to thermal, mechanical, and solvolysis instabilities, attributed to the weak nature of the interactions between the molecular modifier and the support surface [53, 54]. This raises concerns about molecular modifier leaching, either through dissolution or physical abrasion, which is a significant drawback for the long-term stability of MMSs.

The issue of leaching in MMSs can be effectively addressed through the covalent grafting of molecular modifiers to the surface of support materials via chemisorption. This irreversible process relies on the formation of chemical bonds between a functionalized molecular modifier and functionalities present at the surface of the support (*grafted to* method, Fig. 2). The surfaces of the support material may need to be pre-treated to generate the desired reactive sites for bonding, for example through oxidation, etching, or thermal treatment. The formed covalent bonds are inherently stronger than the weak interactions at play in MMSs prepared by physisorption, and thus lead to a more stable attachment of the modifier to the support material. Chemisorption also provides greater selectivity due to its reliance on specific chemical interactions between support materials and molecular modifiers, thus allowing the formation of monolayers.

Several chemisorption strategies exist depending on the functionalities present on the support materials and in the molecular modifier, and we will describe here briefly the most commonly used ones:

– *Silanization (Support-O-Si linkage)* [55]: involves the reaction of alkoxysilane-functionalized molecules with hydroxyl groups present at the surface of oxide supports (e.g., SiO_2, Al_2O_3, TiO_2, ZrO_2) to form one or more Support-O-Si bonds between the support surface and the molecular modifier (Fig. 3a). An example is given for imidazolium-based ionic liquids grafted on silica by silanization, yielding supported ionic liquid phases (SILPs, Fig. 3b) [8]. Carbon materials can become suitable for modification by silanization after undergoing an

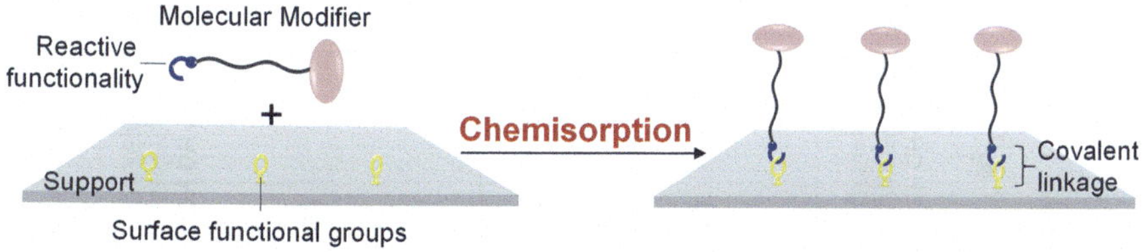

Fig. 2 Molecularly modified surfaces prepared by chemisorption, also sometimes referred to as “Class II” [4, 46] hybrid materials

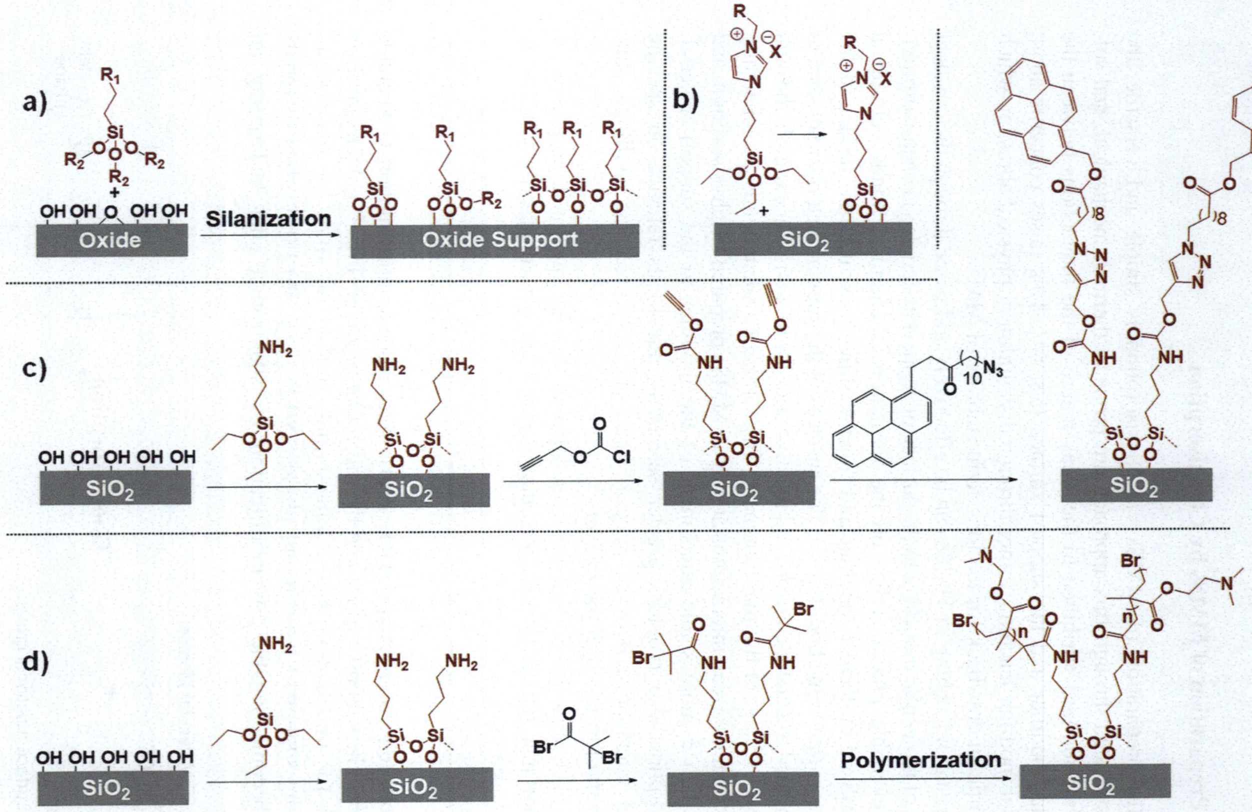

Fig. 3 Preparation of MMSs by silanization. (**a**) Illustration of the grafting method; (**b**) Example for the preparation of an imidazolium-based SILP [8]; (**c**) Silanization of 3-aminopropyltriethoxysilane on SiO_2 for further derivatization by click chemistry [58]; (**d**) Silanization of 3-aminopropyltriethoxysilane on SiO_2 for further derivatization by polymerization [59]

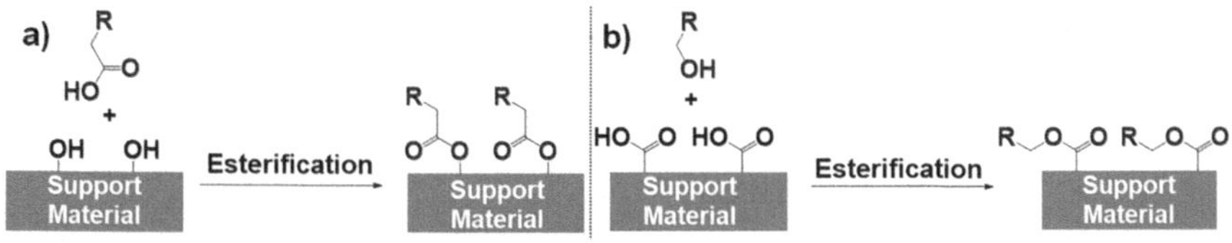

Fig. 4 Preparation of MMSs by esterification

oxidation pretreatment introducing the necessary reactive -OH groups [56, 57]. Silanization is a versatile approach that has been used to prepare a wide variety of MMSs. Interestingly, alkoxysilane-functionalized molecules can also be reacted with molecular tetralkoxysilanes following the molecular imprinting approach [6]. In this case, molecular modifiers are introduced directly during the synthesis of oxide supports and are thus incorporated in their bulk structure.

For molecular modifiers with complex structures, the introduction of alkoxysilane functionalities can be challenging or even impossible. In this case, a simple alkoxysilane-functionalized molecule can be grafted or incorporated first to the support surface, to act as an anchoring point for the introduction of molecular complexity, for example through click chemistry [58] or polymerization [59], as exemplified in Fig. 3c and Fig. 3d, respectively. This technique is referred to as *grafted from* or *ship-in-bottle* [60].

MMSs prepared by silanization are robust and typically used in organic solvents or in the gas phase, the sensitivity of Support-O-Si linkages to hydrolysis preventing their application in aqueous media.

- *Chlorination + reaction with a Grignard reagent (Si-O-C linkage)* [61]: involves the chlorination of SiO_2 support materials, and the subsequent reaction of Grignard reagents with surface Si-Cl groups to generate Si-C bonds possessing excellent stability. However, Grignard reagents are poorly tolerant toward functional groups, severely limiting the variety of surface functionalities that can be introduced following this approach.

- *Esterification (Support-O-CO or Support-CO-O linkage)* [26, 61]: involves the reaction of molecules containing carboxylic acid functionalities with hydroxyl groups at the surface of supports (e.g., oxides, treated carbons, etc.). Alternatively, -OH-functionalized molecules can react at the surface of support materials containing -COOH functionalities, such as oxidized carbon supports (Fig. 4). Note that ester linkages are prone to hydrolysis under basic and/or aqueous conditions.

- *Phosphonation (Support-O-P linkage)* [20, 26]: involves the reaction of phosphonic acid-functionalized molecules ($R\text{-}PO_3H_2$) and their ester derivatives ($RPO_3R'_2$, where R' represents an alkyl or an aryl group) at the surface of oxide supports rich in -OH functionalities, typically metal oxides. The structural flexibility of phosphonates with their three oxygen atoms enables various binding

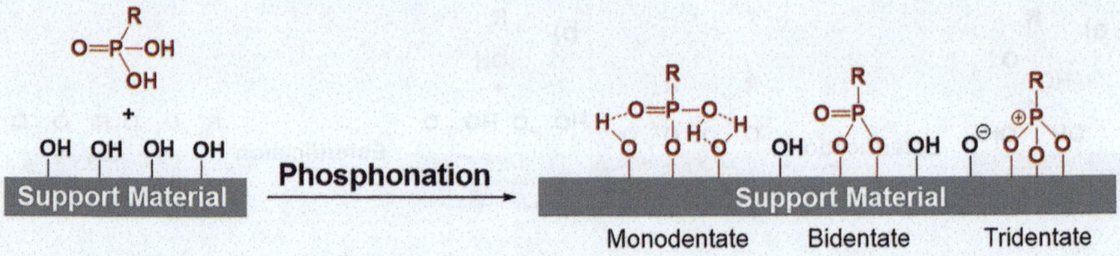

Fig. 5 Preparation of MMSs by phosphonation

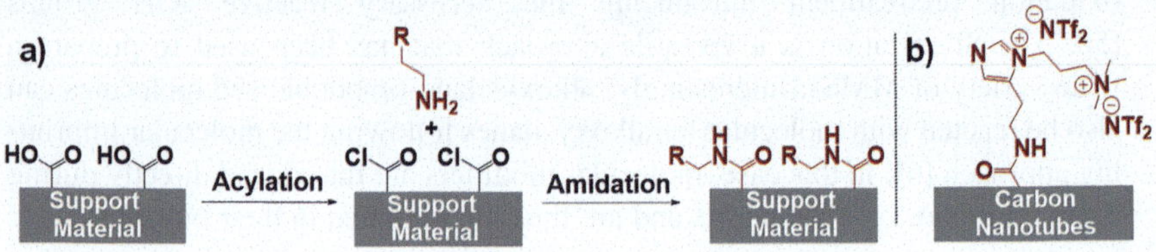

Fig. 6 Preparation of MMSs by amidation. (**a**) General synthesis pathway; (**b**) Example of CNTs molecularly modified by amidation [189]

modes including mono-, bi-, and tri-dentate interactions (Fig. 5) resulting in strong covalent Support-O-P bonds.[1]

– *Amidation (Support-CO-N linkage)* [62]: involves the reaction of amine-functionalized molecules with surface carboxylic acid groups of supports. This approach is widely used to functionalize the surface of oxidized carbon supports (e.g., activated carbon, graphite, graphene, graphene oxide, CNTs, etc.), and typically proceeds through the acylation of -COOH functionalities to produce more reactive acyl chlorides (-COCl), which are then reacted with amine-functionalized molecular modifiers (Fig. 6).

– *Diazotization (Support-C-C linkage)* [63, 64]: has proven to be an effective technique to functionalize carbon supports as well as conductive and semiconducting materials. The surface modification of supports with molecular modifiers containing a diazonium salt is believed to proceed through a free radical pathway involving (a) a single electron transfer from the support to the diazonium ion followed by (b) the concerted release of N_2 to form an aryl radical (c) reacting rapidly at the surface of the support (Fig. 7). Since diazonium salts have a high activity/fast decomposition rate, they are either prepared in situ or freshly synthesized and used. Diazotization can be achieved thermally [65] as well as electrochemically [63] and photochemically [66]. Besides diazotization, nitrenes, carbenes, and arynes can be used to form C-C bonds between molecular modifiers and carbon-based supports [67–69].

The formed C-C bonds between supports and molecular modifiers are very strong, and the resulting MMSs can be used in almost any reaction medium, except in the presence of strongly oxidizing agents/conditions.

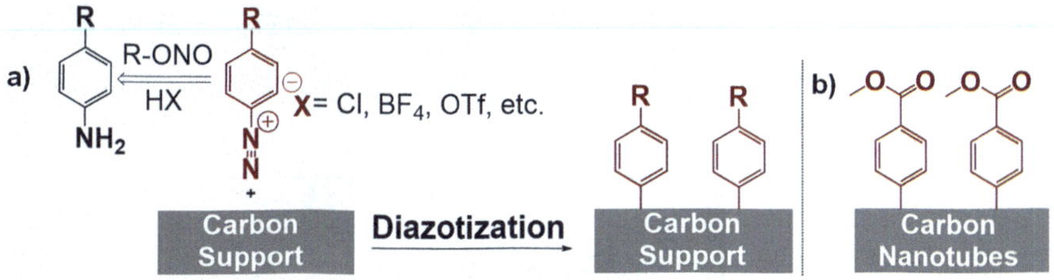

Fig. 7 Preparation of MMSs by diazotization. (**a**) General synthesis pathway; (**b**) Example of CNTs modified molecularly by diazotization [65]

2.1.4 Characterization of MMSs

MMSs can be characterized using various techniques to provide insight into the impact of physisorbed or chemisorbed molecular modifiers on the physicochemical properties of solid supports. Potential changes in textural properties caused by the functionalization can be investigated through physisorption experiments to determine surface area, pore size, and pore volume. After surface molecular modification, a reduction of surface area, pore size, and pore volume is typically observed due to the filling of the pores by the modifier. The loading of molecular modifier on the support as well as the thermal stability of resulting MMSs can be determined by thermogravimetric analysis (TGA). Scanning electron microscopy (SEM), transmission electron microscopy (TEM), and scanning TEM (STEM) give information on the morphology of the materials and reveal changes at the micro-, nano-, and potentially atomic scale. Furthermore, elemental mapping through SEM and STEM coupled with energy-dispersive X-ray spectroscopy (EDX) provide insights into the MMSs composition, particularly into the homogeneity of the molecular modifier layer. This is exemplified in Fig. 8 for an MMS composed of a Cu-NHC (NHC: N-heterocyclic carbene) molecular complex chemisorbed on a SiO_2 support material by silanization (SiO_2-[Cu]) [70].

Potential structural changes in crystalline supports (crystalline phases, crystallite size, etc.) can be monitored by powder X-ray diffraction (XRD) techniques [3, 8, 71]. Molecularly modified surfaces have been shown to typically retain the crystalline structure of the starting support materials, although a lower signal: noises ratio can be observed upon loading of the supports with amorphous species [71].

In the case of MMSs prepared by chemisorption, solid-state NMR can be used to investigate the bonding nature of molecular modifiers to the material surface. Taking the example of MMSs prepared by silanization, ^{29}Si NMR readily distinguishes the connectivity of siloxy groups based on whether they are connected through one, two, or three oxygen atoms [72]. This is illustrated on Fig. 9 for a SiO_2 support modified by silanization using decyltriethoxysilane (Si-Dec). Characterization of the Si-Dec materials using solid-state ^{29}Si CP-MAS (Cross-Polarization Magic-Angle-Spinning) NMR evidenced trifunctionalized signals (T_2: R-Si(OSi)$_2$(OR'); T_1: R-Si(OSi)(OR')$_2$), characteristic of Si atoms from the molecular modifier linked to the SiO_2 surface, thus confirming the covalent attachment of the molecular modifier to the

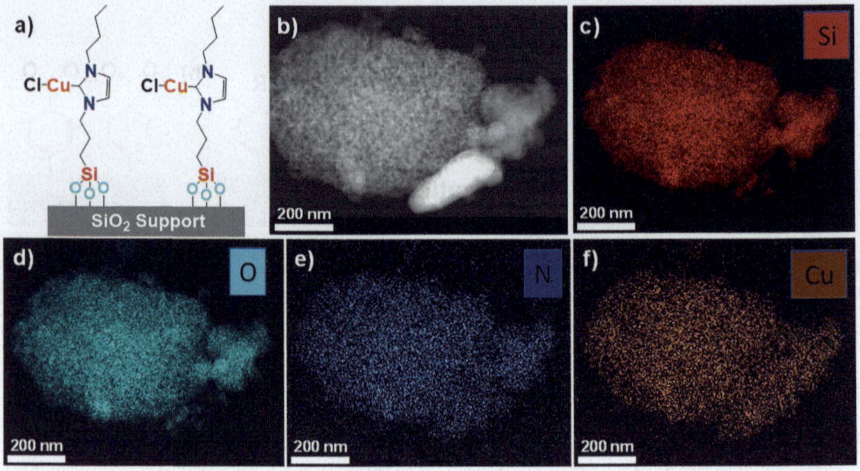

Fig. 8 Characterization of SiO_2-[Cu] by STEM-HAADF-EDX, showing the homogeneous coverage of the SiO_2 support materials by the Cu-NHC molecular modifier. Adapted with permission from Kalsi et al. [70]. Copyright 2022 American Chemical Society

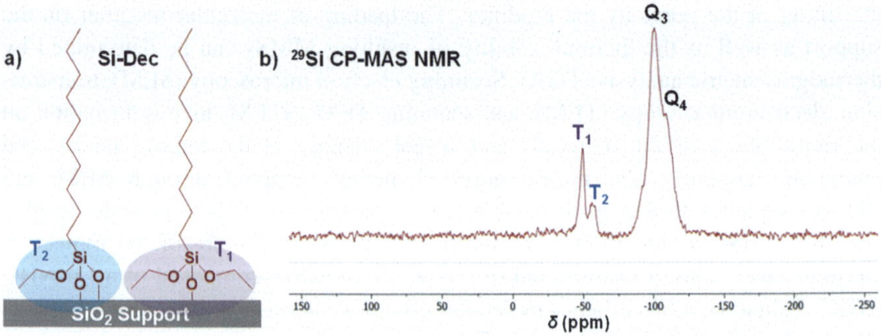

Fig. 9 Characterization of (**a**) Si-Dec by (**b**) solid-state ^{29}Si CP-MAS NMR, showing T_1 and T_2 signals characteristic of the covalent bonding of the molecular modifier at the SiO_2 surface, along with tetrafunctionalized (Q) signals ($Q_4 = Si(OSi)_4$, and $Q_3 = Si(OSi)_3OH$) characteristic of the SiO_2 support. Adapted with permission from Kacem et al. [17]. Copyright 2020 Royal Society of Chemistry

support [17]. Other NMR active nuclei such as naturally abundant ^{31}P and ^{19}F or labeled ^{15}N or ^{13}C in the molecular modifier may also be used for diagnostics [73–75].

Besides solid-state NMR, the presence of molecular modifiers on a support material and the preservation of their structure and functionality can be studied by infrared (IR) spectroscopy (e.g., transmission, ATR, DRIFTS, etc.) through examination of bands characteristic of the molecular modifiers. For example, the DRIFT-FTIR analysis of an imidazolium-based SILP showed C–H stretching from both the imidazolium ring and the aliphatic chain (3,160 and 2,947 cm^{-1}), as well as

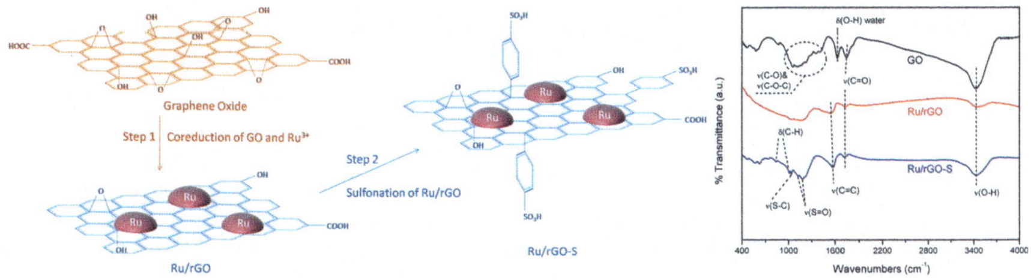

Fig. 10 Characterization of GO, Ru/rGO, and Ru/rGO-S by IR spectroscopy. Adapted with permission from Wang et al. [19]. Copyright 2017 American Chemical Society

symmetric ring stretching of the imidazolium moiety (1,566 and 1,452 cm^{-1}) [18]. As another example, IR spectra demonstrated the functionalization of a Ru/rGO-S with benzenesulfonic acid groups through observation of the expected $\nu_{S=O}$, $\nu_{S\text{-phenyl}}$, and $\delta_{phenyl\text{-}H}$ bands (1,790, 1,121, 1,035, 1,005, and 835 cm^{-1}, Fig. 10) [19].

X-ray photoelectron spectroscopy (XPS) has also been used to investigate the structure of molecular modifiers after chemisorption or physisorption [13, 76–78]. This is exemplified in Fig. 11 for the C 1s, N 1s, O 1s, Br 3s, Br 3p, and Br 3d XPS characterization of an ionic liquid-functionalized polymer material [78].

2.2 *MNPs@MMSs*

Metal nanoparticles (MNPs) possess a high surface-to-volume ratio, making them of great interest in catalysis as they offer increased amounts of active sites as compared to their bulk material counterparts. Size, shape, crystallinity, and composition are key parameters governing the reactivity of MNPs, and the precise control of these parameters allows the development of MNPs with tailor-made activity. MNPs are thermodynamically unstable due to their inherently high surface energy, and they tend to reduce this surface energy through aggregation and coalescence, especially under conditions relevant for catalysis. To prevent such undesirable processes that drastically decrease the efficiency of MNP catalysts, MNPs need to be stabilized, for example by ligands, or by immobilization on support materials. In this context, MMSs were shown to be excellent matrices for the immobilization and stabilization of MNPs that are intended for applications in catalysis [8, 9]. Besides improved MNP stability, MMSs offer various additional benefits for MNPs@MMS catalysis, including for example tunable wettability, enhanced mass transport, possibility to introduce additional functionalities, etc. Thus, the careful selection and association of MNPs@MMS components (support material, molecular modifier, MNPs) provide the opportunity to develop multifunctional catalytic systems with tailor-made reactivity. In this part, we will describe the principal approaches to immobilize MNPs on MMSs.

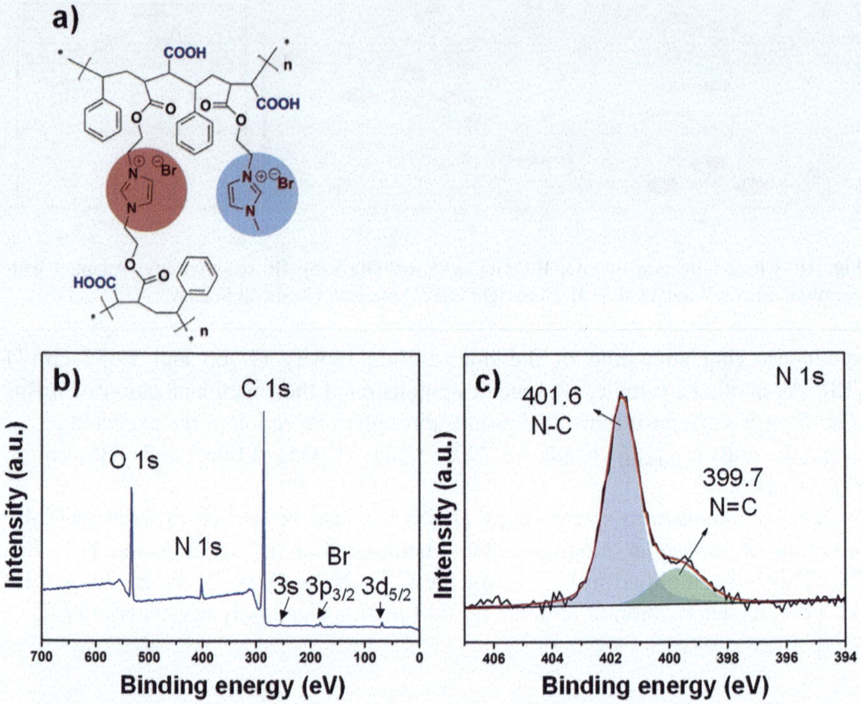

Fig. 11 XPS characterization of porous organic frameworks functionalized with carboxylic acids and imidazolium-based ionic liquid-like structures (CILs-POF). (**a**) Structure of the CILs-POF MMS; (**b**) Survey scan; and (**c**) N 1s XPS spectrum of CILs-POF. Adapted with permission from Yang et al. [78]. Copyright 2017 American Chemical Society

2.2.1 Immobilization of MNPs on MMSs: Ex Situ Approach

The ex situ approach to prepare MNPs@MMSs involves the addition of colloidal MNPs to MMSs, as illustrated in Fig. 12. This approach is widely used in the field as it benefits from the large available expertise on the synthesis of size- and shape-controlled MNPs in solution while excluding any influence of the MMSs.

In this case, MNPs are prepared in solution in the presence of ligands, following one of the many approaches reported for the synthesis of MNPs (organometallic, polyols, etc.) [79–82]. The choice of the ligands is important for subsequent application in catalysis, as there should be an appropriate balance between MNP stabilization and substrate accessibility. Preformed MNPs are immobilized on MMSs in a second step. However, when it comes to depositing colloidal metal MNPs on MMSs, the commonly used wet impregnation method might not be as effective as the presence of ligands on the MNPs and of molecular modifiers on MMSs may impact the deposition process, with unfavorable electrostatic interactions between MNPs and supports. To tackle this challenge, the pH of the impregnation solution can be adjusted to alter electrical charges on MMSs and MNPs. In certain cases, this

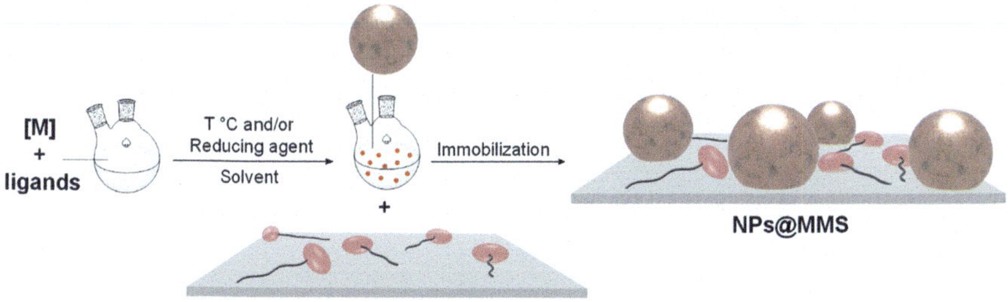

Fig. 12 Preparation of MNPs@MMSs by the ex situ approach, i.e., by immobilizing pre-formed MNPs on MMSs

adjustment leads to the MNPs acquiring a negative charge, while the support becomes positively charged. The resulting electrostatic attraction between these oppositely charged entities facilitates the easy deposition of the negatively charged metal MNPs on the positively charged support [79, 83–85]. In addition, the immobilization of MNPs on MMSs can be facilitated by coordinating interactions between the surface molecular modifiers and the MNPs. For example, Khatri and colleagues investigated the immobilization of Au NPs, which were stabilized by ionic liquids, on a silica support functionalized with 3-mercaptopropyl groups [84]. They found that ionic liquid molecules weakly coordinated to Au NPs could be partially displaced by thiol groups present on the functionalized silica support. This substitution facilitated the formation of strong Au–S interactions, leading to a robust immobilization of the Au NPs on the silica surface.

In the case of physisorbed MMSs, a simple strategy to immobilize MNPs consists in preparing MNPs directly in a solution of the bulk molecular modifier, achieving the direct impregnation of the support material with MNP-loaded molecular modifier solutions. This approach has been particularly used to produce SILPs, benefiting from the extensive knowledge on the synthesis of MNPs in ILs [86–89].

2.2.2 Immobilization of MNPs on MMSs: In Situ Approach

The in situ approach involves the decomposition (via reduction or thermal) of a metal precursor directly in the presence of MMSs to give MNPs@MMSs materials (Fig. 13). MNP formation can be accomplished in solution by mixing metal precursors with suspensions of MMSs, or in the solid state, after impregnation of metal precursors on MMSs and drying. In this case, MMS matrices are in direct interaction with metal precursors and can thus potentially influence metal ions concentration near the support surface, nucleation, and growth processes, and control the morphology of the resulting MNPs.

For example, the immobilization of Rh, Pt, and Ir MNPs on silica supports grafted with 3-aminopropyltriethoxysilane was investigated by several groups [90–92]. Interestingly, the loading of Rh(III), Pt(II), and Ir(III) salts after impregnation

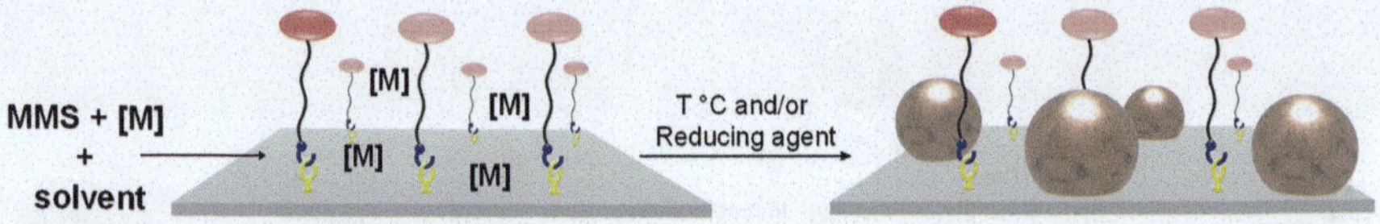

Fig. 13 Preparation of MNPs@MMSs by the in situ approach

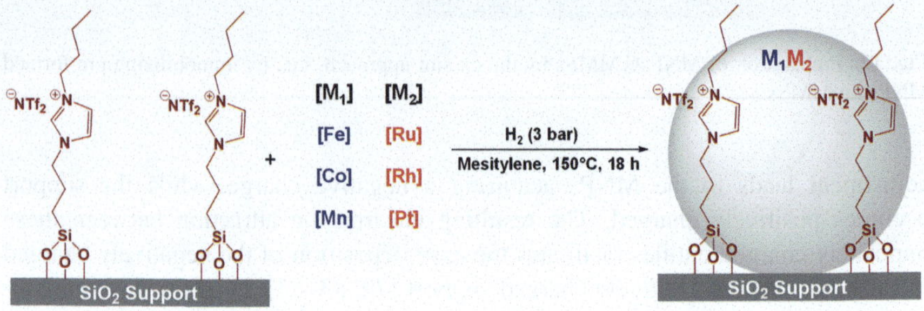

Fig. 14 Organometallic synthesis of metal MNPs on MMSs exemplified for the preparation of bimetallic MNPs on imidazolium-based SILPs

on MMSs was found systematically higher by at least one order of magnitude than on pristine silica. This better impregnation efficiency led to a better dispersion of the metals on the MMSs, and overall, a better utilization of the metals.

The selection of metal precursors displaying favorable interactions with MMSs is thus of critical importance to ensure the successful synthesis of size-controlled MNPs well-dispersed on the support. For decomposition of metal precursors by reduction, the choice of the reducing agent (e.g., sodium borohydride, hydrazine, ethylene glycol, molecular hydrogen, etc.) is also of importance as it impacts on the reaction rate, morphology, and surface chemistry of the resulting MNPs [93].

The organometallic synthesis of MNPs is a bottom-up chemical method providing excellent control over the size, the shape, and the physicochemical properties of MNPs [82, 94]. It relies on the thermal decomposition or hydrogenation with H_2 under relatively mild conditions of high-energy organometallic complexes possessing labile ligands. This approach is simple and versatile while avoiding contamination of MNPs surface with pollutants from precursors or reducing agents, and is thus attractive for the immobilization of MNPs on MMSs [8]. Interestingly, resulting MNPs@MMS materials do not require any activation steps prior to use in catalysis. For example, the group of Bordet and Leitner used carefully selected organometallic precursors (e.g., $\{Fe[N(Si(CH_3)_3)_2]_2\}_2$, [Co(cod)(cyclooctadienyl)], [Ru(cod)(cot)], $[Rh(allyl)_3]$, etc.) to prepare a wide variety of 1–5 nm monometallic (e.g., Fe, Co, Ru, Rh, Pt) [13, 17, 18, 95–103] and bimetallic (e.g., Fe_xRu_{100-x}, Co_xRu_{100-x}, Co_xRh_{100-x}, Fe_xPt_{100-x}) [104–111] MNPs on MMS supports possessing various types of modifiers such as small organic molecules [17, 101], ionic liquids [13, 95–98, 100, 102, 104–111], polymers [99, 103], and organometallic complexes [70] (Fig. 14). In the case of bimetallic systems, small and alloyed bimetallic MNPs were formed on an imidazolium-based SILP but not on unmodified SiO_2, presumably due to the reactivity of the 3d metal precursors used toward surface silanols. Along the same line, Sodreau et al. recently reported a new route to produce transition metal phosphide MNPs under mild conditions. In particular, metal chloride complexes were reacted with tris(trimethylsilyl)phosphine as a phosphorus source, with the release of chloromethylsilane as a byproduct acting as a driving force for the reaction. In this case, 1.6 nm ruthenium phosphide (RuP) MNPs were successfully synthesized on an imidazolium-based SILP but not on pristine silica, presumably because of side-reactions between tris(trimethylsilyl)phosphine and surface silanols [112].

Several groups have investigated the influence that MMSs can have on the formation of MNPs [13–15]. For example, Barsbay and colleagues prepared Pd NPs by radiolytic reduction of $Pd(OAc)_2$ on silica surfaces modified by polyvinylpyrrolidone (PVP) with different brush lengths [14]. The authors observed a decrease in Pd NP size from 333 nm to 31 nm when the PVP brush length increased from no PVP to a PVP shell of 50,425 $g{\cdot}mol^{-1}$. In addition, the PVP surface modification was found to stabilize the Pd NPs used for the hydrogenation 4-nitrophenol, and to prevent Pd leaching.

Along the same line, Bordet and co-workers reduced $[Rh(allyl)_3]$ under H_2 (50 bar, 100°C) to synthesize Rh NPs on a series of imidazolium-based SILPs prepared by silanization on SiO_2. Systematic variations of the IL-like structure (anion, N-alkyl chain length, spacer length) were found to impact the size of the MNPs in the 0.5–2 nm range, as well as their stability under hydrogenation conditions [13]. However, while the benefits of surface molecular modification on the control of MNP size and stability have been clearly observed and documented, efforts are still required to provide fundamental insight into these effects, and bring our fundamental understanding closer to what exists for the synthesis of MNPs in solution.

Sputtering techniques (*top-down physical method*) have also been reported for the deposition of MNPs on MMSs, giving high-purity MNPs under specific conditions (e.g., high vacuum and pure gases) [113]. The bombardment of a target material surface by ions generates a variety of collision events that lead to the ejection of atoms and clusters that can be redeposited as MNPs on solid as well as liquid surfaces. In this case, the nature of the MMSs was found to have little impact on the size and distribution of supported MNPs [114–116].

2.2.3 Characterization of MNPs on MNPs@MMSs

After immobilization of MNPs on MMSs, characterizing the obtained MNPs@MMSs, particularly the formed MNPs, is important. Information about the MNP size and dispersion on the support can be collected using electron microscopy techniques such as SEM, TEM, and STEM (e.g., Fig. 15b, c). In addition, STEM with energy dispersive X-ray spectroscopy (STEM-EDX) or with electron energy loss spectroscopy (STEM-EELS) are powerful techniques to determine MNP

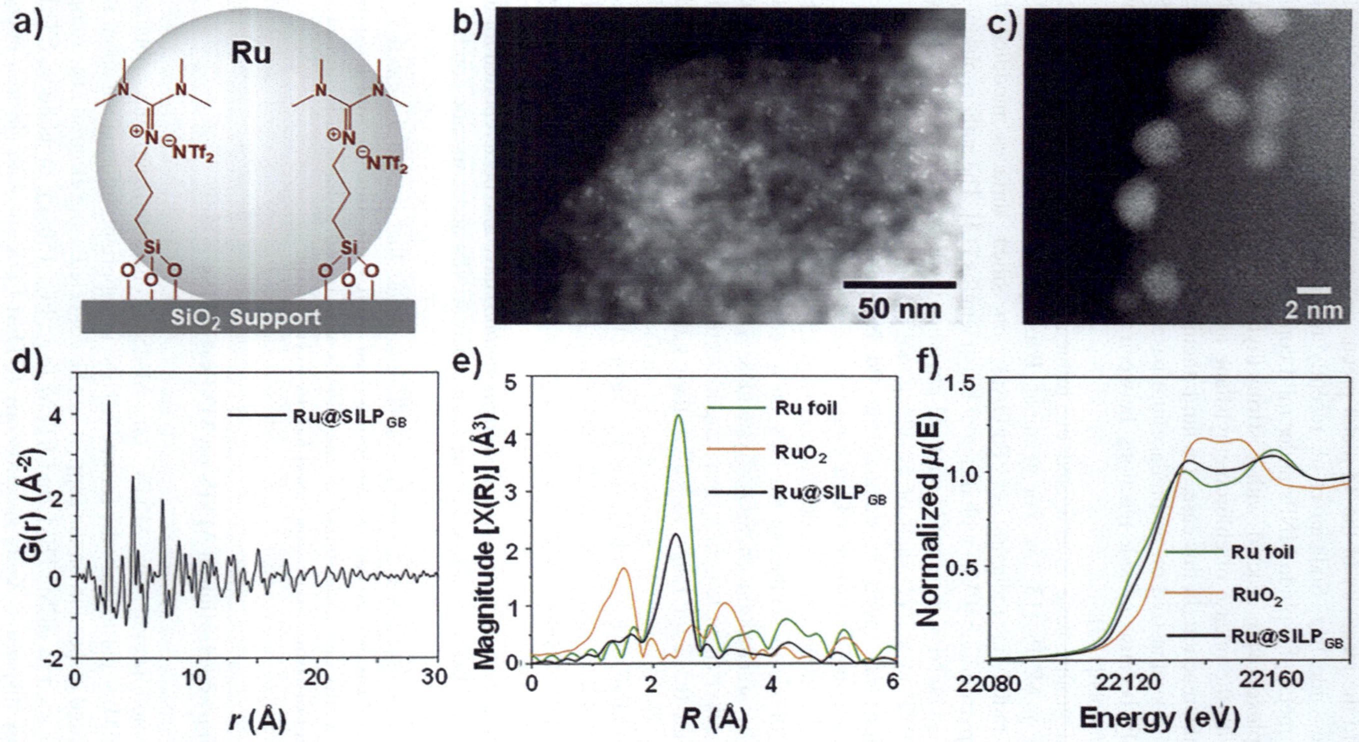

Fig. 15 Characterization of Ru MNPs immobilized on a guanidinium-based SILP (**a**) by (**b–c**) STEM-HAADF, (**d**) PDF analysis, (**e**) EXAFS, (**f**) XANES. Adapted with permission from Zhang et al. [121]

composition, to probe their interaction with molecular modifiers, and to study their local structure.

While electron microscopy techniques offer direct visualization of MNPs@MMSs, their scope is limited to specific local areas. To gain a broader understanding, X-ray spectroscopy techniques prove invaluable. For example, XPS and X-ray absorption near edge spectroscopy (XANES) can be used to study the electronic properties of MNPs@MMSs and determine the oxidation state of the metals (e.g., Fig. 15f). The surface sensitiveness of XPS can also be used to gather information on interactions between MNPs and molecular modifiers [13, 117, 118]. Extended X-ray absorption fine structure (EXAFS) spectroscopy provides insight into the coordination environment of metal atoms (e.g., Fig. 15e), as well as into alloying extent in the case of bimetallic MNPs. In addition, powder XRD is also commonly used to characterize the crystallinity of metal MNPs. In the case of MNPs with very small crystallite size (small or amorphous MNPs), wide-angle X-ray scattering and/or pair distribution function (PDF) analysis can be useful to gain quantitative insight into the structure of materials in which structural order is present only over a few nanometers or even less (e.g., Fig. 15d for the case of 2 nm Ru NPs). In addition, PDF analysis has the potential to reveal interactions between MNPs and surrounding molecules [119], which may be of great interest for the characterization of MNPs@MMSs systems. IR spectroscopy is a relatively simple but powerful technique to characterize selectively the surface physicochemical properties of MNPs and the nature of active sites by studying potential interactions between probe molecules (e.g., CO, acetonitrile, etc.) and metal surfaces [120].

Metal loading (and ratios when applicable) can be determined by inductively coupled plasma (ICP) techniques, as well as SEM-EDX provided that the analysis is performed on many large areas of the material. Potential changes to the textural, morphological, electronic, and structural properties of MNPs@MMS materials upon use in catalysis should be investigated using the same techniques. Particular attention should be paid to leaching (molecular modifier and metal) and MNPs aggregation and growth. Table 1 shows a summary of the techniques that can be used to characterize MNPs@MMS materials, as well as the corresponding information they provide.

3 Applications of MNPs@MMSs in Catalysis

MMSs and MNPs@MMSs materials find applications in various fields, for example in biotechnology as biosensors and bioreactors, in water processing, in the paint industry, as protective coatings in solar cells, optics, electronics, etc. [122]. In addition, the flexible assembly of MNPs@MMSs at the molecular scale provides opportunities to prepare materials with optimized and tailor-made physicochemical properties (e.g., chemical, thermal, and mechanical resistance, wettability, multifunctionality, etc.) that make them of attractive for applications in catalysis. This section will present selected examples of MNPs@MMSs used in catalysis for

Table 1 Summary of characterization techniques suitable for the study of MMSs and MNPs@MMSs materials

Information / Technique	Proof of chemisorption	Textural properties	Elemental composition	Morphological properties	MM characterization	Structural properties	Electronic properties
Solid-state NMR	**X (G)**				**X (G)**		
Physisorption		**X (G)**					
ICP, AAS			**X (G)**				
IR					**X (G)**		
RAMAN					**X (G)**		
TGA	(X) (G)				**X (G)**		
SEM				**X (I)**			
SEM-EDX			**X (I)**				
TEM				**X (L)**			
HRTEM				**X (L)**		**X (L)**	
STEM-EDX/-EELS			**X (L)**	**X (L)**	**X (L)**		**X (L)**
AFM				**X (L)**			
Powder XRD						**X (G)**	
WAXS						**X (G)**	
PDF						**X (G)**	**X (G)**
XPS			**X (G)**		**X (G)**		**X (G)**
XAS						**X (G)**	**X (G)**

Characterization techniques provide information on the global (bulk; **X (G)**), intermediate (ca. 500–0.1 μm; **X (I)**), and local (ca. 500–1 nm; **X (L)**) scale reactions including hydrogenation, hydrodeoxygenation, oxidation, and C-C coupling, as well as their potential for the development of adaptive catalytic systems and electrocatalysts.

3.1 Hydrogenation

Hydrogenation reactions are fundamental pillars of the chemical industry and used across the whole value chain ranging from the production of commodities to fine chemicals, agrochemicals, and pharmaceuticals [123]. While hydrogenation reactions with H_2 typically require metal-containing catalysts of various natures (e.g., molecular metal complexes or conventional heterogeneous catalysts), the modularity of MNPs@MMSs catalytic systems offers unique opportunities to control H_2 activation and transfer processes, and thus activity and selectivity [8]. Indeed, the activation of H_2 at the surface of MNPs can proceed via homolytic, polarized or heterolytic cleavage, giving $H^{\cdot}$–$H^{\cdot}$ (two hydrogen atoms with unpaired electrons), $H^{\delta+}$–$H^{\delta-}$ or H^+–H^- (a hydride and a proton), respectively [123, 124]. These different activation modes of H_2 depend on the catalyst's properties, and influence reaction pathways and selectivities in hydrogenation and hydrodeoxygenation reactions. Hereafter, we present selected examples of MNPs@MMSs catalytic systems used in this context. A summary can also be found in Table 2.

3.1.1 Hydrogenation of C=C and C ≡ C Bonds

The hydrogenation of C=C and C ≡ C bonds present in alkenes, alkynes and aromatics is favored by homolytic cleavage of H_2, and MNPs@MMS catalysts composed of monometallic MNPs and non-functionalized molecular modifiers are particularly suitable for that. In this case, H_2 is activated at the surface of MNPs, while the molecular modification of the support material stabilizes MNPs and prevents aggregation, controls wetting properties, improves mass transport, etc. While alkenes and alkynes can be hydrogenated using a wide variety of metal MNPs (i.e., Co, Ni, Cu, Ru, Rh, Pd or Pt), the hydrogenation of aromatic rings typically requires noble metal surfaces (i.e., Ru, Rh or Pd). For example, Rh NPs used in the hydrogenation of fluoroaromatics to fluorocyclohexanes were found significantly more selective when immobilized on SiO_2 surfaces modified with alkylsilanes (Fig. 16a) [17, 101]. In particular, apolar and hydrophobic environments generated by long alky chains modifiers favored hydrogenation over hydrodefluorination, providing excellent performance while at the same time improving catalyst stability and reusability as compared to state-of-the-art catalysts [125]. Li et al. encapsulated Pt NPs in metal-organic frameworks possessing various functionalities (UiO-66-X, with X = F, H, NH_2, OH), and applied the resulting catalysts to the hydrogenation of phenylacetylene to styrene and ethylbenzene (25°C, 1 bar H_2, Fig. 16b). High

Table 2 Summary of selected examples of MNPs@MMSs used as catalysts in hydrogenation reactions

Support material	Type of MM	MMSs preparation	MNPs	Hydrogenation of	Reference
SiO_2	Alkylsilanes	Chemisorption (silanization)	Rh	Fluoroaromatics	[17, 101]
Mol. sieves	Guad. IL	Physisorption	Pd	Olefins, dienes	[127]
Montmorillonite	Guad. IL	Physisorption	Ru	Arenes	[128]
SiO_2	Im. IL	Chemisorption (silanization)	Pd	Cyclohexadiene	[129]
Al_2O_3	Im. IL	Chemisorption (silanization)	Au	Cinnamaldehyde	[130]
Carbon cloths	Im. and Am. ILs	Physisorption	Pd	Citral	[131]
Carbon fibers	Im. IL	Physisorption	Pd	Acetylene	[132]
SiO_2, Al_2O_3	Im. IL	Chemisorption (silanization)	Rh	Furfuralacetone	[13]
Sepiolite	Guad. IL	Physisorption	Pd	Alkenes	[133]
MOF	Organic molecule	Chemisorption (imprinting)	Pt	Phenylacetylene	[126]
SiO_2	Organic amines	Chemisorption (silanization)	Pd	Phenol	[134]
SiO_2	Organometallic complexes	Chemisorption (silanization)	Ru	Propargylamines	[70]
SiO_2	Im. IL	Chemisorption (silanization)	FeRu CoRu CoRh	Aromatic ketones	[104] [109] [107]
Fe_3O_4	Cysteine Im. IL	Chemisorption (silanization)	CuRu	Nitrophenol	[136]
Graphene oxide	Organosilane	Chemisorption (silanization)	NiCu	Nitrophenol	[137]
Polyionic liquid	Aniline	Chemisorption (imprinting)	Pd	CO_2	[138]
SiO_2	Im. IL	Chemisorption (silanization)	Ru	CO_2	[102]
SiO_2	Im. IL	Physisorption	Ru	CO_2	[114]

Im. IL = imidazolium-based IL; Guad. IL = guanidinium-based IL; Am. IL = ammonium-based IL

hydrophobicity (X = F) proved beneficial to enhance catalytic activity by promoting alkyne solubility on the catalyst [126].

In a pioneer work on the combination of MNPs and SILPs, Han and colleagues reported the immobilization of Pd NPs on molecular sieves modified by physisorbed 1,1,3,3-tetramethylguanidinium-based ILs [127]. Resulting Pd@SILP catalysts with 1–2 nm Pd NPs were applied to the solvent-free hydrogenation of olefins and dienes (20–40°C, H_2), and were found more active and stable for C=C hydrogenation than colloidal Pd NPs with phenanthroline or imidazolium-based ILs ligands. Later, the

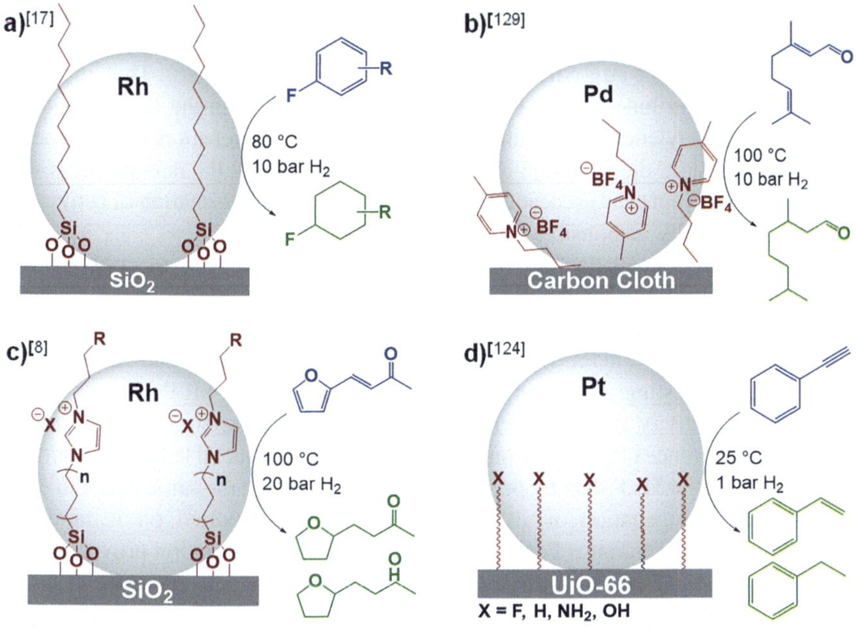

Fig. 16 Examples of MNPs@MMSs used in the hydrogenation of substrates containing C=C and C ≡ C bonds

same group showed that Ru NPs immobilized on a guanidinium-based SILP using montmorillonite as a support material possess enhanced catalytic activity for the hydrogenation of benzene to cyclohexane [128]. Dupont and co-workers reported the sputtering deposition of ca. 2 nm Pd NPs on imidazolium-based SILPs with different anions (prepared by silanization). Interestingly, Pd@SILP catalysts were found more active for the partial hydrogenation of 1,3-cyclohexadiene in the presence of hydrophobic anions [PF_6, NTf_2; TOF (turnover frequency) of 3 s^{-1}] than with more hydrophilic anions (Cl, NO_3; TOF of 1 s^{-1}) [129]. Au NPs immobilized on Al_2O_3-based SILPs were studied by the same group for trans-cinnamaldehyde hydrogenation, particularly focusing on potential confinement effects on catalytic performance. The activation of H_2 was found to be impacted by the dynamic properties of the IL molecular modifiers around the Au NPs, as evidenced by kinetic studies including labeling experiments [130].

Mikkola et al. have used Kynol® active carbon cloths as support materials for the preparation by physisorption of imidazolium- and ammonium-based SILPs [131]. Immobilization of Pd NPs on these SILPs produced Pd@SILP catalytic systems with excellent activity and tunable selectivity for the hydrogenation of citral to dihydrocitronellal (100°C, 10 bar H_2, Fig. 16c), with superior performances observed using the pyridinium-based SILP. Kiwi-Minsker and coworkers studied the hydrogenation of acetylene under continuous flow conditions (150°C, H_2) using Pd NPs immobilized on imidazolium-based SILPs [132]. While the presence of the

IL-like layer did not change the activity of Pd NPs as compared to reference catalysts, the selectivity toward the partially hydrogenated product (ethylene) was found much better for Pd@SILP catalysts (85% over 8 h on stream). This improved selectivity was attributed to the lower solubility of ethylene in the IL-like layer as compared to acetylene, suppressing the consecutive hydrogenation of ethylene to ethane. In addition, the IL cation-anion network surrounding the nanoparticles was suggested to prevent catalyst deactivation by suppressing the formation of active-site ensembles typically responsible for acetylene oligomerization.

More recently, small Rh NPs (0.5–2 nm) prepared on a series of imidazolium-based SILPs with systematically varied IL structure (i.e., anion, N-alkyl chain length, spacer length) and support material (SiO_2, Al_2O_3) were found active in the hydrogenation of biomass-derived furfuralacetone at 100°C and 20 bar H_2 (Fig. 16d). Interestingly, 1.2 nm Rh NPs immobilized on SILP showed better performance (activity and stability) than Rh NPs of the same size immobilized on SiO_2. This difference originated from a better prevention of metal leaching and MNP aggregation by the SILP than by pristine SiO_2 [8, 13]. Along the same line, Liu et al. studied the hydrogenation of alkenes using Pd NPs supported on IL-modified sepiolite as well as reference Pd/C catalysts. The IL functionalization proved important to obtain small and well-dispersed Pd NPs on the support, resulting in better activity and stability of Pd@SILP as compared to reference Pd/C catalysts [133]. Vinod and co-workers reported the synthesis of Pd NPs on periodic mesoporous silica particles possessing secondary amine functionalities (NPMO) and applied the resulting Pd@NPMO catalyst to the aqueous phase hydrogenation of phenol (40°C, 1 bar H_2). Pd NPs were found better stabilized by NPMO than by non-modified support materials (e.g., PMO, TiO_2, etc.), and produced cyclohexanone much more effectively (99% yield). These enhanced catalytic performances were attributed to interactions (e.g., hydrogen bonding) between the phenol molecule and the amine-functionalized silica surface [134].

Kalsi and colleagues recently developed a new type of MNPs@MMS catalytic system combining MNPs and metal complexes on a single support. In particular, Ru NPs were immobilized on a silica support on which Cu-NHC complexes were chemisorbed by silanization [70]. The resulting multifunctional material denoted Ru@SiO_2-[Cu-NHC] catalyzed the one-pot synthesis of various allylamines and alkylamines. The reaction sequence involved the A^3 coupling of amines, formaldehyde, and terminal alkynes (catalyzed by the heterogenized Cu(I)-NHC complex), followed by the selective hydrogenation of the resulting propargyl amines (catalyzed by metallic Ru NPs).

The highlighted selected examples evidence the impact that surface molecular modifiers can have on the formation and stabilization of MNPs. Although often not participating directly in catalysis, these molecular functionalities control the physicochemical properties of the local environment around MNPs (e.g., polarity, stereoelectronic interactions, confinement effects, etc.) and thus play a role in the observed reactivity.

3.1.2 Hydrogenation of Polar Functionalities (e.g., C=O, NO_2, etc.)

The hydrogenation of polar functionalities such as C=O bonds in aldehydes and ketones is typically favored on catalysts offering a polarized activation of H_2, and bimetallic MNPs@MMSs are of great interest in this context. In addition, bimetallic MNPs can possess electronic and structural properties favoring interactions with polar functional groups, while preventing for example hydrogen transfer to arenes. In this context, the preparation of a series of alloy-type bimetallic MNPs composed of noble and 3d metals in various compositions (e.g., Fe_xRu_{100-x}, Co_xRu_{100-x}, Co_xRh_{100-x}, etc.) on imidazolium SiO_2-based SILPs was reported following an organometallic approach [104–111]. To synthesize such MNPs, the selection of organometallic precursors presenting comparable stabilities was found important to prevent the formation of monometallic MNPs and/or core–shell structures. Interestingly the formation and stabilization of small and uniformly dispersed bimetallic MNPs (2–3.5 nm) was achieved on SILP but not on pristine SiO_2 under identical conditions, underlining the importance of the chemisorbed IL-like layer. While monometallic noble metal catalysts gave full hydrogenation of aromatic ketones, bimetallic $Fe_{25}Ru_{75}$, $Co_{70}Ru_{30}$, and $Co_{75}Rh_{25}$ catalysts were found to be highly active and selective for the partial hydrogenation of aromatic ketones to aromatic alcohols (150°C, 50 bar H_2, Fig. 17a) [104, 107, 109]. Interestingly, C=O hydrogenation rates were higher on bimetallic than on monometallic catalysts, while the suppression of aromatic ring hydrogenation was attributed to the incorporation of 3d metal atoms at the MNPs surface disrupting the required ensembles of 3–4 Ru or Rh atoms. Along the same line, Zhao et al. reported that the Pd/Ag ratio in supported Pd_xAg_{100-x} MNP catalysts dictates the selectivity in furfural hydrogenation (25°C, 1 bar H_2) [135]. In particular, partial hydrogenation to furfuryl alcohol was observed at high Ag content (>50%), and full hydrogenation at low Ag content (<20%). The results could be rationalized through DFT calculations, which showed an inversely proportional dependency of the adsorption-free energy of the intermediate furfuryl alcohol on the Pd(111) surface with the Ag content in Pd_xAg_{100-x} NPs.

Gholinejad et al. immobilized CuRu NPs on an MMS composed of Fe_3O_4 NPs as support materials and an L-cysteine-functionalized IL-like molecular modifier. The resulting catalyst was applied to the full hydrogenation of nitrophenol isomers in

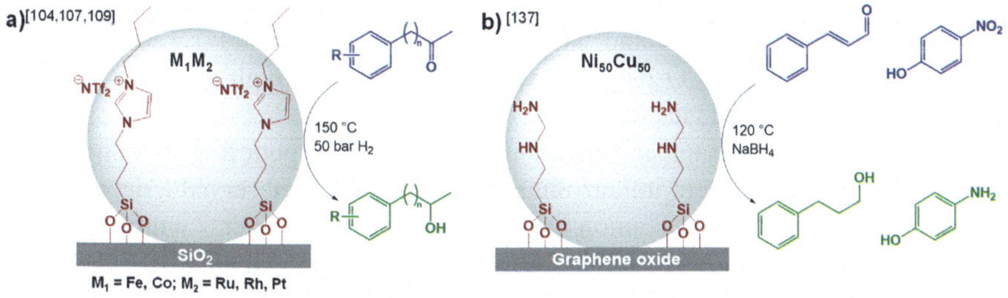

Fig. 17 Examples of bimetallic MNPs@MMSs used in the selective hydrogenation of substrates containing C=O and NO_2 groups

water (25°C, $NaBH_4$), showing superior activity (2–33 times) as compared to reference monometallic Ru and Cu catalysts [136]. Rana and Jonnalagadda reported the hydrogenation of nitrophenol and cinnamaldehyde (120°C, $NaBH_4$) using bimetallic Ni_xCu_{100-x} NPs supported on organo-functionalized graphene oxide prepared by silanization [137]. $Ni_{50}Cu_{50}$ NPs provided selective hydrogenation of nitro and aldehyde functionalities in these substrates while preserving aromaticity, giving aminophenol and cinnamyl alcohol as products, respectively (Fig. 17b). The surface molecular modification of the graphene oxide support material with organic amine groups proved important to stabilize the MNPs and increase their binding capacities toward the substrates.

These examples showcase the potential of bimetallic MNPs@MMS catalysts to control activity and selectivity in hydrogenation reactions. Particularly noteworthy is the synergistic combination of 3d metals and noble metal to boost the hydrogenation of polarized functionalities (e.g., C=O bonds) while suppressing that of aromatic rings.

3.1.3 Hydrogenation of CO_2

MNPs@MMS catalysts have been shown to be productive catalysts for the hydrogenation of CO_2 in the liquid phase, particularly in producing formic acid and formates. For example, Doherty et al. used aniline-functionalized polymerized ionic liquids as support materials for 2–3 nm Pd NPs and applied the resulting catalytic system for the hydrogenation of CO_2 in 0.65 M aqueous K_2CO_3 (100°C, 40 bar, H_2/CO_2 1/1) [138]. Under optimized conditions, TOFs for the formation of potassium formate reached 500 h^{-1}, noticeably higher than for reference Pd-based MNP catalysts. XPS analysis suggested strong interactions between Pd NPs and amine functionalities, resulting in electron-deficient Pd species and electron-rich amines. Louis Anandaraj et al. immobilized 1–3 nm Ru NPs on imidazolium-based SILPs with various molecular structures. Resulting Ru@SILP catalysts proved much more active than a reference Ru@SiO_2 catalyst with Ru NPs of similar size for the hydrogenation of CO_2 in aqueous amine solutions (H_2O/NEt_3) [102]. Sulfonic acid functionalized Ru@SILP(SO_3H) catalysts were found to be particularly promising for this transformation, presumably due to the adjustment of the reaction equilibrium by the presence of ammonium sulfonate species on the SILP support. Under optimized conditions, TONs and TOFs as high as 16,100 and 1,429 h^{-1} were obtained (120°C, 90 bar CO_2/H_2 1/2), which is among the highest so far reported for supported nanoparticle catalysts.

Very recently, the group of Dupont used magnetron sputtering to immobilize 1–2 nm Ru NPs on different imidazolium-based SILPs and observed a dependency of the Ru NP size on the SILP anion (e.g., 1 nm MNPs or "clusters" with BF_4 anion and 1.8 nm MNPs with NTf_2 anion). Applied to the hydrogenation of CO_2 in the ionic liquid 1-butyl-3-methylimidazolium acetate (87°C, 32 bar CO_2/H_2 1/2), Ru@SILP catalysts demonstrated excellent activity, producing free formic acid with TONs as high as 7305 [114]. The authors attributed this remarkable performance to hydrophobic/hydrophilic effects present in the IL-like layers of the SILPs that control formic acid production like catalytic membranes. Mechanistic studies pointed toward a mechanism going through carbonate hydrogenation and not direct CO_2 hydrogenation.

3.2 Hydrodeoxygenation and Hydrogenolysis

Hydrodeoxygenation and hydrogenolysis reactions are typically favored by the presence of acid sites on catalysts, and MNPs@MMS catalytic systems offer various opportunities to incorporate such functions and fine-tune their strength. For example, H^+ can be generated through heterolytic H_2 cleavage, or by the incorporation of Brønsted acid functionalities in molecular modifiers. Certain molecular modifiers or MNPs can also provide Lewis acidity [97]. Catalytic systems containing MNPs and additional functionalities that can take part in catalytic reactions are known as multifunctional catalysts and are attracting tremendous attention to produce various valuable chemicals (e.g., fuels, fine chemicals, or pharmaceuticals), particularly from oxygen-containing renewable carbon feedstocks [139, 140]. In the following, we present selected examples of the use of MNPs@MMS catalytic systems in this context. A summary of MNPs@MMSs can also be found in Table 3.

Luska et al. prepared sulfonic acid-functionalized imidazolium-based SILPs with finely tunable acidity and used them as matrices for the immobilization of 1.6 nm Ru NPs [18]. The resulting bifunctional Ru@SILP-SO_3H catalysts were applied to the hydrodeoxygenation of various substrates (e.g., 4-(2-tetrahydrofuryl)-2-butanol [18], phenol derivatives [95], eucalyptol [141], diaryl ethers [96]) under H_2, evidencing synergistic effects originating from the close proximity between metal and acid sites (Fig. 18a). In these cases, Ru NPs catalyzed hydrogenation steps, while sulfonic acid sites promoted the cleavage of C-O bonds. The ratio between Ru and sulfonic acid functionalities was found important to control activity and selectivity, especially for the conversion of 4-(2-tetrahydrofuryl)-2-butanol to butyltetrahydrofuran (high Ru/acid ratio) or 1-octanol and 1,1-dioctylether (low Ru/acid ratio) [18]. In a comparable approach, Zhu and co-workers reported the immobilization of 2.5 nm Ru NPs on a silica support functionalized with sulfonic acid groups, and used the bifunctional catalytic system for the hydrogenolysis of cellulose, producing sorbitol in good yield (61% at 150°C, 40 bar H_2, 10 h) [142]. In contrast, a Ru@SiO_2 catalyst without acid functionalities gave low conversion (36%) and no sorbitol was formed, evidencing the importance of having both metal and acid sites on the catalyst.

Coan and colleagues explored the potential in hydrodeoxygenation of Pt NPs immobilized on various phosphonic acid-modified support materials (e.g., Al_2O_3, TiO_2, CeO_2, and SiO_2-Al_2O_3 modified by phosphonation). The presence of phosphonic acid functionalities was found beneficial for the hydrodeoxygenation of benzyl alcohol to toluene at 180°C using Pt@Al_2O_3, Pt@TiO_2, and Pt@CeO_2 as it provided new and/or stronger Brønsted acid sites that improved deoxygenation rates (Fig. 18b) [20]. Interestingly, changing the tail functional group on the

Table 3 Summary of selected examples of MNPs@MMSs used as catalysts in hydrodeoxygenation and hydrogenolysis reactions

Support material	Type of MM	MM functionality	MMSs preparation	MNPs	Hydrodeoxygenation of	Reference
SiO_2	Im. IL	$-SO_3H$	Chemisorption (silanization)	Ru	Tetahydrofurylbutanol	[18]
SiO_2	Im. IL	$-SO_3H$	Chemisorption (silanization)	Ru	Phenols	[95]
SiO_2	Im. IL	$-SO_3H$	Chemisorption (silanization)	Ru	Eucalyptol	[141]
SiO_2	Im. IL	$-SO_3H$	Chemisorption (silanization)	Ru	Diaryethers	[96]
SiO_2	Organosilanes	$-SO_3H$	Chemisorption (silanization)	Ru	Cellulose	[142]
Al_2O_3, TiO_2, CeO_2, SiO_2	Organic molecule	$-PO_3H_2$	Chemosorption (phosphonation)	Pt	Benzylalcohol	[20]
MOF@SGO	Organic molecule	$-SO_3H$	Chemosorption (diazotization)	Pd	Monosaccharides	[143]
SiO_2	Im. IL	$-SO_3H$	Chemosorption+Physisorption	FeRu	Aromatic ketones	[105]
SiO_2	Im. IL	–	Chemisorption (silanization)	FeRu	Hydroxyacetophenones	[106, 108]
SiO_2	Im. IL	$-NR_2$	Silanization +Physisorption	FeRu	Aromatic carboxylic acids	[110]
SiO_2	Phos. IL	–	Chemisorption (silanization)	Rh	Aromatic ketones	[98]
SiO_2	Organosilanes	–	Chemisorption (silanization)	Ni_2P	Biphenols	[147]

Im. IL = imidazolium-based IL; Phos. IL = phosphonium-based IL

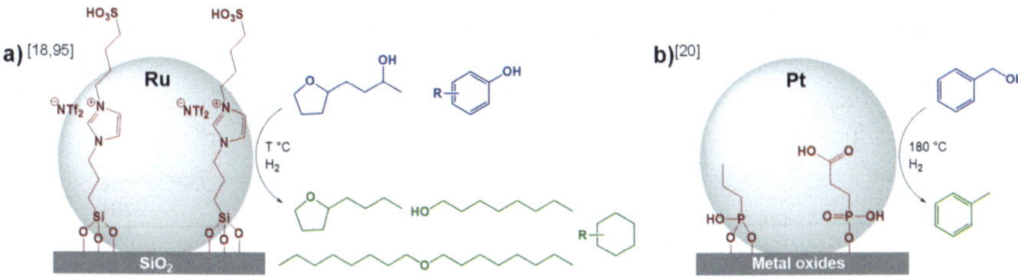

Fig. 18 Examples of bifunctional MNPs@MMSs used in hydrodeoxygenation reactions

phosphonic acid modifier (alkyl chain or carboxylic acid) allowed modulating the strength of the Brønsted acid sites, thereby affecting HDO rates. Kim et al. applied 3.5 nm Pd NPs immobilized on a Zr-based metal-organic framework (UiO-66) itself supported on sulfonated graphene oxide (SGO) as a bifunctional catalyst (Pd/UiO-66@SGO) to produce 2,5-dimethylfuran from monosaccharides [143]. Fructose was deoxygenated to 5-hydroxymethylfurfural by the surface acid sites (e.g., Brønsted acidic –COOH, –OH, and – SO_3H functionalities) while the Pd NPs were responsible for the hydrogenolysis and hydrogenation of 5-hydroxymethylfurfural, giving overall excellent 2,5-dimethylfuran yields (up to 70.5% at 160°C, 10 bar H_2).

With bimetallic MNPs@MMSs and bifunctional MNPs@MMSs found particularly interesting catalysts, the combination of their key features constituted a natural development toward better selectivity control and new reactivity. The potential of this approach was highlighted by Offner-Marko and co-workers, who developed a bimetallic bifunctional catalyst composed of 3 nm $Fe_{25}Ru_{75}$ NPs immobilized on a sulfonic acid-functionalized SILP [105]. In this case, $Fe_{25}Ru_{75}$ NPs were prepared on a neutral imidazolium-based SILP prepared by chemisorption on SiO_2, and further modified by physisorption of a sulfonic acid-functionalized imidazolium-based IL. This post-modification approach is inspired from heterogeneous SCILL catalysts [7], and was found necessary to circumvent side reactions between the Fe precursor and the sulfonic acid functionalities. The combination of chemisorbed and physisorbed IL-like modifiers proved to be robust, providing close contact between the active components while preventing leaching. The resulting $Fe_{25}Ru_{75}$@SILP +IL-SO_3H material was used in the catalytic hydrodeoxygenation of wide variety of aromatic (benzylic and non-benzylic) ketones (175°C, 50 bar H_2), showing excellent activity, selectivity, and stability for the production of alkyl aromatic derivatives (Fig. 19a) typically accessed through classical Clemmensen and Wolff-Kischner reductions. The intimate contact between $Fe_{25}Ru_{75}$ NPs and -SO_3H functionalities generated strong synergistic effects, with a suppression of aromatic hydrogenation activity associated with an enhanced hydrodeoxygenation of C=O bonds. While the selective hydrodeoxygenation of hydroxy- and aminoacetophenone derivatives is of great interest to access alkyl phenols and anilines building blocks [144–146] entering in the synthesis of fine chemicals, polymers, and pharmaceuticals, the $Fe_{25}Ru_{75}$@SILP+IL-SO_3H catalyst proved unsuitable due to various acid-catalyzed

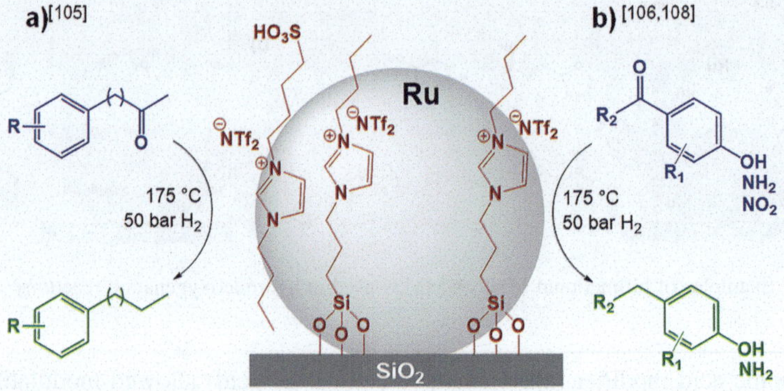

Fig. 19 Examples of bimetallic $Fe_{25}Ru_{75}$@SILP catalyst used in hydrodeoxygenation reactions, (**a**) with [105] and (**b**) without [106, 108] additional sulfonic acid functionalities

side reactions. Interestingly, the corresponding bimetallic acid-free $Fe_{25}Ru_{75}$@SILP catalyst showed excellent activity and selectivity for such transformations, giving a wide range of alkyl phenols and anilines in excellent yields under batch and continuous flow conditions (84–99%, 175°C 50 bar H_2)(Fig. 19b) [106, 108]. The activation and the stabilization of the intermediate cation by mesomeric effects were found beneficial to the hydrodeoxygenation activity, presumably along the Lewis acidity of $Fe_{25}Ru_{75}$ NPs.

Recently, Levin et al. reported another modification of the $Fe_{25}Ru_{75}$@SILP catalyst, this time through the incorporation of tertiary amine functionalities following the post-modification strategy described by Offner-Marko et al. The resulting bimetallic bifunctional $Fe_{25}Ru_{75}$@SILP+IL-NEt_2 catalyst showed excellent performance in the decarboxylation of aromatic carboxylic acids to valuable substituted phenols and anilines [110]. Interestingly, reaction rates were enhanced by conducting the reaction under H_2, and the incorporation of protons from the gas phase in the products through decarboxylation was evidenced by deuterium labeling experiments. Amine functionalities and $Fe_{25}Ru_{75}$ NPs were found to act synergistically, with an activation of the carboxylic acid functionality by the former, and H_2 activation leading to an acceleration of product release by the latter.

Moos et al. showed that catalytic systems composed of 1.2 nm Rh NPs supported on phosphonium-based SILPs hydrodeoxygenate effectively aromatic ketones to cyclohexane derivatives (100–175°C, 50 bar H_2) in the absence of strong acid additive [98]. The NTf_2 anion was found to decompose partially at the surface of Rh NPs, generating Lewis acidic Rh-F species presumably responsible for the HDO activity observed. Interestingly, this behavior was observed only with a SILP possessing triphenylphosphonium-based IL-like structures (Rh@SILP(Ph_3-P-NTf_2)), and other systems showed only hydrogenation activity. Dierks and colleagues reported the preparation of 12–13 nm Ni_2P NPs on silica supports modified with various hydrophobic organosilanes. The resulting MNPs@MMS catalysts were applied to the hydrodeoxygenation of biophenols such as guaiacol at 295°C and under 50 bar H_2 [147]. Ni_2P NPs showed good HDO activity, consistent with their intrinsic BrØnsted and Lewis acidity [148], and produced cyclohexane and benzene in moderate to good yields (25–60%). A correlation between catalytic performance and support surface polarity was identified, with a highly hydrophobic surface being beneficial to prevent fast catalyst deactivation by the water produced in the reaction.

3.3 *Oxidation*

Oxidation reactions are characterized by a loss of electrons or an increase in the oxidation state of a chemical species. In practice, this can happen through a decrease in hydrogen content (e.g., oxidation of alcohols to aldehydes or ketones) or by an increase in oxygen content, as observed in the oxidation of alcohols or aldehydes to carboxylic acids. Oxidation reactions are typically catalyzed by MNPs (e.g., Pd, Ag, Pt, Au) using molecular oxygen, air, or water, as green, cheap, and widely available oxidants. While MNPs@MMSs have been so far studied for reduction reactions, their features make them also of interest for oxidation reactions [149–156]. In the following, we present selected examples of the use of MNPs@MMS catalytic systems in this context. A summary of the application of MNPs@MMSs in oxidation can also be found in Table 4.

Hou et al. investigated the potential of Pd NPs (3 nm) supported on PEG-modified silica surfaces in the oxidation of alcohols with O_2 in supercritical CO_2 as reaction medium (Fig. 20a) [149]. Using Pd@SiO_2-PEG, a variety of aliphatic and benzylic alcohols were selectively (98–99%) oxidized to their corresponding aldehydes and

Table 4 Summary of selected examples of MNPs@MMSs used as catalysts in oxidation reactions

Support material	Type of MM	MMSs preparation	MNPs	Oxidation of	Reference
SiO_2	Polymer	Chemisorption (silanization)	Pd	Alcohols	[149]
Mol.sieves	Organosilanes	Chemisorption (silanization)	Pd	Benzyl alcohol	[150]
TiO_2	APTES	Chemisorption (silanization)	Pd	Benzyl alcohol	[151]
SiO_2	Thiols	Chemisorption (silanization)	Au	Benzyl alcohol	[152]
Fe_3O_4	Chitosan/agarose	Chemisorption (cross linkage)	Pd	Alcohols	[153]
Cellulose	Citric acid	Chemisorption (esterification)	Au	1-indonol	[154]
PIONs	Organic molecule	Chemisorption (imprinting)	Au	Alcohols	[155]
Merrifield resin	Im. IL	Chemisorption (quaternization)	Au	Phenylethanol	[156]

Im. IL = imidazolium-based IL

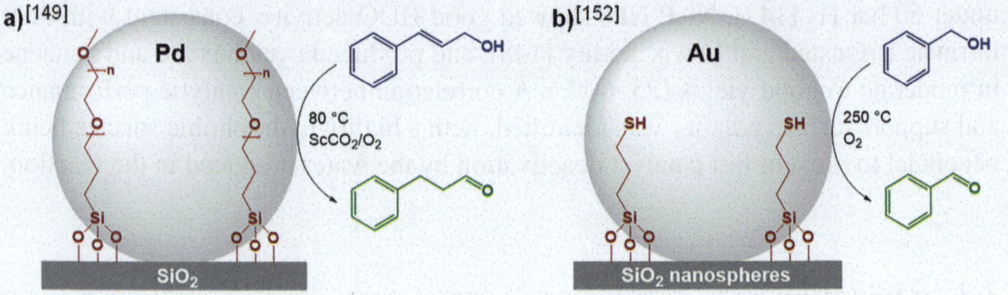

Fig. 20 Examples of MNPs@MMSs used in oxidation reactions

ketones (46–99% yield, 80°C, CO_2/O_2 92/8). Interestingly, the PEG modification of the SiO_2 support material significantly stabilized the Pd NPs, reducing their aggregation and thus deactivation under reaction conditions. Chen's group prepared Pd NPs supported on molecular sieves functionalized with various organosilanes (i.e., with tert-butyl, amine, and thiol functionalities), and used the resulting catalyst for the aerobic solvent-free oxidation of benzyl alcohol to benzaldehyde [150]. The surface molecular modification was found to impact on MNP size (from 1.6 to 17 nm), size distribution and metal-support interactions, which influenced catalytic performance. In particular, superior activity was observed for Pd NPs immobilized on amine-functionalized molecular sieves (TOF of 18,571 h^{-1}, 95% selectivity at 160°C, 20 mL min^{-1} O_2), which provided enhanced metal-support interactions and a modulation of the surface basicity around Pd NPs. Along the same line, Weerachawanasak and colleagues immobilized 3.4 nm Pd NPs on a TiO_2 support functionalized with various amounts of 3-aminopropyltriethoxysilane (APTES) and found that benzaldehyde yields from benzyl alcohol (120°C, 1 bar O_2) were enhanced with increasing APTES coverage, i.e., increasing support basicity [151]. Recently, Zheng et al. reported the synthesis of Au NPs on thiol-modified dendritic mesoporous silica nanospheres (DMSNs, Fig. 20b) [152]. The loading of thiol molecular modifiers on the support surface influenced significantly the MNP size, with the Au NP size decreasing with increasing loading of thiols (i.e., from 20 nm in the absence of thiol to 1.2 nm with 6.0 wt% thiol). The best catalytic activity and lifetime for the gas phase oxidation of benzyl alcohol were obtained for 3.5 nm Au NPs on DMSNs-4.4 wt% SH (91% conversion to benzaldehyde and 820 h lifetime at 250°C). These enhanced performance and stability were attributed to the adsorption and packing of thiols functionalities at the Au NPs surface, which on the one hand prevented metal leaching and MNP aggregation, and on the other hand, provided a favorable environment for the activation and reaction of benzyl alcohol and oxygen.

3.4 C-C Coupling

C-C coupling reactions (e.g., Stille, Suzuki, Sonogashira, Heck, etc. reactions) are important for the chemical sector to produce fine chemicals, agrochemicals, and pharmaceuticals, and this importance was acknowledged with the 2010 Nobel Prize received by Richard Heck, Ei-ichi Negishi, and Akira Suzuki. C-C coupling reactions are typically carried out in the presence of molecular or heterogeneous metal catalysts – very often Pd – for which the definitive identification of the active species remains a topic of debate in the chemical community [157–159]. Taking the example of Pd-catalyzed Heck reactions, the so-called "release-catch mechanism" is commonly accepted and involves the leaching of molecular Pd species from Pd NPs via oxidative addition by aryl halides under reaction conditions [157]. In addition, in cases where $Pd(OAc)_2$ is used as catalyst without additional ligands, Pd NPs are presumably formed in situ, and reactions proceed through the reservoir pathway [157]. Regardless of the exact active species and mechanisms, it is clear that the accessibility of Pd atoms to aryl halides has a strong impact on catalytic performance, and thus that irreversible metal leaching, as well as the growth, aggregation, and coalescence of MNPs, should be prevented to maintain high reaction rates. In this context, the use of MMSs as support for MNPs is attractive, and we will highlight here selected examples of this approach [160–170]. A summary can also be found in Table 5.

Table 5 Summary of selected examples of MNPs@MMSs used as catalysts in coupling reactions

Support material	Type of MM	MMSs preparation	MNPs	Type of coupling	Reference
SiO_2	Im. IL	Chemisorption (silanization)	Pd	Heck-Mizoroki	[160]
Polymer	Guad. IL	Chemisorption (quaternization)	Pd	Suzuki	[161]
SiO_2 or SBA-15	Im. IL	Chemisorption (silanization)	Pd	Suzuki	[162]
SiO_2	Im. IL	Chemisorption (silanization)	Pd	Suzuki	[163]
SiO_2	Im. IL	Chemisorption (silanization)	Pd	Heck	[164]
SiO_2/Fe_3O_4	Organic amines and iminophosphines	Chemisorption (silanization)	Pd	Suzuki	[165]
TiO_2	Mercaptoundecanoic acid	Chemisorption (esterification)	Pd	Ullman	[166]
MOF	Polydopamine	Chemisorption (imprinting)	AgPd	Suzuki-Miyaura	[167]
SiO_2	Im. IL	Chemisorption (silanization)	AuPd	Suzuki	[168]
SBA-15	Kryptofix 23	Physisorption	PdNi	Cross-coupling	[169]
Halloysite clay	Im. IL	Chemisorption (silanization)	Pd	Suzuki	[170]

Im. IL = imidazolium-based IL; Guad. IL = guanidinium-based IL

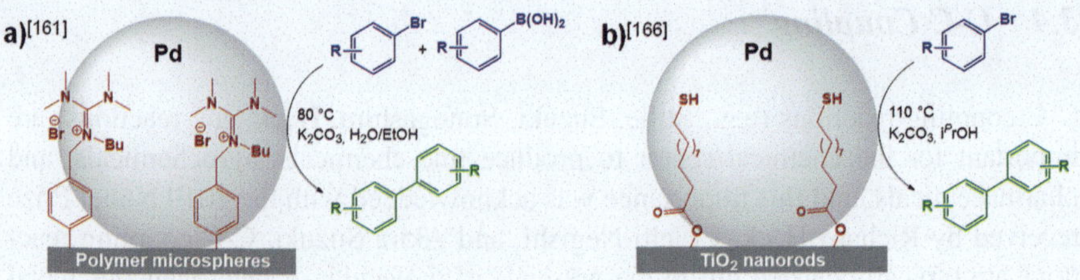

Fig. 21 Examples of MNPs@MMSs used in C=C coupling reactions

In a pioneer work, the group of Corma supported ca. 5 nm Pd NPs on imidazolium-based SILPs prepared by silanization on SiO_2. The resulting Pd@SILP catalysts showed activity in the Heck-Mizoroki coupling of iodobenzene and styrene, giving trans-stilbene in quantitative yield (140°C, 1 h and $^nBu_3N$ as base) [160]. However, hot filtration tests showed activity even after removal of the heterogeneous catalyst, suggesting that Pd NPs served as reservoirs for the leaching of molecular Pd species that were responsible for at least part of the activity observed. Along the same line, Li and co-workers synthesized 3 nm Pd NPs on a guanidinium-based SILP using poly(p-phenylene) microspheres as support materials (Fig. 21a). Catalytic performance was evaluated for the Suzuki-Miyaura cross-coupling reaction between 4-bromoanisole and phenylboronic acid (80°C, K_2CO_3 as base and aqueous ethanol as solvent) [161]. Under optimized conditions, high and stable activity were observed for eight runs. Interestingly, the reaction stopped after the removal of the Pd@SILP catalysts via hot filtration, and the authors concluded that the reaction was heterogeneously catalyzed. Gruttadauria et al. explored the potential of Pd NPs on highly cross-linked imidazolium networks (CLINs) for various C-C coupling reactions (Mizoroki-Miyaura, Suzuki and Heck-Mizoroki reactions) [162–164]. The reduction of $[PdCl_4]^{2-}$ with $NaBH_4$ in the presence of CLINs afforded 2–4 nm Pd NPs, that proved active for the Mizoroki-Miyaura coupling of various aryl iodides and bromides with phenylboronic acid and 4-formylphenylboronic acid (50°C, K_2CO_3 as base and aqueous ethanol as solvent) [163]. Pd@CLINs catalysts were found recyclable for at least four cycles, and hot filtration tests revealed a release-catch mechanism. In this case, Pd NPs acted as reservoirs of Pd molecular species, while the SILP favored the effective capture of these species at the end of the reaction.

Costa et al. prepared SiO_2 supports incorporating magnetic Fe_3O_4 NPs and modified them with amine and iminophosphine functionalities and used these for the synthesis of Pd NPs (2–4 nm). The resulting catalysts were applied to the Suzuki cross-coupling of aryl bromides with phenylboronic acid (100°C, KOH as base and toluene as solvent), showing good performance without the need for additional phosphine ligands [165]. Pd NPs on iminophosphine-functionalized supports were found more active and stable than those on supports possessing amine functionalities, highlighting the crucial role of pendant phosphine groups on catalytic performance. Poisoning experiments with Hg(0) led to an immediate suppression of catalytic activity, and the authors concluded that Pd(0) species are catalyzing the reaction.

Kumar et al. reported that the stability of 25–28 nm Pd NPs supported on TiO_2 nanorods could be significantly enhanced through chemisorption by esterification of mercaptoundecanoic acid on the TiO_2 surface (Fig. 21b). Applied to the Ullman coupling and debromination of aryl bromides (110°C, K_2CO_3 as base and ^{i}PrOH as solvent), this catalyst displayed higher activity and better reusability (over 14 cycles) than reference Pd@TiO_2 catalysts. In particular, the strong affinity of Pd for thiol functionalities was claimed to prevent Pd leaching and MNP aggregation under reaction conditions [166]. Mishra et al. used a polydopamine-modified MOF as MMS for the synthesis of well-dispersed bimetallic 2.2 nm AgPd NPs. The resulting AgPd@MIL-125-NH2-PDA catalyst was tested for the Suzuki-Miyaura coupling reaction of aryl halides with phenylboronic acid (60°C, K_2CO_3 as base and aqueous ethanol as solvent) [167]. Activity and stability were found enhanced for the bimetallic catalyst as compared to monometallic Ag and Pd references, presumably due to strong metal support interactions arising from synergistic stabilization effects from PDA and MOF frameworks on the AgPd NPs.

3.5 *Adaptive Catalysis*

Adaptive catalytic systems are defined as "*capable of adjusting or being adjusted into different states of their performance in response to dynamic changes in the reactive environment*" [171], and typically provide opportunities to adjust product selectivity (i.e., adaptivity in product formation) and/or catalytic activity (i.e., adaptivity to intermittent electricity supply), in a *reversible, rapid, and robust manner (R^3 rule)* (Fig. 22) [171]. The development of such catalysts has attracted increasing interest in recent years, with the application of various stimuli to initiate switches in reactivity, such as temperature [172–177], chemical reactions [99, 103, 121, 178], photochemical irradiation [179–181], and redox switches [182, 183].

In this context, MNPs@MMSs and their flexible molecular design and assembly offer unique opportunities for the development of adaptive catalytic systems. For

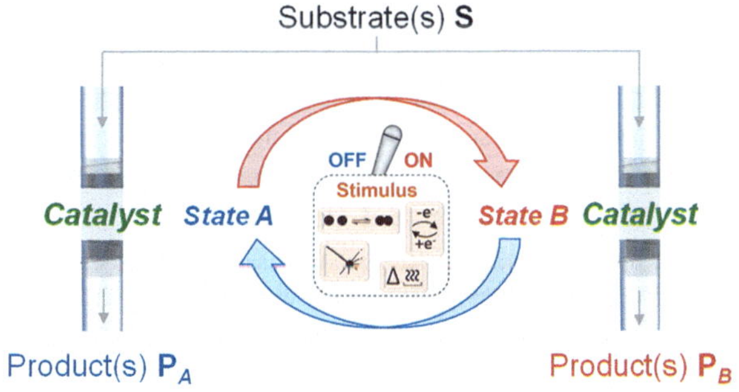

Fig. 22 Adaptive catalytic systems exemplified for the selective conversion of a substrate S to two different products P_A and P_B (adaptivity in product formation). Adapted with permission from Bordet and Leitner [171]. Copyright 2023 Wiley

example, Bordet and co-workers prepared Ru NPs on a tertiary amine-functionalized polymer grafted silica support [99]. The obtained Ru@PGS catalytic system provided full hydrogenation of ketone-containing furan derivatives under H_2 (80°C, 15 bar H_2), while ketone hydrogenation was suppressed under H_2/CO_2 (80°C, 30 bar H_2/CO_2 1/1). This adaptive response to the presence of CO_2 in the feed gas – not observed with reference catalysts such as Ru@SiO_2 – was found to originate from the formation of ammonium formates species at the surface of the Ru NPs under H_2/CO_2, which decomposed upon heating under pure H_2. Recently, Zhang et al. expanded the application of Ru@PGS to the adaptive hydrogenation of bicyclic heteroaromatics (e.g., benzofurans and quinolines) to valuable partially and fully hydrogenated products, for which arene hydrogenation was controlled through the incorporation or removal of CO_2 from the feed gas [103]. Catalyst stability was investigated under batch and continuous flow conditions and found excellent. While the potential of MNPs@MMSs for the development of adaptive catalytic systems has been only hinted until now, these pioneering contributions and proofs of concept pave the way toward a highly dynamic development of this field in the near future.

3.6 Electrocatalysis

The surface molecular modification of electrodes is an important research line in electrochemistry, as the control of the interfacial area between the electrode surface and the electrolyte provides opportunities to tune catalytic activity and selectivity, protect electrode materials from harsh conditions, improve mass transport to active sites, modulate bubble formation (e.g., in H_2 and O_2 evolution reactions), etc. [184, 185]. Thus, the use of MNPs@MMSs as electrocatalysts also attracted significant attention in the past years [186–193].

For example, Wang et al. described the immobilization of Pt NPs on carbon nanotubes (CNTs) modified with polyanions such as poly(styrenesulfonic acid) and poly(acrylic acid sodium). The presence of the polyanions proved beneficial for the activity of supported Pt NPs in electrooxidation of methanol (CVs at room temperature in 0.5 M H_2SO_4 + 0.5 M MeOH at a scan rate of 50 mV/s). These improved performances were attributed to the polyanionic molecular modifiers that transferred charge to the Pt NPs and supplied oxygen-containing species [186]. Along the same line, Zhang and co-workers, prepared Pt NPs (1.4 nm) on a MMS composed of graphene modified with poly(diallyldimethylammonium chloride), giving a Pt/PDDAeG catalyst that was tested for the electrooxidation of methanol (0.5 M H_2SO_4, CVs at a scan rate of 20 mV/s between −0.24 V and 0.96 V vs SCE) [187]. The surface molecular modification of graphene with PDDA contributed to the excellent dispersion and stabilization of Pt NPs, leading to productivity and stability outperforming that of reference catalysts such as Pt/graphene and benchmark Pt/carbon black. Xin and co-workers used the surface modification of graphene with p-phenyl-SO_3H and p-phenyl-NH_2 groups (chemisorption by diazotization) to immobilize Pt NPs and study the impact of NPs support electronic interactions on electrocatalytic performance [188]. Molecular modifiers were found to favor the formation of small MNPs (2 nm, versus 3.5 nm on pristine graphene) and to improve their dispersion on the support. In addition, strong Pt-MMS interactions in the form of electron transfers from the Pt NPs to the MMS supports were observed by XPS and XAS. Interactions were especially strong between Pt NPs and the amine-functionalized graphene support, leading to a minimization of Pt leaching and MNP aggregation/coalescence.

Very recently, Han and colleagues deported an innovative approach to tackle challenges in electrocatalytic hydrogenation associated with substrate solubility and catalytic and Faradaic efficiency [189]. The authors developed MNPs@MMSs composed of 4.5 nm Pd NPs immobilized on CNTs modified with a quaternary ammonium imidazolium salt (chemisorption by amidation) providing a positive charge to the CNTs. The prepared Pd@CNTs(+) acted both as stabilizers for a Pickering emulsion system composed of an aqueous electrolyte and an oil phase containing styrene-type substrates, and as electrocatalysts at the oil-water interface. The system was found extremely efficient for the electrocatalytic hydrogenation of styrene to ethylbenzene, with record faradaic efficiency (95.0%) and mass-specific current density (-148.1 mA·$mg_{Pd}{}^{-1}$). The surface molecular modification of CNTs to produce positively charged emulsion droplets was found to be crucial to assure electrical connection with the negatively charged current collector (Ag electrode), thus allowing the Pd@CNTs(+) to act as an extended cathode.

4 Conclusions and Outlook

In the past decades, the development of MNPs@MMSs has emerged as an attractive and versatile approach for the rational design and modular preparation of tailor-made catalysts bridging molecular ("homogeneous") and materials ("heterogeneous") catalysis. The assembly at the molecular scale of a variety of support materials, molecular modifiers, and MNPs can be achieved using well-documented methods, producing catalytic systems with finely controlled reactivity that showed superior performance in various of transformations such as reductions, oxidations, C-C coupling, and electrocatalytic reactions. Benefits associated with the use of MNPs@MMSs as compared to conventional supported MNP catalysts include inter alia enhanced activity and tunable selectivity, improved MNP stability, and opportunities for multifunctionality. Interestingly, recent developments in the field demonstrated the possibility of introducing reactive functional groups in MNPs@MMSs to generate adaptive catalytic systems capable of changing their activity and/or selectivity in a fully reversible manner.

While many MNPs@MMSs with excellent catalytic properties have been reported in the literature, there is still relatively little fundamental understanding of how MNPs, molecular modifiers and support materials interact (e.g., spatial arrangement, electronic interactions, etc.), and of how these interactions influence their catalytic performance. Future research should investigate this question in detail using advanced materials characterization (including *operando* characterization under catalytic conditions) and computational studies to pave the way toward a facilitated knowledge-based design and development of next-generation MNPs@MMS catalysts. Using the flexibility of the MNPs@MMSs design to prepare adaptive catalytic systems is also of great interest to cope with the dynamics associated with renewable carbon feedstock and energy sources, and significant research efforts should be directed toward the exploration of this strategy.

References

1. Clark JH, Macquarrie DJ (1998) Catalysis of liquid phase organic reactions using chemically modified mesoporous inorganic solids. Chem Commun:853–860
2. Notestein JM, Katz A (2006) Enhancing Heterogeneous Catalysis through Cooperative Hybrid Organic–Inorganic Interfaces. Chem A Eur J 12:3954–3965
3. Kickelbick G (2006) Hybrid materials, synthesis characterization, and applications. Wiley-VCH, Weinheim
4. Díaz U, Brunel D, Corma A (2013) Catalysis using multifunctional organosiliceous hybrid materials. Chem Soc Rev 42:4083–4097
5. Ye R, Zhao J, Wickemeyer BB, Toste FD, Somorjai GA (2018) Foundations and strategies of the construction of hybrid catalysts for optimized performances. Nat Catal 1:318–325
6. Katz A, Davis ME (2000) Molecular imprinting of bulk, microporous silica. Nature 403:286–289
7. Haumann M, Wasserscheid P (2014) SILP and SCILL catalysis. Royal Society of Chemistry
8. Bordet A, Leitner W (2021) Metal nanoparticles immobilized on molecularly modified surfaces: versatile catalytic systems for controlled hydrogenation and hydrogenolysis. Acc Chem Res 54:2144
9. Migowski P, Luska KL, Leitner W (2016) Prechtl MHG (ed) Nanocatalysis in ionic liquids. Wiley VCH, Weinheim
10. Jiang T, Zhou Y, Liang S, Liu H, Han B (2009) Hydrogenolysis of glycerol catalyzed by Ru-cu bimetallic catalysts supported on clay with the aid of ionic liquids. Green Chem 11: 1000–1006
11. van den Berg R, Parmentier TE, Elkjaer CF, Gommes CJ, Sehested J, Helveg S, de Jongh PE, de Jong KP (2015) Support functionalization to Retard Ostwald Ripening in copper methanol synthesis catalysts. ACS Catal 5:4439–4448
12. van Deelen TW, Mejìa CH, de Jong KP (2019) Control of metal-support interactions in heterogeneous catalysts to enhance activity and selectivity. Nat Catal 2:955–970
13. Bordet A, Moos G, Welsh C, Licence P, Luska KL, Leitner W (2020) Molecular control of the catalytic properties of rhodium nanoparticles in supported ionic liquid phase (SILP) systems. ACS Catal 10:13904–13912
14. Barsbay M, Özgür TC, Sütekin SD, Güven A (2020) Effect of brush length of stabilizing grafted matrix on size and catalytic activity of metal nanoparticles. Eur Polym J 134:109811
15. Barlocco I, Bellomi S, Delgado JJ, Chen X, Prati L, Dimitratos N, Roldan A, Villa A (2021) Enhancing activity, selectivity and stability of palladium catalysts in formic acid decomposition: effect of support functionalization. Catal Today 382:61–70
16. Boniface KJ, Dykeman RR, Cormier A, Wang H-B, Mercer SM, Liu G, Cunningham MF, Jessop PG (2016) CO_2-switchable drying agents. Green Chem 18:208–213
17. Kacem S, Emondts M, Bordet A, Leitner W (2020) Selective hydrogenation of fluorinated arenes using rhodium nanoparticles on molecularly modified silica. Cat Sci Technol 10:8120–8126

18. Luska KL, Julis J, Stavitski E, Zakharov DN, Adams A, Leitner W (2014) Bifunctional nanoparticle–SILP catalysts (MNPs@SILP) for the selective deoxygenation of biomass substrates. Chem Sci 5:4895–4905
19. Wang Y, Rong Z, Wang Y, Wang T, Du Q, Wang Y, Qu J (2017) Graphene-based metal/acid bifunctional catalyst for the conversion of levulinic acid to γ-valerolactone. ACS Sustain Chem Eng 5:1538–1548
20. Coan PD, Griffin MB, Ciesielski PN, Medlin JW (2019) Phosphonic acid modifiers for enhancing selective hydrodeoxygenation over Pt catalysts: the role of the catalyst support. J Catal 372:311–320
21. Vansant EF, Van Der Voort P, Vrancken KC (1995) Characterization and chemical modification of the silica surface.1st edn. Elsevier, p 93
22. Notestein JM, Katz A (2006) Enhancing heterogeneous catalysis through cooperative hybrid organic–inorganic interfaces. Chem A Eur J 12:3954–3965
23. Jung H-S, Moon D-S, Lee J-K (2012) Quantitative analysis and efficient surface modification of silica nanoparticles. J Nanomat:593471
24. Kickelbick G, Schubert U (2003) Organic functionalization of metal oxide nanoparticles. In: Baraton MI (ed) Synthesis, functionalization and surface treatment of nanoparticles. Stevenson Ranch, p 91
25. Neouze M-A, Schubert U (2008) Surface modification and functionalization of metal and metal oxide nanoparticles by organic ligands. Monatsh Chem 139:183–195
26. Pujari SP, Scheres L, Marcelis ATM, Zuilhof H (2014) Covalent surface modification of oxide surfaces. Angew Chem Int Ed 53:6322–6356
27. Martinez C, Corma A (2011) Inorganic molecular sieves: preparation, modification and industrial application in catalytic processes. Coord Chem Rev 255:1558–1580
28. de Gennaro B (2019) Surface modification of zeolites for environmental applications. In: Modified clay and zeolite nanocomposite materials. Elsevier, pp 57–85
29. Bahmanzadegan F, Ghaemi A (2024) Modification and functionalization of zeolites to improve the efficiency of CO_2 adsorption: a review. Case Studies in Chemical and Environmental Engineering 9:100564
30. Fang W, Zhang Y, Kang L, DeBeer S, Leitner W, Bordet A, Riisager A. Molecularly modified aluminum phosphates as support materials for Ru nanoparticles in selective hydrogenation. Under review
31. Tasis D, Tagmatarchis N, Bianco A, Prato M (2006) Chemistry of carbon nanotubes. Chem Rev 106:1105–1136
32. Yu W, Sisi L, Haiyan Y, Jie L (2020) Progress in the functional modification of graphene/graphene oxide: a review. RSC Adv 10:15328–15345
33. Rathinavel S, Priyadharshini K, Panda D (2021) A review on carbon nanotube: an overview of synthesis, properties, functionalization, characterization, and the application. Mater Sci Eng B 268:115095
34. Wang S, Morris W, Liu Y, McGuirk CM, Zhou Y, Hupp JT, Farha OK, Mirkin CA (2015) Surface-specific functionalization of nanoscale metal–organic frameworks. Angew Chem Int Ed 54:14738–14742
35. Forgan RS (2019) The surface chemistry of metal–organic frameworks and their applications. Dalton Trans 48:9037–9042
36. Zhang W, Yang G, Deng F, Liang J, Huang Q, Dou J, Liu M, Cao Q-Y, Zhang Y, Wei Y (2022) Direct grafting of cellulose nanocrystals with poly(ionic liquids) via gamma-ray irradiation and their utilization for adsorptive removal of CR. Int J Biol Macromol 194: 1029–1037
37. Singh A, Singh N, Kaur N, Jang DO (2022) Gold nanoparticles supported on ionic-liquid-functionalized cellulose (au@CIL): a heterogeneous catalyst for the selective reduction of aromatic nitro compounds. Appl Organomet Chem 36:e6855
38. Ikram M, Haider A, Imran M, Haider J, Naz S, Ul-Hamid A, Shahzadi A, Moeen S, Nazir G, Nabgan W, Bashir A, Ali S (2023) Cellulose grafted poly acrylic acid doped manganese oxide

nanorods as novel platform for catalytic, antibacterial activity and molecular docking analysis. Surf Interfaces 37:102710
39. Hetemi D, Pinson J (2017) Surface functionalization of polymers. Chem Soc Rev 46:5701–5713
40. Yadav R, Baskaran T, Kaiprathu A, Ahmed M, Bhosale SV, Joseph S, Al-Muhtaseb AH, Singh G, Sakthivel A, Vinu A (2020) Recent advances in the preparation and applications of organo-functionalized porous materials. Chem Asian J 15:2588–2621
41. Xin B, Hao J (2014) Imidazolium-based ionic liquids grafted on solid surfaces. Chem Soc Rev 43:7171–7187
42. Mévellec V, Roussel S, Tessier L, Chancolon J, M. Mayne-L'Hermite, G. Deniau, P. Viel, S. Palacin. (2007) Grafting polymers on surfaces: a new powerful and versatile diazonium salt-based one-step process in aqueous media. Chem Mater 19:6323–6330
43. Blacha A, Krukiewicz K, Zak J (2011) The covalent grafting of polymers to the solid surface. Chemik 65:11–19
44. Coperet C, Comas-Vives A, Conley MP, Estes DP, Fedorov A, Mougel V, Nagae H, Núnez-Zanur F, Zhizhko PA (2016) Surface organometallic and coordination chemistry toward single-site heterogeneous catalysts: strategies, methods, structures, and activities. Chem Rev 116:323–421
45. Samantaray MK, Pump E, Bendjeriou-Sedjerari A, D'Elia V, Pelletier JDA, Guidotti M, Psaro R, Basset J-M (2018) Surface organometallic chemistry in heterogeneous catalysis. Chem Soc Rev 47:8403–8437
46. Sanchez C, Ribot F, Lebeau B (1999) Molecular design of hybrid organic–inorganic nanocomposites synthesized via sol–gel chemistry. J Mater Chem 9:35–44
47. Hutchings GJ, Védrine JC (2003) Heterogeneous catalyst preparation. In: Basic principles in applies catalysis. Springer, Berlin. GmbH
48. de Jong KP (2009) Synthesis of solid catalysts. Wiley-VCH, Weinheim
49. Bénard J, Berthier Y, Delarnare F, Hondros E, Huber M, Marcus P, Masson A, Oudar J, Rhead GE (1983) Equilibrium adsorption and surface segregation. Stud Surf Sci Catal 13:44–99
50. Cremer T, Killian M, Gottfried JM, Paape N, Wasserscheid P, Maier F, Steinrück H-P (2008) Physical vapor deposition of [EMIM][Tf_2N]: a new approach to the modification of surface properties with ultrathin ionic liquid films. ChemPhysChem 9:2185–2190
51. Sobota M, Schmid M, Happel M, Amende M, Maier F, Steinruck HP, Paape N, Wasserscheid P, Laurin M, Gottfried JM, Libuda J (2010) Ionic liquid based model catalysis: interaction of [BMIM][Tf_2N] with Pd nanoparticles supported on an ordered alumina film. Phys Chem Chem Phys 12:10610–10621
52. Obst M, Arnauts G, Cruz AJ, Gonzalez MC, Marcoen K, Hauffman T, Ameloot R (2021) Chemical vapor deposition of ionic liquids for the fabrication of Ionogel films and patterns. Angew Chem Int Ed 60:25668–25673
53. Gaspard JP (1982) Physisorption and chemisorption. In: Mutaftschiev B (ed) Interfacial aspects of phase transformations. NATO advanced study institutes series, vol 87. Springer, Dordrecht
54. Bellmann C (2008) Surface modification by adsorption of polymers and surfactants. In: Stamm M (ed) Polymer surfaces and interfaces. Springer, Berlin
55. Brzoska JB, Azouz IB, Rondelez F (1994) Silanization of solid substrates: a step toward reproducibility. Langmuir 10:4367–4373
56. Velasco-Santos C, Martínez-Hernández AL, Castano VM (2010) Silanization of carbon nanotubes: surface modification and polymer nanocomposites. IntechOpen
57. Speranza G (2019) The role of functionalization in the applications of carbon materials: an overview. J Carbon Res 5:84
58. Ranjan R, Brittain WJ (2011) Combination of living radical polymerization and click chemistry for surface modification. J Mater Chem 21:16717–16734

59. Panzarasa G, Aghion S, Marra G, Wagner A, Liedke MO, Elsayed M, Krause-Rehberg R, Ferragut R, Consolati G (2017) Probing the impact of the initiator layer on grafted-from polymer brushes: a positron annihilation spectroscopy study. Macromolecules 50:5574–5581
60. Wight AP, Davis ME (2002) Design and preparation of organic-inorganic hybrid catalysts. Chem Rev 102:3589–3614
61. Angloher S, Bein T (2005) Organic functionalization of mesoporous silica. Stud Surf Sci Catal 158(Part B):2017–2026
62. Ban C, George SM (2016) Molecular layer deposition for surface modification of lithium-ion battery electrodes. Adv Mater Interfaces 3:1600762
63. Allongue P, Delamar M, Desbat B, Fagebaume O, Hitmi R, Pinson J, Savéant J-M (1997) Covalent modification of carbon surfaces by aryl radicals generated from the electrochemical reduction of diazonium salts. J Am Chem Soc 119:201–207
64. Chehimi MM (2012) Aryl diazonium salts. In: New coupling agents in polymer and surface science. Wiley-VCH, Weinheim
65. Dyke CA, Tour JM (2003) Unbundled and highly functionalized carbon nanotubes from aqueous reactions. Nano Lett 3:1215–1218
66. Busson M, Berisha A, Combellas C, Kanoufi F, Pinson J (2011) Photochemical grafting of diazonium salts on metals. Chem Commun 47:12631
67. Delgado JL, de la Cruz P, Langa F, Urbina A, Casado J, López Navarrete JT (2004) Microwave-assisted sidewall functionalization of single-sall carbon nanotubes by Diels–Alder cycloaddition. Chem Commun:1734–1735
68. Ismaili H, Geng D, Sun AX, Kantzas TT, Workentin MS (2011) Light-activated covalent formation of gold nanoparticle–graphene and gold nanoparticle–glass composites. Langmuir 27:13261–13268
69. Criado A, Melchionna M, Marchesan S, Prato M (2015) The covalent functionalization of graphene on substrates. Angew Chem Int Ed 54:10734–10750
70. Kalsi D, Loui Anandaraj SJ, Durai M, Weidenthaler C, Emondts M, Nolan SP, Bordet A, Leitner W (2022) One-pot multicomponent synthesis of allyl and alkylamines using a catalytic system composed of ruthenium nanoparticles on copper N-heterocyclic carbene-modified silica. ACS Catal 12:14902–14910
71. Jiang Y, Wang Z, Zhu Z, He M, Zhou P (1800) Effective removal of humic acid by Zr-MOFs with surface modification. Water 2022:14
72. Lelli M, Gajan D, Lesage A, Caporini MA, Vitzthum V, Miéville P, Héroguel F, Rascaon F, Roussey A, Thieuleux C, Boualleg M, Veyre L, Bodenhausen G, Coperet C, Emsley L (2011) Fast characterization of functionalized silica materials by silicon-29 surface-enhanced NMR spectroscopy using dynamic nuclear polarization. J Am Chem Soc 133(7):2104–2107
73. Macdonald PM, Damha MJ, Ganeshan K, Braich R, Zabarylo SV (1996) Phosphorus 31 solid state NMR characterization of oligonucleotides covalently bound to a solid support. Nucleic Acids Res 24:2868–2876
74. Zhang G, Tsujimoto M, Packwood D, Duong NT, Nishiyama Y, Kadota K, Kitagawa S, Horike S (2018) Construction of a hierarchical architecture of covalent organic frameworks via a postsynthetic approach. J Am Chem Soc 140:2602–2609
75. Tavera-Méndez C-L, Bergen A, Trzeciak S, Heinemann FW, Graf R, Zahn D, Meyer K, Hartmann M, Wisser D (2023) Self-assembled supported ionic liquids. Chem A Eur J: e202303673
76. Lovelock KRJ, Villar-Garcia IJ, Maier F, Steinrück H-P, Licence P (2010) Photoelectron spectroscopy of ionic liquid-based interfaces. Chem Rev 110:5158–5190
77. Steinrück H-P (2012) Recent developments in the study of ionic liquid interfaces using X-ray photoelectron spectroscopy and potential future directions. Phys Chem Chem Phys 14:5010–5029
78. Yang Y, Guo Y, Yuan J, Xie H, Gao C, Zhao T, Zheng Q (2022) Agile construction of porous organic frameworks pending carboxylic acids and imidazolium-based ionic liquids for the efficient fixation of CO_2 to cyclic carbonates. ACS Sustain Chem Eng 10(24):7990–8001

79. Jia C-J, Schüth F (2011) Colloidal metal nanoparticles as a component of designed catalyst. Phys Chem Chem Phys 13:2457–2487
80. Fiévet F, Ammar-Merah S, Brayner R, Chau F, Giraud M, Mammeri F, Peron J, Piquemal J-Y, Sicard L, Viau G (2018) The polyol process: a unique method for easy access to metal canoparticles with tailored sizes, shapes and compositions. Chem Soc Rev 47:5187–5233
81. Guan H, Harris C, Sun S (2023) Metal–ligand interactions and their roles in controlling nanoparticle formation and functions. Acc Chem Res 56:1591–1601
82. Amiens C, Chaudret B, Ciuculescu-Pradines D, Collière V, Fajerwerg K, Fau P, Kahn M, Maisonnant A, Soulantica K, Philippot K (2013) Organometallic approach for the synthesis of nanostructures. New J Chem 37:3374–3401
83. Rossi LM, Fioro JL, Garcia MAS, Ferraz CP (2018) The role and fate of capping ligands in colloidally prepared metal nanoparticle catalysts. Dalton Trans 47:5889–5915
84. Khatri OP, Adachi K, Murase K, Okazaki K, Torimoto T, Tanaka N, Kuwabata S, Sugimura H (2008) elf-Assembly of ionic liquid (BMI-PF6)-stabilized gold nanoparticles on a silicon surface: chemical and structural aspects. Langmuir 24:7785–7792
85. Costa NJS, Rossi LM (2012) Synthesis of supported metal nanoparticle catalysts using ligand assisted methods. Nanoscale 4:5826–5834
86. Wasserscheid P, Welton T (2008) Ionic liquids in synthesis. Wiley-VCH Verlag GmbH & Co. KGaA
87. Prechtl MHG, Campbell PS (2013) Metal oxide and bimetallic nanoparticles in ionic liquids: synthesis and application in multiphase catalysis. Nanotechnol Rev 2:577–595
88. Wegner S, Janiak C (2017) Metal nanoparticles in ionic liquids. Top Curr Chem 375:1–32
89. Seitkalieva MM, Samoylenko DE, Lotsman KA, Rodygin KS, Ananikov VP (2021) Metal nanoparticles in ionic liquids: synthesis and catalytic applications. Coord Chem Rev 445: 213982
90. Jacinto MJ, Kiyohara PK, Masunaga SH, Jardim RF, Rossi LM (2008) Recoverable rhodium nanoparticles: synthesis, characterization and catalytic performance in hydrogenation reactions. Appl Catal A: General 338:52–57
91. Jacinto MJ, Landers R, Rossi LM (2009) Preparation of supported Pt(0) nanoparticles as efficient recyclable catalysts for hydrogenation of alkenes and ketones. Catal Commun 10: 1971–1974
92. Jacinto MJ, Silva FP, Kiyohara PK, Landers R, Rossi LM (2012) Catalyst recovery and recycling facilitated by magnetic separation: iridium and other metal nanoparticles. ChemCatChem 4:698–703
93. Mourdikoudis S (2021) Reducing agents in colloidal nanoparticle synthesis. Nanoscience & nanotechnology series. Royal Society of Chemistry
94. Martinez-Prieto LM, Chaudret B (2018) Organometallic ruthenium nanoparticles: synthesis, surface chemistry, and insights into ligand coordination. Acc Chem Res 51:376–384
95. Luska KL, Migowski P, El Sayed S, Leitner W (2015) Synergistic interaction within bifunctional ruthenium nanoparticle/SILP catalysts for the selective hydrodeoxygenation of phenols. Angew Chem Int Ed 54:15750–15755
96. Rengshausen S, Etscheidt F, Großkurth J, Luska K, Bordet A, Leitner W (2019) Catalytic hydrogenolysis of substituted diaryl ethers by using ruthenium nanoparticles on an acidic supported ionic liquid phase (Ru@SILP-SO_3H). Synlett 30:405–412
97. El Sayed S, Bordet A, Weidenthaler C, Hetaba W, Luska KL, Leitner W (2020) Selective hydrogenation of benzofurans using ruthenium nanoparticles in Lewis acid-modified ruthenium-supported ionic liquid phases. ACS Catal 10:2124–2130
98. Moos G, Emondts M, Bordet A, Leitner W (2020) Selective hydrogenation and hydrodeoxygenation of aromatic ketones to cyclohexane derivatives using a Rh@SILP catalyst. Angew Chem Int Ed 59:11977–11983
99. Bordet A, El Sayed S, Sanger M, Boniface KJ, Kalsi D, Luska KL, Jessop PG, Leitner W (2021) Selectivity control in hydrogenation through adaptive catalysis using ruthenium nanoparticles on a CO_2-responsive support. Nat Chem 13:916–922

100. Zenner J, Moos G, Luska KL, Bordet A, Leitner W (2021) Rh MNPs immobilized on phosphonium-based supported ionic liquid phases (Rh@SILPs) as hydrogenation catalysts. Chimia 75:724–732
101. Kacem S, Qiao Y, Wirtz C, Theyssen N, Bordet A, Leitner W (2022) Supercritical carbon dioxide as reaction medium for selective hydrogenation of fluorinated arenes. Green Chem 24: 8671–8676
102. Louis Anandaraj SJ, Kang L, DeBeer S, Bordet A, Leitner W (2023) Catalytic hydrogenation of CO_2 to formate using ruthenium nanoparticles immobilized on supported ionic liquid phases. Small:2206806
103. Zhang Y, El Sayed S, Kang L, Sanger M, Wiegand T, Jessop PG, DeBeer S, Bordet A, Leitner W (2023) Adaptive catalysts for the selective hydrogenation of bicyclic heteroaromatics using ruthenium nanoparticles on a CO_2-responsive support. Angew Chem Int Ed 62:e202311427
104. Luska KL, Bordet A, Tricard S, Sinev I, Grünert W, Chaudret B, Leitner W (2016) Enhancing the catalytic properties of ruthenium nanoparticle-SILP catalysts by dilution with iron. ACS Catal 6:3719–3726
105. Offner-Marko L, Bordet A, Moos G, Tricard S, Rengshausen S, Chaudret B, Luska KL, Leitner W (2018) Bimetallic nanoparticles in supported ionic liquid phases as multifunctional catalysts for the selective hydrodeoxygenation of aromatic substrates. Angew Chem Int Ed 57: 12721–12726
106. Goclik L, Offner-Marko L, Bordet A, Leitner W (2020) Selective hydrodeoxygenation of hydroxyacetophenones to ethyl-substituted phenol derivatives using a FeRu@SILP catalyst. Chem Commun 56:9509–9512
107. Rengshausen S, Van Stappen C, Levin N, Tricard S, Luska KL, DeBeer S, Chaudret B, Bordet A, Leitner W (2021) Organometallic synthesis of bimetallic cobalt-rhodium nanoparticles in supported ionic liquid phases (Co_xRh_{100-x}@SILP) as catalysts for the selective hydrogenation of multifunctional aromatic substrates. Small 17:e2006683
108. Goclik L, Walschus H, Bordet A, Leitner W (2022) Selective hydrodeoxygenation of acetophenone derivatives using a $Fe_{25}Ru_{75}$@SILP catalyst: a practical approach to the synthesis of alkyl phenols and anilines. Green Chem 24:2937–2945
109. Sisodiya-Amrute S, Van Stappen C, Rengshausen S, Han C, Sodreau A, Weidenthaler C, Tricard S, DeBeer S, Chaudret B, Bordet A, Leitner W (2022) Bimetallic MxRu100-x nanoparticles (M = Fe, Co) on supported ionic liquid phases (MxRu100-x@SILP) as hydrogenation catalysts: influence of M and M:Ru ratio on activity and selectivity. J Catal 407:141–148
110. Levin N, Goclik L, Walschus H, Antil N, Bordet A, Leitner W (2023) Decarboxylation and tandem reduction/decarboxylation pathways to substituted phenols from aromatic carboxylic acids using bimetallic nanoparticles on supported ionic liquid phases as multifunctional catalysts. J Am Chem Soc 145:22845–22854
111. Marchenko N, Lacroix L-M, Ratel-Ramon N, Leitner W, Bordet A, Tricard S (2023) Bimetallic Fe_xPt_{100-x} nanoparticles immobilized on supported ionic liquid phases as hydrogenation and hydrodeoxygenation catalysts: influence of the metal content on activity and selectivity. ACS Appl Nano Mater 6:20231–20239
112. Sodreau A, Zahedi HG, Dervisoglu R, Kang L, Menten J, Zenner J, Terefenko N, DeBeer S, Wiegand T, Bordet A, Leitner W (2023) A simple and versatile approach for the low-temperature synthesis of transition metal phosphide nanoparticles from metal chloride complexes and $P(SiMe_3)_3$. Adv Mater 35:2306621
113. Wender H, Migowski P, Feil AF, Teixeira SR, Dupont J (2013) Sputtering deposition of nanoparticles onto liquid substrates: recent advances and future trends. Coord Chem Rev 257: 2468–2483
114. Qadir MI, Castegnaro MV, Selau FF, Baptista DL, Chacon G, Pontes RB, Lisbôa AM, Eberhardt D, Dupont J (2024) Dynamic tuning of naked ruthenium clusters/nanoparticles in ionic liquids cages to boost CO_2 hydrogenation to formic acid. Appl Catal B 341:123315

115. Foppa L, Luza L, Gual A, Weibel DE, Eberhardt D, Teixeira SR, Dupont J (2015) Sputtering-deposition of Ru nanoparticles onto Al_2O_3 modified with imidazolium ionic liquids: synthesis, characterisation and catalysis. Dalton Trans 44:2827–2834
116. Luza L, Gual A, Eberhardt D, Teixeira SR, Chiaro SSX, Dupont J (2013) Imprinting catalytically active Pd nanoparticles onto ionic-liquid-modified Al_2O_3 supports. ChemCatChem 5:2471–2478
117. Qiu L, Liu F, Zhao L, Yang W, Yao J (2006) Evidence of a unique electron donor – acceptor property for platinum nanoparticles as studied by XPS. Langmuir 22:4480–4482
118. Cerra S, Carlini L, Salamone TA, Hajareh F, Mercurio M, Pennacci B, Sappino C, Battocchio C, Nottola S, Fratoddi I (2023) Noble metal nanoparticles networks stabilized by rod-like organometallic bifunctional thiols. ChemistrySelect 8:e202300874
119. Thomä SLJ, Krauss SW, Eckardt M, Chater P, Zobel M (2019) Atomic insight into hydration shells around facetted nanoparticles. Nat Commun 10:995
120. Doherty S, Knight JG, Backhouse T, Tran TST, Paterson R, Stahl F, Alharbi HY, Chamberlain TW, Bourne RA, Stones R, Griffiths A, White JP, Aslam Z, Hardare C, Daly H, Hart J, Temperton RH, O'Shea JN, Rees NH (2022) Highly efficient and selective aqueous phase hydrogenation of aryl ketones, aldehydes, furfural and levulinic acid and its ethyl Ester catalyzed by phosphine oxide-decorated polymer immobilized ionic liquid-stabilized ruthenium nanoparticles. Cat Sci Technol 12:3549–3567
121. Zhang Y, Levin N, Kang L, Müller F, Zobel M, DeBeer S, Leitner W, Bordet A (2024) Design and understanding of adaptive hydrogenation catalysts triggered by the H_2/CO_2-formic acid equilibrium. J Am Chem Soc In press
122. Mobin R, Rangreez TA, Chisti HTN, Rezakazemi IM (2019) Organic-inorganic hybrid materials and their applications. In: Functional polymers. Springer
123. Zhang L, Zhou M, Wang A, Zhang T (2020) Selective hydrogenation over supported metal catalysts: from nanoparticles to single atoms. Chem Rev 120:683–733
124. Cano I, Martínez-Prieto LM, van Leeuwen PWNM (2021) Heterolytic cleavage of dihydrogen (HCD) in metal nanoparticle catalysis. Catal Sci Tech 11:1157–1185
125. Wiesenfeldt MP, Nairoukh Z, Li W, Glorius F (2017) Hydrogenation of fluoroarenes: direct access to all-cis-(multi)-fluorinated cycloalkanes. Science 357:908–912
126. Guo M, Wang F, Zhang M, Wang L, Zhang X, Li G (2023) Modulating Pt nanoparticles confined in UiO-66 by linker functional groups for controllable hydrogenation. J Catal 424: 221–235
127. Huang J, Jiang T, Gao HX, Han BX, Liu ZM, Wu WZ, Chang YH, Zhao GY (2004) Pd nanoparticles immobilized on molecular sieves by ionic liquids: heterogeneous catalysts for solvent-free hydrogenation. Angew Chem Int Ed 43:1397–1399
128. Miao SD, Liu ZM, Han BX, Huang J, Sun ZY, Zhang JL, Jiang T (2006) Ru nanoparticles immobilized on montmorillonite by ionic liquids: a highly efficient heterogeneous catalyst for the hydrogenation of benzene. Angew Chem Int Ed 45:266–269
129. Luza L, Gual A, Rambor CP, Eberhardt D, Teixeira SR, Bernardi F, Baptista DL, Dupont J (2014) Hydrophobic effects on supported ionic liquid phase Pd nanoparticle hydrogenation catalysts. Phys Chem Chem Phys 16:18088–18091
130. Luza L, Gual A, Fernandes JA, Eberhardt D, Dupont J (2019) Tunneling effects in confined gold nanoparticle hydrogenation catalysts. Phys Chem Chem Phys 21:16615–16622
131. Mikkola JPT, Virtanen PP, Kordas K, Karhu H, Salmi TO (2007) SILCA – supported ionic liquid catalysts for fine chemicals. Appl Catal, A 328:68–76
132. Ruta M, Laurenczy G, Dyson PJ, Kiwi-Minsker L (2008) Pd nanoparticles in a supported ionic liquid phase: highly stable catalysts for selective acetylene hydrogenation under continuous-flow conditions. J Phys Chem C 112:17814–17819
133. Tao R, Miao S, Liu Z, Xie Y, Han B, An G, Ding K (2009) Pd nanoparticles immobilized on Sepiolite by ionic liquids: efficient catalysts for hydrogenation of alkenes and heck reactions. Green Chem 11:96–101

134. Betsy KJ, Bajpai P, Lazar A, Vanka K, Vinod CP (2023) Pd nanoparticles supported on N-incorporated hybrid Organosilica as an active and selective low-temperature phenol hydrogenation catalyst. ACS Appl Nano Mater 6:11500–11512
135. Wu Z-L, Wang J, Wang S, Zhang Y-X, Bai G-Y, Ricardez-Sandoval L, Wang G-C, Zhao B (2020) Controllable chemoselective hydrogenation of furfural by PdAg/C bimetallic catalysts under ambient operating conditions: an Interesting Ag Switch. Green Chem 22:1432–1442
136. Gholinejad M, Iranpanah M, Karimi S, Sansano JM (2024) Cysteine and ionic liquid modified magnetic nanoparticles supported RuCu as a new bimetallic catalyst in reduction reactions. J Mol Struct 1298:137100
137. Rana S, Jonnalagadda SB (2017) A facile synthesis of cu–Ni bimetallic nanoparticle supported organo functionalized graphene oxide as a catalyst for selective hydrogenation of *p*-nitrophenol and cinnamaldehyde. RSC Adv 7:2869–2879
138. Paterson R, Fahy LE, Arca E, Dixon C, Wills CY, Yan H, Griffiths A, Collins SM, Wu K, Bourne RA, Chamberlain TW, Knight JG, Doherty S (2023) Amine-modified polyionic liquid supports enhance the efficacy of PdNPs for the catalytic hydrogenation of CO_2 to formate. Chem Commun 59:13470–13473
139. Robinson AM, Hensley JE, Medlin JW (2016) Bifunctional catalysts for upgrading of biomass-derived oxygenates: a review. ACS Catal 6:5026–5043
140. Fang Z, Smith Jr RL, Li H (2017) Production of biofuels and chemicals with bifunctional catalysts. Springer
141. Luska KL, Migowski P, El Sayed S, Leitner W (2016) Bifunctional ruthenium nanoparticle-SILP catalysts (RuNPs@ SILP) for the hydrodeoxygenation of eucalyptol under batch and continuous flow conditions. ACS Sustain Chem Eng 4:6186–6192
142. Zhu W, Yang H, Chen J, Chen C, Guo L, Gan H, Zhao X, Hou Z (2014) Efficient hydrogenolysis of cellulose into sorbitol catalyzed by a bifunctional catalyst. Green Chem 16:1534–1542
143. Insyani R, Verma D, Kim SM, Kim J (2017) Direct one-pot conversion of monosaccharides into high-yield 2,5-dimethylfuran over a multifunctional Pd/Zr-based metal−organic Framework@Sulfonated graphene oxide catalyst. Green Chem 19:2482–2490
144. Hutchins RO, Hutchins MK, Fleming I (1991) Comprehensive organic synthesis, vol 8. Elsevier, Oxford, p 327
145. Yu S, Oh J, Li F, Kwon Y, Cho H, Shin J, Lee SK, Kim S (2017) New scaffold for angiogenesis inhibitors discovered by targeted chemical transformations of Wondonin natural products. ACS Med Chem Lett 8:1066–1071
146. Bhattacharya and V. Kashaw. (2019) Design, synthesis and anticonvulsant potential of (E)-3-(5-(substituted aminomethyl)-1,3,4-thiadiazol-2-yl)-2-substituted styrylquinazolin-4-(3H)-one. J Drug Delivery Ther 9:591
147. Dierks M, Cao Z, Manayil JC, Akilavasan J, Wilson K, Schüth F, Rinaldi R (2018) Impact of hydrophobic organohybrid silicas on the stability of Ni_2P catalyst phase in the hydrodeoxygenation of biophenols. ChemCatChem 10:2219–2231
148. Wang Z, Feng X, Yang S, Xiao L, Wu W (2020) Influence of acidity on the catalytic performance of Ni_2P in liquid-phase hydrodeoxygenation of furfural to 2-methylfuran. J Nanopart Res 22:67
149. Hou Z, Theyssen N, Leitner W (2007) Palladium nanoparticles stabilised on PEG-modified silica as catalysts for the aerobic alcohol oxidation in supercritical carbon dioxide. Green Chem 9:127–132
150. Chen Y, Guo Z, Chen T, Yang Y (2010) Surface-functionalized TUD-1 mesoporous molecular sieve supported palladium for solvent-free aerobic oxidation of benzyl alcohol. J Catal 275: 11–24
151. Weerachawanasak P, Hutchings GJ, Edwards JK, Kondrat SA, Miedziak PJ, Prasertham P, Panpranot J (2015) Surface functionalized TiO_2 supported Pd catalysts for solvent-free selective oxidation of benzyl alcohol. Catal Today 250:218–225

152. Zheng L-X, Peng B, Zhou J-F, Shan B-Q, Xue Q-S, Zhang K (2022) High efficient and stable thiol-modified dendritic mesoporous silica nanospheres supported gold catalysts for gas-phase selective oxidation of benzyl alcohol with ultra-long lifetime. Microporous Mesoporous Mater 342:112140
153. Shahriari M, Sedigh MA, Mahdavian Y, Mahdigholizad S, Pirhayati M, Karmakar B, Veisi H (2021) In situ supported Pd NPs on biodegradable chitosan/agarose modified magnetic nanoparticles as an effective catalyst for the ultrasound assisted oxidation of alcohols and activities against human breast cancer. Int J Biol Macromol 172:55–65
154. Chutimasakul T, Uetake Y, Tantirungrotechai J, Asoh TA, Uyama H (2020) Sakurai size-controlled preparation of gold nanoparticles deposited on surface-fibrillated cellulose obtained by citric acid modification. ACS Omega 5:33206–33213
155. Zhang P, Qiao ZA, Jiang X, Veith GM, Dai S (2015) Nanoporous ionic organic networks: stabilizing and supporting gold nanoparticles for catalysis. Nano Lett 15:823–828
156. Restrepo J, Lozano P, Burguete MI, García-Verdugo E, Luis SV (2015) Gold nanoparticles immobilized onto supported ionic liquid-like phases for microwave phenylethanol oxidation in water. Catal Today 255:97–101
157. de Vries JG (2006) A unifying mechanism for all high-temperature heck reactions. The role of palladium colloids and anionic species. Dalton Trans 3:421–429
158. Phan NTS, Van Der Sluys M, Jones CW (2006) On the nature of the active species in palladium catalyzed Mizoroki–Heck and Suzuki–Miyaura couplings – homogeneous or heterogeneous catalysis, a critical review. Adv Synth Catal 348:609–679
159. Pagliaro M, Pandarus V, Ciriminna R, Béland F, Carà PD (2012) Heterogeneous versus homogeneous palladium catalysts for cross-coupling reactions. ChemCatChem 4:432–445
160. Corma A, Garcia H, Leyva A (2005) Catalytic activity of palladium supported on Single Wall carbon nanotubes compared to palladium supported on activated carbon: study of the heck and Suzuki couplings, aerobic alcohol oxidation and selective hydrogenation. J Mol Catal A Chem 230:97–105
161. Li SH, Wang J, Kou Y, Zhang S (2010) Ionic-liquid-grafted rigid poly(*p*-phenylene) microspheres: efficient heterogeneous media for metal scavenging and catalysis. Chem A Eur J 16: 1812–1818
162. Gruttadauria M, Liotta LF, Salvo AMP, Giacalone F, La Parola V, Aprile C, Noto R (2011) Multi-layered, covalently supported ionic liquid phase (mlc-SILP) as highly cross-linked support for recyclable palladium catalysts for the Suzuki reaction in aqueous medium. Adv Synth Catal 353:2119–2130
163. Pavia C, Ballerini E, Bivona LA, Giacalone F, Aprile C, Vaccaro L, Gruttadauria M (2013) Palladium supported on cross-linked imidazolium network on silica as highly sustainable catalysts for the Suzuki reaction under flow conditions. Adv Synth Catal 355:2007–2018
164. Pavia C, Giacalone F, Bivona LA, Salvo AMP, Petrucci C, Strappaveccia G, Vaccaro L, Aprile C, Gruttadauria M (2014) Evidences of release and catch mechanism in the heck reaction catalyzed by palladium immobilized on highly cross-linked-supported imidazolium salts. J Mol Catal A Chem 387:57–62
165. Costa NJS, Kiyohara PK, Monteiro AL, Coppel Y, Philippot K, Rossi LM (2010) A single-step procedure for the preparation of palladium nanoparticles and a phosphine-functionalized support as catalyst for Suzuki cross-coupling reactions. J Catal 276:382–389
166. Kumar S, Singh S, Mathur N, Roy P, Joshi H (2023) Titania Nanorod-supported mercaptoundecanoic acid-grafted palladium nanoparticles as a highly reusable heterogeneous catalyst for substrate-dependent Ullmann coupling and debromination of aryl bromides. Inorg Chem 62:3993–4002
167. Mishra B, Ghosh D, Tripathi BP (2022) Finely dispersed AgPd bimetallic nanoparticles on a polydopamine modified metal organic framework for diverse catalytic applications. J Catal 411:1–14

168. Shaikh NM, Sawant AD, Bagihalli GB, Challa M, Adimule VM (2022) Highly active mixed Au–Pd nanoparticles supported on RHA silica through immobilised ionic liquid for Suzuki coupling reaction. Top Catal
169. Aalinejad M, Pesyan Noroozi N, Alamgholiloo H (2022) Stabilization of Pd–Ni alloy nanoparticles on Kryptofix 23 modified SBA-15 for catalytic enhancement. Colloids Surf A Physicochem Eng Asp 634:127956
170. Massaro M, Riela S, Cavallaro G, Gruttadauria M, Milioto S, Noto R, Lazzara G (2014) Eco-friendly functionalization of natural Halloysite clay nanotube with ionic liquids by microwave irradiation for Suzuki coupling reaction. J Organomet Chem 749:410–415
171. Bordet A, Leitner W (2023) Adaptive catalytic systems for chemical energy conversion. Angew Chem Int Ed 62:e202301956
172. Bordet A, Lacroix L-M, Fazzini P-F, Carrey J, Soulantica K, Chaudret B (2016) Magnetically induced continuous CO_2 hydrogenation using composite iron carbide nanoparticles of exceptionally high heating power. Angew Chem Int Ed 55:15894–15898
173. Kreissl H, Jin J, Lin S-H, Schüette D, Störtte S, Levin N, Chaudret B, Vorholt A, Bordet A, Leitner W (2021) Commercial $Cu_2Cr_2O_5$ decorated with iron carbide nanoparticles as multifunctional catalyst for magnetically induced continuous flow hydrogenation of aromatic ketones. Angew Chem Int Ed 60:26639–26646
174. Lin S-H, Hetaba W, Chaudret B, Leitner W, Bordet A (2022) Copper-decorated iron carbide nanoparticles heated by magnetic induction as adaptive multifunctional catalysts for the selective Hydrodeoxygenation of aldehydes. Adv Energy Mater 12:2201783
175. Zhang W, Wang L, Wang K, Khan MU, Wang M, Li H, Zeng J (2016) Integration of Photothermal effect and heat insulation to efficiently reduce reaction temperature of CO_2 hydrogenation. Small 13:1602583
176. Li Y, Hao J, Song H, Zhang F, Bai X, Meng X, Zhang H, Wang S, Hu Y, Ye J (2019) Selective light absorber-assisted single nickel atom catalysts for ambient sunlight-Diven CO_2 Methanation. Nat Commun 10:2359–2368
177. Stuart MAC, Huck WTS, Genzer J, Mller M, Ocer C, Stamm M, Sukhorukov GB, Szleifer I, Tsukruk VV, Urban M, Winnik F, Zauscher S, Luzinov I, Minko S (2010) Emerging applications of stimuli-responsive polymer materials. Nat Mater 9:101–113
178. Poole DA, Mathew S, Reek JNH (2021) Just add water: modulating the structure-derived acidity of catalytic hexameric resorcinarene capsules. J Am Chem Soc 143:16419–16427
179. Koumura N, Zijlstra RWJ, van Delden RA, Harada N, Feringa BL (1999) Light-driven monodirectional molecular rotor. Nature 401:153–155
180. Wei Y, Han S, Kim J, Soh S, Grzybowski BA (2010) Photoswitchable catalysis mediated by dynamic aggregation of nanoparticles. J Am Chem Soc 132:11018–11020
181. Kean ZS, Akbulatov S, Tian Y, Widenhoefer RA, Boulatov R, Craig SL (2014) Photomechanical actuation of ligand geometry in enantioselective catalysis. Angew Chem Int Ed 53: 14508–14511
182. Peris E (2018) Smart N-heterocyclic carbene ligands in catalysis. Chem Rev 118:9988–10031
183. Ruiz-Zambrana C, Gutiérrez-Blanco A, Gonell S, Poyatos M, Peris E (2021) Redox-switchable cycloisomerization of alkynoic acids with napthalenediimide-derived N-heterocyclic carbene complexes. Angew Chem Int Ed 60:20003–20011
184. Snell KD, Keenan AG (1979) Surface modified electrodes. Chem Soc Rev 8:259–282
185. Wrighton MS (1986) Surface functionalization of electodes with molecular reagents. Science 231:32–37
186. Wang S, Yang F, Jiang SP, Chen S, Wang X (2010) Tuning the electrocatalytic activity of Pt nanoparticles on carbon nanotubes via surface functionalization. Electrochem Commun 12: 1646–1649
187. Zhang M, Xie J, Sun Q, Yan Z, Chen M, Jing J (2013) Enhanced electrocatalytic activity of high Pt-loadings on surface functionalized graphene nanosheets for methanol oxidation. Int J Hydrogen Energy 38:16402–16409

188. Xin L, Yang F, Rasouli S, Qiu Y, Li Z-F, Uzunoglu A, Sun C-J, Liu Y, Ferreira P, Li W, Ren Y, Stanciu LA, Xie J (2016) Understanding Pt nanoparticle anchoring on graphene supports through surface functionalization. ACS Catal 6:2642–2653
189. Han C, Zenner J, Johny J, Kaeffer N, Bordet A, Leitner W (2022) Electrocatalytic hydrogenation of alkenes with Pd/Carbon nanotubes at an oil–water interface. Nat Catal 5:1110–1119
190. Zagoraiou E, Paloukis F, Daletou MK (2021) Correlations between properties and activity of platinum supported on modified carbon nanotubes with pyridines. Appl Surf Sci 567:150749
191. Chen Y, Ma Y, Zhou Y, Huang Y, Li S, Chen Y, Wang R, Tang J, Wu P, Zhao X, Chen C, Zhu Z, Chen S, Cheng K, Lin D (2022) Enhanced methanol oxidation on PtNi nanoparticles supported on silane-modified reduced graphene oxide. Int J Hydrogen Energy 47:6638–6649
192. Li S, Liang X, Shen S, Yang H, Wu CL (2021) Surface engineering of flower-like ionic liquid-functionalized graphene anchoring palladium nanocrystals for a boosted ethanol oxidation reaction. Inorg Chem 60:17388–17397
193. Kim JH, Cheon JY, Shin TJ, Park JY, Joo SH (2016) Effect of surface oxygen functionalization of carbon support on the activity and durability of Pt/C catalysts for the oxygen reduction reaction. Carbon 101:449–457

Top Organomet Chem (2024) 75: 47–104
https://doi.org/10.1007/3418_2024_120
© The Author(s), under exclusive license to Springer Nature Switzerland AG 2024
Published online: 27 September 2024

Metal Nanoparticles on Polymeric Membranes Applied in Catalytic Hydrogenations

Rosa Pich, Jean-François Lahitte, Jean-Christophe Remigy, Daniel Pla, and Montserrat Gómez

Contents

Abstract Polymeric catalytic membranes, adapted for the immobilization of metal nanoparticles (MNPs) with the objective of being applied in hydrogenation reactions, are reviewed in this chapter, highlighting the works including a rational design and full characterization of the catalytic membrane reactors. The most relevant

R. Pich
Laboratoire Hétérochimie Fondamentale et Appliquée, UMR CNRS 5069, Toulouse, Cedex 09, France

Laboratoire de Génie Chimique, Université de Toulouse; INPT, UPS, UMR CNRS 5503, Toulouse, Cedex 09, France

J.-F. Lahitte and J.-C. Remigy
Laboratoire de Génie Chimique, Université de Toulouse; INPT, UPS, UMR CNRS 5503, Toulouse, Cedex 09, France

D. Pla and M. Gómez (✉)
Laboratoire Hétérochimie Fondamentale et Appliquée, UMR CNRS 5069, Toulouse, Cedex 09, France
e-mail: montserrat.gomez-simon@univ-tlse3.fr

contributions concern zero-valent MNPs (mono- and bimetallic systems) of late transition metals (groups 8–10), involving a significant variety of (un)functionalized polymers. Both batch and continuous-flow processes are discussed, with the latter being of particular interest for large-scale applications. The literature analysis carried out evidences that there is still room for developing innovative polymeric catalytic membranes, regarding both design and applications. Thus, for instance, well-defined stereo-controlled polymeric membranes for enantioselective hydrogenations or multi-step processes involving a series of polymeric catalytic reactors would be of interest, particularly for the production of fine chemicals under smooth conditions.

Keywords Catalytic polymeric membranes · Flow chemistry · Hydrogenation · Metal nanoparticles · Molecular hydrogen activation · Nanocatalysis · Process intensification

1 Introduction

According to the IUPAC definition, membrane reactors (MRs) are devices for simultaneously carrying out a reaction and a membrane-based separation in the same apparatus [1, 2]. Thus, in particular, catalytic membrane reactors (CMRs) allow the implementation of both catalytic transformations and separation of reagents/products from the catalyst, joining different unit operations in line with process intensification strategies leading to low energy consumption, small footprint, and improved efficiency [3]. Moreover, for some specific applications, CMRs enable coupling endothermic and exothermic reactions, for instance in dehydrogenation/hydrogenation processes using palladium membranes [4]. Conventionally, inorganic membranes have been applied in gas phase processes that require high temperatures [5, 6]. In contrast, the more versatile polymeric catalytic membranes (porous or dense) find applications in processes working at relatively low temperature (<150 °C) [7], especially those transformations taking place in liquid phases for both fine and bulk chemical synthesis including bio-catalytic transformations, where the catalyst provides conversion and the polymeric membrane is responsible for transfer phenomena [8–11]. Other than catalytic processes, it is important to highlight that polymer-based membranes have been largely applied in separation processes, gas or liquid permeation, and also water purification [12], aspects outside the scope of this chapter.

In this frame, metal-based catalytic hydrogenations represent an important field permitting to work under relatively low pressure and temperature, keeping high catalytic performance. Concerning the nature of metals, those from group 10 (Ni, Pd, and Pt) are among the most efficient, with palladium occupying a privileged place [13]. In the 1960s, Gryaznov introduced for the first time the concept of CMR using non-porous palladium membranes [14]. However due to their high cost, in the 1970s and 1980s, composite membranes were firstly developed by the same group

[15, 16]. Later on, mainly at the beginning of the twenty-first century, well-defined metal-based nanoparticles immobilized on polymeric membranes have been described for different types of applications [17], in particular in catalysis [18].

In this contribution, we focus on catalytically active membrane reactors coming from the immobilization of metal-based nanoparticles on polymeric membranes for hydrogenation processes (molecular hydrogen activation). Consequently, those containing heterogenized organometallic complexes are outside the scope of this review. More specifically, this chapter contains a section dedicated to the design and principles of catalytic polymeric membranes for hydrogenation purposes, followed by an analysis of the reported works classified by type of metal, highlighting the most relevant contributions related to group 10 elements (mainly Pd and Pt) and other transition metals (mostly from groups 8 and 9), including bimetallic-based polymeric membrane reactors. Contributions involving reduction processes using borohydride derivatives instead of molecular hydrogen have not been considered [19]. Some perspectives are also addressed as promising research outcomes.

The polymers discussed in this chapter are summarized in Table 1, including their names, acronyms, and structures.

2 General Aspects of Polymeric Catalytic Membranes

2.1 Concept and Design of Polymeric Catalytic Membranes

Drioli and coworkers classified catalytic membranes into three main categories depending on their role in catalytic reactivity [10, 18], as follows:

- Active catalytic membrane: the catalyst is integrated into the membrane.
- Intrinsically catalytic membrane: the membrane material exhibits inherent catalytic properties (e.g., acid-promoted catalysis).
- Inert catalytic membrane: the membrane retains the catalyst in the reactor, also able to extract a reactant (for example to shift a reaction equilibrium or avoid side reactions).

For the two first configurations, a reactive separation involves both catalytic reaction and separation as described by Stankiewicz [20]. In each of these three cases, the membrane can improve the contact between the different phases, particularly in filtration mode by convection, forcing the different reactants to be encountered. This last aspect can explain the higher efficiency of flow-based CMRs compared to batch reactors in some processes.

Miachon et al. proposed another classification based on the three different roles that a CMR can show [21], as follows (Fig. 1):

- Extractor: the membrane selectively separates the products from the reagents thanks to its permselectivity with respect to one of the products causing the shift in the reaction equilibrium and increasing the reaction yields.

Table 1 Polymers (names, acronyms, and structures) discussed in this chapter

Acronym	Polymer name	Polymer structure
CA	Cellulose acetate	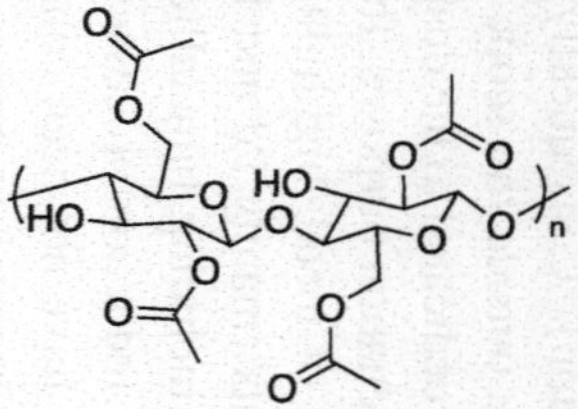
CS	Chitosan	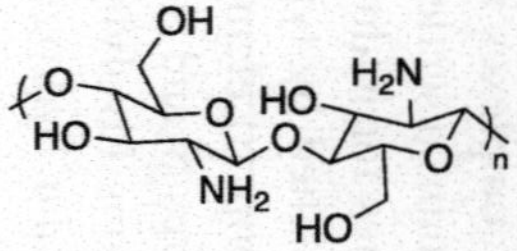
EC	Ethylcellulose	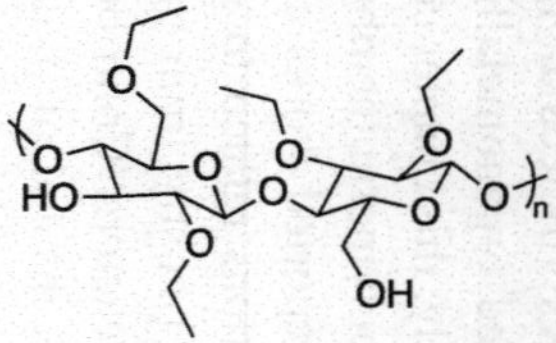
ePTFE	Expanded porous polytetrafluoroethylene	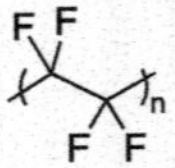

MTD-co-Cp[N]	Methyltetracyclododecene-co-*endo*-2-(cyclopentadienylmethyl)norborn-5-ene	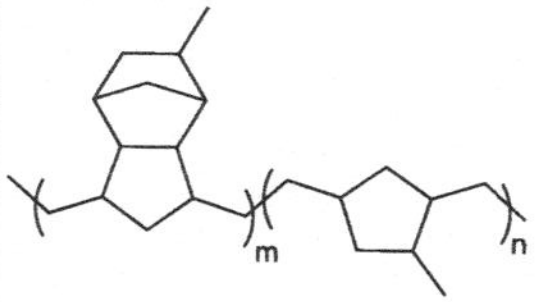
Monobor	Styrene-divinylbenzene polymer bearing triphenylborate sodium groups	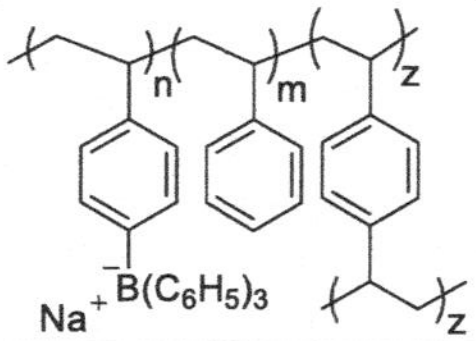
Nafion 117	Tetrafluoroethylene-perfluoro-3,6-dioxa-4-methyl-7-octenesulfonic acid copolymer	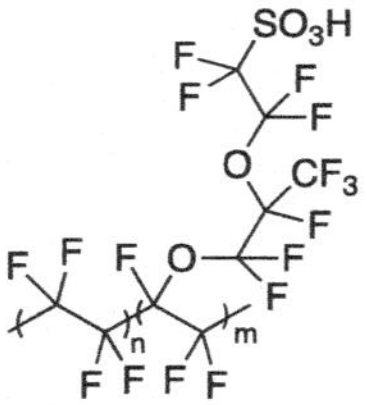
PSPI	Photosensitive polyimide	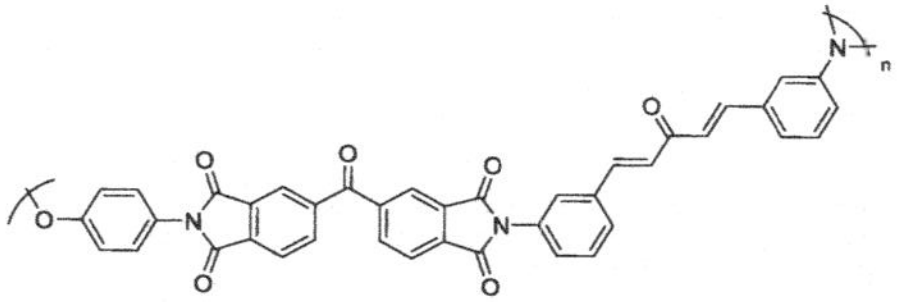
PAA	Poly(acrylic acid)	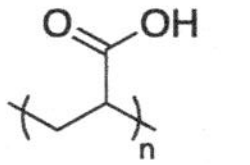

(continued)

Table 1 (continued)

Acronym	Polymer name	Polymer structure
PAI	Poly(amide imide)	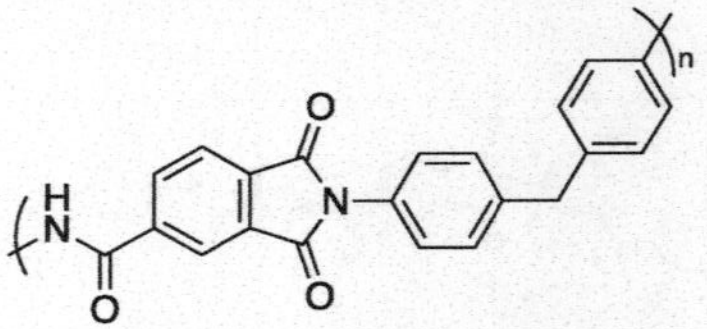
PAI-F	Poly(amide imide)-co-hexafluoroisopropylidene	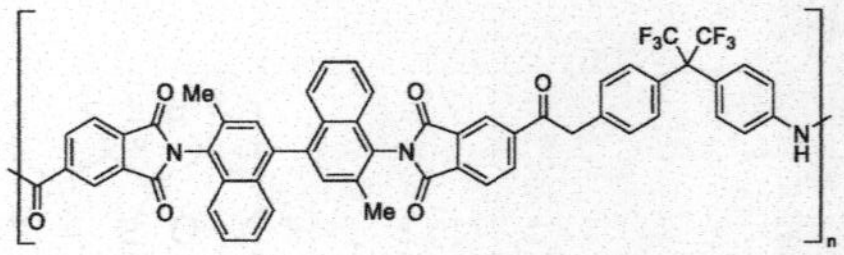
PAH	Poly(allylamine hydrochloride)	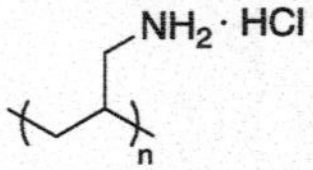
PAM	Poly(acrylamide)	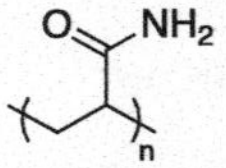
PAMAM	Poly(amidoamine)	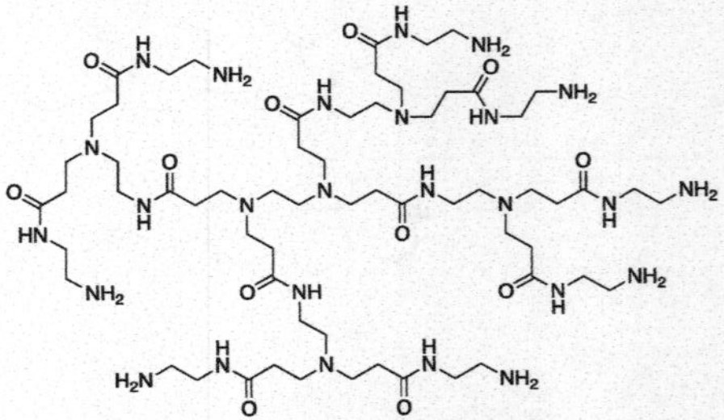

PANI	Polyaniline	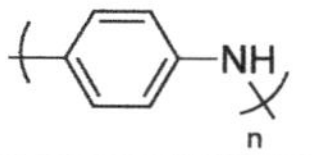
PCS	*N*-Propylchitosan	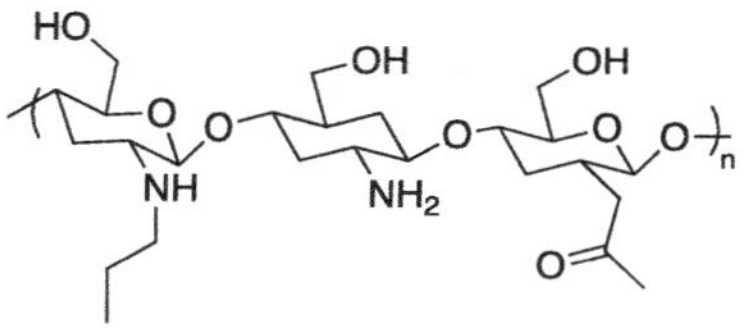
PDA	Polydopamine	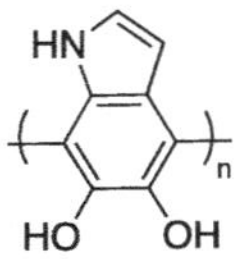
PDMS	Poly(dimethylsiloxane)	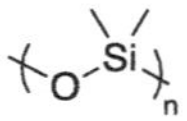
P(DVD-DIIL)	Poly(divinylbenzene-co-dication imidazole) bromide	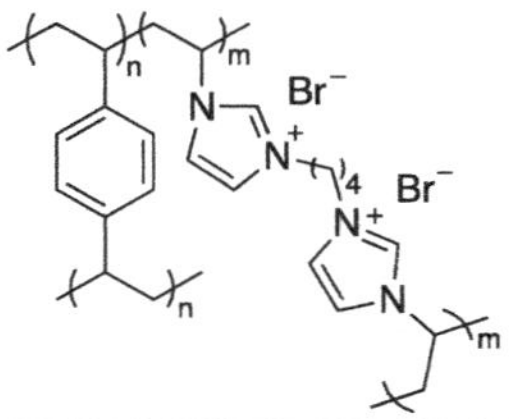

(continued)

Table 1 (continued)

Acronym	Polymer name	Polymer structure
PEEK-WC	Polyetheretherketone with carbo(lactone) groups	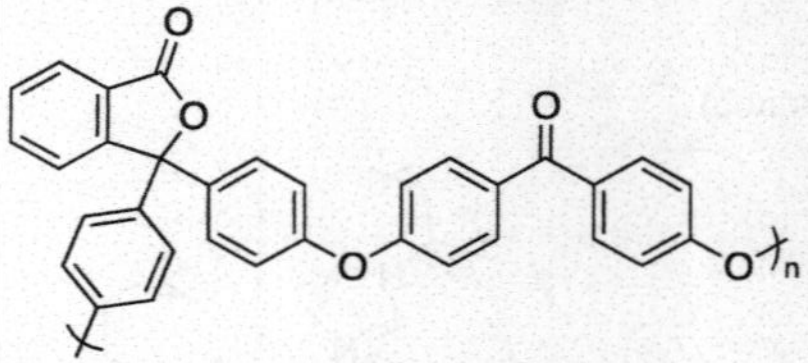
PEI	Polyethylenimine	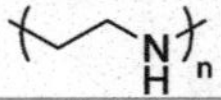
PEID	Polyetherimide	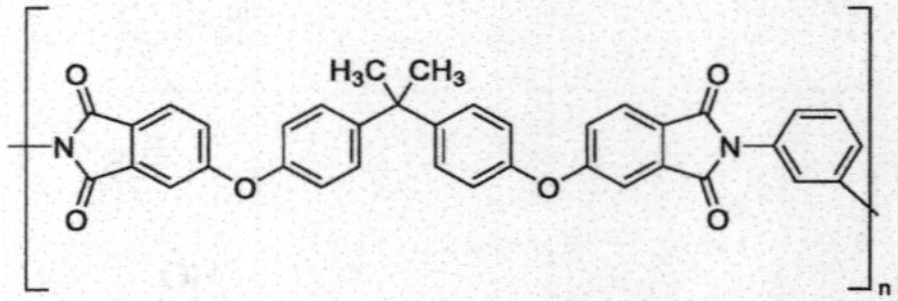
PEO	Polyethylene oxide	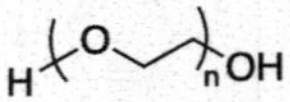
PES	Polyethersulfone	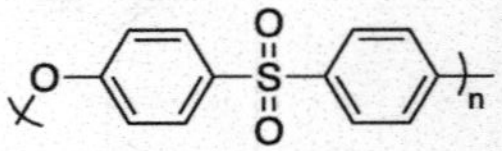
PET	Polyethyleneterephthalate	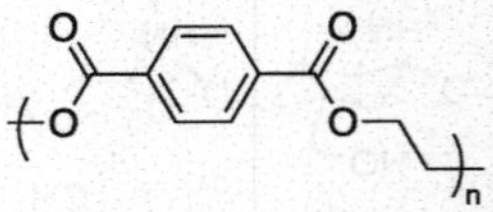

PLA	Polylactic acid	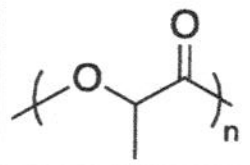
PLA-b-PS	Polylactic-b-polystyrene diblock copolymer	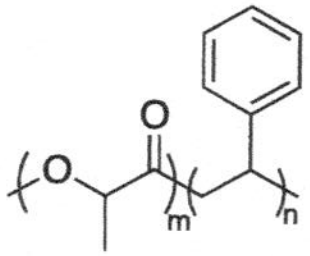
PLA-CTA	Polylactic chain transfer agent	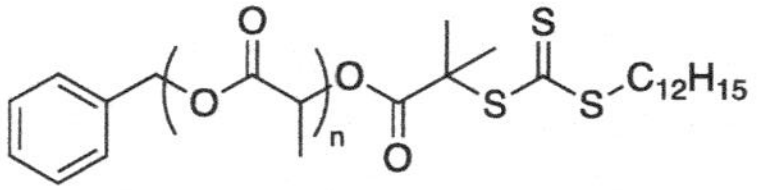
PP	Polypropylene	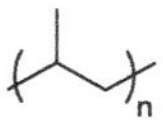
PPS	Polyphenylene sulfide	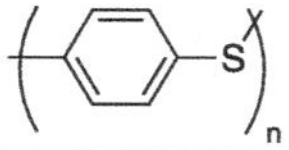
PS	Polystyrene	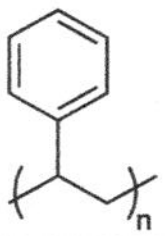
PSF	Polysulfone	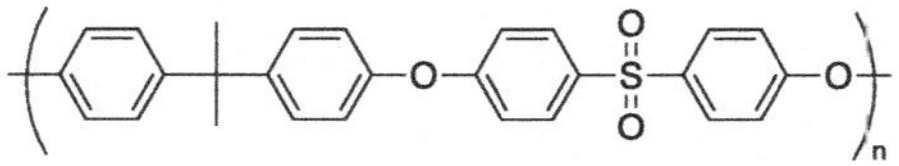

(continued)

Table 1 (continued)

Acronym	Polymer name	Polymer structure
PSS	Polystyrene sulfonate	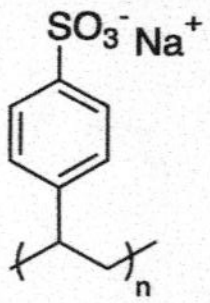
PTFE	Polytetrafluoroethylene	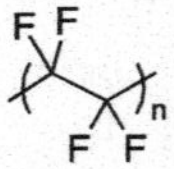
PVDF	Polyvinylidene fluoride	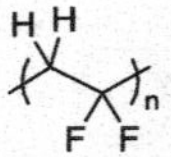
PVP	Polyvinylpyrrolidone	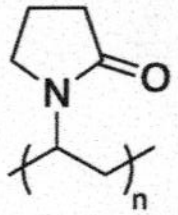
AF-2400	Poly[4,5-difluoro-2,2-bis(trifluoromethyl)-1,3-dioxole-co-tetrafluoroethylene]	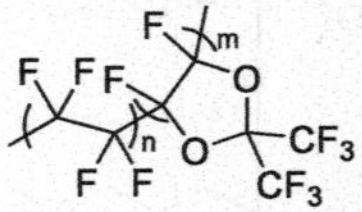

Tosflex	Tetrafluoroethylene fluoropolymer-copolymer	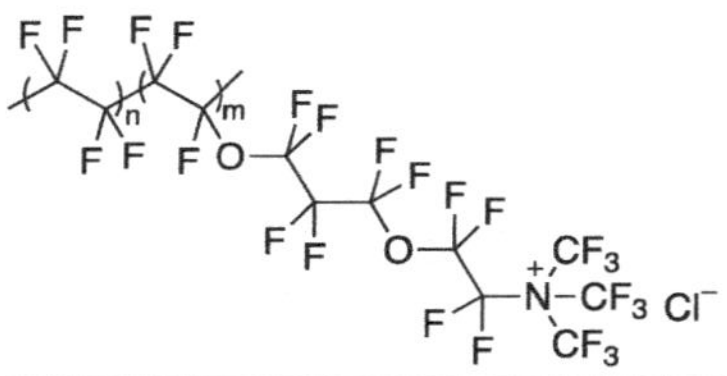
PS-co-PVBAH	*N*-(4-Vinylbenzyl)-*N*,*N*-dibutylamine hydrochloride-co-styrene	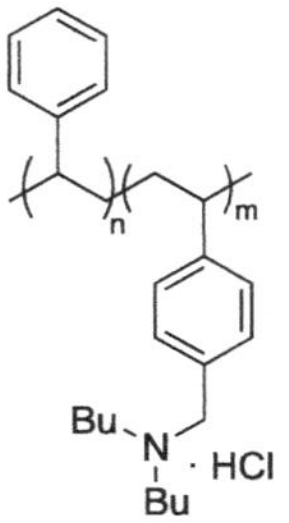
TentaGel-NH_2	*O*-(2-Aminoethyl)polyethylene glycol polymer-bound	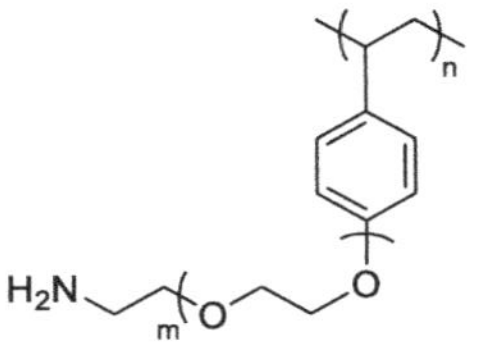

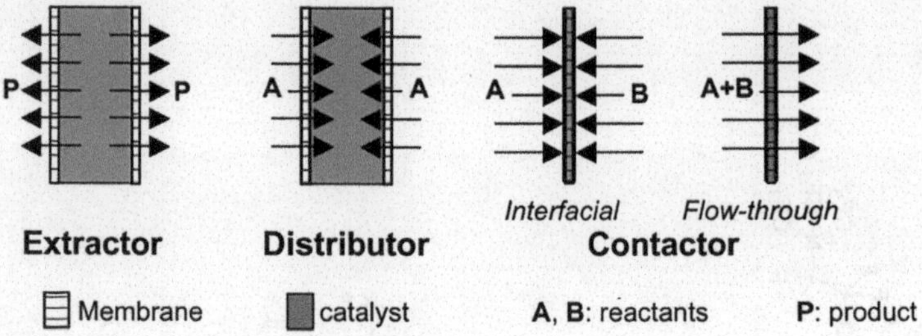

Fig. 1 Schematic representation of the different catalytic membrane reactors. Reprinted from [21]. Copyright (2004), with permission from Springer Nature

- Distributor: the membrane controls the quantity of reagents according to their properties and transfer conditions; this operating mode is particularly suitable for gas/liquid contactors.
- Contactor: the membrane promotes contact between the reagents and the catalyst; the reaction takes place in the membrane. This working mode can be divided into two sub-categories:
  - Interfacial contactor: the reagents are introduced from each side of the membrane.
  - Flow-through contactor: this is the closest mode to traditional filtration; reactants are forced to pass through the active pores of the membrane. Preferentially, the membrane must not exhibit retention with respect to one of the products.

One example of a contactor membrane has been described by Remigy, Lahitte, Gómez, and co-workers using the flow-through contactor mode for hydrogenation reactions with polymeric membranes [22] (Fig. 2).

Alternatively, Venezia et al. used the distributor contactor mode for cyclohexene hydrogenation, feeding the inside of a hollow fiber with a solution of nitrobenzene and the outside with H_2 (Fig. 3) [23].

It should be noted that in some reported works the use of a membrane simply consists of an addition of a piece of catalytic membrane to a batch reactor without taking advantage of the configuration cited above [24]. Thus, the term catalytic membrane should not be used in these cases to avoid misunderstandings.

Regarding the design of polymeric catalytic membranes, most metal-based catalysts involve the immobilization of MNPs, in particular for hydrogenation purposes. As known, metal-based nanoparticles tend to aggregate due to their high-energy surfaces, often losing their catalytic properties; for that reason, it is necessary to stabilize them in order to increase their lifetime and efficiency [25]. Moreover, metal leaching should be avoided during the filtration process [26]. The main strategy developed in the literature to achieve these goals is based on an electrosteric stabilization of MNPs by polyelectrolytes deposited inside or at the surface of the membrane, or even using a membrane constituted of polyelectrolytes [27]. The presence of functional groups in the membranes often permits to immobilize metal

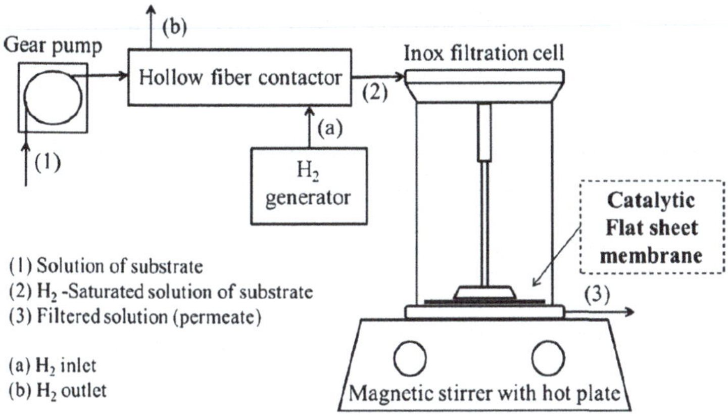

Fig. 2 Schematic representation corresponding to the setup for continuous-flow hydrogenation reactions using flat-sheet membranes in flow-through configuration. Reprinted from [22], Copyright (2021), with permission from Elsevier

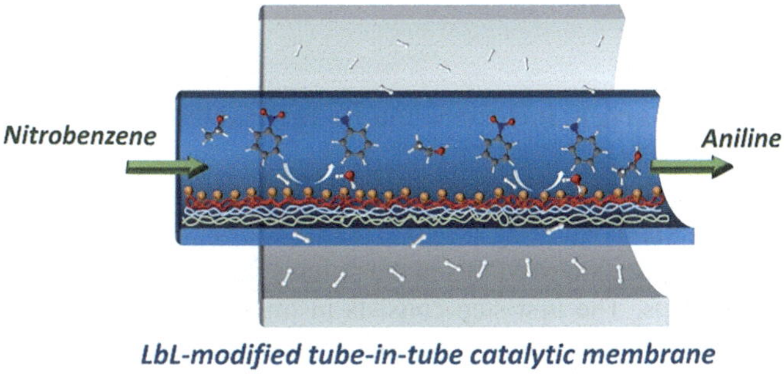

Fig. 3 Distributor contactor for cyclohexene hydrogenation. Reprinted from [23], Copyright (2021), with permission from Elsevier

precursors by ion-pair interactions (e.g., SO_3^- or COO^- and NR_3^+) or by interaction with Lewis bases, such as amino, pyridyl, benzimidazoyl, thiol, methylsulfide, or nitrile groups [28]. Regarding the immobilization of MNPs on polymeric membranes, two synthetic ways are commonly applied. The so-called ex situ route, which consists of preforming MNPs and subsequently incorporating them into the polymer matrix, also known as direct blending. The in situ approach is based on carrying out the synthesis of MNPs directly in the presence of the polymeric membrane from the corresponding metal precursor [29]. In the case of ionic polymeric membranes (presence of polyelectrolytes), an ion exchange between the metallic salt precursor and the ionic function present in the membrane takes place. This approach is known as intermatrix synthesis (IMS) [29].

Hereafter, the two main approaches to immobilize MNPs on polymeric membranes are detailed, i.e., direct blending and intermatrix synthesis.

2.1.1 Metal Nanoparticle Incorporation by Direct Blending

A first approach to obtain catalytic polymeric membranes consists in incorporating the nanocatalyst during the membrane preparation process. MNPs are first prepared in solution and incorporated into the polymer solution. The solvents of the nanoparticle solution and of the polymer solution must be compatible to avoid aggregation or polymer precipitation during the blending process. Hereafter, we highlight some representative examples of this approach. Peinmann et al. obtained a catalytic polymeric membrane for catalytic nitrate reduction by hydrogen incorporating MNPs (4.45 wt.% palladium and 0.95 wt.% copper) with a size of 2.5 μm in a polyetherimide solution [30]. The quantity of catalyst represented 27% of the mass of the material. On the other hand, Oliveira et al. synthesized Rh and Pd nanoparticles in the ionic liquid [BMI]BF_4 (BMI = 1-*n*-butyl-3-methylimidazolium) and directly incorporated them in a solution of cellulose acetate to be applied in cyclohexene hydrogenation [24]. Groschel et al. incorporated PdNPs in a block copolymer, constituted of poly(acrylic acid) (PAA) and a difunctional epoxide acting as crosslinker; this catalytic material was efficiently applied in the hydrogenation of propyne [31].

2.1.2 Metal Nanoparticle Incorporation by Intermatrix Synthesis

IMS is based on the ion exchange between the salt precursor and the ionic function present in the material containing the ionogenic groups coming from polyelectrolytes [29]. The incorporation of MNPs in the polymeric catalytic membranes is often achieved in two steps. The first step consists in impregnating the matrix with the metallic salt, followed by the reduction of the metallic salt toward zero-valent MNPs. This latter step generally consists of treating the material with a reducing agent (e.g., $NaBH_4$ or H_2), or less often, it can also be obtained by thermo-degradation [32]. Ruiz et al. proposed the following stages corresponding to ion exchange and reduction reaction, from the metallic salt to nanoparticle formation [33]:

$$2\,Na^+\,R-SO_3{}^- + \left[Pd(NH_3)_4\right]^{2+} \rightarrow \left[Pd(NH_3)_4\right](R-SO_3{}^-)_2 + 2\,Na^+ \quad (1)$$

$$\left[Pd(NH_3)_4\right](R-SO_3{}^-)_2 + 2\,Na^+ \rightarrow 2\,Na^+R-SO_3{}^- + \left[Pd(NH_3)_4\right]^{2+} \quad (2)$$

$$\left[Pd(NH_3)_4\right]^{2+} + 2\,BH_4{}^- + 6\,H_2O \rightarrow Pd^0 + 7\,H_2 + 2\,B(OH)_3 + 4\,NH_3 \quad (3)$$

This method permits to obtain small spherical MNPs with narrow size and homogeneous distribution in the matrix, preventing the formation of aggregates. Domènech et al. explained this behavior by the dual function of the matrix, stabilizing MNPs and also avoiding their aggregation by steric effects of the polymer, which provides a confined environment for the design of nanoreactors [34].

2.2 Functionalization of Polymeric Membranes

As discussed above, the polymeric material must contain ionogenic functions when the IMS strategy is applied. MNPs should be localized on the surface of the open and accessible pores of the membrane to avoid diffusion issues that could prevent the desired reactivity. Thus, a general strategy consists of placing the ionogenic groups mainly on the membrane surface at the direct interface between the polymeric membrane and the solution where the catalytic reaction takes place. In this regard, several operating modes can be applied, as detailed in the following subsections.

2.2.1 Native Complexing Polymer

In this method, the polymeric membrane is made of a native charged or complexing polymer. Poly(acrylic acid) is often used due to its ease of polymerization affording a highly charged matrix. A neutral polymer can also be modified at a later stage. For instance, sulfonation strategies have been largely explored in order to graft ionogenic groups to the polymeric chains of the membrane [26]. However, it should be taken into account that charged polymers are highly soluble in water, which can complicate the preparation of membrane supports with convenient mechanical properties. To overcome this limitation, crosslinking steps or low functionalization of the parent matrixes are commonly applied [35, 36].

Chitosan is a very effective polymer for the incorporation of MNPs. Indeed, this amino polysaccharide polymer is characterized by its great affinity for metal ions, which can be easily adsorbed either by chelation or by ion exchange [37, 38]. The presence of amino functions induces chelating properties for cationic metals, and in its protonated form; this polymer acts as an anion exchanger.

Under harsh reaction conditions (mainly concerning, temperature and pH), the stability of the polymeric matrix shall be taken into account. Indeed, it may be worth using polymeric membranes that present a relative robustness. Volkov et al. functionalized hollow polypropylene fibers with PdNPs by treatment with an alkaline sodium solution (25 wt.%) followed by a classical impregnation/reduction step [39]. The nanoparticles exhibited a diameter between 10 and 300 nm with aggregates between 200 nm and 10 mm. The fibers were then used to deoxygenate water by hydrogenation of dissolved oxygen. In contrast, the MNPs are directly incorporated into neutral polymers, where the formation of aggregates is generally observed [40–42].

2.2.2 Coating of a Polymeric Layer/Host Layer

The surface of polymer membranes is generally neutral and consequently a functionalization of their surfaces is often required. For instance, the coating using a polyelectrolyte chain permits to charge the surface of the pores. Two ways of coating are reported in the literature to achieve this goal:

– Polyelectrolyte grafting
– Layer-by-layer polyelectrolyte deposition

Polyelectrolyte Grafting

Polyelectrolyte grafting consists in initiating polymerization from the surface of the membrane to obtain brush-like structures or polymeric gels. The covalent immobilization of the polyelectrolyte at the membrane surface prevents water solubility issues. The most common grafting methods are chemical grafting, plasma induction, high energy radiation, and UV-initiated grafting polymerization (for a recent review, see [43]). Among them, UV-initiated grafting polymerization is a very powerful tool to achieve an efficient surface modification of the polymeric membrane, showing the following advantages [44]:

– Equipment simplicity.
– Possibility to control the grafting of polymer chains with a desired density and precise localization onto the membrane surface.
– High versatility, since there is a large choice of grafting monomers to target different applications.
– High stability: the covalent attachment of polymer chains onto the membrane surface avoids desorption and provides a long-term chemical stability.

The light irradiation of the membrane surface encompasses the formation of radical species due to the homolytic cleavage of chemical bonds in the polymeric chain. These radicals can then initiate a free-radical polymerization process with the monomer in solution. The adjustment of the irradiation energy is crucial to favor the formation of radicals on the surface of the membrane, avoiding the direct polymerization of monomers in solution (the use of filters prevents this type of problem) [45–47]. Using an irradiation pilot with continuous scrolling of a hollow fiber, Remigy and co-workers showed that this technique is particularly suitable for quickly modifying the external surface of hollow fibers in large quantities. The different operating parameters (nature and concentration of the monomer, solvent, viscosity, flow speed, O_2 concentration, and crosslinker amount) make it possible to tune the thickness and porosity of the resulting polymeric materials [45, 46, 48, 49].

Due to its photosensitive nature, polysulfone and polyethersulfone, polymers widely used for water filtration membranes, are particularly suited to surface modification by photografting [50]. Remigy, Lahitte, Gómez, and co-workers used UV-photografting to deposit a grafted layer of polyacrylic acid (PAA) at the surface of polyethersulfone (PES) membrane by IMS with the aim of incorporating PdNPs to be applied in hydrogenation reactions [22], obtaining a grafted layer at the surface of the membrane and well-dispersed PdNPs (Fig. 4).

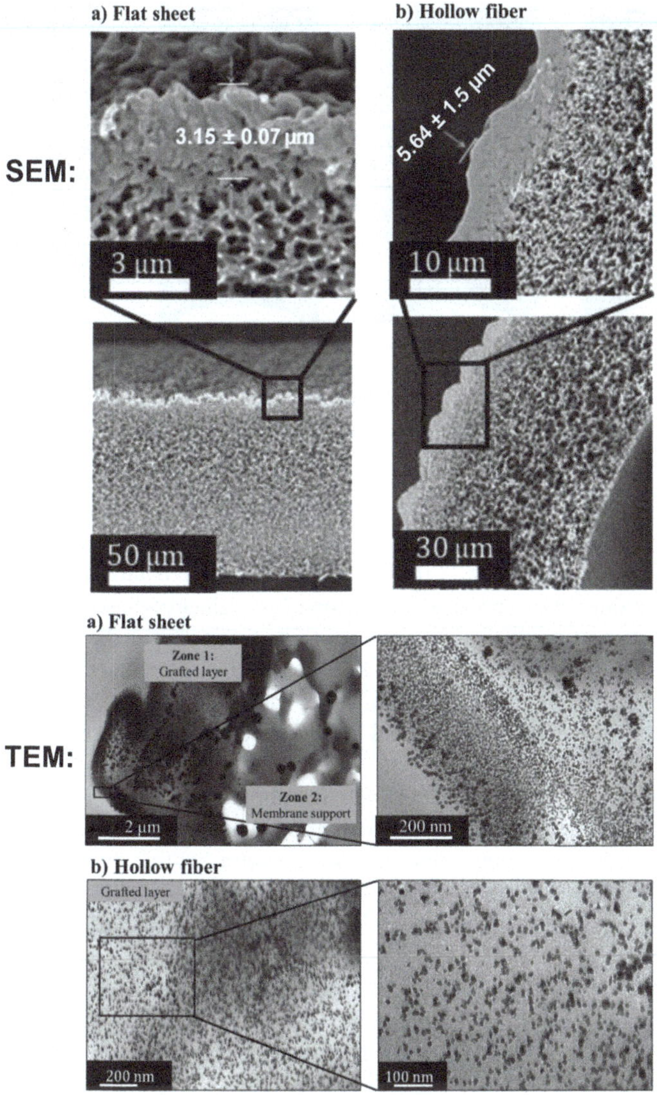

Fig. 4 Top: SEM cross-section micrographs of (**a**) functionalized flat-sheet membrane and (**b**) functionalized hollow fibers. Bottom: TEM micrographs of PdNPs in the grafted PAA layer on (**a**) flat-sheet membrane and (**b**) hollow fiber. Reprinted from [22], Copyright (2021), with permission from Elsevier

Layer-by-Layer Deposition

The layer-by-layer (LBL) method developed by Decher et al. consists in the sequential deposition of oppositely charged polyelectrolytes (PAA), poly(allylamine

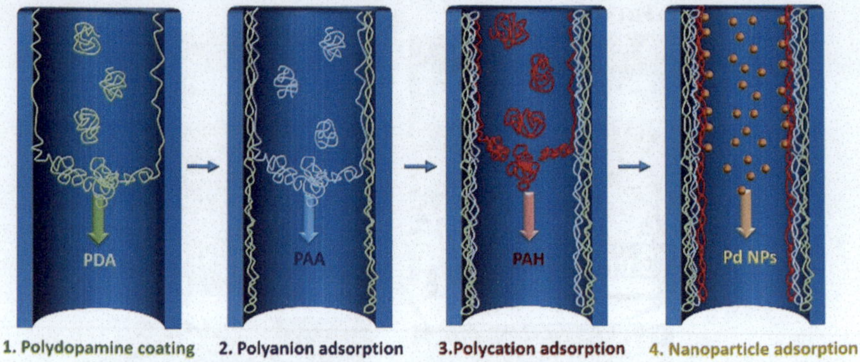

Fig. 5 Graphical representation of the PDA/PAA/PAH layer-by-layer modification of a Teflon AF-2400 tubular membrane, followed by adsorption of PdNPs. Reprinted from [23], Copyright (2021), with permission from Elsevier

hydrochloride) (PAH), or polystyrenesulfonate (PSS)) layers by alternating dipping in polyelectrolyte solutions [51]. This method is simple and allows the surface of the pores to be modified by successive filtration of polyelectrolyte solution. Lahitte and co-workers demonstrated that the number of filtration cycles is directly related to the number of layers and therefore the thickness of the deposited layer [52]. It is suitable for the modification of both flat membranes and hollow-fiber ones, including the modification of the internal channel of the fibers, generally not accessible by photochemical methods [52, 53].

Venezia et al. developed a catalytic membrane based on a modified Teflon AF-2400 membrane for nitrobenzene hydrogenation. PdNPs were incorporated after the layer-by-layer deposition approach, starting with a first coating of PDA (Fig. 5) [23].

Blending of Polymers

Fischer et al. designed nanocomposite materials containing NiNPs immobilized on a PES-based membrane functionalized with ionomers, such as Fumion or sodium dodecyl sulfate (SDS); this functionalization not only avoided NiNPs agglomeration, but also permitted to tune the hydrophobicity/hydrophilicity of the membranes. These composite catalytic materials found applications in the H_2 generation from ammonia-borane in water, which was in situ applied in the 4-nitrophenol reduction to 4-aminophenol [54, 55].

2.3 Efficiency of Catalytic Membrane Reactors

Process intensification using polymeric membranes represents a modern methodology in chemical reaction engineering toward the development of sustainable and cost-effective chemical processes (low energy consumption and limited waste generation), leading to processes on the order of seconds under continuous-flow conditions. The contact time between the reagents and the catalyst is therefore very short in comparison to the time generally needed under batch conditions. In addition, the quantities of metals that are incorporated are generally low. Despite these considerations, high conversion rates are observed, demonstrating the efficiency of the CMR systems.

Remigy, Lahitte, Gómez, and co-workers demonstrated, combining experimental and modeling data, that the efficiency of a CMR resides in three key points [47, 56]. The first one is the locally high concentration of MNPs without aggregations. This concentration can be a thousand times higher than that obtained in a batch reactor using colloidal MNPs; the reaction rate is consequently favored by this same factor. Secondly, thanks to the permeation flux, the transport of reagents toward the catalyst is ensured by a strong convection, which in turn promotes high rates. The small distance between the particles increases the probability for a reactant molecule to be adsorbed and then react on the MNP surface. Finally, as products are continuously separated, the consumption of reagents increases according to Le Chatelier principle [57].

This literature analysis evidences that the strategy of including MNPs within a polymeric membrane is a valuable approach for catalytic purposes. While most research is limited to studying membranes in discontinuous mode, the continuous filtration mode allows a clear improvement in catalytic performance. The conversion rate and selectivity of a reaction are mainly determined by the residence time of the reagents in the membrane. The improvement of the transport of the reactants toward the catalyst increases the performance compared to batch reactor processes, generally limited by mass transfer diffusion. The way in which MNPs are incorporated influences the reactivity of the catalytic membrane: the larger the incorporated amount of MNP, the higher the local concentration of MNPs. An essential aspect is to ensure the stability of MNPs on the membrane by preventing their aggregation and/or their release into the reaction solution.

3 Metal-Based Nanoparticles Immobilized on Polymeric Membranes for Hydrogenation Applications

3.1 Group 10 Elements

3.1.1 Monometallic Nanoparticles of Pd and Ni

Palladium nanoparticles (PdNPs) have attracted significant attention as versatile catalysts due to their unique properties, such as high surface area, tunable size, and remarkable activity for a plethora of catalytic reactions. In particular, the integration of PdNPs into polymeric membrane supports has gained prominence due to their inherent advantages over traditional catalysts, including tunable pore size, enhanced catalyst stability, and simplified catalyst recovery with potential for integration in separation processes, making them suitable for various applications. Thus, PdNPs supported on polymeric membranes have emerged as a promising class of catalysts for a diverse range of chemical reactions [58, 59], although only hydrogenation will be herein discussed. This section provides an overview focusing on bottom-up strategies for the immobilization of PdNPs on polymeric membranes, organized depending on the polymer nature.

Among the polymeric membranes, hydrophobic polypropylene (PP) hollow-fiber membranes have been studied as organic supports for the immobilization of PdNPs using a method consisting of the development of adlayers upon pretreatment of the membrane surface with an organic solvent and an alkaline aqueous solution before the immobilization of the metal precursor. Solvothermal PdNP syntheses were carried out using a PP hollow-fiber membrane (featuring an outer diameter of 300 μm, a wall thickness of 40 μm, and an average pore diameter of 30 nm with 40% porosity) that was treated with a solution of $Pd(OAc)_2$ in MeOH at 25–35 °C for 5–25 h. The sizes of the PdNPs ranged from 10 to 500 nm, although larger aggregates (from 200 nm to tens of microns) were also obtained as estimated by PXRD and observed by SEM [39], paving the way for the elucidation of key stabilization factors for efficient Pd nanocatalysts supported on polymeric membranes.

Ciebien et al. prepared PdNPs immobilized on diblock copolymer films of methyltetracyclododecene-co-*endo*-2-(cyclopentadienylmethyl)norborn-5-ene (MTD-co-Cp^N) (Scheme 1). The resulting polymeric materials present molecular

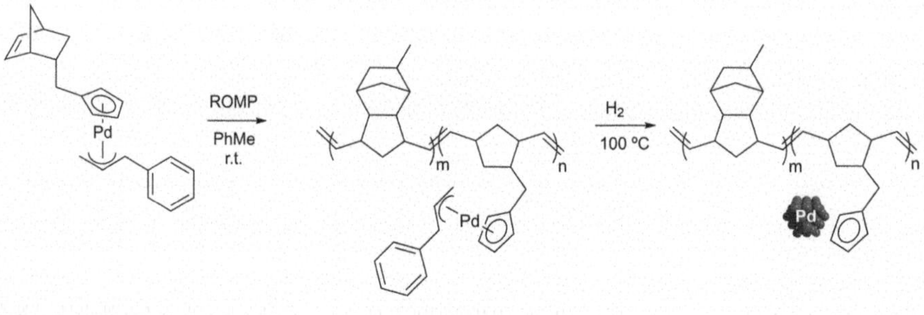

Scheme 1 Synthesis of $[MTD]_{113}[Pd(CpN)PA]_{50}$ [60, 61]

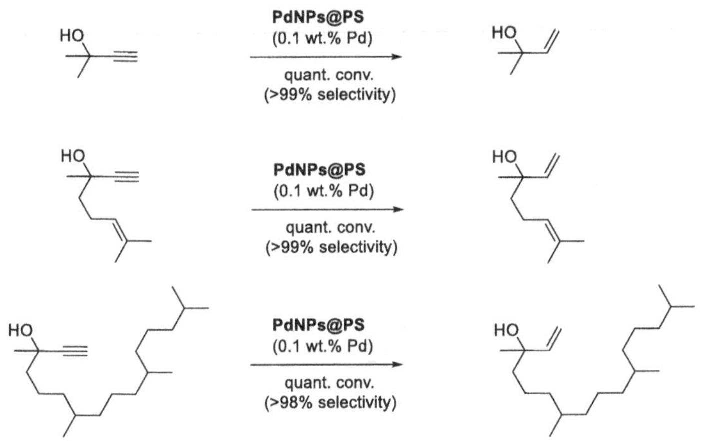

Scheme 2 Selective semi-hydrogenation of 3,7-dimethyloct-6-en-1-yn-3-ol, 2-methyl-3-butyn-2-ol, and 3,7,11,15-tetramethyl-1-hexadecyn-3-ol using a PdNPs@PS catalytic membrane under batch conditions [62]

weights ranging from 19,000 to 31,000 g/mol with a polydispersity index of 1.10 [60, 61]. The as-prepared catalytic materials containing well-dispersed PdNPs of 1.4–3 nm average sizes throughout the polymer matrix were used in the hydrogenation of ethylene, propylene, and 1,3-butadiene. The polymeric matrix plays a dual role, serving as a support for PdNPs while acting as a stabilizer against aggregation, as only a slight increase in PdNP size was observed after 10 catalytic runs [60, 61].

Besides, the development of polymeric polystyrene (PS) matrixes has classically yielded hydrophobic membranes, presenting high crosslinking density and tunable porosity. Given that, the affinity of metal precursors toward the membrane surfaces determines not only their adsorption profiles on the membrane surface but also impacts on the metal particle formation upon decomposition of the former. Sulman et al. have recently reported the impregnation of a solution of $Na_2[PdCl_4]$ in $THF/MeOH/H_2O$ (4:1:1) on the hydrophobic micro/mesoporous pores of polystyrene (PS), followed by Na_2CO_3 treatment and subsequent drying at 75 °C, affording PdNPs ranging from 22.2 to 34.9 nm average particle sizes [62]. The as-prepared catalytic composites were evaluated in the semi-hydrogenation of C5, C10, and C20 propargyl alcohols, such as 3,7-dimethyloct-6-en-1-yn-3-ol, 2-methyl-3-butyn-2-ol, and 3,7,11,15-tetramethyl-1-hexadecyn-3-ol to the corresponding allyl alcohols, namely, linalool, 2-methyl-3-buten-2-ol, and 3,7,11,15-tetramethyl-1-hexadecen-3-ol), providing high conversions and high selectivity while preventing the total hydrogenation of the substrates (Scheme 2). Despite the detrimental steric factors observed for the substrates of higher molecular weight, these novel Pd-based nanocatalyst composites using a PS matrix represent an environmentally reliable alternative to classical Lindlar catalyst that contains lead [62].

In quest of overcoming the hydrophobic properties of PS-based membranes, aryl groups with polar functional groups on their surface such as amino [63], sulfonate

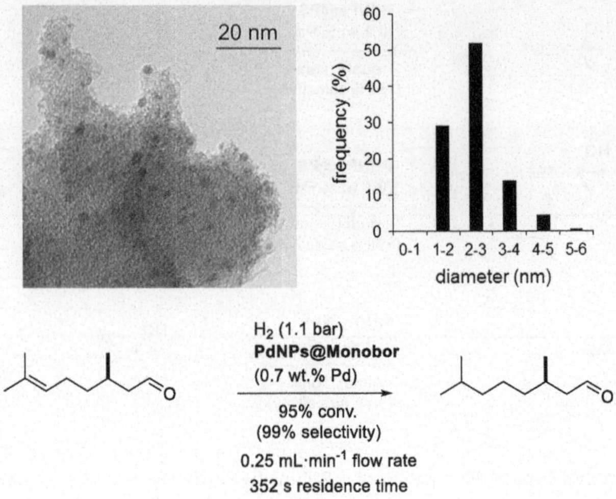

Fig. 6 TEM micrograph and nanoparticle size distribution chart of PdNPs supported on polystyrene-borate monoliths (top). Hydrogenation of (*R*)-citronellal catalyzed by PdNPs@Monobor (bottom). Reprinted from [65], Copyright (2014), with permission from Elsevier

[63, 64], or triphenyl borate groups [65, 66] have been explored. The choice of amine vs. sulfonate groups on the polymeric membrane can affect the structural and electronic properties of the MNPs, which ultimately influences the selectivity and stability of the corresponding catalytic materials. Nemygina et al. observed different behaviors between the functionalization of PS with amino or sulfonate groups [63]. Whereas the amino-functionalized PS matrix interacted with Pd(II) ions from the $Pd(OAc)_2$ precursor, thus preventing the formation of metal aggregates inside the polymeric matrix, the presence of sulfonate groups not only retarded the formation of Pd clusters, but also poisoned the active surface of PdNPs by the in situ formation of reduced sulfur species. Therefore, the difference in the behavior of PdNPs and their interaction with the surface functional groups triggered different activities of the amine versus sulfonate catalysts in two benchmark reactions, i.e., the selective hydrogenation of 2-methyl-3-butyne-2-ol to 2-methyl-3-butene-2-ol and Suzuki cross-coupling of 4-bromoanisole with phenylboronic acid [63]. Besides, a catalyst system encompassing PdNPs within the polymeric triphenylborate monolith described by Liguori et al., featuring 2.5 ± 0.8 nm PdNPs (determined by TEM, Fig. 6), was employed in the continuous-flow partial hydrogenation of (*R*)-citronellal to (*R*)-3,7-dimethyloctanal with high conversion (95%) and selectivity (99%), operating at 1.1 bar H_2 and 0.25 mL·min^{-1} flow rate with 352 s residence time [65, 66].

In addition to surface functionalization of PS, alternative encapsulation strategies of ligand-free MNPs within hyper-crosslinked hollow porous polymeric nanosphere frameworks (HPPNFs), featuring yolk–shell structured Pd nanocatalysts, have been achieved by a molecular self-assembly approach in solution. This method involved the use of a polylactic chain transfer agent (PLA-CTA), styrene, and an initiator for the reversible addition fragmentation chain transfer (RAFT) polymerization at 70 °

C for 24 h, furnishing the polylactic-b-polystyrene (PLA-b-PS) diblock copolymer support. Subsequently, hyper-crosslinking of the PS fragments was achieved by a Lewis acid treatment ($FeCl_3$ at 90 °C), encompassing the formation of hollow cavities within the polymeric framework via degradation of PLA fragments, which was confirmed by FT-IR analysis (disappearance of the characteristic absorption bands related to carbonyl stretching at 1758 cm^{-1} of PLA). The hollow polymeric framework was then suspended in a solution of $Pd(OAc)_2$ in DMF prior to $NaBH_4$ reduction, which resulted in the formation of encapsulated PdNPs inside the hollow polymeric nanoframeworks (PdNPs@HPPNFs), as characterized by (HR)TEM (Fig. 7) [67]. The PdNPs@HPPNF-based catalyst system showed a high TON of 2,500 for the hydrogenation of styrene to ethylbenzene at 0.2 mol% Pd loading, 1 bar of H_2, and 30 min reaction time without a significant decrease in catalytic activity over 15 cycles, which represents a superior catalyst efficiency in comparison to commercial Pd/C catalysts [67]. Despite swelling of the nanoreactors being observed depending on solvent polarity, no significant changes in either the structure of the polymeric material or Pd NP morphology were observed after catalysis [67].

The exploration of alternative polymeric structures with embedded p-block elements has been widely studied given their capacity not only to stabilize PdNPs but also to prevent non-desired leaching processes from the polymeric matrixes. Crosslinking strategies for the preparation of block copolymers with PAA [68, 69] in combination with PS [31, 70], polyethylene oxide (PEO) [70, 71], or polyethylene imine (PEI) [72] have been reported by Bruening, Schomäcker, and collaborators. These catalytic systems were mainly applied to the (semi)hydrogenation of propyne to propene and cyclohexene to cyclohexane in continuous flow at mild reaction conditions (r.t., 1 bar H_2) [70, 71]. Schmidt et al. showed that compared to a fixed fluidized bed, the selectivity for the desired compound (geraniol) could be improved by 40%, mainly due to an improved transport to the catalyst and without any influence of the nature of the support material [69]. The catalytic activity of the prepared membranes depends on both the synthesis method of PdNPs and structural factors of the polymeric matrix. While slow solvothermal methods may result in PdNPs with relatively large sizes and poor catalytic activities, the use of $NaBH_4$ or $LiAlH_4$ for the reduction of Pd(II) salts dispersed in the PAA matrixes resulted in small PdNPs (2–6 nm average diameter) that are highly active toward hydrogenation. Moreover, high porosities of the membrane matrix can result in relatively low catalytic activity as substrates and catalyst may hardly come into contact [31, 70, 71]. Indeed, reaction kinetics correlation with transport studies on PdNPs@PAA-co-PEI system suggests that the catalytic selectivity strongly depends on the diffusion through the membrane toward the catalyst sites, with steric effects being crucial for the catalytic performance [72]. Alternative developments using double emulsion water-in-oil-in-water methods for the synthesis of micron-sized polycaprolactone membrane reactors with embedded PdNPs of 4–20 nm in diameter have been prepared by Robello et al., although their application in the hydrogenation of 2-butyne-1,4-diol was inferior in comparison with colloidal PdNPs systems [73], highlighting once more the impact of structural factors of polymeric matrixes and the need of tailoring them toward the development of catalytic membranes with enhanced catalytic properties.

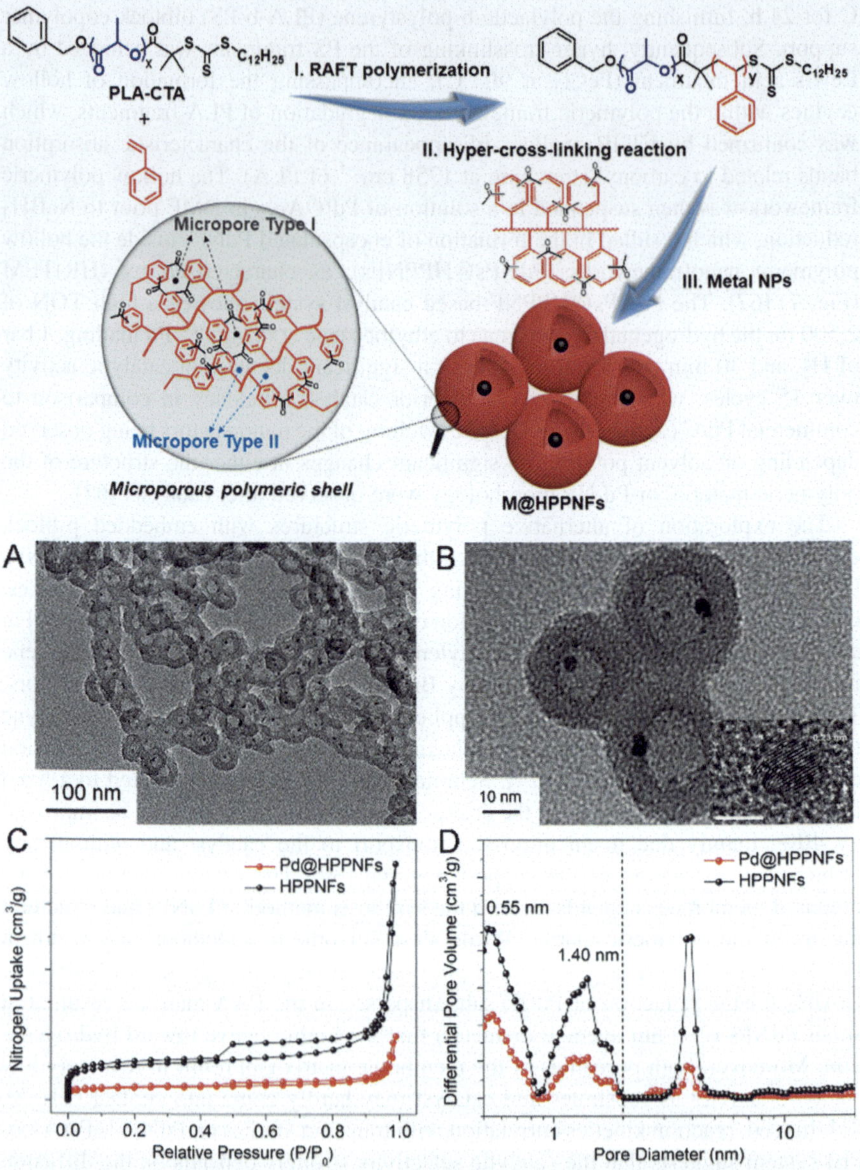

Fig. 7 Synthesis of PdNPs@HPPNFs and (HR)TEM micrographs of PdNPs@HPPNFs (**a**, **b**). Inset (**b**) shows the HRTEM micrograph of PdNPs encapsulated in HPPNFs (Scale bar: 2 nm). (**c**) Nitrogen absorption isotherms. (**d**) Differential pore volume (DFT) distribution for HPPNFs and PdNPs@HPPNFs. Reprinted with permission from [67]. Copyright (2019) American Chemical Society

A number of copolymerization strategies have successfully been employed to enhance the physicochemical properties of the resulting polymeric materials. For instance, TentaGel resins consisting of a crosslinked polystyrene matrix on which poly(ethylene glycol) is strongly grafted via ether linkages confer both hydrophobic and hydrophilic properties. Two different methods have recently been reported for the generation of PdNPs onto this support. On the one hand, Goszewska et al. described the preformation of PdNPs of approximately 4 nm from palladium (II) acetylacetone via sodium borohydride reduction in the presence of trioctylphosphine oxide as stabilizer and their subsequent grafting onto TentaGel-NH_2 beads (90 μm featuring a pore volume 0.13 cm^3/g), bearing amino groups [74]. Kowalewski et al. also described the immobilization of PdNPs (2.2 wt.% Pd) onto TentaGel-NH_2 beads of 130 μm by a similar reduction protocol but in the absence of phosphine oxide. In this case, the resulting PdNPs were also well-dispersed on the polymer surface with a slightly narrower particle size distribution (~2 nm). The polymeric surface functionalization of TentaGel-NH_2 not only allowed the efficient immobilization and distribution of the PdNPs onto the resin, but also facilitated their use as catalysts [74, 75]. Concerning the catalytic behavior using the as-prepared catalytic systems in continuous flow, the former system selectively hydrogenated C=C over C=O bonds in 6-methyl-5-hepten-2-one operating at 85 °C and 10 bar H_2 pressure and 0.5 mL/min flow rate [74], whereas the latter exhibited excellent performance in the semi-hydrogenation of 2-methyl-3-butyn-2-ol and nitrocyclohexane toward 2-methyl-3-buten-2-ol operating at 25 °C and 5 bar H_2 and cyclohexanone oxime at 40 °C and 10 bar H_2, respectively (Scheme 3) [75]. Full

Scheme 3 Partial Pd-catalyzed hydrogenations of 6-methyl-5-hepten-2-one, 2-methyl-3-butyn-2-ol, and nitrocyclohexane in continuous-flow conditions

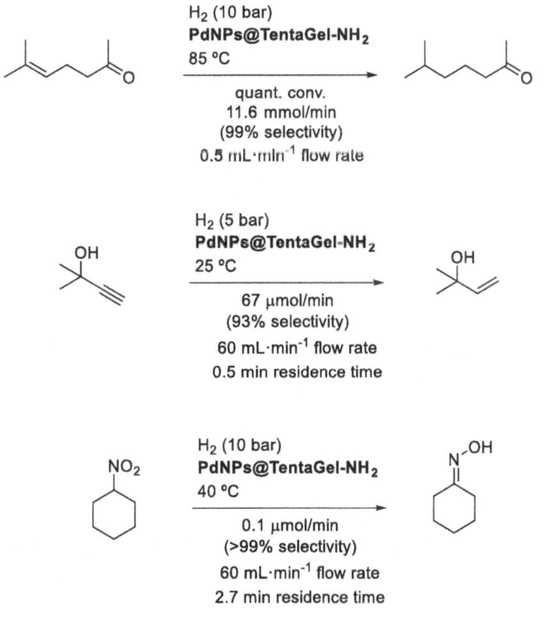

hydrogenation to 2-methyl-2-butanol and cyclohexylamine was achieved at higher reaction temperatures [75].

In an analogous manner, Ni-NPs supported on Tentagel-NH_2 (NiNPs@Tentagel-NH_2) have been reported by Sá et al. via copolymerization of PS with a 1% of divinylbenzene matrix [76]. More specifically, Ni-NPs were first synthesized through $NaBH_4$ reduction of nickel(II) acetylacetonate in the presence of trioctylphosphine oxide as stabilizing agent, followed by their immobilization onto the resin in the second step. (HR)TEM analysis showed Ni-NPs of less than 10 nm in size on both the polymer surface and inside the pores (Fig. 8a, b), although to a certain extent agglomeration was observed [76]. Given the ease of oxidation of Ni (0), Ni(II) species was mostly found by XPS (Fig. 8c). Notably, the authors attributed the main signal of Ni $2p_{3/2}$ at 856.1 eV to $Ni(OH)_2$, even though the presence of Ni(0) was also detected (Ni $2p_{3/2}$ 852.9 eV, together with the Ni satellite at 860.4 eV) [76], which is in good agreement with literature reports [77–80]. This catalytic polymeric membrane was used for the selective hydrogenation of cinnamaldehyde to hydrocinnamaldehyde in continuous flow using 0.7 wt.% loading of catalyst. The system was stable for 12 h of reaction with a conversion up to 80% and high selectivity (93%) without leaching. The same system was applied for the hydrogenation of 3-methyl-2-butenal with a selectivity of 95% albeit in low conversion (37%). Overall, this strategy is economically more competitive since it employs nearly tenfold less metal loading than analogous catalysts immobilized on inorganic supports [81–85].

The exploration of other crosslinking strategies has afforded a number of hybrid polymeric materials with enhanced catalytic properties. For example, Jiao et al. prepared a supported ionic liquid polymeric material via crosslinking, featuring poly(divinylbenzene-co-dication imidazole) units that served as precursors, upon a basic treatment, to generate bis-*N*-heterocyclic carbenes (NHCs), which in turn acted as a reticulation matrix to facilitate the immobilization of PdNPs (Scheme 4) [86]. Indeed, NHCs are ideal stabilizers for PdNPs due to their electron-donor properties, allowing them to easily form strong ligand-metal bonds [87, 88]. Characterization of the catalytic materials by TEM confirmed that PdNPs ranging from 1 to 10 nm were embedded into the polymer matrixes, although the metal loading had an impact on the resulting particle size dispersions. Remarkably, the higher the metal loading, the bigger the particle size. XPS analysis revealed the presence of two sets of peaks in the Pd 3d region that were attributed to Pd(0) (Pd $3d_{5/2}$ and $3d_{3/2}$ at 335 eV and 340 eV, respectively) and Pd(II) species (Pd $3d_{5/2}$ and $3d_{3/2}$ at 337 eV and 343 eV, respectively; Fig. 9), matching literature reports for PdNPs systems [89, 90]. Despite no clear correlations drawn among the catalyst loading, particle sizes, and ratios of Pd(0) vs Pd(II), the presence of oxidized species indicates a strong interaction between the NHC moieties and palladium. Notably, the immobilization of NHC moieties in the polymeric matrix helps to avoid leaching processes, observed with molecular systems. The as-prepared systems showed high activity for the hydrogenation of nitroarenes under mild conditions (1 bar of H_2, r.t., 1 to 7 h reaction time), affording the corresponding aniline derivatives in excellent yields (conversions up to 97 99% and selectivities ranging from 75 to 99%)

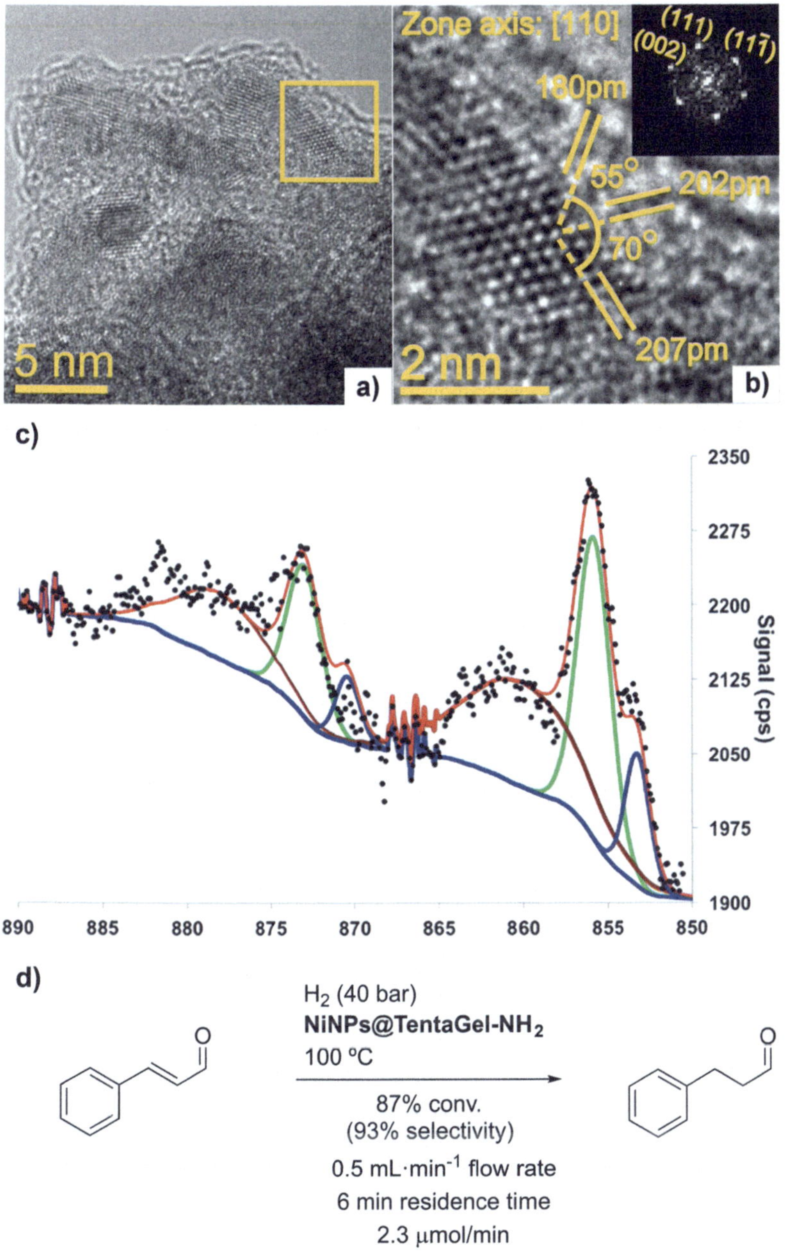

Fig. 8 HRTEM overall view (**a**) and magnification of the selected area with FFT image insert (**b**), showing Ni(0)NPs in the polymer matrix exhibiting a fcc structure (left); XPS analysis of Ni-NPs supported on TentaGel-NH_2 with magnification of Ni 2p region (**c**); hydrogenation of cinnamaldehyde to hydrocinnamaldehyde catalyzed by NiNPs@TentaGel-NH_2 (0.7 wt.% Ni loading) in continuous flow (**d**). Reprinted from [76]. Copyright (2017), with permission from Elsevier

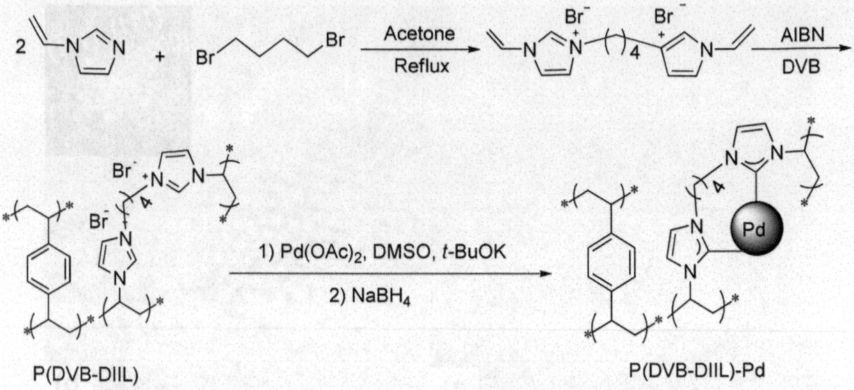

Scheme 4 Synthesis of PdNPs immobilized on crosslinked membranes featuring NHC functional groups [86]. Reprinted with permission from [86]. Copyright (2017) with permission from Wiley

[86]. Furthermore, the best catalyst of the series was highly stable and could be reused at least 10 times under optimized conditions without any significant loss of activity. These findings emphasize the valuable role of crosslinked polymeric ionic liquid materials as an excellent catalyst support.

Alternatively, poly(acryl amide) (PAM) matrixes have also been crosslinked to ionic liquid (imidazolium)polymer microgel network [91]. An IMS strategy was used to successfully introduce $[PdCl_4]^{2-}$ as counteranions of the imidazolium fragments, followed by a $NaBH_4$ reduction. Size control of PdNPs in this polymer was achieved, with an optimum average size of 3 nm. The physicochemical properties of the material (swelling behavior) in organic solvents had an impact on the catalytic activity. Therefore, the selective hydrogenation of C=C and C ≡ C bonds (versus C=O bonds) was observed in MeOH but not in THF, the latter precluding microgel particles swelling. This polymeric catalytic membrane was reused 10 times without activity loss [91].

Cage-like porous polymeric microspheres ranging from 300 to 600 nm with pore volumes of 0.31 cm^3/g and pH-tunable properties were conveniently prepared via a one-stage synthesis through the emulsion copolymerization of the surfactant monomer *N*-(4-vinylbenzyl)-*N*,*N*-dibutylamine hydrochloride and styrene in water (PS-co-PVBAH). The trialkylammonium groups present at acidic pH can reversibly provide trialkylamino moieties under basic medium, favoring the accessibility of reactants [92]. The polymeric PS-co-PVBAH membrane was then added to a solution of $PdCl_2$ in ethanol, followed by $NaBH_4$ addition for the synthesis of the corresponding PdNPs embedded on the support (PdNPs@PS-co-PVBAH, Fig. 10). Powder XRD characterization revealed crystallites from 6.4 to 7.6 nm depending on the metal loading [92]. The as-prepared systems showed excellent catalytic results with TOF values up to 578 h^{-1} for the hydrogenation of nitrobenzene to aniline in water at 30 °C under bubbling H_2 at atmospheric pressure [92].

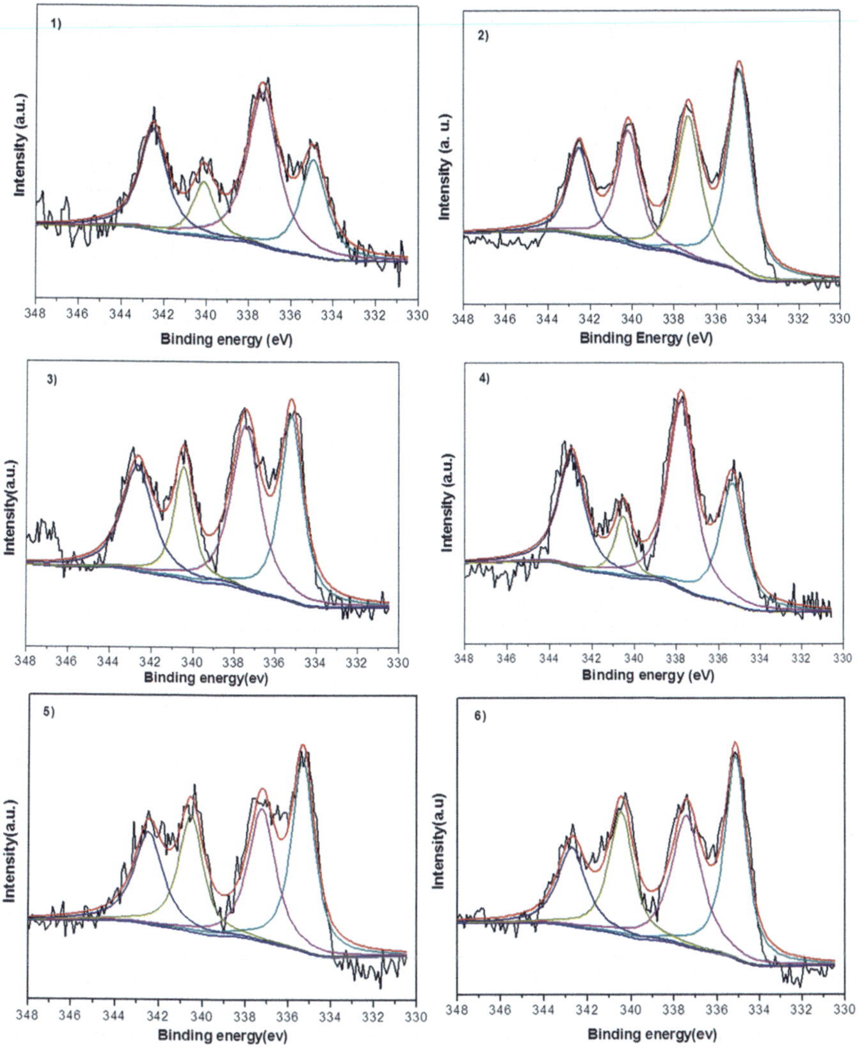

Fig. 9 XPS Pd 3d spectra of PdNPs immobilized on crosslinked materials featuring NHC stabilizing groups at (1) 2.5 wt.%, (2) 6.7 wt.%, (3) 9.3 wt.%, (4) 11.3 wt.%, (5) 17.7 wt.%, (6) 21.3 wt. %. Reprinted with permission from [86]. Copyright (2017) with permission from Wiley

To highlight the key role of nitrogen-based functions in the stabilization of PdNPs, a number of synthetic methods for the preparation of Pd nanocomposites in alternative nitrogenated polymeric matrixes have been reported, namely, polyaniline (PANI) [93], polyvinylpyrrolidone (PVP) [94–97], poly(acrylamide) PAM [91, 98, 99], poly(amide imide) (PAI) [100, 101], and aza-biopolymers such as chitosan [102] and collagen [103]. Thus, the synthesis of the PdNPs on both PANI thin films (previously prepared by thermal evaporation) and membranes (obtained

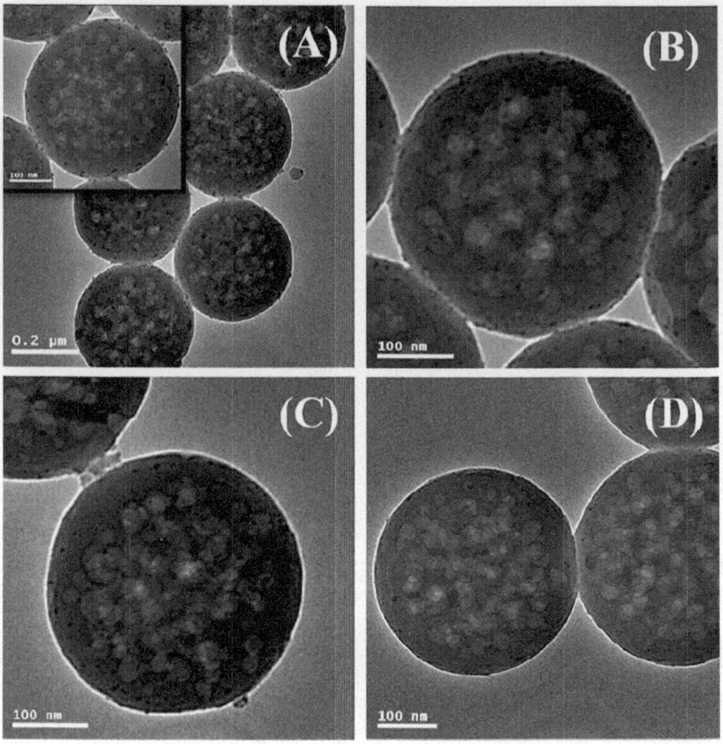

Fig. 10 TEM images of PdNPs@PS-co-PVBAH catalysts at 10 wt.% (**a**), 5 wt.% (**b**), 2 wt.% (**c**), and 1 wt.% (**d**). Reprinted with permission from [92]. Copyright (2011) American Chemical Society

by phase inversion) was achieved by the in situ reduction of $Pd(NO_3)_2$ on their surfaces, resulting in particles with an average size of 13 nm together with larger agglomerates. These catalytic polymeric membranes exhibited efficient catalytic activity toward full hydrogenation of alkynes (diphenylacetylene and ethyl 3-phenylpropynoate toward 1,2-diphenylethane and ethyl 3-phenylpropionate, respectively) with high selectivity. In addition, the reduction of cinnamaldehyde to hydrocinnamaldehyde showed conversions higher than 90% with 81% selectivity (see reaction scheme in Fig. 8d). The recyclability of PdNPs@PANI was also investigated and the systems were reused for more than 7 runs without loss of activity; authors did not comment on metal leaching or MNPs aggregations, plausibly triggering catalyst deactivation.

Given the successful application of PVP polymers in providing the stabilization of Pd nanocatalysts in liquid media [58, 89, 104–107], PVP has also been implemented on more complex hollow-fiber reactors and composite nanofibers with enhanced capability for the immobilization of PdNPs in such polymeric supports [94–97]. HRTEM analysis of the nanocomposite materials synthesized (at a 20:1 M ratio of PVP vs $PdCl_2$) showed that PdNPs were well-distributed on

PVP nanofibers, with average diameters ranging from 3 to 10 nm [94]. Among the applications in catalysis, these systems were mostly applied in the partial hydrogenation of butadiene toward 1-butene with 97% selectivity (at 40 °C and 1 bar pressure of a gas mixture of H_2 and butadiene), although isomerization by-products were also observed [97]. However, the use of the analogous bimetallic Pd–Co nanocomposite enhanced the selectivity of the process overriding isomerization processes, evidencing a synergistic effect between Pd and Co [97] (see Sect. 3.1.3 for further details). Overall, the as-prepared polymeric catalytic nanocomposites showed relatively high catalytic activity toward hydrogenation, simple reaction product separation, and excellent recyclability properties.

Polycondensation of hexamethylenediamine and adipoyl chloride resulted in the formation of an ultrathin nylon membrane, which underwent layer-by-layer self-assembly after impregnation of $H_2[PdCl_4]$ and further reduction by $NaBH_4$. This process was repeated three times to ensure sufficient catalyst loading. The chemical composition and surface morphology of the catalyst layer were characterized by XPS and SEM, which confirmed the successful generation of Pd(0)NPs (Pd $3d_{3/2}$ and $3d_{5/2}$ were 341.1 and 335.7 eV, respectively), below 100 nm on the nylon membrane, although the presence of Pd-O oxidized species could also be detected by XPS evidencing the interaction with the matrix support (Fig. 11). The immobilization on a nylon polymeric matrix not only avoids nanoparticle aggregation but also enables substrate diffusion, resulting in a promising catalyst for hydrogenation reactions in continuous flow [64].

The entrapment of metal clusters could also be successfully achieved thanks to the stabilization provided by the PAI polymeric membrane, enabling high metal dispersion even at high metal concentrations (up to 15 wt.%), as described by Fritsch and co-workers [101]. The catalytic membrane reactors based on PAI were used for the immobilization of both Pd and Ag NPs, as well as bimetallic Pd/Ag, Pd/Cu, Pd/Co, and Pd/Pb NPs, under H_2 at 27 °C. MNPs on the membrane exhibited narrow size distributions (i.e., 1–3 nm), although larger particle aggregates up to 30 nm were also observed in some of the as-prepared films. Particle sizes and distribution critically depended on the solvents used (being *N*-methyl-2-pyrrolidone (NMP) better than THF) and stirring times. For bimetallic NPs systems, no clear evidence of alloy formation could be concluded [101]. The same authors also described the use of TiO_2 as an inorganic filler suspended in the membrane casting solution to enhance the low thermal stability of the resulting PAI polymeric membranes. Despite the porosity changes of PAI membranes with and without the additive (Fig. 12), the TiO_2-modified PAI one was successfully used as support in the development of catalytically active Pd nanocomposites (i.e., PdNPs@PAI-TiO_2) exhibiting enhanced stability up to 200 °C. For instance, the hydrogenation of propene to propane could be observed with quantitative conversions and up to 98% selectivity toward propane with PdNPs@PAI-TiO_2 at 30 °C. The selective hydrogenation of propyne to propene could also be observed with full conversion and 94% selectivity at a flux of 0.23 $m^3 \cdot m^{-2} \cdot h^{-1} \cdot bar^{-1}$ at 0.9 bar pressure using PdNPs@PAI-TiO_2 as catalyst, although analogous polyacrylonitrile-based

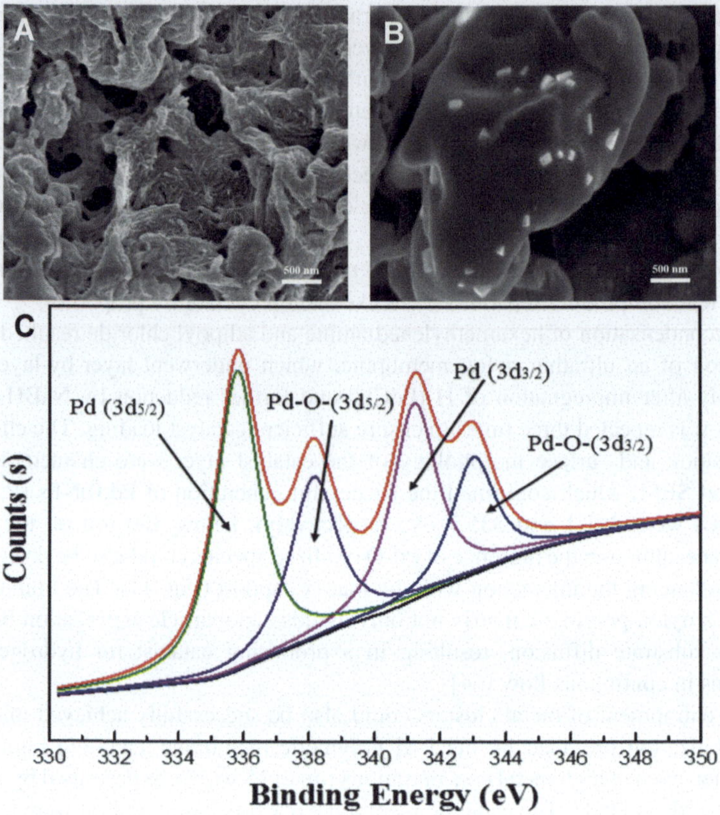

Fig. 11 SEM (top) and XPS (bottom) characterization of ultrathin nylon membrane and Pd-based nanocomposite. SEM images of the (**a**) pristine ultrathin nylon membrane and (**b**) nylon membrane with embedded PdNPs; (**c**) XPS Pd 3d core-level spectra of the catalyst layer. Reprinted with permission from [64]. Copyright (2020) American Chemical Society

membraned offered slightly better results (full conversion and 99% selectivity) under the optimized flux conditions of 0.08 $m^3/m^2 \cdot h \cdot bar$ at 0.9 bar [100].

Alternatively, the use of bio-sourced polymeric materials for the immobilization of PdNPs has been explored with chitosan [102] and collagen fibers (CF) [103]. For example, both unmodified CF and functionalized CF with polyphenol epigallocatechin-3-gallate (EGCG) were used as supports for the preparation of PdNPs via the $NaBH_4$ reduction of Pd(II) salts previously impregnated onto these supports (Fig. 13) [103]. The latter presented better PdNPs distribution, as well as better activity and selectivity in the hydrogenation of allyl alcohol (selectivity up to 89% and TOF of 9,713 $mol \cdot mol^{-1} \cdot h^{-1}$ vs 72% selectivity and TOF of 7,283 $mol \cdot mol^{-1} \cdot h^{-1}$ for the parent unfunctionalized catalytic material), even after recycling for five times thanks to the enhanced coordinating/stabilizing effects of polyphenolic gallate units [103].

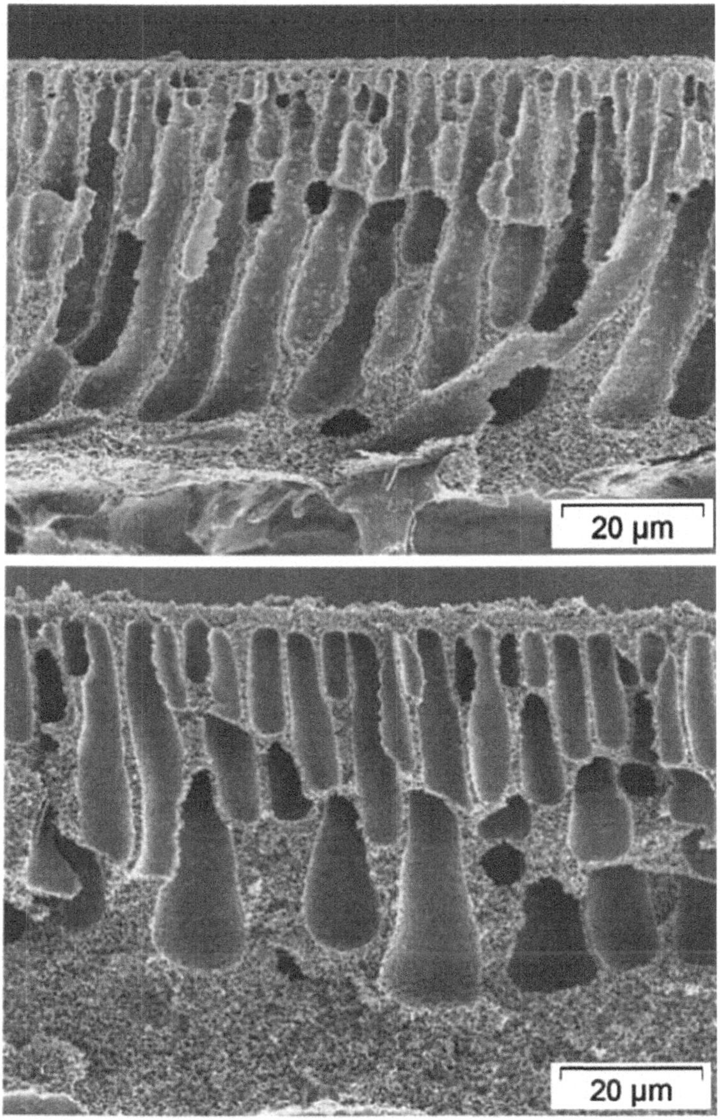

Fig. 12 SEM of the cross-sections of PAI-F polymeric membranes with PdNPs: (top) with TiO_2 inorganic filler; (bottom) without TiO_2. Reprinted from [100]. Copyright (2001), with permission from Elsevier

On the other hand, the strong affinity of chitosan fibers for Pd was exploited by Guibal and co-workers using a glutaraldehyde crosslinked strategy. After the impregnation of the support with $H_2[PdCl_4]$ at pH 2 in HCl solution, it was reduced in situ using generated hydrogen gas (Fig. 13) [102]. At high Pd loadings (100 mg of Pd per g of support), evenly dispersed PdNPs of average diameter size of 3.9 nm

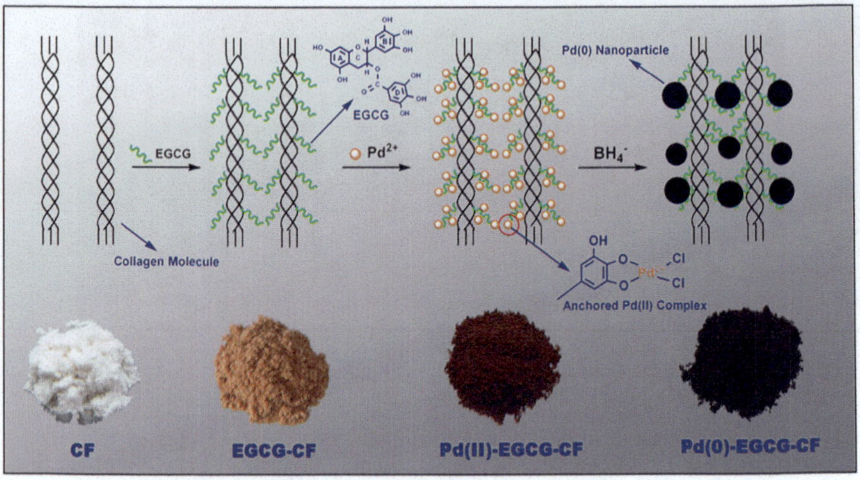

Fig. 13 Reaction scheme (top) and photographs (down) of CF, EGCG-CF, Pd(II)-EGCG-CF, and PdNPs@EGCG-CF. Reprinted from [102], Copyright (2010), with permission from Elsevier

were obtained throughout the material, whereas Pd was mainly located at the lumen of the hollow fiber when Pd loadings were lower (33 mg of Pd per g of support), due to their smaller size (1.7–2.9 nm). The as-prepared PdNPs@EGCG-CF were evaluated in the hydrogenation of *o*-nitrotoluene to *o*-toluidine in MeOH at 1 bar of H_2 with TOF ranging from 1,500 to 2,500 $mmol^{-1}{\cdot}h^{-1}$ [102].

A number of Pd nanocomposites based on sulfur-containing polymeric catalytic membranes have been prepared [22, 36, 44, 108, 109]. In particular, polymeric PES flat-sheet membranes and PES hollow fibers have been widely used after their functionalization under UV-photografting involving acrylic acid [22] and sodium *p*-styrene sulfonate [45, 46] as vinyl monomers (see Sect. 2). Alternatively, the direct sulfonation of the PES matrix has also been described under concentrated sulfuric acid conditions [36]. Moreover, the preparation of composite polysulfone (PSF) matrixes with good dispersions of both PdNPs and thiol-functionalized MWCNTs has been reported by Mahdavi et al. [108]. The as-prepared systems showed better sorption control for polar reactants and enhanced thermal stability of the catalytic membranes, important aspects for their application as CMRs.

Besides, dynamic metal-polymer interactions have been exploited using polyphenylene sulfide (PPS) polymeric matrixes for the design of highly chemoselective and long-lived hydrogenation nanocatalytic systems for the semi-hydrogenation of acetylene [110]. It has been proposed that the fluxional polyphenylene sulfide chains in the glass transition temperature (80 °C) cover PdNPs surface via strong metal-polymer interactions, although the strong binding affinity of acetylene can disturb the Pd-PPS interface and permit the H_2 activation. Once acetylene is hydrogenated to ethylene, exhibiting a much weaker binding affinity for Pd, the readsorption of PPS on the Pd surface precludes the full hydrogenation to ethane even in an ethylene rich stream, thus suppressing catalyst deactivation due to coke formation. These results highlight the unique properties of harnessing dynamic metal-polymer systems for designing engineered long-lived catalytic systems with excellent chemoselectivities [110].

Organosilicon polymeric catalytic membrane reactors based on poly (dimethylsiloxane) (PDMS) matrixes have been successfully developed for hydrogenation purposes, operating under several feed compositions and flow rates withstanding moderate temperatures (35 °C) [32, 111–113]. A PDMS catalytic membrane exhibiting an average thickness of 4.4 μm loaded with 5 Pd wt.% and featuring PdNPs of 9 nm was efficiently employed toward propyne hydrogenation, yielding to different propylene/propane ratios depending on the initial H_2/propyne molar feed ones [111, 112]. Brandao et al. also incorporated Pd nanoclusters in crosslinked PDMS for propylene hydrogenation [30]. Moreover, Rownaghi and co-workers designed catalytically active membranes for 4-nitrophenol reduction by crosslinking PDMS thin layers onto a porous PAI hollow-fiber support, followed by grafting of aminosilane and salicylic aldehyde linkers prior to the formation of PdNPs [113].

Perfluorinated polymeric membrane supports have been broadly used due to the inertness of the C–F bonds they withstand harsh reaction conditions. For example, polyvinylidene fluoride (PVDF), polytetrafluoroethylene (PTFE), and the amorphous fluoropolymer poly[4,5-difluoro-2,2-bis(trifluoromethyl)-1,3-dioxole-co-tetrafluoroethylene] (AF-2400) with enhanced permeability for low-molecular-weight gases have been used as fluorinated membranes. Bottino et al. prepared two membranes by a phase inversion method [114], based on PVDF and a copolymer membrane composed of PVDF and PVP with a ratio 1:0.5 with respect to the monomers. With the membranes in hand, an impregnation with $PdCl_2$ and a subsequent reduction of the salt by $NaBH_4$ were performed to achieve the incorporation of the metallic nanoparticles on the membrane. TEM analysis showed different nanoparticle size distribution depending on the nature of the membrane. In the case of the PVDF membrane, well-dispersed PdNPs were observed (4.5–3.5 nm) while for the $PVDF_1PVP_{0.5}$ membrane some aggregates were also identified. Both systems were tested in the hydrogenation of methylenecyclohexane to methylcyclohexane, obtaining TON values of $7.6 \cdot 10^{-3}$ and $1.0 \cdot 10^{-3}$, respectively. The lower activity of the copolymer was justified by the presence of metal aggregates. It is important to highlight the high selectivity toward methylcyclohexane observed in comparison to ceramic Pd systems on γ-Al_2O_3, which exhibit poor selectivities (i.e., methylcyclohexane and methylcyclohexene mixtures in 6:4 ratios).

In another example, Venezia et al. [23] functionalized a tubular Teflon AF-2400 membrane with PDA in basic medium. Afterward, an acidic solution with PAA was pumped through the membrane. Finally, a polycation absorption into the membrane was achieved by pumping a poly(allylamine hydrochloride) solution. Once the membrane was obtained, the preformed PdNPs showing different shapes (spheres, octahedra, and nanocubes) were adsorbed on the membrane by flushing NP solutions through the doped Teflon AF-2400 membrane. The triangular-shaped palladium supported nanoparticles were tested in the hydrogenation of nitrobenzene to afford

Table 2 Full hydrogenation of alkynes and alkenes to alkanes with a PDVF-PAMAM catalytic membrane with embedded Pt(0) NPs and catalytic control experiments [115]

Entry	Catalyst	Catalyst (Pt mol%)	Substrate	Time (h)	Conversion (%)	Selectivity (%)
1	PtNPs@PDVF-PAMAM	0.1	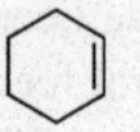	12	98	100
2	PtNPs@PDVF-PAMAM	0.1	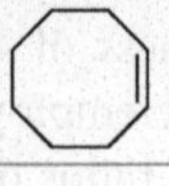	12	95	100
3	Pt@ZIF-8	1.0	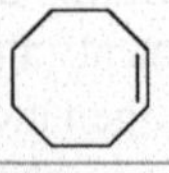	24	0	–
4	Pt/C	1.0	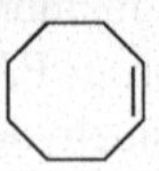	24	7.6	100
5	PtNPs@PDVF-PAMAM	0.1	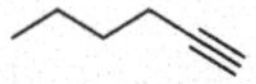	12	98	100
6	Pt@ZIF-8	1.0	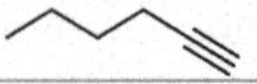	24	40	8[a]
7	Pt/C	1.0	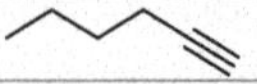	24	90	60[a]
8	PtNPs@PDVF-PAMAM	0.1	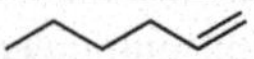	12	96	100
9	PtNPs@PDVF-PAMAM	0.1	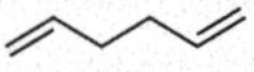	12	95	100
10	PtNPs@PDVF-PAMAM	0.1		24	0	–

Reaction conditions: 8 mmol of substrate, the appropriate metal loading, 4 mL of methanol, 1 bar of H_2 at r.t.

[a] Hex-1-ene intermediate obtained in ca. 30%

aniline. However, after a 6–7 h run, the three catalytic systems showed an activity drop due to metal leaching.

3.1.2 Monometallic Nanoparticles of Pt

In 2018, Diallo's group synthesized a poly(amidoamine) dendrimer (PAMAM) with amine and amide functional groups anchored on a polyvinylidene fluoride membrane (PVDF) by thermally induced phase separation (TIPIS) and non-solvent-induced phase separation (NIPS) processes [115]. This membrane was cast into PET supports for its subsequent impregnation with Pt(IV) by pumping a solution of the metal precursor $H_2[PtCl_6]\cdot 6H_2O$ through the membrane. The reduction of the metal was then carried out with sodium borohydride, obtaining PtNPs of 2–3 nm of average size. The catalytic activity of the as-prepared catalytic membrane was evaluated for the full hydrogenation of alkenes and alkynes as shown in Table 2.

Table 3 Benzene hydrogenation catalyzed by polymeric catalytic membranes containing PtNPs [118]

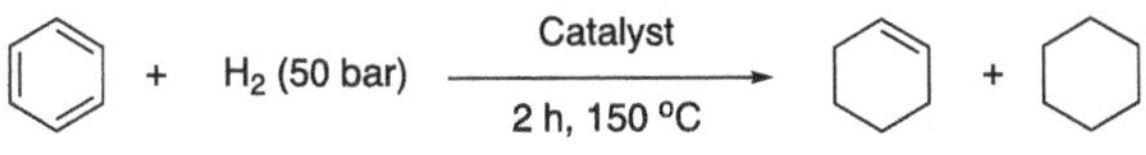

Entry	Catalyst	Catalyst (Pt mol%)	Swelling degree (%)	Conversion (%)	Selectivity to cyclohexene (%)
1	PtNPs	0.04	–	1.08	0
2	PtNPs/PSPI	0.1	0.42	2.80	72
3	PtNPs/CS	0.1	1.06	0.54	58
4	PtNPs/PCS	0.1	1.43	0.69	85

Reaction conditions: 0.34 mol of benzene (30 mL), the appropriate metal loading, 50 bar of H_2, 2 h of reaction at 150 °C

For the hydrogenation of cycloalkenes, the polymeric catalytic membrane was more active than PtNPs supported on classical inorganic materials (entries 1–2 vs 3–4, Table 2). It is worth mentioning that for the hydrogenation of hex-1-yne two different behaviors were observed. When the polymeric catalytic membrane was used, full hydrogenation was achieved (entry 5, Table 2), but the inorganic supported catalysts gave a mixture of the fully hydrogenated product and the corresponding alkene (entries 6–7, Table 2). Moreover, the polymeric catalytic membrane also exhibited enhanced catalytic performance for the hydrogenation of alkenes such as hex-1-ene and hexa-1,5-diene, providing full hydrogenation toward the saturated product (entries 8–9, Table 2). However, internal alkenes were not reduced under the same reaction conditions (entry 10, Table 2). Reusability tests on the cyclohexene hydrogenation using the PtNPs supported on this polymeric catalytic membrane showed that the catalytic system exhibited full conversion during 5 consecutive runs.

Cyclohexene is a relevant building block exhibiting wide use in the synthesis of pharmaceutical active compounds [116, 117]. The partial hydrogenation of benzene to cyclohexene is kinetically more favorable than the full hydrogenation to cyclohexane. This is why research in catalyst enhancing this reaction is a key point. In this line, Zhang et al. synthesized a photosensitive polyimide membrane (PSPI) with PtNPs, previously synthesized in polyimide acid [118], and used it in the partial hydrogenation of benzene. It is important to highlight that for the synthesis of MNPs a new strategy was used in that case. First, an ionic exchange was performed and then a reduction of the metal was achieved via a solvothermal method using ethylene glycol under microwave irradiation. Three polymeric supports were used, namely neat CS, PCS, and PSPI [119]. Moreover, colloidal PtNPs solutions in ethylene glycol were also prepared as catalyst control. Then, the three different membrane reactors were used for the partial hydrogenation of benzene (Table 3).

For the membrane-free catalytic system, full hydrogenation of benzene was observed (entry 1, Table 3). The more selective catalytic polymeric membrane was

1.97 ± 0.36 nm 2.62 ± 1.31 nm 16.91 ± 4.87 nm

Fig. 14 TEM images of PtNPs on chitosan membrane using methanol (left), sodium borohydride (middle), and hydrazine (right) as reducing agents. Reprinted from [125]. Copyright (2004), with permission from Elsevier

that one based on the PCS membrane (entry 4, Table 3). This behavior could be explained by a combination of its higher hydrophilicity and a more appropriate swelling degree leading to enhanced contacting times of the substrate with this catalytic system compared to the other two polymers evaluated (entries 2–3, Table 3).

On the other hand, saturated cyclic alkanes are building blocks for the synthesis of ketone-alcohol (KA) oil, a mixture of cyclohexanone and cyclohexanol which are the main precursors of ε-caprolactam and adipic acid, used for the production of nylon-6 and nylon-6,6, respectively [120, 121]. In addition, KA oil has direct applications in lacquers, alkyd resins, oils, varnishes, gums, and dyes [122]. In 1990, Hodges et al. used two perfluorinated membranes, namely Nafion 117 membrane and Nafion on a PTFE membrane, for the immobilization of PtNPs with a size range of 2–6 nm, generated by an ion exchange process followed by a reduction with sodium borohydride [123]. The activity of the catalytic membranes toward the hydrogenation of cyclohexene was compared with commercially available catalysts such as Pt/C and Pt on Triton Kaowool, achieving better TOF values (ranging from 0.14 to 0.07 s^{-1} for the polymer supported catalysts at 10 bar of H_2 at 20 °C with 1 wt.% of Pt).

Taking into account the encouraging results using Nafion membranes, in 2002 H. Byun and T. Lee compared the activity of Nafion and Tosflex membranes containing PtNPs for the hydrogenation of cyclohexene [124]. They evaluated two reducing agents, $NaBH_4$ and molecular hydrogen, observing full reduction of the complex when $NaBH_4$ was used, and a partial reduction when H_2 was employed. In addition, the effect of the wt.% metal loading in the membrane (from 0 to 3 wt.%) was studied. Both membranes showed similar catalytic activities (TOF of 0.23 s^{-1}), when 0.5 wt.% of Pt was used, indicating that the nature of the polymeric support does not affect the cyclohexene hydrogenation. For both membranes, higher metal loading (from 0.5 to 3 wt.%) induced a drop of the catalytic activity which was explained by the porosity loss of the membrane, affecting the diffusion of reactants.

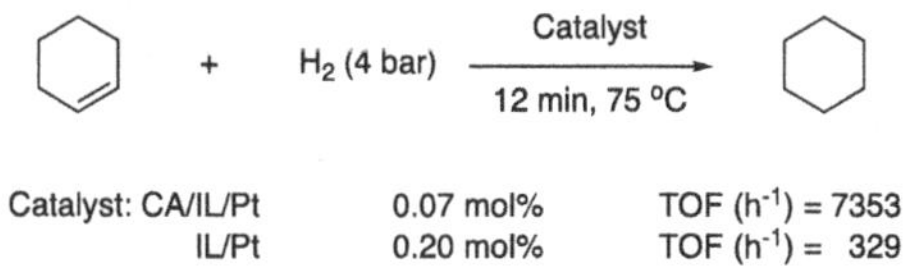

Scheme 5 Cyclohexene hydrogenation using PtNPs stabilized by IL in cellulose (CA/IL/Pt) and PtNPs stabilized by IL (IL/Pt) as catalysts [126]. IL = [BMI](NTf$_2$)

A greener catalytic approach was used by Adlim et al. that prepared PtNPs on a chitosan membrane [125]. In this work three reducing agents were used (methanol, sodium borohydride, and hydrazine; see TEM images of the PtNPs obtained in Fig. 14). Despite the use of hydrazine provided particle aggregates, small and well-distributed PtNPs (1.9–2.2 nm) on the membrane were obtained when MeOH or $NaBH_4$ was used. For the hydrogenation of cyclooctene, the best catalytic results were obtained for the polymeric catalytic membrane obtained using methanol as reducing agent in comparison to the system obtained via $NaBH_4$ reduction (TOFs of 93,808 h^{-1} and 4,486 h^{-1}, respectively).

Gelesky et al. prepared a functionalized cellulose membrane covalently modified by an imidazolium-based ionic liquid with the aim of immobilizing PtNPs [126]. The PtNPs were first synthesized by an organometallic approach using [BMI](NTf$_2$) as ionic liquid which also acts as stabilizing agent. Then, this IL phase containing the NPs was either covalently attached to cellulose or directly deposited, forming the corresponding catalytic films. The catalytic properties of the membranes and the PtNPs stabilized by IL were tested for the hydrogenation reaction of cyclohexene, proving that cellulose enhanced the reaction rate (Scheme 5).

Concerning the hydrogenation of fatty acids and their derivatives, Singh et al. developed an anisotropic polyetherimide (PEID) membrane with PtNPs for the partial hydrogenation of soybean oil, mostly constituted of unsaturated fatty esters [127]. The PEID membrane was synthesized using the phase inversion process; this process consists in the solidification of a thermodynamically stable solution of the polymer [128]. After the synthesis of the membrane, preformed PtNPs were added to the membrane by sputtering. The sputtering time was evaluated at 3, 6, and 9 s, yielding PtNPs ranging from 4 nm (at shorter times) to 10 nm average sizes (at longer times). Both systems of nanoparticles prepared at shorter and longer times were applied in the hydrogenation reaction, evidencing that the former (with smaller PtNPs) were more active toward the partial hydrogenation of glycerides present in vegetable oil (0.9 μg cm^{-2} of catalyst loading).

Hybrid (in)organic catalytic membranes for hydrogenation at high temperatures have been developed using alumina-supported polymeric membranes of polyester and polyphenylenesulfide, showing high permeability for viscous liquids [129]. Pt nanocomposite membranes offered enhanced selectivity over Pd ones, which avoid side isomerization of oleic acid toward elaidic acid [129].

Table 4 Selective hydrogenation of butadiene in but-1-ene using monometallic PdNPs and bimetallic Pd/Co NPs [130]

Catalytic fiber	Product (wt.%)				
	C_4H_{10}	But-1-ene	*trans*-2-C_4H_6	*cis*-2-C_4H_6	C_4H_8
PVP-Pd/CA[a]	0.5	96.8	1.3	1.2	<0.001
PVP-Pd/CA[b]	0.5	97.1	1.1	1.1	<0.001
PVP-Pd-Co/CA[a]	0.5	97.4	1.0	0.9	<0.001
PVP-Pd-Co/CA[b]	0.3	99.0	0.3	0.2	<0.001

Reaction conditions: but-1-ene feed inside fibers, C_2 0.2%, butane 0.3%, but-1-ene 99.0%, butadiene 0.5%, 30 mL/min, $H_2 + N_2$ outside the fibers, 0.7 mL/min, H_2:N_2 1:1, H_2/C_6H_4 2.3:1, 40 °C, 0.1 MPa

[a] NH_2NH_2 as reducing agent in the NPs synthesis

[b] $NaBH_4$ as reducing agent in the NPs synthesis

Table 5 Selective hydrogenation of cyclopentadiene to cyclopentene [130]

Entry	Catalytic fiber	Conversion (%)	Selectivity (%)
1	PVP-PdNPs/CA	92.7	93.4
2	PVP-CoNPs/CA	0	0
3	PVP-PdCoNPs/CA[b]	97.5	98.5

Reaction conditions: cyclopentadiene 3 mL/min, H_2/N_2 2:1, $H_2 + N_2$ 6 mL/min, H_2/cyclopentadiene 1.33/1, 40 °C, 0.1 MPa

3.1.3 Bimetallic Metallic Nanoparticles

Xu et al. used CA or PSF as hollow fibers for the synthesis of PdNPs and Pd/Co bimetallic NPs [130], although the structure of the latter bimetallic systems was not described. The generation of the NPs in the membrane was achieved by first pumping a solution of the stabilizing agent and the metal precursor through the membrane. Afterward, a solution of the reducing agent was pumped through the membrane. The nature of the membrane and three different stabilizers for the NPs [PVP, ethyl cellulose (EC), and melamine-formaldehyde resin (AR)] were evaluated in the selective hydrogenation of butadiene. The most selective system toward the formation of but-1-ene was that one based on the CA membrane with Pd/Co NPs stabilized by PVP (Table 4). The results obtained suggested a synergistic effect between Pd and Co in the bimetallic systems, especially on those reduced with $NaBH_4$. The bimetallic NPs inhibited the isomerization and reduced the quantity of trans isomer in the product crude.

The same group also studied the selective hydrogenation of cyclopentadiene to cyclopentene [130]. They started by evaluating the effect of the stabilizing (PVP, EC, AR) and the reducing agent ($NaBH_4$, NH_2NH_2) in PdNPs immobilized on CA. The results were in agreement with those previously reported for the selective hydrogenation of butadiene. The most active and selective system was the one using PVP as stabilizer and $NaBH_4$ as reducing agent. To improve the conversion and selectivity, bimetallic Pd/Co NPs were prepared with a Pd:Co ratio of 1:0.5. The results of the monometallic PdNPs and bimetallic Pd/Co systems toward the hydrogenation of cyclopentadiene are shown in Table 5. Slightly higher activity was

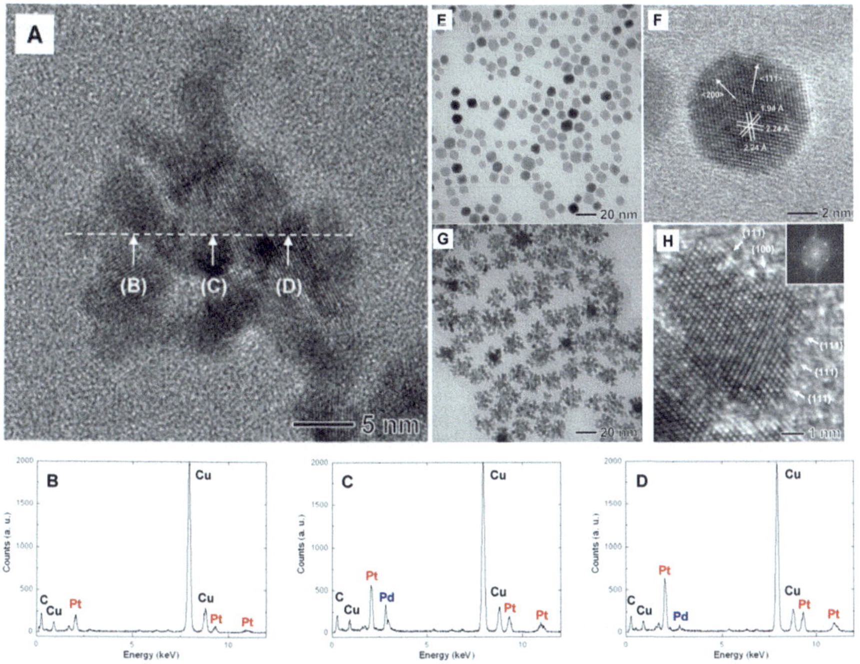

Fig. 15 (**a**) HRTEM image and (**b–d**) EDS spectra obtained from three different regions along the dashed line on the HRTEM image; (**e**) TEM image of the truncated octahedral PdNPs; (**f**) HRTEM image of a single truncated octahedral PdNP; (**g**) TEM image of Pd-Pt nanodendrites; (**h**) HRTEM image of the Pt-branches on a Pd core [131]. Reprinted with permission from AAAS, license number OP-00153480

achieved for the hydrogenation of cyclopentadiene using bimetallic Pd/Co polymeric catalytic membranes (entries 1 and 3, Table 5); however, it is worth to mention that monometallic CoNPs were inactive. The catalytic materials maintained the activity after 300 h of reaction, suggesting that no leaching of the metal from the polymeric membrane occurred.

Gavriidis et al. functionalized a tubular Teflon AF-2400 membrane with PDA in basic medium [23]. Afterward, an acidic solution with polyacrylic acid was pumped through the membrane. A polycation absorption into the membrane was then carried out by pumping a poly(allylamine hydrochloride) solution. Once the membrane was obtained, the preformed monometallic PdNPs exhibiting different shapes (spheres, octahedra, and nanocubes) [23], were adsorbed on it by flushing the nanoparticle solutions through the doped Teflon AF-2400 membrane. For the bimetallic Pt-Pd system, a dendritic Pt deposition onto PdNPs was achieved according to Lim et al.'s methodology [131], encompassing the reduction of a Pt(II) salt in the presence of the preformed PdNPs. A HRTEM-EDS analysis confirmed a core-dendrimer structure with Pt dendrimers onto PdNPs (Fig. 15). The as-prepared polymeric catalytic

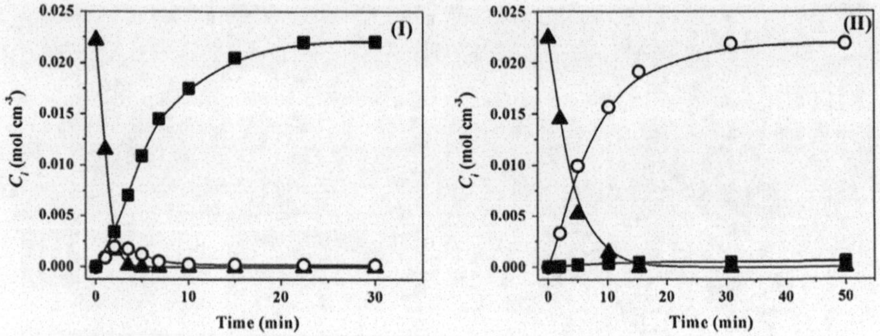

Fig. 16 Concentrations of 3-nitrostyrene (*filled triangle*), 3-vinylaniline (*open circle*), and 3-ethylaniline (*filled square*) as a function of time using the PtNPs@HPS (left) and Pt/ZnNPs@HPS (right) catalyst at 348 K, 10 bar of H_2, [3-nitrostyrene] = $2.2 \cdot 10^{-3}$ mol/cm^{-3}, 3-nitrostyrene/Pt = 10^2. Reprinted from [132], Copyright (2015), with permission from Elsevier

membranes were tested in the hydrogenation of nitrobenzene obtaining aniline in 6–7 h.

Kiwi-Minsker and co-workers prepared bimetallic Pt/Zn NPs by wetness impregnation into hypercrosslinked polystyrene beads (HPS) [132]. The Pt/Zn NPs were synthesized after an impregnation with a solution of acetonitrile containing Zn (II) and Pt(II) in an equimolar ratio. The catalysts were activated using a flow of H_2 and heating the sample up to 573 K, obtaining small NPs of 4.7 nm average size. To gain structural insights of the Pt-Zn bimetallic system, the polymeric catalytic membrane was treated with CO and studied by DRIFTS, concluding that an alloy was plausibly formed. The authors used the monometallic and bimetallic systems for the selective hydrogenation of 3-nitrostyrene into 3-vinylaniline. The monometallic-based catalyst hydrogenated the nitro and vinyl functional groups affording 3-ethylaniline; meanwhile the bimetallic catalytic membrane preferentially hydrogenated the nitro substituent to give 3-vinylaniline (Fig. 16) [132].

The lack of activity of the bimetallic system toward the hydrogenation of the C=C bonds could be explained by the electronic effect between the d-band of Pt and the molecular orbitals of the substrate. The increase of the electron density of the Pt due to the interaction with Zn triggered a lower strength of the coordination of the alkene bond to the nanoparticle surface. This agrees with the hydrogenation product distribution: when the molar ratio of Pt/Zn increased from 1 to 5, the conversion toward 3-vinylaniline increased and the 3-ethylaniline product formation decreased. The bimetallic catalyst was reused four times without observing a drop in catalytic activity.

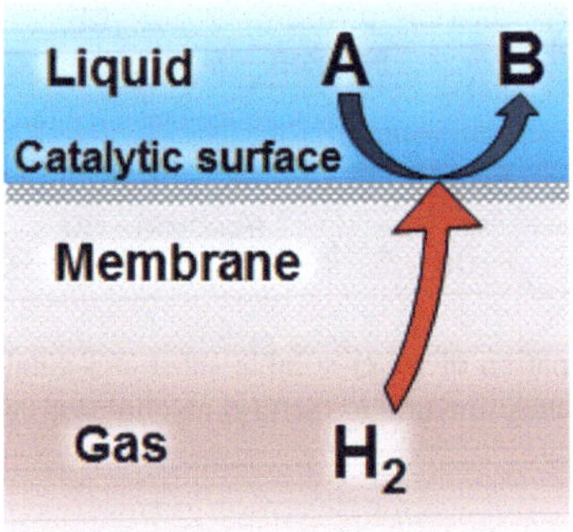

Fig. 17 Scheme illustrating the catalytic membrane performing as a liquid/gas phase contactor where the metal-based catalyst is coated on the surface of the membrane. Hydrogenation reaction takes place as hydrogen permeates the membrane reaching the catalytically active sites. Reprinted from [133]. Copyright (2016), with permission from Elsevier

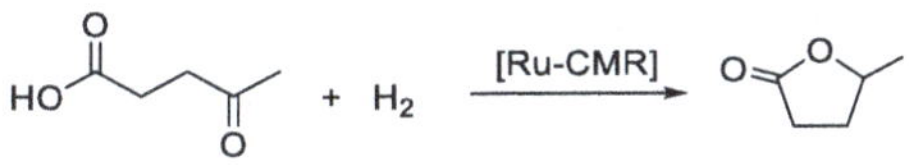

Scheme 6 Levulinic acid hydrogenation to γ-valerolactone catalyzed by RuNPs supported on ePTFE-based membranes [133]

3.2 Groups 8 and 9 Elements

3.2.1 Metal Nanoparticles of Group 8

CMRs have been successfully applied in the hydrogenation of water-soluble substrates such as those from biomass, despite the low solubility of H_2 in water but taking advantage of the high permeability of polymeric membranes to hydrogen. Moreover, CMRs give an efficient methodology for three-phase processes, where the membrane appears as a liquid/gas contactor and as a support for the metal-based catalyst. In this context, Rezac and co-workers conceived a CMR based on a porous expanded polytetrafluoroethylene membrane (ePTFE) containing ruthenium nanoparticles (RuNPs), leading to a gas/liquid phase contactor (Fig. 17) [133]. The hydrophobic ePTFE membrane maintains the aqueous phase on one side permitting H_2 to permeate from the opposite side and react with RuNPs at the interface between the catalytic surface and the liquid phase (Fig. 17). In this work, both unmodified (commercially available) and modified (by adding polyimide polymers by spin-coating) membranes were used. The CMRs were prepared by one-step synthesis mixing $RuCl_3$ and the corresponding membrane in ethanol; after evaporation of ethanol, the ruthenium-impregnated membrane was treated under a H_2 flow at 150 ° C for 2 h (the reduction of Ru(III) to Ru(0) was established by X-ray diffraction analysis). TEM analyses revealed the presence of RuNPs exhibiting an average particle size of 2–3 nm. The full-characterized CMRs were applied in the hydrogenation of levulinic acid to γ-valerolactone (Scheme 6), obtaining the best catalytic

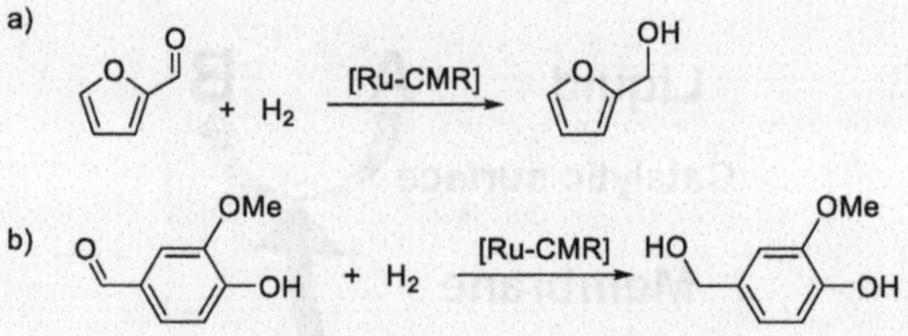

Scheme 7 Selective hydrogenation of water-soluble substrates catalyzed by RuNPs supported on sulfonated PEEK-WC membranes: furfural to furfuryl alcohol (top) and vanillin to vanillin alcohol (bottom) [26]

results with the membrane derived from pristine ePTFE membrane, working under relatively smooth conditions (0.07–5.6 bar H_2; 50–90 °C) and showing a high stability (no decrease in catalytic activity was observed over 200 h).

Galiano, Sanna, and co-workers reported the immobilization of RuNPs on PES [134] and sulfonated polyetheretherketone containing a spiro lactone (PEEK-WC) for the selective hydrogenation of furfural to furfuryl alcohol [26]. PES was modified by photochemical polymerization of acrylic acid in the presence of an appropriate photo-initiator leading to hydrophilic membranes. Several parameters were optimized (i.e., membranes showing pore sizes from 50 to 800 nm, time of UV irradiation, and nature of photo-initiator) before immobilizing the metal-based catalyst. The modified PES membranes were impregnated with $[Ru(NH_3)_6]Cl_2$ and further treated with $NaBH_4$ permitting the reduction of the ruthenium complex to Ru (0) NPs. After characterization of the different CMRs, the modified PES exhibiting 220 nm pore size (UV polymerization at 7.26 J cm^{-2}) and a Ru content of 10.9 μg cm^{-2} was privileged for catalytic purposes, mainly due to its more rigid structure compared to the other CMRs, thus avoiding deformation under catalytic conditions. This CMR showed an excellent selectivity toward the formation of furfuryl alcohol precluding the formation of other hydrogenated products (such as tetrahydrofurfural or tetrahydrofurfuryl alcohol) or hydrodeoxygenated products (i.e., 2-methylfuran, furan, 2-methyltetrahydrofuran, 1-butanol, 2-pentanol), showing a high turnover frequency (48,000 h^{-1}) under smooth conditions (7 bar H_2, 70 °C, H_2/furfural ratio = 1/1, liquid flow rate of ca. 0.5 mL min^{-1}), although conversion was relatively low (ca. 26% after 4 h). Unfortunately, after 30 min of reaction, ca. 50% of ruthenium was present in solution (determined by ICP-EOS analysis). The authors attributed this significant leaching to the presence of ruthenium clusters not well-adsorbed on the acrylic acid functional groups. These metal-based clusters appeared to be catalytically inactive, suggested by the preserved CMR activity up to 3 h and the presence of RuNPs on the modified PES membrane after catalysis (post-reaction TEM analysis).

With the aim of overcoming the stability limitation of the Ru-based PES catalytic membrane, sulfonated PEEK-WC membranes were used and efficiently applied in the selective hydrogenation of aldehydes derived from biomass in aqueous phase, such as furfural and vanillin (Scheme 7) [26]. After sulfonation of the pristine PEEK-

WC by H_2SO_4 (different concentrations were used in order to obtain different sulfonation degrees), hydrated $RuCl_3$ salt was added to an aqueous suspension of the corresponding membrane followed by addition of $NaBH_4$. SEM analyses of the Ru-based CMR evidenced the presence of RuNPs not homogenously distributed on the sulfonated PEEK-WC membrane exhibiting diameters comprised between ca. 30 and 200 nm (Fig. 18). These CMRs were selective in the hydrogenation of furfural and vanillin to the corresponding alcohols, exhibiting up to 75% and 33% conversion, respectively (18 bar H_2 and 85 °C). In contrast to the PES-based membrane, Ru-based sulfonated PEEK-WC catalytic membrane was stable for 90 h and negligible Ru leaching was observed (25.20000 μg cm^{-2} of Ru before catalysis versus 25.19998 μg cm^{-2} of Ru after catalysis). However, the catalytic activity was lower than using the PES-based CMR probably due to the worse dispersion of RuNPs on the sulfonated PEEK-WC surface.

Grosu and co-workers have recently reported the immobilization of OsNPs on CA (hydrophilic membrane), PSF (slightly hydrophilic membrane), and PP (hydrophobic membrane) hollow-fiber membranes, for the hydrogenation of 5-nitrobenzimidazole to 5-aminobenzimidazole in water, which can be applied in the synthesis of drugs (Scheme 8) [42]. The Os-based CMRs were prepared in a permeation cell by reaction of OsO_4 in *t*-butylalcohol under H_2 pressure (Fig. 19). SEM analysis evidenced the formation of spherical OsNPs with diameters ranging between 15 and 55 nm in the three types of CMRs, both on the surface and inside the corresponding membrane. The Os surface content was determined by energy-dispersive spectroscopy (EDS), ca. 4–6.5 atomic%. However, the surface state of the osmium-based nanoparticles was not studied, and in consequence, it was not possible to conclude on the oxidation state of the metal. The three CMRs were catalytically active, with the Os-based propylene hollow fiber being the most active, probably due to its larger active surface area (10,000 cm^2) than the other two membranes, CA and PSF (both showing a working surface of 100 cm^2).

3.2.2 Metal Nanoparticles of Group 9

J. Dupont and co-workers prepared RhNPs and PtNPs in the ionic liquid [BMI](BF_4) before their immobilization on CA by spreading the IL homogeneous solution containing MNPs and CA over a glass plate [126]. The resulting CMR membranes, RhNPs/IL@CA and PtNPs/IL@CA, were fully characterized including their mechanical properties. SEM analyses clearly evidenced the morphological changes of CA in the presence and in the absence of IL: in the presence of [BMI](BF_4), a microstructural order of CA was observed for both CMRs (Fig. 20). For both materials, RhNPs/IL@CA and PtNPs/IL@CA, the surface area was higher than for CA/IL (96 for RhNPs/IL@CA and 113 for PtNPs/IL@CA versus 38 m^2 g^{-1} for CA/IL), indicating that the presence of small MNPs in the CA/IL membrane (3–5 nm determined by TEM) causes a surface area increase. Mechanical studies showed that the presence of IL in CA induces an increase of the elasticity, probably due to an increase of the distance between cellulose units, along with a better

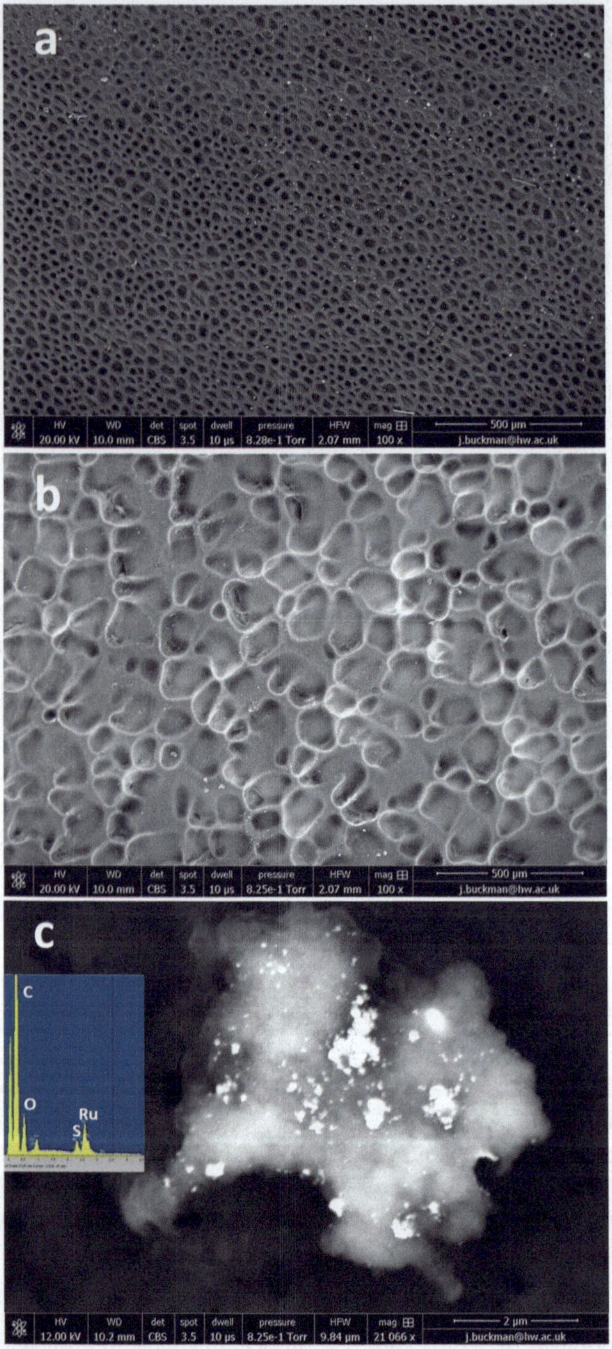

Fig. 18 SEM and EDS analyses: (**a**) sulfonated PEEK-WC membrane; (**b**) Ru-sulfonated PEEK-WC membrane; (**c**) magnification of Ru-sulfonated PEEK WC membrane (21,066x resolution). Reprinted from [26]. Copyright (2021), with permission from Elsevier

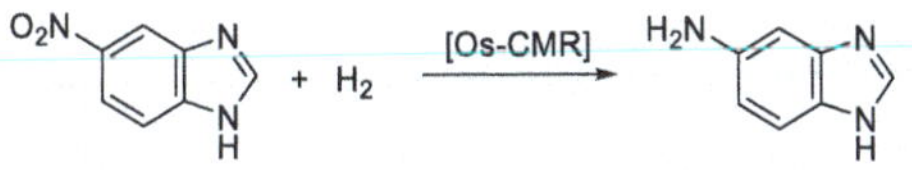

Scheme 8 5-Nitrobenzimidazole hydrogenation to yield 5-aminobenzimidazole, catalyzed by OsNPs supported on CA, PSF, and PP hollow-fiber membranes [42]

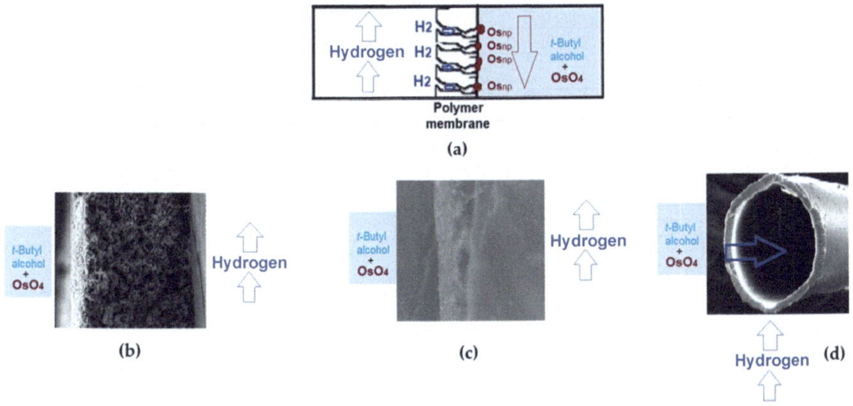

Fig. 19 (**a**) General illustration representing the OsNPs immobilized on different polymer membranes: (**b**) cellulose acetate, (**c**) polysulfone, and (**d**) polypropylene hollow fiber [42]. Reprinted permission pending from [42]. Copyright (2021) MDPI

formability of the CA-based material. Both CMRs were applied in the cyclohexene hydrogenation under batch conditions (75 °C, 4 bar H_2), proving that MNPs/IL@CA were more active than MNPs in IL.

C. W. Scheeren reported zero-valent IrNPs, preformed in the ionic liquid [BMI](NTf$_2$) and immobilized on CA membrane following a similar methodology than that reported for Rh- and Pt-based systems [126], to be applied in the hydrogenation of benchmark substrates (hex-1-ene, cyclohexene, benzene) [41]. IrNPs were homogenously distributed on the CA membrane, preserving their morphology (spherical nanoparticles of mean size of 2.2 nm) (Fig. 21). EDS analysis evidenced the presence of the IL on the CMR, which influences the CA structure, triggering a fibrous-like structure. The surface area of the CA/IL membrane clearly decreased comparing with the pristine CA (24 $m^2\ g^{-1}$ vs 192 $m^2\ g^{-1}$), suggesting that IL occupies the free pores of the CA membrane (its presence was evidenced by IR spectroscopy). On the other hand, the surface area increased up to 92 $m^2\ g^{-1}$ for the catalytic membrane IrNPs/IL@CA. It is also important to highlight that the presence of the IL on the membrane resulted in a better flexibility (reduction of the Young's modulus in comparison with the pristine CA), probably due to the intermolecular interactions between the NTf$_2$ anions and the hydrogen bond network of CA. The hydrogenation study (batch conditions: 75 °C, 4 bar H_2, 0.03–1.5 h) of the chosen substrates proved the synergy effects between the membrane, ionic liquid, and IrNPs. For the three reactions, the highest conversions were obtained with IrNPs/

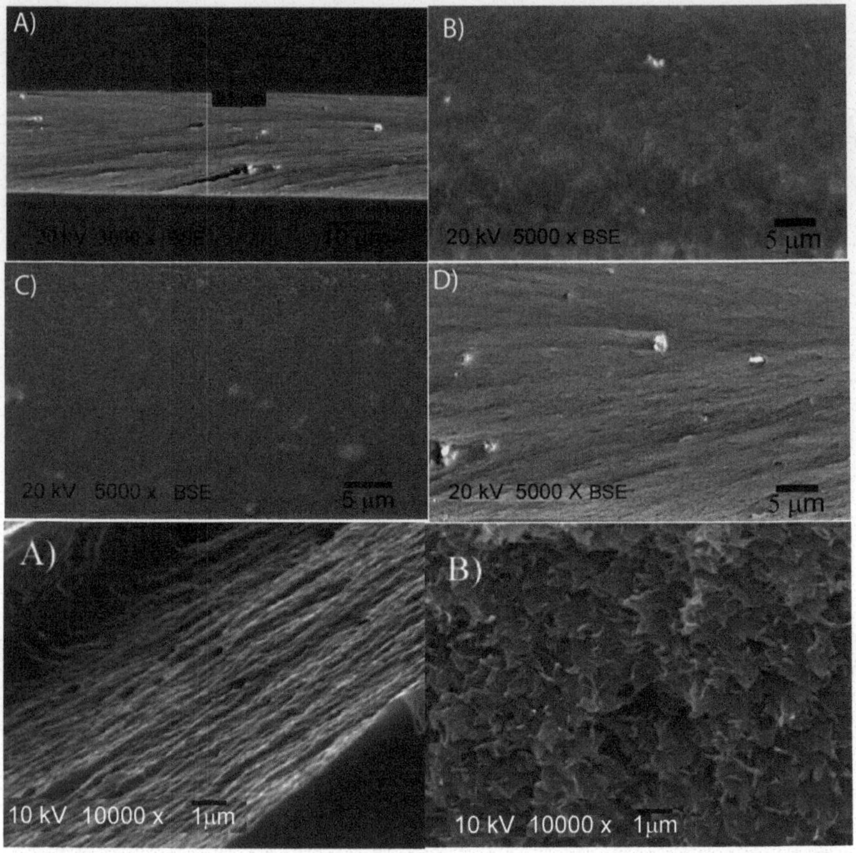

Fig. 20 Top: SEM images of (**a**) PtNPs/IL@CA; (**b**) compact membrane of IL-free PtNPs@CA; (**c**) PtNPs/IL@CA; (**d**) PtNPs/IL@CA exhibiting the longitudinal order of the membrane. Bottom: SEM images of (**a**) RhNPs/IL@CA; (**b**) IL-free RhNPs@CA [126]. Reprinted with permission from [126]. Copyright (2009) American Chemical Society

IL@CA (in comparison with unsupported Ir(0)NPs and IL-free IrNPs on CA), along with a higher stability (IrNPs/IL@CA was efficiently recycled up to 7 runs).

Liguori and Barbaro reported the immobilization of Rh NPs and PdNPs on polystyrene-based borate monolith (Monobor) for hydrogenation purposes under continuous-flow conditions (Fig. 22) [65, 66]. Both materials were prepared by one-pot synthesis methodology from the corresponding metal precursors (Pd$(NO_3)_2$ and $[Rh(nbd)_2](BF_4)$, nbd = norbornadiene) and Monobor placed in a glass column and treated with a H_2 stream. TEM analysis showed the formation of small MNPs (ca. 2.5 and 4 nm for PdNPs and Rh NPs, respectively). MNPs@Monobor catalytic reactors were applied in the hydrogenation of cyclohexene and several carbonyl derivatives (*trans*-4-phenyl-3-buten-2-one, isophorone, (*R*)-carvone, (*R*)-citronellal) under smooth conditions (r.t., 1–2 bar H_2,

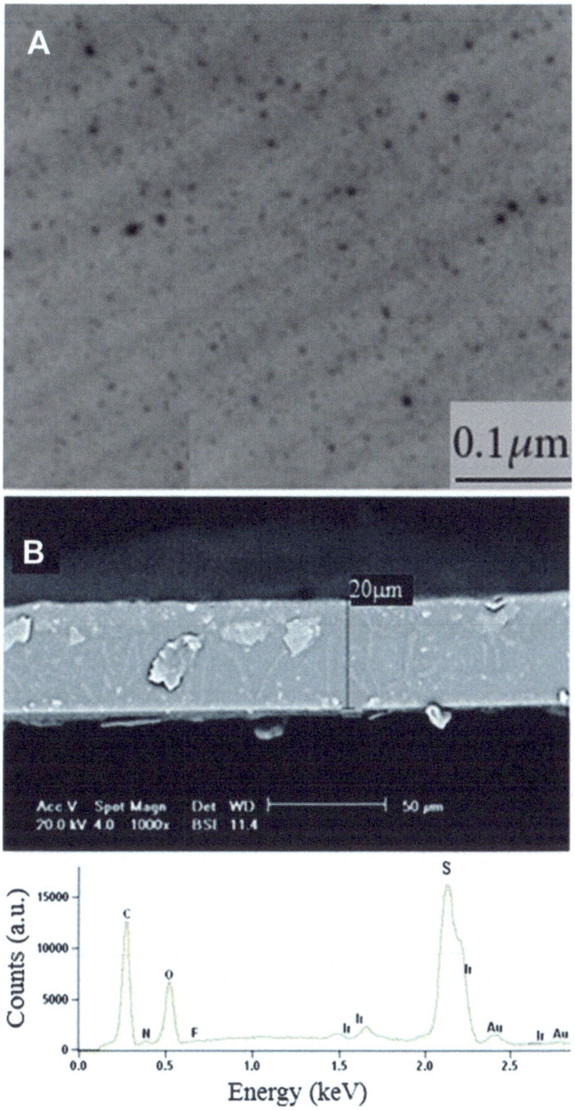

Fig. 21 (**a**) TEM image of IrNPs/IL on CA polymeric membrane; (**b**) SEM image of IrNPs/IL@CA along with EDS analysis, exhibiting the presence of Ir, S, and F. Reprinted with permission from [41]. Copyright (2015) Royal Society of Chemistry

H_2 flow 0.25–1.00 mL min^{-1}), showing high selectivity and durability. No rational conclusions could be established to explain the differences in selectivity and reactivity observed for both catalytic systems, highly dependent on the nature of the substrate.

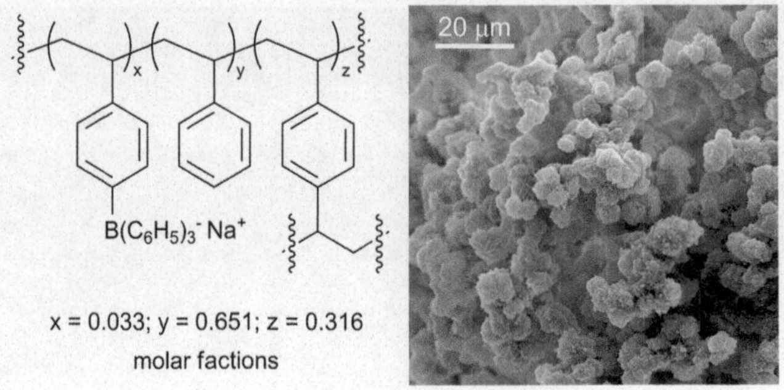

Fig. 22 Schematic illustration (left) and SEM image (right) of Monobor . Reprinted from [132], Copyright (2014), with permission from Elsevier

4 Conclusions and Perspectives

Catalytic polymeric membranes are easily adaptable materials to different module geometries, such as flat or spiral sheets, or hollow fibers. This feature is an important aspect of implementing membrane modules on an industrial-scale process, where highly specific surfaces are required to produce the required amount of a specific target product. The design of original CMRs can be greatly enhanced in combination with process modeling applied in intensification strategies. Hollow-fiber modules are of special industrial interest, since they present the highest specific surface area, leading to size reduction of modules. They also present some other advantages such as low cost and easy processability. In order to obtain better stability and enhanced immobilization of MNPs within the polymeric network, crosslinking steps can be performed prior to or after the incorporation of MNPs. Functionalization strategies have also been explored, as well as the incorporation of inorganic solids by casting within the polymeric matrixes as an effective strategy to improve the physicochemical and thermal stability of the resulting membranes.

An important element of catalytic polymeric membranes is that they can be assembled into a small module. Tuning the number of modules makes it possible to adapt and achieve production objectives. The potential for industrial-scale production of a CMR has been proven using a parallel configuration, exhibiting a better performance than a series one [27]. Based on reported kinetic studies, the production capacity of CMRs is comparable to that of microreactors and makes it possible to achieve a production of ca. 10,000 tons·year^{-1} m^{-3}).

As discussed in this chapter through the several works of polymeric CMRs containing MNPs for hydrogenation purposes using molecular hydrogen as a reagent, these reactors represent an efficient and sustainable approach for both fundamental research and industrial applications, mainly in flow conditions, considering their relatively low energy consumption, low footprint, and in the main of cases, improved efficiency. Design, modeling, and suitable preparations of MNPs on polymeric CMRs are well-established mostly for noble metals of groups 8–10 (Ru,

Os, Rh, Ir, Pd, Pt), evidencing a lack of contributions related to first-row transition metals (except for nickel-based CMRs and some contributions involving cobalt). It is also important to mention the importance of conceiving metal nanoparticles-based CMRs for stereoselective purposes toward the production of high added value compounds, in particular for fine chemicals applications. The versatility in terms of functionalization of polymeric membranes permits the immobilization of different types of nanocatalysts and therefore envisages multi-processes using polymeric CMRs in series or parallel configurations depending on the envisaged reactivity (tandem, cascade, or sequential processes), combining different types of reactivity (C-C or C-heteroatom couplings, oxidation or reduction steps, etc.). The valuable fluxionality of polymeric membranes offers advantageous capabilities to finely tune the chemoselectivity of hydrogenation reactions, as evidenced for the semi-hydrogenation of alkynes via efficiently tuning the adsorption energies of alkene products versus the polymer matrix featuring coordinating groups. Alternative activation modes, such as photochemical-promoted dehydrogenations, deserve to be developed to obtain simultaneously both fine chemicals (e.g., carbonyl or nitrogen derivatives depending on the substrates used from alcohols or amines, respectively) and H_2 (for a recent review on photocatalytic membrane reactors, see [135]).

References

1. Koros WJ, Ma YH, Shimidzu T (1996) Terminology for membranes and membrane processes. Pure Appl Chem 68:1479–1489
2. Abdullah N, Rahman MA, Dzarfan Othman MH, Jaafar J, Ismail AF (2018) Chapter 2 – membranes and membrane processes: fundamentals. In: Basile A, Mozia S, Molinari R (eds) Current trends and future developments on (bio-) membranes. Elsevier, pp 45–70
3. Sitter S, Chen Q, Grossmann IE (2019) An overview of process intensification methods. Curr Opin Chem Eng 25:87–94
4. Itoh N, Wu T-H (1997) An adiabatic type of palladium membrane reactor for coupling endothermic and exothermic reactions. J Membr Sci 124:213–222
5. Dong X, Jin W, Xu N, Li K (2011) Dense ceramic catalytic membranes and membrane reactors for energy and environmental applications. Chem Commun 47:10886–10902
6. Zhang G, Jin W, Xu N (2018) Design and fabrication of ceramic catalytic membrane reactors for green chemical engineering applications. Engineering 4:848–860
7. Qing W, Liu F, Yao H, Sun S, Chen C, Zhang W (2020) Functional catalytic membrane development: a review of catalyst coating techniques. Adv Colloid Interf Sci 282:102207
8. Vankelecom IFJ (2002) Polymeric membranes in catalytic reactors. Chem Rev 102:3779–3810
9. Qing W, Li X, Shao S, Shi X, Wang J, Feng Y, Zhang W, Zhang W (2019) Polymeric catalytically active membranes for reaction-separation coupling: a review. J Membr Sci 583:118–138
10. Ozdemir SS, Buonomenna MG, Drioli E (2006) Catalytic polymeric membranes: preparation and application. Appl Catal A 307:167–183
11. Brunetti A, Zito PF, Giorno L, Drioli E, Barbieri G (2018) Membrane reactors for low temperature applications: an overview. Chem Eng Process 124:282–307
12. Robeson LM (2012) 8.13 – polymer membranes. In: Matyjaszewski K, Möller M (eds) Polymer science: a comprehensive reference. Elsevier, Amsterdam, pp 325–347

13. Volkov VV, Petrova IV, Lebedeva VI, Roldughin VI, Tereshchenko GF (2011) Palladium-loaded polymeric membranes for hydrogenation in catalytic membrane reactors. In: Basile A, Gallucci F (eds) Membranes for membrane reactors: preparation, optimization and selection. Wiley, p 531
14. Gryaznov VM (1963) Mechanism of the catalytic redistribution of hydrogen in unsaturated cyclic hydrocarbons. Russ Chem Rev 32:188
15. Gryaznov VM, Smirnov VS, Vdovin VM, Ermilova MM, Gogua LD, Pritula NA, Litvinov IA (1979) Method of preparing a hydrogen-permeable membrane catalyst on a base of palladium or its alloys for the hydrogenation of unsaturated organic compounds, U.S. Patent
16. Gryaznov VM, Smirnov VS, Vdovin VM, Ermilova MM, Gogua LD, Pritula NA, Fedorova GK (1983) Membrane catalyst for hydrogenation of organic compounds and method for preparing same, U.S. Patent
17. Ng LY, Mohammad AW, Leo CP, Hilal N (2013) Polymeric membranes incorporated with metal/metal oxide nanoparticles: a comprehensive review. Desalination 308:15–33
18. Buonomenna MG, Choi SH, Drioli E (2010) Catalysis in polymeric membrane reactors: the membrane role. Asia Pac J Chem Eng 5:26–34
19. Kudaibergenov SE, Dzhardimalieva GI (2020) Flow-through catalytic reactors based on metal nanoparticles immobilized within porous polymeric gels and surfaces/hollows of polymeric membranes. Polymers 12:572
20. Stankiewicz A (2003) Reactive separations for process intensification: an industrial perspective. Chem Eng Process 42:137–144
21. Miachon S, Dalmon J-A (2004) Catalysis in membrane reactors: what about the catalyst? Top Catal 29:59–65
22. López-Viveros M, Favier I, Gómez M, Lahitte J-F, Remigy J-C (2021) Remarkable catalytic activity of polymeric membranes containing gel-trapped palladium nanoparticles for hydrogenation reactions. Catal Today 364:263–269
23. Venezia B, Panariello L, Biri D, Shin J, Damilos S, Radhakrishnan ANP, Blackman C, Gavriilidis A (2021) Catalytic Teflon AF-2400 membrane reactor with adsorbed ex situ synthesized Pd-based nanoparticles for nitrobenzene hydrogenation. Catal Today 362:104–112
24. Oliveira DGM, Rosa GR, Scheeren CW (2017) Rh and Pd nanoparticles supported in a polymeric membrane: evaluation in hydrogenation and Suzuki-Miyaura reactions. J Nanosci Nanotechnol 17:9295–9301
25. Roucoux A, Schulz J, Patin H (2002) Reduced transition metal colloids: a novel family of reusable catalysts? Chem Rev 102:3757–3778
26. Bagnato G, Figoli A, Garbe R, Russo F, Galiano F, Sanna A (2021) Development of Ru-PEEK-WC catalytic membrane using a more sustainable solvent for stable hydrogenation reactions. Fuel Process Technol 216:106766
27. Gu Y (2016) Membranes polymères fonctionnalisées par des poly(liquide ionique)s et des nanoparticules de palladium : applications au captage de CO_2 et aux membranes catalytiques, Laboratoire de Génie Chimique (LGC), UMR 5503. Université Toulouse III Paul Sabatier, Toulouse
28. Chen B, Li F, Huang Z, Lu T, Yuan G (2014) Stability or flexibility: metal nanoparticles supported over cross-linked functional polymers as catalytic active sites for hydrogenation and carbonylation. Appl Catal A 481:54–63
29. Bastos-Arrieta J, Munoz J, Stenbock-Fermor A, Munoz M, Muraviev DN, Cespedes F, Tsarkova LA, Baeza M (2016) Intermatrix synthesis as a rapid, inexpensive and reproducible methodology for the in situ functionalization of nanostructured surfaces with quantum dots. Appl Surf Sci 368:417–426
30. Ludtke K, Peinemann K-V, Kasche V, Behling R-D (1998) Nitrate removal of drinking water by means of catalytically active membranes. J Membr Sci 151:3–11

31. Groeschel L, Haidar R, Beyer A, Coelfen H, Frank B, Schomaecker R (2005) Hydrogenation of propyne in palladium-containing polyacrylic acid membranes and its characterization. Ind Eng Chem Res 44:9064–9070
32. Goyal A, Mohl M, Kumar A, Puskas R, Kukovecz A, Konya Z, Kiricsi I, Ajayan PM (2011) In situ synthesis of catalytic metal nanoparticle-PDMS membranes by thermal decomposition process. Compos Sci Technol 71:129–133
33. Ruiz P, Munoz M, Macanas J, Turta C, Prodius D, Muraviev DN (2010) Intermatrix synthesis of polymer stabilized inorganic nanocatalyst with maximum accessibility for reactants. Dalton Trans 39:1751–1757
34. Domènech B, Bastos-Arrieta J, Alonso A, Macanás J, Muñoz M, Muraviev DN (2012) Bifunctional polymer-metal nanocomposite ion exchange materials. In: Ayben KL (ed) Ion exchange technologies. IntechOpen, Rijeka. Ch. 3
35. Gu Y, Emin C, Remigy J-C, Favier I, Gómez M, Noble RD, Gin DL, Macanás J, Domènech B, Lahitte J-F (2016) Hybrid catalytic membranes: tunable and versatile materials for fine chemistry applications. Mater Today Proc 3:419–423
36. Domènech B, Muñoz M, Muraviev DN, Macanas J (2011) Polymer-stabilized palladium nanoparticles for catalytic membranes: ad hoc polymer fabrication. Nanoscale Res Lett 6
37. Vincent T, Spinelli S, Guibal E (2003) Chitosan-supported palladium catalyst. II. Chlorophenol Dehalogenation Ind Eng Chem Res 42:5968–5976
38. Vincent T, Guibal E (2004) Chitosan-supported palladium catalyst. 5. Nitrophenol degradation using palladium supported on hollow chitosan fibers. Environ Sci Technol 38:4233–4240
39. Volkov VV, Lebedeva VI, Petrova IV, Bobyl AV, Konnikov SG, Roldughin VI, van Erkel J, Tereshchenko GF (2011) Adlayers of palladium particles and their aggregates on porous polypropylene hollow fiber membranes as hydrogenization contractors/reactors. Adv Colloid Interf Sci 164:144–155
40. Chen Y, Fan S, Qiu B, Chen J, Mai Z, Wang Y, Bai K, Xiao Z (2021) Cu-ag bimetallic Core-shell nanoparticles in pores of a membrane microreactor for enhanced synergistic catalysis. ACS Appl Mater Interfaces 13:24795–24803
41. Faria VW, Brunelli MF, Scheeren CW (2015) Iridium nanoparticles supported in polymeric membranes: a new material for hydrogenation reactions. RSC Adv 5:84920–84926
42. Nechifor AC, Goran A, Grosu V-A, Pirtac A, Albu PC, Oprea O, Grosu AR, Pascu D, Pancescu FM, Nechifor G, Tanczos S-K, Bungau SG (2021) Reactional processes on osmium-polymeric membranes for 5-nitrobenzimidazole reduction. Membranes 11:633
43. Durmaz EN, Sahin S, Virga E, de Beer S, de Smet LCPM, de Vos WM (2021) Polyelectrolytes as building blocks for next-generation membranes with advanced functionalities. ACS Appl Polym Mater 3:4347–4374
44. Bhattacharya A, Misra BN (2004) Grafting: a versatile means to modify polymers. Techniques, factors and applications. Prog Polym Sci 29:767–814
45. Macanas J, Ouyang L, Bruening ML, Munoz M, Remigy JC, Lahitte JF (2010) Development of polymeric hollow fiber membranes containing catalytic metal nanoparticles. Catal Today 156:181–186
46. Emin C, Gu Y, Remigy JC, Lahitte JF (2015) Polyethersulfone hollow fiber modified with poly(styrenesulfonate) and Pd nanoparticles for catalytic reaction. Eur Phys J Spec Top 224: 1843–1848
47. Gu Y, Favier I, Pradel C, Gin DL, Lahitte J-F, Noble RD, Gómez M, Remigy J-C (2015) High catalytic efficiency of palladium nanoparticles immobilized in a polymer membrane containing poly(ionic liquid) in Suzuki-Miyaura cross-coupling reaction. J Membr Sci 492: 331–339
48. Goma-Bilongo T, Akbari A, Clifton MJ, Remigy JC (2006) Numerical simulation of a UV photografting process for hollow-fiber membranes. J Membr Sci 278:308–317
49. Akbari A, Desclaux S, Remigy JC, Aptel P (2002) Treatment of textile dye effluents using a new photografted nanofiltration membrane. Desalination 149:101–107

50. Yamagishi H, Crivello JV, Belfort G (1995) Evaluation of photochemically modified poly (arylsulfone) ultrafiltration membranes. J Membr Sci 105:249
51. Decher G (1997) Fuzzy nanoassemblies: toward layered polymeric multicomposites. Science 277:1232–1237
52. Ouyang L, Dotzauer DM, Hogg SR, Macanas J, Lahitte J-F, Bruening ML (2010) Catalytic hollow fiber membranes prepared using layer-by-layer adsorption of polyelectrolytes and metal nanoparticles. Catal Today 156:100–106
53. Hogg SR, Muthu S, O'Callaghan M, Lahitte J-F, Bruening ML (2012) Wet air oxidation of formic acid using platinum nanoparticle modified polysulfone hollow fiber gas liquid contactor. ACS Appl Mater Interfaces 4:1440–1448
54. Fischer L, Volz A, Hagemann U, Ulbricht M (2023) Polymeric multi-composites with a tailored nickel microenvironment as catalytic flow-through membrane reactors for efficient p-nitrophenol degradation. Chem Eng J 463:142437
55. Fischer L, Hesaraki SAH, Volz A, Ulbricht M (2023) Highly nickel loaded nanocomposite membranes for catalytic H2 production and hydrogenation of p-nitrophenol using ammonia borane. ACS Appl Nano Mater 6:14910–14920
56. Gu Y, Bacchin P, Lahitte JF, Remigy JC, Favier I, Gómez M, Gin DL, Noble RD (2017) Catalytic membrane reactor for Suzuki-Miyaura C–C cross-coupling: explanation for its high efficiency via modeling. AICHE J 63:698–704
57. Hashim SM, Mohamed AR, Bhatia S (2011) Catalytic inorganic membrane reactors: present research and future prospects. Rev Chem Eng 27:157–178
58. Favier I, Pla D, Gómez M (2020) Palladium nanoparticles in polyols: synthesis, catalytic couplings and hydrogenations. Chem Rev 120:1146–1183
59. Liu X, Astruc D (2018) Development of the applications of palladium on charcoal in organic synthesis. Adv Synth Catal 360:3426–3459
60. Ciebien JF, Cohen RE, Duran A (1998) Catalytic properties of palladium nanoclusters synthesized within diblock copolymer films: hydrogenation of ethylene and propylene. Supramol Sci 5:31–39
61. Ciebien JF, Cohen RE, Duran A (1999) Membrane catalysts for partial hydrogenation of 1,3-butadiene: catalytic properties of palladium nanoclusters synthesized within diblock copolymer films. Mater Sci Eng C C7:45–50
62. Sulman E, Sulman M, Tyamina I, Doluda V, Nikoshvili L, Lyubimova N, Sidorov A, Matveeva V (2015) Adsorption processes for the synthesis of catalytically active metal nanoparticles in polymeric matrices. Chem Eng Technol 38:683–689
63. Nemygina NA, Nikoshvili LZ, Matveeva VG, Sulman MG, Sulman EM, Kiwi-Minsker L (2016) Pd-nanoparticles confined within hollow polymeric framework as effective catalysts for the synthesis of fine chemicals. Top Catal 59:1185–1195
64. Liu M, Zhu X, Liao Q, Chen R, Ye D, Chen G, Wang K (2020) Catalytic membrane microreactors with an ultrathin freestanding membrane for nitrobenzene hydrogenation. ACS Appl Mater Interfaces 12:9806–9813
65. Liguori F, Barbaro P (2014) Green semi-hydrogenation of alkynes by Pd@borate monolith catalysts under continuous flow. J Catal 311:212–220
66. Liguori F, Barbaro P (2014) Continuous flow synthesis of Rh and Pd nanoparticles onto ion-exchange borate monoliths: application to selective catalytic hydrogenation of unsaturated carbonyl compounds under flow conditions. Cat Sci Technol 4:3835–3839
67. Xu Y, Yao Y, Yu H, Shi B, Gao S, Zhang L, Miller AL, Fang J-C, Wang X, Huang K (2019) Nanoparticle-encapsulated hollow porous polymeric nanosphere frameworks as highly active and tunable size-selective catalysts. ACS Macro Lett 8:1263–1267
68. Liu Y, Li M, Chen G (2013) A new type of raspberry-like polymer composite sub-microspheres with tunable gold nanoparticles coverage and their enhanced catalytic properties. J Mater Chem A 1:930–937
69. Schmidt A, Haidar R, Schomaecker R (2005) Selectivity of partial hydrogenation reactions performed in a pore-through-flow catalytic membrane reactor. Catal Today 104:305–312

70. Groeschel L, Haidar R, Beyer A, Reichert K-H, Schomaecker R (2004) Characterization of palladium nanoparticles adsorbed on polyacrylic acid particles as hydrogenation catalyst. Catal Lett 95:67–75
71. Beyer A, Schomaecker R, Reichert K-H (2003) Synthesis and characterization of palladium containing membranes based upon polyacrylic acid. Colloid Polym Sci 281:862–868
72. Kidambi S, Bruening ML (2005) Multilayered polyelectrolyte films containing palladium nanoparticles: synthesis, characterization, and application in selective hydrogenation. Chem Mater 17:301–307
73. Robello DR, Mis MR, Nair M (2015) Micron-sized membrane reactors: multicompartment semipermeable polymer particles containing palladium nanoparticles. J Appl Polym Sci 132: 42021/42021-42021/42010
74. Goszewska I, Gizinski D, Zienkiewicz-Machnik M, Lisovytskiy D, Nikiforov K, Masternak J, Srebowata A, Sa J (2017) A novel nano-palladium catalyst for continuous-flow chemoselective hydrogenation reactions. Catal Commun 94:65–68
75. Kowalewski E, Zawadzki B, Matus K, Nikiforow K, Srebowata A (2021) Continuous-flow hydrogenation over resin supported palladium catalyst for the synthesis of industrially relevant chemicals. React. Kinet Mech Catal 132:717–728
76. Gizinski D, Goszewska I, Zielinski M, Lisovytskiy D, Nikiforov K, Masternak J, Zienkiewicz-Machnik M, Srebowata A, Sa J (2017) Chemoselective flow hydrogenation of α,β – unsaturated aldehyde with nano-nickel. Catal Commun 98:17–21
77. Biesinger MC, Lau LW, Gerson AR, Smart RSC (2012) The role of the Auger parameter in XPS studies of nickel metal, halides and oxides. Phys Chem Chem Phys 14:2434–2442
78. Biesinger MC, Payne BP, Lau LWM, Gerson A, Smart RSC (2009) X-ray photoelectron spectroscopic chemical state quantification of mixed nickel metal, oxide and hydroxide systems. Surf Interface Anal 41:324–332
79. Mamontova E, Trabbia C, Favier I, Serrano-Maldonado A, Ledeuil J-B, Madec L, Gómez M, Pla D (2023) Novel catalyst composites of Ni- and co-based nanoparticles supported on inorganic oxides for fatty acid hydrogenations. Nanomaterials 13:1435
80. Pérez Alonso A, Mauriés S, Ledeuil J-B, Madec L, Pham Minh D, Pla D, Gómez M (2022) Nickel nanoparticles immobilized on pristine halloysite: an outstanding catalyst for hydrogenation processes. ChemCatChem 14:e2022007
81. Xie R, Fan G, Ma Q, Yang L, Li F (2014) Facile synthesis and enhanced catalytic performance of graphene-supported Ni nanocatalyst from a layered double hydroxide-based composite precursor. J Mater Chem A 2:7880–7889
82. Gryglewicz S, Sliwak A, Cwikla J, Gryglewicz G (2014) Performance of carbon nanofiber and activated carbon supported nickel catalysts for liquid-phase hydrogenation of cinnamaldehyde into hydrocinnamaldehyde. Catal Lett 144:62–69
83. Bhogeswararao S, Pavan Kumar V, Chary KVR, Srinivas D (2013) Noble metal promoted CeO_2-ZrO_2-supported Ni catalysts for liquid-phase hydrogenation of cinnamaldehyde. Catal Lett 143:1266–1276
84. Malobela LJ, Heveling J, Augustyn WG, Cele LM (2014) Nickel-cobalt on carbonaceous supports for the selective catalytic hydrogenation of cinnamaldehyde. Ind Eng Chem Res 53: 13910–13919
85. Ceatra L, Parvulescu OC, Rodriguez Ramos I, Dobre T (2016) Preparation, characterization, and testing of a carbon-supported catalyst obtained by slow pyrolysis of nickel salt impregnated vegetal material. Ind Eng Chem Res 55:1491–1502
86. Jiao N, Li Z, Xia C, Liu J (2017) Palladium nanoparticles immobilized on cross-linked polymeric ionic liquid material: application as efficient and recoverable catalyst for the hydrogenation of nitroarenes. ChemistrySelect 2:4545–4556
87. Copéret C, Comas-Vives A, Conley MP, Estes DP, Fedorov A, Mougel V, Nagae H, Núñez-Zarur F, Zhizhko PA (2016) Surface organometallic and coordination chemistry toward single-site heterogeneous catalysts: strategies, methods, structures, and activities. Chem Rev 116:323–421

88. Astakhov AV, Khazipov OV, Chernenko AY, Pasyukov DV, Kashin AS, Gordeev EG, Khrustalev VN, Chernyshev VM, Ananikov VP (2017) A new mode of operation of Pd-NHC systems studied in a catalytic Mizoroki-heck reaction. Organometallics 36:1981–1992
89. Favier I, Toro M-L, Lecante P, Pla D, Gómez M (2018) Palladium-mediated radical homocoupling reactions: a surface catalytic insight. Cat Sci Technol 8:4766–4773
90. Duarte TAG, Favier I, Pradel C, Martins LMDRS, Carvalho AP, Pla D, Gómez M (2020) Tetraalkylammonium functionalized hydrochars as efficient supports for palladium nanocatalysts. ChemCatChem 12:2295–2303
91. Zhang Y, Quek X-Y, Wu L, Guan Y, Hensen EJ (2013) Palladium nanoparticles entrapped in polymeric ionic liquid microgels as recyclable hydrogenation catalysts. J Mol Catal A Chem 379:53–58
92. Xu J, Chen G, Yan R, Wang D, Zhang M, Zhang W, Sun P (2011) One-stage synthesis of Cagelike porous polymeric microspheres and application as catalyst scaffold of Pd nanoparticles. Macromolecules 44:3730–3738
93. Gao Y, Chen C-A, Gau H-M, Bailey JA, Akhadov E, Williams D, Wang H-L (2008) Facile synthesis of polyaniline-supported Pd nanoparticles and their catalytic properties toward selective hydrogenation of alkynes and cinnamaldehyde. Chem Mater 20:2839–2844
94. Guo L, Bai J, Li C, Meng Q, Liang H, Sun W, Li H, Liu H (2013) A novel catalyst containing palladium nanoparticles supported on PVP composite nanofiber films: synthesis, characterization and efficient catalysis. Appl Surf Sci 283:107–114
95. Gao H, Xu Y, Liao S, Liu R, Liu J, Li D, Yu D, Zhao Y, Fan Y (1995) Catalytic polymeric hollow-fiber reactors for the selective hydrogenation of conjugated dines. J Membr Sci 106: 213
96. Liu C, Xu Y, Liao S, Yu D, Zhao Y, Fan Y (1997) Selective hydrogenation of propadiene and propyne in propene with catalytic polymeric hollow-fiber reactor. J Membr Sci 137:139–144
97. Liu C, Xu Y, Liao S, Yu D (1998) Mono- and bimetallic catalytic hollow-fiber reactors for the selective hydrogenation of butadiene in 1-butene. Appl Catal A 172:23–29
98. Gancheva T, Virgilio N (2019) Tailored macroporous hydrogel-nanoparticle nanocomposites for monolithic flow-through catalytic reactors. React Chem Eng 4:806–811
99. Yamada YMA, Watanabe T, Ohno A, Uozumi Y (2012) Development of polymeric palladium-nanoparticle membrane-installed microflow devices and their application in hydrodehalogenation. ChemSusChem 5:293–299
100. Ziegler S, Theis J, Fritsch D (2001) Palladium modified porous polymeric membranes and their performance in selective hydrogenation of propyne. J Membr Sci 187:71–84
101. Troeger L, Huennefeld H, Nunes S, Oehring M, Fritsch D (1997) Structural characterization of catalytically active metal nanoclusters in poly(amide imide) films with high metal loading. J Phys Chem B 101:1279–1291
102. Peirano Blondet F, Vincent T, Guibal E (2008) Hydrogenation of nitrotoluene using palladium supported on chitosan hollow fiber, catalyst characterization and influence of operative parameters studied by experimental design methodology. Int J Biol Macromol 43:69–78
103. Wu H, Wu C, He Q, Liao X, Shi B (2010) Collagen fiber with surface-grafted polyphenol as a novel support for Pd(0) nanoparticles: synthesis, characterization and catalytic application. Mater Sci Eng C 30:770–776
104. Long NV, Nguyen DC, Hirata H, Ohtaki M, Hayakawa T, Nogami M (2010) Chemical synthesis and characterization of palladium nanoparticles. Adv Nat Sci Nanosci 1:035012
105. Harraz FA, El-Sheikh SM, Mohamed RM, Ibrahim IA (2009) Synthesis and characterization of palladium nanoparticles for catalytic applications. Int J Nanopart 2:361–371
106. Astruc D, Lu F, Aranzaes JR (2005) Nanoparticles as recyclable catalysts: the frontier between homogeneous and heterogeneous catalysis. Angew Chem Int Ed 44:7852–7872
107. Narayanan R, El-Sayed MA (2003) Effect of catalysis on the stability of metallic nanoparticles: Suzuki reaction catalyzed by PVP-palladium nanoparticles. J Am Chem Soc 125:8340–8347

108. Mahdavi H, Rahimi A, Shahalizade T (2016) Catalytic polymeric membranes with palladium nanoparticle/multi-wall carbon nanotubes as hierarchical nanofillers: preparation, characterization and application. J Polym Res 23:1–12
109. Zhang N, Wu Y, Yuen G, de Lannoy C-F (2023) Ultrafiltration Pd-immobilized catalytic membrane microreactors continuously reduce nitrophenol study of catalytic activity and simultaneous separation. Sep Purif Technol 312:123318
110. Lee S, Shin S-J, Baek H, Choi Y, Hyun K, Seo M, Kim K, Koh D-Y, Kim H, Choi M (2020) Dynamic metal-polymer interaction for the design of chemoselective and long-lived hydrogenation catalysts. Sci Adv 6:eabb7369
111. Brandao L, Madeira LM, Mendes AM (2007) Propyne hydrogenation in a continuous polymeric catalytic membrane reactor. Chem Eng Sci 62:6768–6776
112. Brandão L, Fritsch D, Mendes AM, Madeira LM (2007) Propylene hydrogenation in a continuous polymeric catalytic membrane reactor. Ind Eng Chem Res 46:5278–5285
113. He Y, Cheshomi N, Lawson SM, Itta AK, Rezaei F, Kapila S, Rownaghi AA (2021) PDMS/PAI-HF composite membrane containing immobilized palladium nanoparticles for 4-nitrophenol reduction. Chem Eng J 410:128326
114. Bottino A, Capannelli G, Comite A, Di Felice R (2002) Polymeric and ceramic membranes in three-phase catalytic membrane reactors for the hydrogenation of methylenecyclohexane. Desalination 144:411–416
115. Kotte MR, Kuvarega AT, Talapaneni SN, Cho M, Coskun A, Diallo MS (2018) A facile and scalable route to the preparation of catalytic membranes with in situ synthesized supramolecular dendrimer particle hosts for Pt(0) nanoparticles using a low-generation PAMAM dendrimer (G1-NH_2) as precursor. ACS Appl Mater Interfaces 10:33238–33251
116. Büker J, Alkan B, Chabbra S, Kochetov N, Falk T, Schnegg A, Schulz C, Wiggers H, Muhler M, Peng B (2021) Liquid-phase cyclohexene oxidation with O_2 over spray-flame-synthesized $La_{1-x}Sr_xCoO_3$ perovskite nanoparticles. Chem Eur J 27:16912–16923
117. Dintzner MR, Kinzie CR, Pulkrabek K, Arena AF (2012) The cyclohexanol cycle and synthesis of nylon 6,6: green chemistry in the undergraduate organic laboratory. J Chem Educ 89:262–264
118. Long S, Zhang A, Liu H, Li L, Ding L (2009) Fabrication of a Pt nano-cluster/photosensitive polyimide hybrid membrane reactor and its partial hydrogenation of benzene. Chinese J Catal 30:276–278
119. Zhai H, Zhang A, Li L, Long S (2012) Partial hydrogenation of benzene catalyzed by Pt/N-n-propyl chitosan hybrid membrane. J Appl Polym Sci 123:2140–2146
120. Hereijgers BPC, Weckhuysen BM (2010) Aerobic oxidation of cyclohexane by gold-based catalysts: new mechanistic insight by thorough product analysis. J Catal 270:16–25
121. Schuchardt U, Cardoso D, Sercheli R, Pereira R, da Cruz RS, Guerreiro MC, Mandelli D, Spinacé EV, Pires EL (2001) Cyclohexane oxidation continues to be a challenge. Appl Catal A 211:1–17
122. Musser MT (2011) Cyclohexanol and cyclohexanone, Ullmann's encyclopedia of industrial chemistry. Wiley
123. Hodges AM, Linton M, Mau AWH, Cavell KJ, Hey JA, Seen AJ (1990) Perfluorinated membranes as catalyst supports. Appl Organomet Chem 4:465
124. Byun H, Lee T (2002) Hydrogenation of cyclohexene with perfluorinated cation- and anion-exchange membranes supported platinum. Desalination 144:419–424
125. Adlim M, Abu Bakar M, Liew KY, Ismail J (2004) Synthesis of chitosan-stabilized platinum and palladium nanoparticles and their hydrogenation activity. J Mol Catal A Chem 212:141–149
126. Gelesky MA, Scheeren CW, Foppa L, Pavan FA, Dias SLP, Dupont J (2009) Metal nanoparticle/ionic liquid/cellulose: new catalytically active membrane materials for hydrogenation reactions. Biomacromolecules 10:1888–1893

127. Singh D, Rezac ME, Pfromm PH (2010) Partial hydrogenation of soybean oil using metal-decorated integral-asymmetric polymer membranes: effects of morphology and membrane properties. J Membr Sci 348:99–108
128. Holda AK, Vankelecom IFJ (2015) Understanding and guiding the phase inversion process for synthesis of solvent resistant nanofiltration membranes. J Appl Polym Sci 132:42130/42131
129. Fritsch D, Bengtson G (2006) Catalytic polymer membranes for high temperature hydrogenation of viscous liquids. Adv Eng Mater 8:386–389
130. Liu C, Xu Y, Liao S, Yu D (2000) Selective hydrogenation of cyclopentadiene in mono- and bimetallic catalytic hollow-fiber reactors. J Mol Catal A Chem 157:253–259
131. Lim B, Jiang M, Camargo PHC, Cho EC, Tao J, Lu X, Zhu Y, Xia Y (2009) Pd-Pt bimetallic nanodendrites with high activity for oxygen reduction. Science 324:1302–1305
132. Yarulin A, Berguerand C, Yuranov I, Cardenas-Lizana F, Prokopyeva I, Kiwi-Minsker L (2015) Pt-Zn nanoparticles supported on porous polymeric matrix for selective 3-nitrostyrene hydrogenation. J Catal 321:7–12
133. Stanford JP, Soto MC, Pfromm PH, Rezac ME (2016) Aqueous phase hydrogenation of levulinic acid using a porous catalytic membrane reactor. Catal Today 268:19–28
134. Bagnato G, Figoli A, Ursino C, Galiano F, Sanna A (2018) A novel Ru-polyethersulfone (PES) catalytic membrane for highly efficient and selective hydrogenation of furfural to furfuryl alcohol. J Mater Chem A 6:4955–4965
135. Molinari R, Lavorato C, Argurio P (2020) Visible-light photocatalysts and their perspectives for building photocatalytic membrane reactors for various liquid phase chemical conversions. Catalysts 10:1334

Top Organomet Chem (2024) 75: 105–132
https://doi.org/10.1007/3418_2024_116
© The Author(s), under exclusive license to Springer Nature Switzerland AG 2024
Published online: 16 October 2024

Influence of Capping Ligands on Metal-Nanoparticle-Driven Hydrogen Evolution and CO_2 Reduction Reactions

Gerard Martí, Álvaro Lozano-Roche, Nuria Romero, Laia Francàs, Karine Philippot, Roger Bofill, Jordi García-Antón, and Xavier Sala

Contents

Abstract This chapter looks into the influence of the functionalization of the surface of metal nanoparticles (MNPs) with ligands on their catalytic performance in two crucial reactions, the hydrogen evolution reaction (HER) and the CO_2 reduction reaction (CO_2RR). The birth of the surface functionalization of MNPs with ligands was greatly inspired by the advanced knowledge accumulated on molecular metal catalysts, whose physical and chemical properties can be tuned by coordinating appropriate ligands to the metal active center. This field of research has profited from the progress in the synthetic methods for the generation of MNPs and the evolution of the spectroscopic and analytical techniques for their exhaustive characterization, as well as the improvement of the computational techniques. Relevant examples will illustrate how the functionalization of the surface of MNPs allow to modulate the energy profile of catalytic intermediate species through electronic effects and regulate the product selectivity through hydrophobic/

G. Martí, Á. Lozano-Roche, L. Francàs, R. Bofill (✉), J. García-Antón (✉), and X. Sala (✉)
Departament de Química, Universitat Autònoma de Barcelona, Bellaterra, Catalonia, Spain
e-mail: roger.bofill@uab.cat; jordi.garciaanton@uab.es; xavier.sala@uab.cat

N. Romero and K. Philippot (✉)
CNRS, LCC (Laboratoire de Chimie de Coordination), UPR8241, Université de Toulouse, UPS, INPT, Toulouse, Cedex 4, France
e-mail: karine.philippot@lcc-toulouse.fr

hydrophilic effects. Another key parameter is the tuning of the number of surface-active sites as a function of the metal to ligand ratio and the chemical nature of the coordinating ligand. Furthermore, in some peculiar cases, second coordination sphere effects have even been described in analogy to biological enzymes.

Keywords CO_2 reduction reaction · Electronic effects · Hydrogen evolution reaction · Hydrophobic/hydrophilic effects · Ligand-capped metal nanoparticles · Second coordination sphere effects · Surface-functionalized metal nanoparticles

1 Introduction

Metal nanoparticles (MNPs), with dimensions comprised between 1 and 100 nm, show characteristic electronic properties in between those of molecular metal complexes and bulk metals [1, 2]. In the field of catalysis, MNPs are very attractive materials because of the high abundance of coordinatively unsaturated atoms at their surface, which gives them unique reactivity features for numerous reactions [3, 4].

The use of H_2 as an energy carrier is pivotal to overcome global warming provided the development of green and sustainable technologies for its production. Another strategy to mitigate global warming is the reduction of the atmospheric concentration of CO_2. This can be partially achieved by using this greenhouse gas as the starting material to produce synthetic fuels (methane, ethane, propane, methanol, ethanol, propanol, etc.) and chemical feedstocks (syngas, formic acid, formaldehyde, ethylene, etc.), aiming to attain carbon neutrality in transportation and industry. These strategies require the development of efficient and robust catalysts able to perform the hydrogen evolution reaction (HER) and the CO_2 reduction reaction (CO_2RR) together with the use of green electricity (electrocatalysis), solar light (photocatalysis) or both (photoelectrocatalysis) for their activation. In this way, MNPs have been shown to be relevant candidates in both the fields of electrocatalysis [5–7] and photo(electro)catalysis [8].

One feature of MNPs is the ratio between the number of atoms constituting their core and the number of atoms present at their surface, the latter being exposed to the catalytic substrate and solvent in liquid catalysis. This ratio can be calculated according to the so-called magic number rule, that depends on the type of crystalline structure of the metal [9]. In the case of very small MNPs ($\leq$1 nm), the number of surface atoms is greater than 90% of the total number of atoms in the MNPs, thus granting a high amount of potential active sites. Although sub-nanometric MNPs are often amorphous, larger MNPs can be crystalline, presenting exposed atoms at different locations (corners, edges or faces) with different coordination numbers, which may induce distinctive reactivities [10]. This implies that the reactivity of MNPs will both depend on their size and shape and the nature of metal. Here is where surface functionalization becomes important, since the catalytic performance of MNPs can vary depending on the nature and number of ligands (L) coordinated to

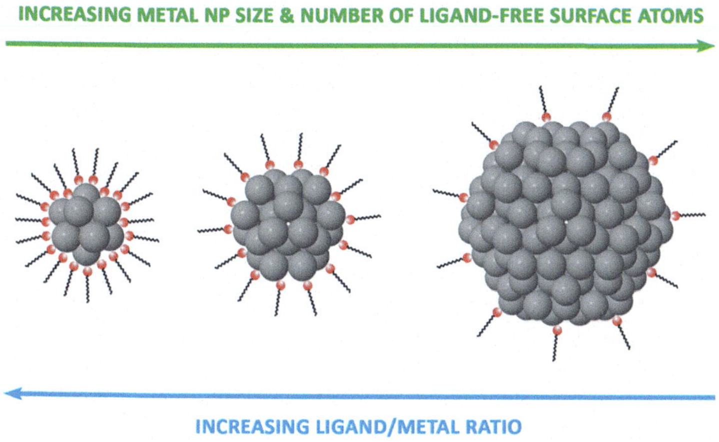

Fig. 1 Schematic representation of the expected effects of the L:M ratio on the MNP size and surface coverage

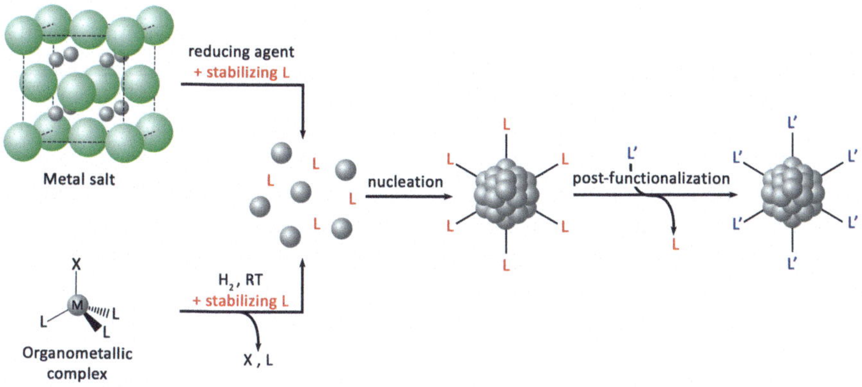

Fig. 2 Schematic representation of the different chemical bottom-up synthetic approaches to obtain MNPs

their surface. Ligands are usually small organic molecules [11], which a priori have two main roles: to stabilize the MNPs (avoiding their aggregation, sintering or coalescence) and to tune their reactivity by tailoring their surface properties via steric, hydrophilic/hydrophobic and/or electronic/electrostatic effects [12]. When these capping ligands are employed for the synthesis of MNPs, the stochiometric L to metal ratio (L:M) used may affect the average size of the MNPs, where it is usually inversely proportional to the L:M ratio (Fig. 1). Additionally, a post-functionalization process can be carried out by a ligand-exchange procedure to tune the surface properties [11, 13] (see Fig. 2, right, in Sect. 2). This can be of interest to compare the effect of various ligands at the metal surface at a fixed MNP size. However, in any of the two cases, the number of coordinated ligands must be controlled in order to have MNPs with enough free surface sites to react, being those generally fewer at higher ligand concentrations (Fig. 1). A compromise between a small size (to achieve a high number of potential active sites) and a low surface coverage (to leave enough free metal sites) is thus often needed to have good performing nanocatalysts.

The chemical nature of the ligands used so far in the literature to synthesize and/or functionalize MNPs is wide and diverse, encompassing amines [14–25], alkynes [26], carboxylates [15, 19, 22, 24, 27, 28], thiolates [19, 24, 25, 29–31], carbenes [32–34], phosphines [24, 31], phosphonates [24, 35], polypyridyl [36, 37], and porphyrin-derived [38] molecules, among others. The choice of the ligand depends on its relative affinity towards the metals of study [12]. Furthermore, the electron density of the MNP surface can be modified by using capping-ligands of different electron-donor/electron-withdrawing properties or different σ-donor/π-acceptor character, as will be exemplified later [22, 25, 28, 32, 33, 39].

The ligand-capped MNPs can be used as catalysts either directly used as colloidal suspensions or after their deposition onto a solid support, the latter being the most common case. The use of supported nanocatalysts presents benefits from both classical homogeneous and heterogeneous catalysis, namely high reactivity and selectivity typical from the former, and high recyclability and long-term stability typical from the latter [40].

The number of works devoted to the functionalization of MNPs [41] has incessantly increased during the last decade owing to the advanced knowledge gained on molecular catalysts, which provide inspiration for the development of nanocatalysts. The field has benefited as well from the advances in the synthesis of MNPs, that has given rise to efficient and reproducible methods to obtain well-defined ligand-capped MNPs. Additionally, the progress made in the spectroscopic and analytical methods allow their thorough characterization, and the development of theoretical techniques for their modelling and surface reactivity have also boosted this field of research.

This chapter aims to be a critical analysis of the most remarkable ligand-capped MNP systems in HER and CO_2RR catalysis, focusing on the influence of surface ligands on their catalytic properties. Thus, this chapter is not intended to be comprehensive nor to provide an exhaustive comparison of catalytic performances.

2 Synthesis of Ligand-Capped Metal Nanoparticles

The best synthetic strategies to obtain crystalline ligand-capped MNPs in a reproducible manner with narrow size distribution, well-defined shape, and good dispersibility are the chemical methods in solution [42, 43]. These liquid-phase bottom-up methods are more versatile and easier to implement than the physical top-down methods [44]. The principle of the chemical approach consists in the decomposition of either a metallic salt [45] or an organometallic complex [46–49] to release naked metallic atoms in solution that will nucleate to generate the MNPs.

Their growth is controlled by the stabilizer added in the reaction medium (Fig. 2). The nature of the metal source will condition which solvent, reducing agent, stabilizing ligand, pressure, temperature, etc., should be used.

The reduction of a commercial metal salt in aqueous medium in the presence of a stabilizing agent, either at room temperature or under temperature-controlled assisted conditions, is a widely used method to synthesize MNPs (Fig. 2, upper left). Several reducing agents can be used, namely surfactants such as long alkyl amines [50], alcohols [51], diols or polyols [52], hydrogen gas under pressure [33] or hydrides such as $NaBH_4$ [53] and hydrazine [44], to name a few. The stabilizer must be soluble and stable in water. Many different stabilizing agents have been used so far, such as polyacrylamide [54], polyvinylpyrrolidone [55], polyvinyl alcohol [56] and ammonium-based surfactants [57], as well as amine, phosphine, citrate, and thiolate ligands [58]. If this method is advantageous in terms of cost, ease of implementation, and scalability, the main drawbacks are the potential contamination of the resulting MNPs with species arising from the reactants, which may adsorb to the MNPs surface, such as halide counterions, and the potential oxidation of the MNP surface, especially when working with oxyphilic metals [59].

Another efficient strategy consists in the decomposition of an organometallic precursor in an organic solvent under hydrogen (1-3 bar) at mild temperature (ranging from room temperature to below 150°C) and in the presence of a stabilizer (Fig. 2, bottom left). This methodology yields well-defined, small (ca. 1–10 nm) MNPs with a narrow size distribution and a clean surface (free of contaminants). This clean surface is easy to achieve with precursors containing olefinic complexes, since the volatile alkanes released under H_2 are inert towards the MNP surface [60]. The choice of the stabilizer added to control the MNP growth will also functionalize their surface and affect their catalytic properties [61]. However, in some cases, the solvent itself (THF/MeOH [62] or heptanol [63, 64]) can act as a stabilizer, with no need of an additional ligand. The major limitation of the organometallic approach may be the synthesis of the organometallic precursor if it is not commercially available, or its handling, which in some cases entails the use of a glove box under inert conditions, especially for the case of oxyphilic metals. Finally, the risk of oxidation can be an issue for the very small MNPs obtained by the organometallic approach due to the high reactivity of their surface.

The dispersion of MNPs onto solid supports for their use as electrodes can be performed by two different methods, namely the in situ-synthesis in the presence of the support [20, 65–77] or their ex situ-synthesis [69, 78, 79]. For the former, the support (carbon nanotubes [65, 66, 74], metal organic frameworks [71, 72, 74], covalent organic frameworks [70, 72, 74], carbon nitride [67, 69, 74], (reduced) graphene oxide [20, 65, 66], fluorine-doped tin oxide [75], carbon and carbon quantum dots [73, 74], quantum dots [74], TiO_2 [68], Ni foam [76], carbon paper [77] or carbon fibers [79], among others) is inserted inside the synthesis medium. For the latter, the MNPs are spin-coated onto the support [78] or alternatively the support is soaked into a MNP dispersion [69, 79]. In both cases, the support can interact with the MNPs and thus affect their electronic properties and stability.

3 Ligand Effects in the Hydrogen Evolution Reaction

The hydrogen evolution reaction (HER) is a two-electron reduction process (Eq. 1) [80].

$$2\,H^{+} + 2\,e^{-} \rightarrow H_{2(g)} \qquad (E^{0}(H_2O/H_2) = -0.41\ \text{V vs.NHE at pH 7}) \qquad (1)$$

Historically, Pt has been the metal of choice in acidic media owing to its ideal Pt-H adsorption energy, leading to a very low overpotential and a very high catalytic activity [81]. This model behavior is related to the Sabatier principle [82] proposed in heterogeneous catalysis and electrocatalysis [83], which relates the catalytic activity with the Gibbs free energy (ΔG_{ads}) of the substrate adsorption to the surface of the catalyst. According to Sabatier, an optimal catalyst should not show neither too negative nor too positive ΔG_{ads} towards the substrate. Thus, if the interaction is too strong (negative ΔG_{ads}), the desorption of the product from the catalyst surface will be slow, becoming the rate determining step (rds) of the whole process. On the other hand, if the interaction is not favorable (positive ΔG_{ads}), the rds of the catalytic process will be the interaction between the substrate and the catalyst.

The in silico-application of the Sabatier principle in the case of HER is translated into the so-called Volcano plots [84], in which the DFT calculated Gibbs free energy of H-adsorption, ΔG_{H*} [85–88], is plotted versus the experimental electrocatalytic activity (log exchange current density, log j_O), as shown in Fig. 3 (left). In this plot, the catalysts that lay closer to the top, such is the case of Pt, show a higher HER performance. However, even though Pt shows ideal Pt-H adsorption energy, Pt-based catalysts are not stable under neutral or alkaline conditions due to corrosion [89]. This has impelled researchers to move towards other transition metals, such as Ru (among others). This metal shows strong stability under both acidic and basic conditions and only a bit lower HER catalytic activity (slightly stronger metal-H bond than Pt, Fig. 3, left) [90, 91]. In this context, the use of capping ligands in nanocatalysis can play a major role in the optimization of the binding of key *H reaction intermediate for improving the overall HER electrocatalytic activity. Thus, the electronic effects exerted by the ligands, analogously to those induced by the ligands in organometallic or coordination complexes, will allow the rational modification of ΔG_{H*}.

Interestingly, a realistic computational model that rationalizes the effect of the mono-N-dentate 4-phenylpyridine (4PP) ligand (Chart 1) onto the ΔG_{H*} values at acidic pH for 1.5 nm Ru NPs synthesized by the organometallic method was built [36, 94]. The initial model was a 55-atom hcp Ru NP [95]. According to the experimental ratios measured (elemental analysis (EA), inductively coupled plasma optical emission spectroscopy (ICP-OES), thermogravimetric analysis (TGA) and hydride titration [96]), a chemical formula can be calculated with 55 Ru atoms, 53 hydride ligands, 4 4PP ligands, and 2 THF solvent molecules (Fig. 3, right). The DFT calculations allowed to establish the influence of the different coordination modes (σ, π or σ-π) of the 4PP ligands as well as that of the THF molecules onto the

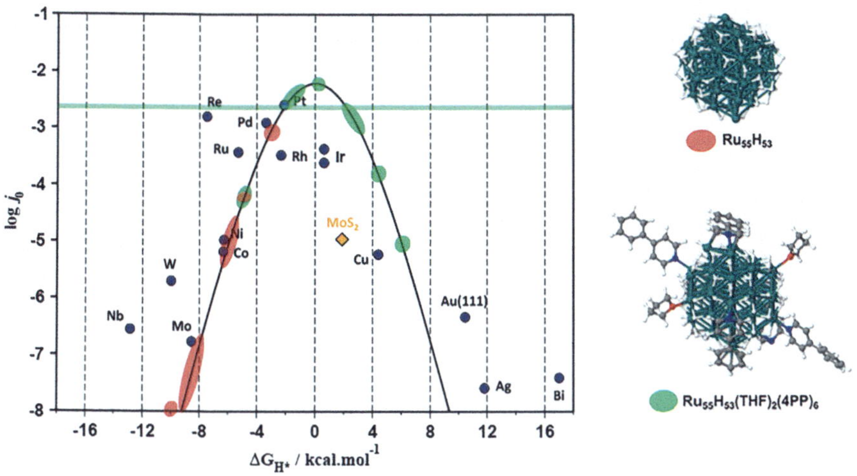

Fig. 3 Left: volcano plot of the log of the experimental exchange current density j_0 (A/cm^2) vs. the DFT calculated Gibbs free energy of H-adsorption, ΔG_{H*} (kcal/mol), for metals [92] (blue circles) and nanoscale MoS_2 (orange diamond) [93] at acidic pH. The horizontal green line is the experimental $\log j_0$ value for Ru-4PP [36]. The results from our DFT calculations for ΔG_{H*} in the different hydride sites for $Ru_{55}H_{53}$ (salmon) and $Ru_{55}H_{53}(THF)_2(4PP)_6$ (green) are shown as oval shapes. Right: Representation of the DFT calculated models for $Ru_{55}H_{53}$ and $Ru_{55}H_{53}(THF)_2(4PP)_6$. Adapted with permission [94]. Copyright 2021, John Wiley and Sons

ΔG_{H*} values of their surrounding hydride ligands [36]. This evidenced that 4PP coordination causes the shift of the ΔG_{H*} values from the left of the volcano plot (Fig. 3, salmon oval shapes), where strong Ru-H bonds dominate for bare Ru NPs, to values closer to the top of the volcano plot, with more moderate ΔG_{H*} values (Fig. 3, green oval shapes). This ligand effect on ΔG_{H*} induces the extraordinary HER electrocatalytic performance of the Ru-4PP NPs (η_{10} = 20 mV and Tafel slope (b) = 29 mV/dec, where η_{10} stands for the required overpotential to reach the standard benchmark current density of 10 mA·cm^{-2}). Moreover, the theoretical calculations have demonstrated that the weakly coordinating THF molecules exert smaller electronic influence than the 4PP ligand, thus only marginally decreasing the ΔG_{H*} values of the hydrides standing at the same Ru sites or nearby. This may explain the inferior HER performance (η_{10} = 83 mV and b = 46 mV/dec) of Ru NPs when stabilized by THF and MeOH solvent molecules only [62]. In summary, the excellent electrocatalytic activity of the Ru-4PP NPs might be explained by the beneficial electronic modulation exerted by the 4PP ligand. Also, the moderate Ru-4PP bond strength ensures the presence of enough accessible surface sites for the hydride substrate. Further validation of these theoretical results through the use of advanced surface-sensitive characterization techniques, such as Time-of-Flight Secondary Ion Mass Spectrometry (ToF-SIMS) [97] and in situ-XPS, is expected to provide insights on the ligand coordination modes and binding strength throughout the electrocatalytic process.

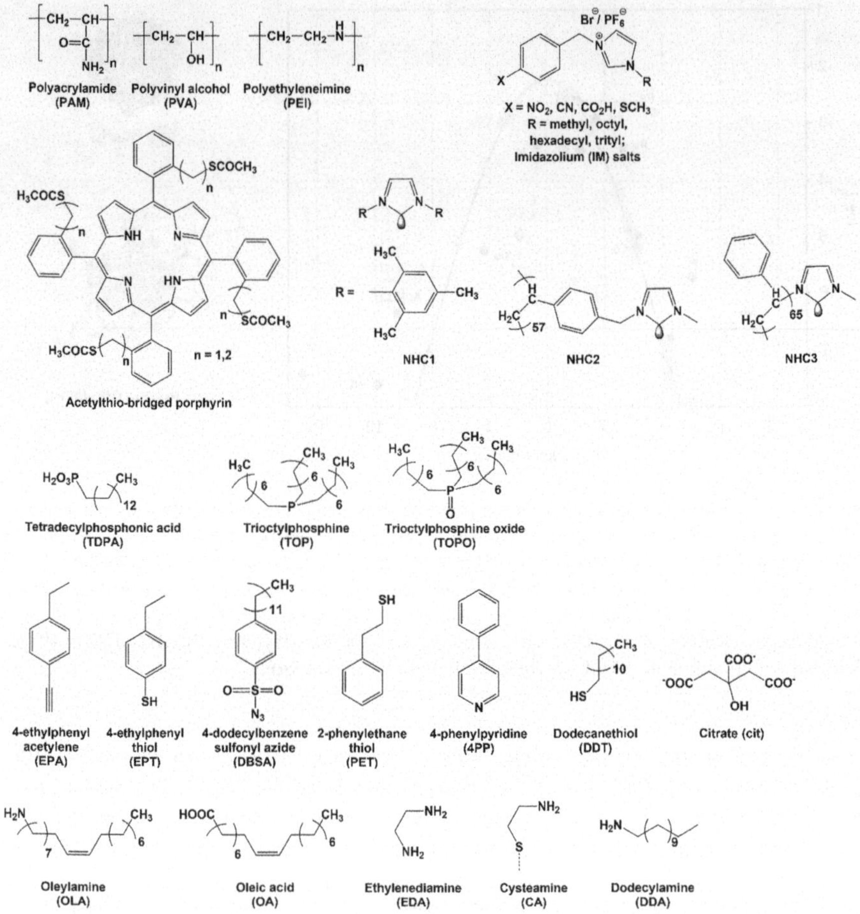

Chart 1 Illustration of the most representative capping ligands described in this chapter

Analogously to Ru, other metals forming stronger M-H bonds than Pt, thus laying on the left branch of the volcano plot (Fig. 3), may also benefit after weakening their M-H bond. This is the case of W, as the HER electrocatalytic activity of dodecylamine-functionalized (DDA, Chart 1) WSe_2 nanocrystals (NCs) increased (η_{10} decreased 118 mV) after removing the stabilizing ligands from the surface. This ligand stripping is achieved by using the Meerwein's reagent ($(Et_3O)(BF_4)$) [14], an alkylating reagent used for the release of coordinating ligands from nanocrystal thin films or dispersions to yield bare or BF_4^-/DMF-passivated surfaces. Ultraviolet photoelectron spectroscopy (UPS) evidenced that ligand stripping populates the antibonding states and thus weakens the W-H bond [14]. However, apart from the electronic effects modulating the W-H bond strength, other effects may intervene. In fact, an increase in the electrocatalytically active surface area (ECSA) was observed after ligand stripping, indicating that the number of available active sites increased

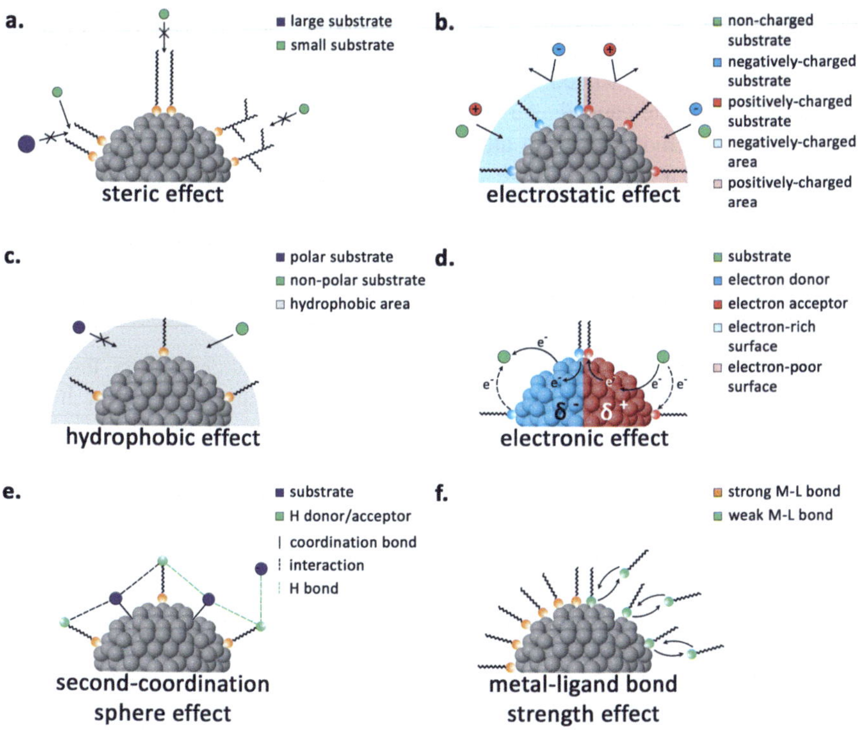

Fig. 4 Schematic overview of the different effects induced by metal NP capping in electrocatalytic HER and CO_2RR

after ligand removal, thus potentially increasing the HER performance of the nanocatalyst. To sum up, the authors attributed the lower electrocatalytic HER activity of the DDA-capped WSe_2 NCs to a combination of the steric hindrance induced by DDA, which blocks the catalytic centers (Fig. 4a), and the unfavorable electronic effects caused by the same ligand, that shifts ΔG_{H*} towards more negative values [14].

The same authors have also demonstrated the beneficial effect of ligand stripping with Meerwein's reagent when using CoP NPs stabilized with either carboxylate (acetate, octanoate, oleate) or amine (oleylamine, octylamine, dioctylamine, trioctylamine, butylamine) ligands (Chart 1) [15]. This work establishes the effect that the different Co-L bond strengths and L chain lengths have on the overall HER electrocatalytic activity. With the carboxylate ligands, η_{10} increased 251 mV from acetate to oleate, showing that the higher steric hindrance and hydrophobic character of the oleate-capped MNPs prevent catalysis by inhibiting the access of the substrate to the surface active sites (Fig. 4a, c). In contrast, with amine ligands, the same HER activity was measured independently of the chain length of the amine. Subsequent ECSA and TGA measurements showed lower density of ligands onto the NP surface as compared to their carboxylate analogues. This was assigned to a weaker coordinating capacity of the amine ligands, allowing an easier access of the substrate to the catalytic sites [15] (Fig. 4f).

The advantages associated to a moderate M-L bond strength have also been observed for citrate-capped Au NPs (Chart 1) [27]. Their excellent HER electrocatalytic activity has been attributed to the weak bonds between the citrate ligands and the Au NPs, which are strong enough to stabilize the MNPs but weak enough to leave enough free surface-active sites. Furthermore, this work showed that the highly negative surface charge density provided by the electrostatic interactions of the citrate ligands induced an increase in the amount of hydronium ions present at the interface (Fig. 4b, **left**), thus decreasing the activation energy barrier of the HER electrocatalysis.

Similarly to Ru, for which the M-H bond energy decrease eases the HER performance (Fig. 3), an increased HER activity has been described for Ir NPs capped with 4-ethylphenylacetylene (EPA, Chart 1) (Ir-C≡) compared to commercial Ir NPs and analogous mercapto (R-SH, 4-ethylphenyl thiol, EPT) and nitrene (R-·N·, 4-dodecylbenzene sulfonyl azide, DBSA) surface-capped derivatives (Chart 1) [26]. The Ir-H bond weakening in the Ir-C ≡ NPs was assigned to an intraparticle charge delocalization induced by the conjugated interfacial Ir-C ≡ C- bonds, which resulted in a charge transfer from the Ir core to the acetylene ligands [26].

In contrast, for metals on the right part of the volcano plot (Fig. 3), the HER activity will increase when the M-H bond is strengthened, as it is below the optimal values. This has been demonstrated in the case of porphyrin-modified Au nanoclusters (Au NCs) through an acetylthio (CH_3-CO-S-) bridge (Chart 1) [38]. XPS analysis indicated that the binding of porphyrin onto the Au surface induced a movement of charge from the 5d state of the Au core to the Fermi level. This emptied the antibonding state of the Au-H* intermediate, thus increasing the Au-H* bond strength [38].

As conclusion of this section, the coordination abilities of the capping ligands at the NP surface are crucial to determine the extent to which surface-active sites are available. Moreover, both the chemical nature and the volume of the capping ligands can influence the mass transfer process from the electrolyte to the surface of the MNPs by different interactions (steric, electrostatic and/or hydrophobic/hydrophilic). In fact, ligands can produce several effects simultaneously, being thus difficult to distinguish separately one from another. However, in general, MNPs with moderate M-L bond strength, low steric hindrance and high proton affinity/hydrophilicity are the best candidates for catalyzing the HER.

4 Ligand Effects in the CO_2 Reduction Reaction

The reduction of CO_2, an extremely thermodynamically inert molecule with a fully oxidized C atom, is a complex process involving several electrons and protons (at least two, but even 18 or more can be involved), which gives rise to low reaction rates, high overpotentials and low selectivities [98, 99]. As shown in Eqs. 2a–2i, the

CO_2RR can lead to several partially or fully reduced products, all formed at close standard reduction potentials at pH 7 vs. the normal hydrogen electrode (NHE), ranging from single C products (CO, formic acid, formaldehyde, methanol, methane) to C_{2+} products (ethylene, ethanol, ethane, 1-propanol, etc.):

$$CO_2 + 2\,H^+ + 2\,e^- \rightarrow CO + H_2O \qquad \left(E^0 = -0.52\ \mathrm{V, pH\ 7}\right) \tag{2a}$$

$$CO_2 + 2\,H^+ + 2\,e^- \rightarrow HCOOH \qquad \left(E^0 = -0.61\ \mathrm{V, pH\ 7}\right) \tag{2b}$$

$$CO_2 + 4\,H^+ + 4\,e^- \rightarrow HCHO + H_2O \qquad \left(E^0 = -0.51\ \mathrm{V, pH\ 7}\right) \tag{2c}$$

$$CO_2 + 6\,H^+ + 6\,e^- \rightarrow CH_3OH + H_2O \qquad \left(E^0 = -0.38\ \mathrm{V, pH\ 7}\right) \tag{2d}$$

$$CO_2 + 8\,H^+ + 8\,e^- \rightarrow CH_4 + 2\,H_2O \qquad \left(E^0 = -0.24\ \mathrm{V, pH\ 7}\right) \tag{2e}$$

$$2\,CO_2 + 12\,H^+ + 12\,e^- \rightarrow CH_2CH_2 + 4\,H_2O \qquad \left(E^0 = -0.34\ \mathrm{V, pH\ 7}\right) \tag{2f}$$

$$2\,CO_2 + 12\,H^+ + 12\,e^- \rightarrow CH_3CH_2OH + 3\,H_2O \qquad \left(E^0 = -0.33\ \mathrm{V, pH\ 7}\right) \tag{2g}$$

$$2\,CO_2 + 14\,H^+ + 14\,e^- \rightarrow CH_3CH_3 + 4\,H_2O \qquad \left(E^0 = -0.27\ \mathrm{V, pH\ 7}\right) \tag{2h}$$

$$3\,CO_2 + 18\,H^+ + 18\,e^- \rightarrow CH_3CH_2CH_2OH + 5\,H_2O \qquad \left(E^0 = -0.32\ \mathrm{V, pH\ 7}\right) \tag{2i}$$

The similarity of the standard potentials of all these reductive processes (at pH 7, CO and formaldehyde are formed at −0.5 V vs. NHE, and ethylene, ethane, ethanol and 1-propanol at −0.3 V vs. NHE) leads to the formation of mixtures of products. The formation of multiple products is costly as purification steps are required before commercialization. In addition, the competitive HER process occurs at a median potential compared to those of the processes listed above (−0.41 V vs. NHE at pH 7) [100]. Finally, another limiting parameter for a successful CO_2RR process is the low solubility of CO_2 gas in aqueous electrolytes. As a consequence, the development of cheap and selective catalysts is needed for an economically viable electrocatalytic CO_2RR process. MNPs, with their unique properties, are promising candidates to achieve this aim. Among the surface-functionalized MNPs that reduce CO_2 to CO with high selectively, Au and Ag NPs stand out as the most active systems. The decoration of these MNPs with thiolate ligands, such as 2-phenylethanethiol (PET), cysteamine (CA) or dodecanethiol (DDT) (Chart 1), can be used to disfavor the competing HER process [16–18, 29–31]. The PET ligand gave rise to diverse electrostatic and electronic effects. For instance, in $[Au_{25}(PET)_{18}]^-$ atomically precise clusters, the anionic nature of the ligand shell promoted the adsorption of CO_2 [30] (Figs. 4b and 5, **left**). These extremely small clusters show a perfectly defined surface structure and differ from larger MNPs because their energy levels become quantized and possess electronic structures like molecules. Thus, they are suitable platforms for studying catalytic processes by representing the gap between complexes and larger MNPs.

In this work, partial electron density was transferred from the thiolate donating group to the O atoms of CO_2 (Fig. 4d, **left**), which activated the C=O bonds and

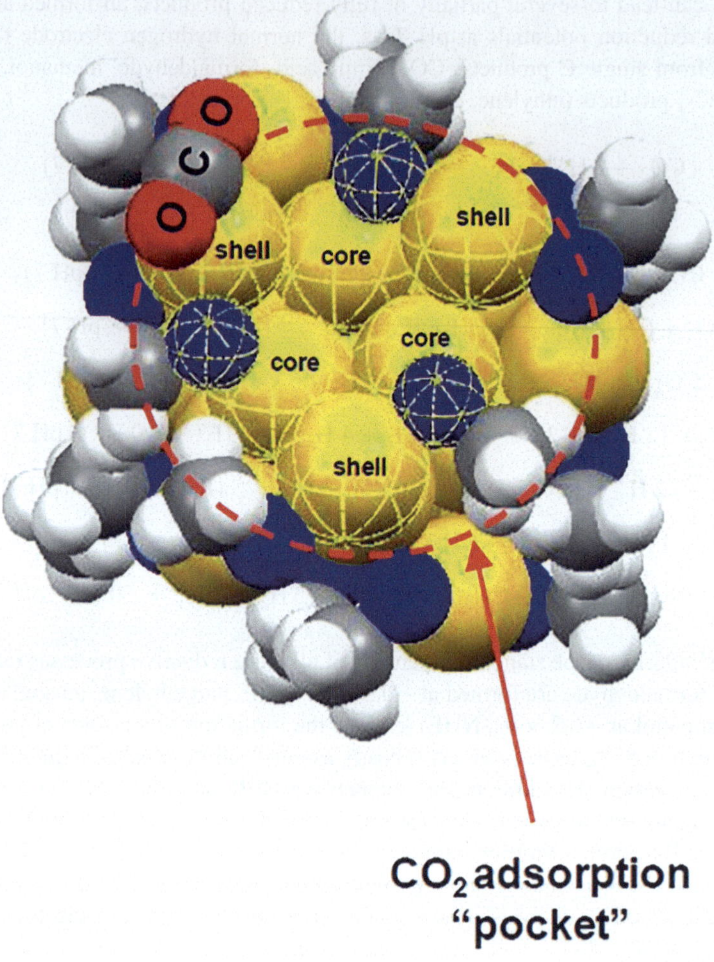

Fig. 5 Space-filled DFT calculated model for the Au_{25}-CO_2 binding site formed by three core Au atoms, three shell Au atoms and three shell S atoms (highlighted with yellow crosshatches). Color code: Au, yellow; S, blue; O, red; C, grey; H, white. The PET ligands were replaced by SCH_3 ligands in the DFT calculations. Reproduced from Ref. [30] with permission from the American Chemical Society

generated adsorbed $*CO_2^-$ entities [30]. This polarized the Au-S bonds and induced electronic changes similar to oxidation within the core of the gold clusters. Then, after a proton-coupled electron transfer, the initially adsorbed $*CO_2^-$ intermediates were converted into *COOH species which finally split into CO and water. The effect of the PET ligand on the selective CO formation has been studied by DFT calculations for 1.5 nm Au/Cu (51% Au, 49% Cu) bimetallic NPs [29]. The stabilization of the *CO intermediate by the PET ligand and the suppression of the competing HER reaction due to a disfavored *H intermediate formation were proven. Also, theoretical simulations of gradual loss of the thiolate ligand indicated a progressive *CO intermediate stabilization making its desorption difficult, with a concomitant decrease in the global CO selectivity vs. HER.

Analogous electronic effects to those of PET, yielding to the stabilization of the key *COOH radical intermediate, have also been described for the CA ligand (Fig. 4d, **left**). DFT calculations on ligand-free and CA-capped Ag NPs indicated that the CA ligand induced a change in the electronic spin density of the surface of the MNPs [16, 17]. Upon coordination of the CA ligand, the starting symmetry of the system was destroyed, because the unpaired electrons were progressively localized at the NP surface. This favored the stabilization of the *COOH radical intermediate since the covalent character of the Ag-COOH bond increased. Oppositely, the Ag-CO bond did not experience any stabilization upon the appearance of the unpaired electrons, and thus CO desorption was favorable.

Regarding Au NPs, EXAFS measurements have shown that the binding of CA induces a decrease of the Au-Au coordination numbers, thus creating more vacancies and reactive sites that will drive the CO_2RR catalysis, which leads to a mass activity 110 times superior to that of the ligand-free counterparts [18]. Second coordination sphere effects have also been described for face-centered cubic 4.2 nm Au NPs [18] and analogous 10 nm Ag NPs [17], both stabilized by a self-assembled monolayer (SAM) of CA. The dangling amino groups of the CA monolayer promoted the chemisorption of CO_2 at the NP surface through H bonds (Fig. 4e). This eased the activation of CO_2 (overpotentials for the ligand-free counterparts were ca. 300 mV higher) and blocked the competitive HER process. This second coordination sphere effect was lost when the alkyl chain was extended (due to an increased distance between the substrate and the NP surface), or when a non-coordinating group substituted the terminal amino group.

Dodecanethiol (DDT, Chart 1) has also been used as a capping ligand for 2.0–2.1 nm Au NPs. The electrocatalytic performance of DDT-capped Au NPs has been compared with those of PPh_3-capped and ligand-free analogues as well as those of DDT-capped and bare polycrystalline Au surfaces [31]. It has been demonstrated that DDT forms a stable hydrophobic monolayer on top of the Au NPs, helping the CO_2 to reach the catalytic active sites while avoiding the access of hydrophilic poisoning species, such as small (ca. 0.53 nm) solvated Zn^{2+} ions, a common contaminant in sodium bicarbonate-based electrolytes (Fig. 4c). In coherence with this hypothesis, the ligand-free Au NPs were quickly poisoned in the presence of aqueous Zn^{2+}. Under a negative potential, it was shown that DDT partially desorbed from the surface of Au NPs during the first minutes, but a part remained attached for longer times. In contrast, under the same conditions, all DDT and PPh_3 ligands were desorbed from the flat surface of polycrystalline Au and the surface of PPh_3-capped Au NPs, respectively. This is a consequence of a decreased ligand-Au bond strength (Fig. 4f). This work highlights the crucial role of capping ligand- electrodesorption from the MNP surface under the negative potentials used in electrocatalytic CO_2RR, as it affects their overall catalytic output.

Aside from thiolates, amines have also been used as stabilizing ligands for Ag [19] and Au [20] NPs as well as CuO NPs [21] for selective CO_2 electroreduction into CO. The higher activity and selectivity towards CO formation displayed by 9 nm CuO NPs decorated with oleylamine (OLA, Chart 1) as compared to ligand-free MNPs was assigned to the presence of OLA, which favored the reactivity of surface oxide sites towards the adsorption of CO_2 and its conversion into CO [21]. Considering Ag NPs, a comparative study has been carried out with a series of ligands like oleylamine (OLA), ethylenediamine (EDA), oleic acid (OA), dodecanethiol (DDT) and cysteamine (CA) (Chart 1) [19]. The CO selectivity displayed the trend amines > carboxylate > thiolate. According to DFT calculations, the superior performance of the amine-stabilized Ag NPs is due to their inferior HER electroactivity compared to that of the thiolate-capped Ag NPs, whereas the CO mass activity is not significantly different. This is due to the ability of amine-capped Ag NPs to both stabilize the *COOH intermediate at under-coordinated corner surface sites and destabilize the *H intermediate. In contrast, the thiolate ligands favor the stabilization of both *COOH and *H intermediates at the NP surface. Regarding carboxylate ligands, they present intermediate CO_2RR and HER electrocatalytic activities because in both processes there is a charge transfer from the carboxylate ligands to the 4.9–5.0 nm Ag NPs. Concerning Au NPs of ca. 2.5 nm, the advantageous effects of using moderate amounts of linear amines (propylamine, hexylamine, oleylamine (OLA) and ethylenediamine (EDA), Chart 1) have been reported, with an increase of 23% in Faradaic Efficiency (FE) towards CO evolution for the OLA-capped MNPs compared to ligand-free counterparts [20]. However, when larger Au NPs (ca. 11.7 nm) were functionalized with OLA, the selectivity towards CO decreased. Given their larger size, these MNPs have a lower number of undercoordinated sites at their surface. The amine ligands coordinate preferentially to these surface sites, thus blocking the favorable sites for *COOH formation [19]. As a result, the amine-functionalized large Au NPs display similar FE to their ligand-free counterparts. An analogous decrease in selectivity was observed for the small 2.5 nm mean size Au NPs when capped with the bulky polyethyleneimine ligand (PEI, Chart 1), or with much larger amounts of linear alkylamines, which decreased the amount of free reactive sites. These results indicate that the formation of CO can be inhibited by either using amines with multiple functional groups (PEI), that lead to concomitant stronger Au-N bonds (Fig. 4f, **left**), or by employing higher amounts of linear amines. This CO_2RR inhibition is either due to steric hindrance effects, which limit the CO_2 mass transfer (Fig. 4a), or the blockage of the catalytically active sites (Fig. 4f, **left**), decreasing in both cases the selectivity to CO against the competing HER process.

Interestingly, the use of positively charged imidazolium (IM) salts (derivatized-phenyl substituents with various anchoring groups (X = NO_2, CN, COOH, SCH_3) and different IM rings N-functionalized by alkyl tails of different length and bulkiness (R = methyl, octyl, hexadecyl, 1-triphenylmethyl or trytil), Chart 1) as stabilizers for Ag NPs (ca. 13 nm) has been proven advantageous compared to using the neutral oleylamine (OLA) ligand [22], yielding a superior FE for CO evolution. The electrostatic attraction between the positively charged IMs and the CO_2 substrate induced the approach of CO_2 to the MNP surface (Fig. 4b, **right**). The current density for CO evolution was found to follow the trend -NO_2 > -CN > -COOH >

-SCH_3, suggesting that the higher the electron-withdrawing capacity of the functional group in the IM, the higher the CO activity (Fig. 4d, **right**). Such electronic effect appeared to contradict previous literature data indicating that electron-rich surfaces promote the CO_2RR (Fig. 4d, **left**) [32, 33]. However, the previous literature data concerned neutral ligands, and therefore the different effect observed may result from electronic influences coming from both the anchoring groups and the cationic IM groups. Considering the optimal anchoring group (-NO_2), the study of the hydrophobic and steric effects of the different alkyl tails on the IM ring on the FE for CO evolution was carried out. The results showed that as the length increased, the ligand shell permeability to protons and CO_2 decreased, leading to an overall decrease in the catalytic CO and H_2 formation (Fig. 4a). However, when the hydrophobicity of the shell increased up to a certain extent, the HER was disfavored against the CO_2RR (Fig. 4c). These results indicate that an optimal selectivity of CO_2 over protons and a good overall permeability are required. The optimal situation was observed with a medium-length octyl chain (92% FE for CO), whereas both shorter (methyl, 74% FE), bulkier (trityl, 81% FE) or longer chains (hexadecyl, 43% FE) showed inferior performances. This work proves the significance of achieving a correct equilibrium between ligand hydrophobicity and steric hindrance in order to optimize the solid/liquid interface in electrocatalytic CO_2RR.

N-heterocyclic carbene (NHC) ligands, bearing a strong sigma-donor character, have also been shown to favor CO evolution in the electrocatalytic CO_2RR through electronic effects when used to functionalize Au NPs [33, 34] or nanoclusters [101], as well as Pd NPs [34]. A first relevant example concerns 6.9 nm Au NPs functionalized with NHC1 (Chart 1) [33], whose electron richness provided by the carbene increased the reductive abilities of the MNPs (Fig. 4d, **left**). Furthermore, NHC1 coordination decreased the strength of the Au-Au bonds, thus leading to a reorganization of the NP surface and an increased number of defects, being the latter a crucial parameter to improve the CO_2RR. This induced a dramatic decrease of the Tafel slope from 138 $mV \cdot dec^{-1}$ (ligand-free MNPs) to 72 $mV \cdot dec^{-1}$ (NHC1-capped MNPs), hence changing the reaction mechanism from a single-electron transfer rds (from $^*CO_2$ to $^*CO_2^{\cdot -}$) to a fast electron transfer to CO_2 before a chemical rds [33]. A similar reaction-mechanism change has been described for the OLA-capped Au NPs, with a decrease of the Tafel slope from 252 $mV \cdot dec^{-1}$ for the ligand-free MNPs to 78 $mV \cdot dec^{-1}$ for the OLA-capped MNPs [23]. In another work on Au NPs (14 nm) and Pd NPs (1–2 nm), a higher CO selectivity against the competitive HER electrocatalytic process was reported in the presence of NHC ligands compared to thiolates [34]. The polymeric NHC2 and NHC3 ligands (Chart 1) showed two combined effects, a strong σ-donor electronic effect due to the carbene function (Fig. 4d, **left**) and a hydrophobic effect induced by the polymeric nature of the ligands (Fig. 4c). The hydrophobicity reduces the ability of the positively charged hydronium ions to diffuse through the polymeric shell surrounding the MNPs, thus controlling the surface accessibility in an analogous way as metalloenzymes do through secondary interactions between the amino acids of their protein framework [102]. Furthermore, the polymeric shell stabilizes the MNPs against sintering under negative potentials [34].

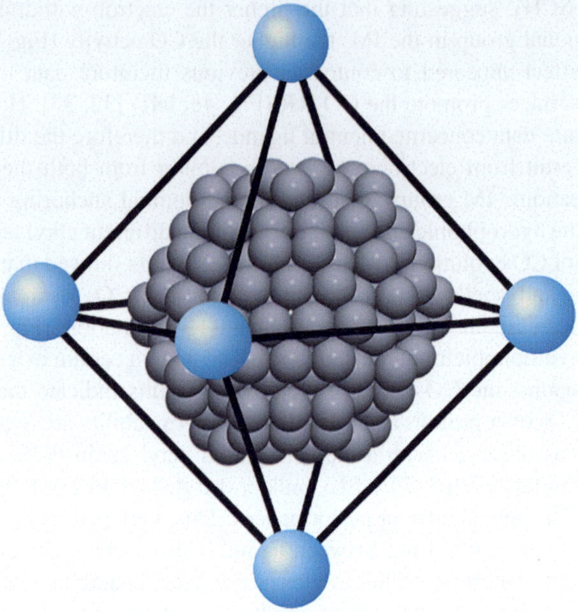

Fig. 6 Representation of a Pd NP encapsulated into a bpy-derived MOF. Color code: $Zr_6O_4(OH)_4$ secondary building units, blue; linear 2,2′-bipyridine-5,5′-dicarboxlate bridging ligands, black; Pd NP, grey. Adapted with permission [37]. Copyright 2021, American Chemical Society

An electronic modulation of the electrocatalytic CO_2RR capacity of Pd NPs encapsulated into a 2,2'-bipyridine (bpy)-derived metal organic framework (MOF) [37] has been reported (Fig. 6). In situ X-ray diffraction and Raman spectroscopies, as well as DFT calculations, evidenced that the electronic effect of the bpy linkers at the Pd NPs surface allowed not only the stabilization of the *CO_2 and *COOH intermediates, but also the destabilization of the *H intermediate, in contrast to a non-coordinating biphenyl-based MOF. Furthermore, the bpy ligands limited the formation of HCOOH (Eq. 2b) at low overpotentials and avoided the potential CO poisoning at higher overpotentials owing to the destabilization of the *CO intermediate. This yielded a significant FE (≥80%) over a wide range of potentials (−0.3 to −1.2 V vs. NHE).

The use of H-bond forming polymeric stabilizers has been demonstrated to be useful for increasing the FE towards CO formation in the case of polyacrylamide (PAM) stabilized Ag NPs [103] and polyvinyl alcohol (PVA) stabilized Au NPs [104]. The PAM and PVA polymers (Chart 1) have a stabilizing effect through second coordination sphere interactions, as the result of the formation of H bonds between their functional group (amide NH_2 for PAM, OH for PVA) and the O atoms of the CO_2 substrate (Fig. 4e). According to *operando*-ATR-IR measurements and DFT calculations, the H bond network eases the absorption of CO_2 and the stabilization of the *COOH intermediate species that is formed prior to CO evolution [103].

Two works have reported that tetradecylphosphonic acid (TDPA, Chart 1), a long-chain linear ligand with a polar phosphonate group in one end, desorbs from

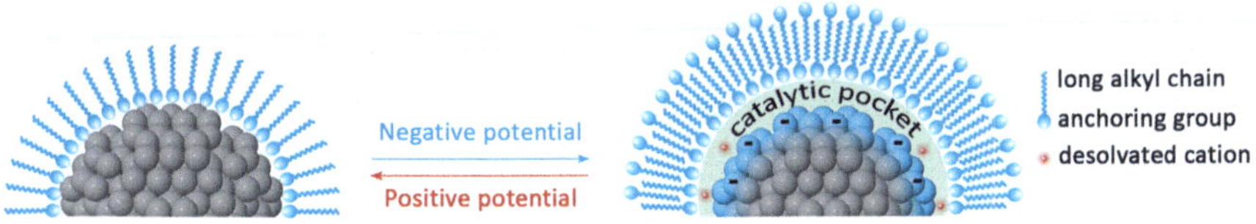

Fig. 7 Generation of a NP/ordered-ligand interlayer (NOLI) state from ligand-capped Ag NPs upon application of a negative potential during electrocatalytic CO_2RR (0.1 M $KHCO_3$ saturated with 1 atm CO_2) [35]

TDPA-stabilized MNPs under reductive potentials during electrocatalytic CO_2RR, and that the desorbed molecules stay close to the MNPs by forming an ordered bilayer [35, 105]. The space between this bilayer and the NP surface, referred to as the NP/ordered-ligand interlayer (NOLI) state by the authors, is shown to be crucial for the overall catalytic performance (Fig. 7). The NOLI state, stabilized by the intermolecular enzyme-like interactions that exist between the chains of the decoordinated ligands that are now physisorbed, offers a highly selective catalytic microenvironment with unique pseudocapacitive characteristics. This catalytic pocket allows the presence of desolvated cations and the electrostatic stabilization of intermediates. In addition, it creates an active-site configuration that enhances the intrinsic CO_2 to CO turnover by 2 orders of magnitude over the ligand-free analogues, while disfavoring the competing HER process.

To produce more reduced multi-carbon (C_{2+}) compounds, such as ethylene (Eq. 2f) or ethanol (Eq. 2g), Cu NPs [106–108] and Cu nanocrystals (NCs) [24, 109] are appropriate catalysts, being able to promote C-C bond formation. However, the tuning of their catalytic properties by ligand functionalization has been less explored. A recent study reports the post-functionalization of the Cu atoms of a micrometric layered tandem Ag-Cu surface with aromatic heterocycles [25]. The selectivity towards C_{2+} products was found to increase as the average Cu oxidation state rises. The use of electron-withdrawing ligands, which favor the formation of $Cu^{\delta+}$ sites increasing their average oxidation state, caused a three-fold increment of the FE for the formation of C_{2+} products (Fig. 4d**, right**). This was attributed to the positive partial charge on the $Cu^{\delta+}$ sites that enabled the adsorption of CO, subsequently enhancing the formation of C_{2+} products. This work opened the field of molecular surface-capping of Cu-based nano catalysts. At the nanoscale, the number of surface sites can be maximized, and their electronic properties can potentially be tuned at will by using organic ligands of varied chemical nature. A recent example with polymer-stabilized Cu NPs (50 nm) successfully demonstrated how the presence of COOH and -CF_2 functional groups affected the distribution of diverse C_1 (CO, HCOOH, CH_4) and C_{2+} (C_2H_4) products [28].

Analogously to the previously discussed Ag NPs [22], the use of positively charged IMs (with $X = NO_2$ and R = octyl, Chart 1) has been evaluated for cubic Cu NCs and spheric Cu NCs [109]. The stable cubic NCs mainly produced ethylene (C_2H_4), whereas the less stable spheric ones underwent a surface reconstruction upon the application of negative potentials, causing the removal of the IM stabilizers. This loss of ligand induced the formation of H_2 as the main product through the competitive HER process. Overall, this work emphasizes the importance of producing structurally stable ligand-capped nanocatalysts with the proper morphology.

A comparative study was published on the ligand binding strength and lability among different amine, thiolate, carboxylate, phosphine, phosphine oxide, and phosphonate ligands functionalizing Cu NCs. In all cases, formate ($HCOO^-$) was the main product formed during electrocatalytic CO_2RR, and ethylene, methane, and CO the minor products [24]. The combination of experimental surface-characterization results and DFT theoretical calculations allowed to determine that dodecanethiol (DDT, Chart 1) was the most strongly bound ligand under application of a negative bias, being thus the least labile of all ligands tested. Consequently, DDT-capped Cu NCs showed the lowest ECSA, current density and FE for the reduced products. In contrast, the Cu NCs stabilized with labile ligands that easily electro-desorb from the surface like amine (oleylamine), carboxylate (oleic acid), phosphonate (tetradecylphosphonic acid), and phosphine oxide (trioctylphosphine oxide) ligands (Chart 1), showed similar electrocatalytic activities than the ligand-free Cu NCs. The phosphine ligand (trioctylphosphine, TOP, Chart 1), with an intermediate labile character, stood in the middle. This work demonstrates that the metal-ligand bond strength (Fig. 4f) allows to predict the fate of the capping ligands under electrocatalytic turnover. Therefore, weakly-bound ligands should be chosen for the catalytic processes that do not require the influence of the stabilizing ligands. On the contrary, strongly bound ligands will affect the catalytic output of the nanocatalysts, as they will tune their electrocatalytic properties.

This section discussed how the use of hydrophobic ligands (i.e., alkylamines, alkylthiolates) or surrounding-ligand pockets with alkylphosphonic acids (i.e., NOLI effect) [35, 105] can be beneficial to increase the selectivity towards CO_2 reduced products by blocking the diffusion of the competing H^+/H_2O substrates onto the surface of the MNPs [20, 22, 34]. However, the bulkiness of the ligands [20, 22, 33] and the degree of surface coverage [20] should always be taken into account. The use of bulky ligands or long linear alkyl chain ones, as well as the presence of a high density of ligands at the NP surface, was shown to limit the CO_2 substrate from reaching the active sites. The crucial role of capping ligands within the solid/liquid interface during electrocatalytic CO_2RR was also highlighted, affecting both the overall activity and the reduced products distribution, especially in the case of C_{2+} products. The delicate balance between interrelated parameters such as M-L bond strength, ligand coverage, ECSA, steric hindrance, and hydrophilicity/hydrophobicity of the ligands affects the final electrocatalytic output of the nanocatalysts, as also described for HER electrocatalysis (Fig. 4). The combination of multiscale microkinetic modelling techniques [110–113] with *in operando*-surface-sensitive techniques, such as Time-Resolved Surface-Enhanced Raman Spectroscopy (TR-SERS) [114] or Time-of-Flight Secondary Ion Mass Spectrometry (ToF-SIMS) [97], should help to study their reaction dynamics [115, 116]. This will allow to define the exact role of the capping ligands and to rationalize the design of improved ligand-capped nanometric CO_2RR electrocatalysts. In addition, the NOLI strategy appears as a promising approach, which could be extended to C_{2+}-promoting metals as well, such as copper. This is likely to be a promising area of research in the near future.

5 Conclusions and Future Directions

Through relevant examples from the literature, this chapter emphasizes how the functionalization of MNPs is a powerful strategy to rationally modify their catalytic behavior in the HER and CO_2RR processes. It has also been shown how the interface between the NP surface and the electrolyte can affect many crucial parameters of catalysis, such as substrate concentration, intermediate stabilization and mass transfer, a subject whose analysis has largely increased over the last 5 years [117].

The development of the functionalization of MNPs has benefited from the mechanistic knowledge accumulated on homogeneous molecular catalysts, that has been partially transferred to the nanocatalysts. This transfer has been possible not only by the improvements made on the advanced characterization and computational techniques, but also by the development of reproducible synthetic methodologies for the formation of homogeneous in size MNPs with a clean surface. The almost unlimited number of ligand/metal combinations offers the possibility to access countless ligand-capped MNPs with diverse catalytic properties.

The strategy of surface-functionalization of MNPs in the electrocatalytic HER has been already explored in a substantial number of works. The accumulated knowledge clearly indicates that the free Gibbs energy of hydrogen adsorption (ΔG_{H*}) is a useful descriptor for predicting the catalytic behavior of ligand-capped MNPs. The control over the electronic effects of the capping ligands is thus crucial for achieving a ΔG_{H*} value close to zero and, therefore, an optimal HER electrocatalytic activity. However, to predict trustable ΔG_{H*} values, the construction of realistic theoretical representations that match the analytical, electrochemical, and spectroscopic data is compulsory. Furthermore, this combined experimental/theoretical methodology allows to consider other significant properties of the capping ligands, such as the bulkiness, M-L bond strength, hydrophilicity/hydrophobicity, etc. This enables the design, in an even more sensible way, of ligand-capped MNPs with improved performance in the HER.

Regarding the CO_2RR, its inherent more complex mechanism compared to the HER makes the synthesis and study of ligand-capped MNPs even more compelling. The various effects of the capping ligands and the broad range of reduced products that can be formed render the rational design of selective electrocatalysts more difficult compared to the HER. Nonetheless, huge advances have been made in the electronic tuning of the stability of the reaction-intermediate species, which determines the overall selectivity of the CO_2RR. It is well-known that the noble metals Au and Ag are highly selective towards CO formation through a 2-electron process, and their systematic study has allowed to face more complex situations, such as the formation of C_{2+} reduced products by ligand-capped Cu NPs. Interestingly, nickel oxygenate-derived electrocatalysts bearing $Ni^{\delta+}$ surface sites have been recently reported to have a similar reactivity than that of Cu sites [39]. Thus, by favoring the partially positively charged Ni sites by the use of electron-withdrawing ligands, it is possible to produce C_{n+} products from CO_2.

Important advances have been achieved to prevent the competing HER process during the CO_2RR by the introduction of second coordination sphere effects, which induces a local increase in the concentration of CO_2 and the concomitant decrease in the concentration of hydronium ions at the NP surface. Furthermore, in view of determining the right product selectivity, it is important to reach a correct balance between the hydrophobicity, bulkiness, and extent of metal NP surface coverage by the capping ligands. The latter parameter not only depends on the intrinsic bulkiness of the ligands but also on the M-L bond strength. The bond strength must be moderate, neither too strong nor too weak, in order to reach a good balance between NP stability and the total amount of free active sites. Moreover, one crucial point is the fate of the capping ligands under electrocatalytic turnover conditions. In this sense, controlled ligand expulsion has been demonstrated to be advantageous in the presence of a NOLI state.

Finally, an important field in which the use of ligand-capped MNPs is starting to emerge is photo(electro)catalysis. Although most of the published works until now deal with the oxygen evolution reaction (OER) [64, 118–121], the results can be extrapolated to the reductive HER [122, 123] and CO_2RR [124] processes. MNPs can have three different roles in photo(electro)catalysis: they can act as light absorbers [125], as catalysts [68] or as both light absorbers and catalysts [126]. Once the light is absorbed, the photogenerated electrons and holes need to find the appropriate reacting centers before charge recombination takes place, which makes the desired catalytic processes less efficient. In order to promote catalysis, the favorable charge-transfer processes need to be accelerated whereas the unfavorable recombination events require to be slowed down. This is perfectly known for solid metal oxide materials able to perform the photoelectrocatalytic OER [121]. It is very important to modulate the charge-transfer processes occurring between the photoabsorbing units and the catalysts. This has been shown for hybrid HER photocatalysts consisting of a molecular Co catalyst bonded onto the surface of a semiconductor (TiO_2) [123], and for ligand-capped Cu_2O NPs also deposited onto TiO_2 [68]. In these HER catalytic systems, analogously to the pure electrocatalytic CO_2RR processes, the product distribution depends on the capping ligand employed. Lastly, hybrid dyad-metal NP systems constitute promising alternative photo (electro)catalytic systems that incorporate ligand-capped MNPs. For example, such a dyad system has been synthesized via the covalent anchoring of an organophosphorus polypyridyl Ru molecular complex that acts as the light absorber (or photosensitizer, PS) onto the surface of Ru NPs [122]. This type of hybrid covalently bonded PS-NP systems can perform photo(electro)catalytic reduction reactions when a sacrificial electron-donor agent is added to the colloidal suspension, or when the system is supported onto an electrode. The main advantage of these hybrid systems is that the distance between the light absorbing units and the catalytic centers can be tuned by changing the length of the chemical linkers. Additionally, by modifying the nature of the binding functional group of the PS, the degree of electronic coupling between the PS and the catalytic MNPs can be modulated, directly affecting the yield of the electron transfer process.

To sum up, this chapter has shown that the use of ligands to synthesize and stabilize MNPs may strongly affect their final electrocatalytic performance. This offers plenty of room to rationally tune the activity (HER, CO_2RR) and selectivity (CO_2RR) of these nanocatalysts. However, the fine understanding of the role of the ligand is still difficult in many cases, since the observed effects can be a mixture of electronic, steric, hydrophobic/hydrophilic, and second coordination sphere effects all at once. Thus, for a complete picture of the role and fate of the capping ligands during catalysis, correlation studies, including catalyst benchmarking, pioneering *in operando*-characterization tools and realistic multiscale theoretical models are needed [110–113]. Although electrocatalyst benchmarking is well-established and accepted by the research community, it is not yet usual for photo(electro)catalysts [127, 128]. This is thus an important objective that the scientific community needs to address. Additionally, it is advisable to exploit the synergies between theoretical modelling and advanced surface-sensitive characterization tools, such as ToF-SIMS [97] or TR-SERS [114], among others, to understand which are the effects of the coordination modes of the ligands before and during catalysis, as well as to disclose the dynamics of the reaction intermediate species [116]. Up to date, the implementation of more complex theoretical computational methods that go beyond the standard atomistic DFT models is still missing. In this context, the use of microkinetic methods [110–113] can be very useful to detect reaction intermediate species and their rds steps. Also, the use of fluid dynamics methodologies that take the real parameters within the reaction into consideration [129] can also be extremely useful to optimize the design of performant nanocatalysts. The outcomes from a multidisciplinary approach should draw clear lines for the rational design of more active and selective ligand-capped MNPs for both the HER and the CO_2RR in the upcoming years.

Acknowledgements

Sustained support from MICINN, FEDER, AGAUR, CNRS, the University Paul Sabatier-Toulouse, and the IRN-HC3A Franco-Catalan action are gratefully acknowledged. Recent research projects and grants include PID2019-104171RB-I00-, PID2021-128197NA-I00, TED2021-129237B-I00, IDEX UNITI Emergence project (UFTMIP: 2015-209-CIF-d-DRD-127185), CTP2013-0016 and CTP2013-Midi-Pyrénées13053026. Á.L.-R. and G.M. acknowledge the AGAUR and MICINN for an FI and a FPU predoctoral grant, respectively. L.F. is indebted to the Ramón y Cajal Program (RYC2018-025394-I Fellowship) and to the Royal Society of Chemistry (R20-8077 Research Fund). X.S. thanks ICREA for the ICREA Academia prize 2020.

Declaration of Competing Interest The authors declare that they have no known competing financial interests or personal relationships that could have appeared to influence the work reported in this chapter.

References

1. Corain B, Schmid G, Toshima N (2008) Metal nanoclusters in catalysis and materials science: the issue of size control.1st edn. Elsevier Science, Amsterdam
2. Schmid G (2010) Nanoparticles from theory to application.2nd edn. Wiley-VCH
3. Astruc D (2020) Introduction: nanoparticles in catalysis. Chem Rev 120:461–463. https://doi.org/10.1021/acs.chemrev.8b00696
4. Rocoux A, Philippot K (2021) Nanoparticles in catalysis: advances in synthesis and applications. Wiley-VCH Verlag GmbH & Co. KGaA, Weinheim
5. Zhu J, Hu L, Zhao P et al (2020) Recent advances in electrocatalytic hydrogen evolution using nanoparticles. Chem Rev 120:851–918. https://doi.org/10.1021/acs.chemrev.9b00248
6. Trindell JA, Duan Z, Henkelman G, Crooks RM (2020) Well-defined nanoparticle electrocatalysts for the refinement of theory. Chem Rev 120:814–850. https://doi.org/10.1021/acs.chemrev.9b00246
7. Wu L, Xi Z, Sun S (2017) Well-defined metal nanoparticles for electrocatalysis. Stud Surf Sci Catal 177:123–148. https://doi.org/10.1016/B978-0-12-805090-3.00004-8
8. Wang Q, Domen K (2020) Particulate photocatalysts for light-driven water splitting: mechanisms, challenges, and design strategies. Chem Rev 120:919–985. https://doi.org/10.1021/acs.chemrev.9b00201
9. Kaatz FH, Bultheel A (2019) Magic mathematical relationships for nanoclusters. Nanoscale Res Lett 14:150. https://doi.org/10.1186/s11671-019-2939-5
10. Zhou K, Li Y (2012) Catalysis based on nanocrystals with well-defined facets. Angew Chem Int Ed 51:602–613. https://doi.org/10.1002/anie.201102619
11. Lu L, Zou S, Fang B (2021) The critical impacts of ligands on heterogeneous nanocatalysis: a review. ACS Catal 11:6020–6058. https://doi.org/10.1021/acscatal.1c00903
12. Heuer-Jungemann A, Feliu N, Bakaimi I et al (2019) The role of ligands in the chemical synthesis and applications of inorganic nanoparticles. Chem Rev 119:4819–4880. https://doi.org/10.1021/acs.chemrev.8b00733
13. Dong A, Ye X, Chen J et al (2011) A generalized ligand-exchange strategy enabling sequential surface functionalization of colloidal nanocrystals. J Am Chem Soc 133:998–1006. https://doi.org/10.1021/ja108948z
14. Henckel DA, Lenz O, Cossairt BM (2017) Effect of ligand coverage on hydrogen evolution catalyzed by colloidal WSe_2. ACS Catal 7:2815–2820. https://doi.org/10.1021/acscatal.7b00074
15. Ung D, Cossairt BM (2019) Effect of surface ligands on CoP for the hydrogen evolution reaction. ACS Appl Energy Mater 2:1642–1645. https://doi.org/10.1021/acsaem.9b00240
16. Kim C, Jeon HS, Eom T et al (2015) Achieving selective and efficient electrocatalytic activity for CO_2 reduction using immobilized silver nanoparticles. J Am Chem Soc 137:13844–13850. https://doi.org/10.1021/jacs.5b06568
17. Wang Z, Wu L, Sun K et al (2018) Surface ligand promotion of carbon dioxide reduction through stabilizing chemisorbed reactive intermediates. J Phys Chem Lett 9:3057–3061. https://doi.org/10.1021/acs.jpclett.8b00959
18. Wang Z, Sun K, Liang C et al (2019) Synergistic chemisorbing and electronic effects for efficient CO_2 reduction using cysteamine-functionalized gold nanoparticles. ACS Appl Energy Mater 2:192–195. https://doi.org/10.1021/acsaem.8b01699
19. Kim C, Eom T, Jee MS et al (2017) Insight into electrochemical CO_2 reduction on surface-molecule-mediated Ag nanoparticles. ACS Catal 7:779–785. https://doi.org/10.1021/acscatal.6b01862
20. Zhao Y, Wang C, Liu Y et al (2018) Engineering surface amine modifiers of Ultrasmall gold nanoparticles supported on reduced graphene oxide for improved electrochemical CO_2 reduction. Adv Energy Mater 8:1801400. https://doi.org/10.1002/aenm.201801400

21. Kauffman DR, Ohodnicki PR, Kail BW, Matranga C (2011) Selective electrocatalytic activity of ligand stabilized copper oxide nanoparticles. J Phys Chem Lett 2:2038–2043. https://doi.org/10.1021/jz200850y
22. Pankhurst JR, Guntern YT, Mensi M, Buonsanti R (2019) Molecular tunability of surface-functionalized metal nanocrystals for selective electrochemical CO_2 reduction. Chem Sci 10: 10356–10365. https://doi.org/10.1039/c9sc04439f
23. Gao M, Zhu Y, Liu Y et al (2020) The role of adsorbed oleylamine on gold catalysts during synthesis for highly selective electrocatalytic reduction of CO_2 to CO. Chem Commun 56: 7021–7024. https://doi.org/10.1039/d0cc01088j
24. Pankhurst JR, Iyengar P, Loiudice A et al (2020) Metal-ligand bond strength determines the fate of organic ligands on the catalyst surface during the electrochemical CO_2 reduction reaction. Chem Sci 11:9296–9302. https://doi.org/10.1039/d0sc03061a
25. Wu H, Li J, Qi K et al (2021) Improved electrochemical conversion of CO_2 to multicarbon products by using molecular doping. Nat Commun 12:7210. https://doi.org/10.1038/s41467-021-27456-5
26. Peng Y, Liu Q, Lu B et al (2021) Organically capped iridium nanoparticles as high-performance bifunctional electrocatalysts for full water splitting in both acidic and alkaline media: impacts of metal-ligand interfacial interactions. ACS Catal 11:1179–1188. https://doi.org/10.1021/acscatal.0c03747
27. Alba-Molina D, Puente Santiago AR, Giner-Casares JJ et al (2019) Tailoring the ORR and HER electrocatalytic performances of gold nanoparticles through metal-ligand interfaces. J Mater Chem A Mater 7:20425–20434. https://doi.org/10.1039/c9ta05492h
28. Chang Q, Lee JH, Liu Y et al (2022) Electrochemical CO_2 reduction reaction over cu nanoparticles with tunable activity and selectivity mediated by functional groups in polymeric binder. J Am Chem Soc 2:214–222. https://doi.org/10.1021/jacsau.1c00487
29. Kauffman DR, Alfonso DR, Tafen DN et al (2018) Selective electrocatalytic reduction of CO_2 into CO at small, thiol-capped au/cu nanoparticles. J Phys Chem C 122:27991–28000. https://doi.org/10.1021/acs.jpcc.8b06234
30. Kauffman DR, Alfonso D, Matranga C et al (2012) Experimental and computational investigation of Au_{25} clusters and CO_2: a unique interaction and enhanced electrocatalytic activity. J Am Chem Soc 134:10237–10243. https://doi.org/10.1021/ja303259q
31. Shang H, Wallentine SK, Hofmann DM et al (2020) Effect of surface ligands on gold nanocatalysts for CO_2 reduction. Chem Sci 11:12298–12306. https://doi.org/10.1039/d0sc05089j
32. Cao Z, Derrick JS, Xu J et al (2018) Chelating N-heterocyclic carbene ligands enable tuning of electrocatalytic CO_2 reduction to formate and carbon monoxide: surface organometallic chemistry. Angew Chem Int Ed 57:4981–4985. https://doi.org/10.1002/anie.201800367
33. Cao Z, Kim D, Hong D et al (2016) A molecular surface functionalization approach to tuning nanoparticle electrocatalysts for carbon dioxide reduction. J Am Chem Soc 138:8120–8125. https://doi.org/10.1021/jacs.6b02878
34. Zhang L, Wei Z, Thanneeru S et al (2019) A polymer solution to prevent nanoclustering and improve the selectivity of metal nanoparticles for electrocatalytic CO_2 reduction. Angew Chem Int Ed 131:15981–15987. https://doi.org/10.1002/ange.201909069
35. Yu S, Kim D, Qi Z et al (2021) Nanoparticle assembly induced ligand interactions for enhanced electrocatalytic CO_2 conversion. J Am Chem Soc 143:19919–19927. https://doi.org/10.1021/jacs.1c09777
36. Creus J, Drouet S, Suriñach S et al (2018) Ligand-capped Ru nanoparticles as efficient electrocatalyst for the hydrogen evolution reaction. ACS Catal 8:11094–11102. https://doi.org/10.1021/acscatal.8b03053
37. Liang Y, Zhao J, Zhang H et al (2021) Bias-adaptable CO_2-to-CO conversion via tuning the binding of competing intermediates. Nano Lett 21:8924–9832. https://doi.org/10.1021/acs.nanolett.1c02719

38. Eguchi D, Sakamoto M, Teranishi T (2017) Ligand effect on the catalytic activity of porphyrin-protected gold clusters in the electrochemical hydrogen evolution reaction. Chem Sci 9:261–265. https://doi.org/10.1039/c7sc03997b
39. Zhou Y, Martín AJ, Dattila F et al (2022) Long-chain hydrocarbons by CO_2 electroreduction using polarized nickel catalysts. Nat Catal 5:545–554. https://doi.org/10.1038/s41929-022-00803-5
40. Somorjai GA, Li Y (2010) Major successes of theory-and-experiment-combined studies in surface chemistry and heterogeneous catalysis. Top Catal 53:311–325. https://doi.org/10.1007/s11244-010-9449-0
41. Martí G, Mallón L, Romero N et al (2023) Surface-functionalized nanoparticles as catalysts for artificial photosynthesis. Adv Energy Mater 13:2300282. https://doi.org/10.1002/aenm.202300282
42. Jin R, Pei Y, Tsukuda T (2019) Controlling nanoparticles with atomic Precision_Acc Chem Res 2019.pdf. Acc Chem Res 1:1. https://doi.org/10.1021/acs.accounts.8b00631
43. Du Y, Sheng H, Astruc D, Zhu M (2020) Atomically precise Noble metal nanoclusters as efficient catalysts: a bridge between structure and properties. Chem Rev 120:526–622. https://doi.org/10.1021/acs.chemrev.8b00726
44. Cushing BL, Kolesnichenko VL, O'Connor CJ (2004) Recent advances in the liquid-phase syntheses of inorganic nanoparticles. Chem Rev 104:3893–3946. https://doi.org/10.1021/cr030027b
45. Roucoux A, Schulz J, Patin H (2002) Reduced transition metal colloids: a novel family of reusable catalysts? Chem Rev 102:3757–3778. https://doi.org/10.1021/cr010350j
46. Philippot K, Chaudret B (2003) Organometallic approach to the synthesis and surface reactivity of noble metal nanoparticles. C R Chim 6:1019–1034
47. Philippot K, Chaudret B (2007) Organometallic derived metals, colloids and nanoparticles. In: Crabtree RH, Mingos MP (eds) Comprehensive organometallic chemistry III. Elsevier, pp 71–99
48. Lara P, Philippot K, Lacroix L-M et al (2013) Organometallic Nanoparticles. In: Advances in organometallic chemistry and catalysis. Wiley, pp 421–436
49. Amiens C, Chaudret B, Ciuculescu-Pradines D et al (2013) Organometallic approach for the synthesis of nanostructures. New J Chem 37:3374–3401. https://doi.org/10.1039/c3nj00650f
50. Meffre A, Lachaize S, Gatel C et al (2011) Use of long chain amine as a reducing agent for the synthesis of high quality monodisperse iron(0) nanoparticles. J Mater Chem 21:13464–13469. https://doi.org/10.1039/c1jm12127h
51. Toshima N, Yonezawa T (1998) Bimetallic nanoparticles—novel materials for chemical and physical applications. New J Chem 22:1179–1201. https://doi.org/10.1039/a805753b
52. Fievet F, Ammar-Merah S, Brayner R et al (2018) The polyol process: a unique method for easy access to metal nanoparticles with tailored sizes, shapes and compositions. Chem Soc Rev 47:5187–5233. https://doi.org/10.1039/c7cs00777a
53. Sanyal U, Jagirdar BR (2012) Metal and alloy nanoparticles by amine-borane reduction of metal salts by solid-phase synthesis: atom economy and green process. Inorg Chem 51:13023–13033. https://doi.org/10.1021/ic3021436
54. Sabadasch V, Dachwitz S, Hannappel Y et al (2022) Acrylamide-based Pd-nanoparticle carriers as smart catalysts for the Suzuki-Miyaura cross-coupling of amino acids. Synthesis 54:3180–3192. https://doi.org/10.1055/a-1782-4224
55. Koczkur KM, Mourdikoudis S, Polavarapu L, Skrabalak SE (2015) Polyvinylpyrrolidone (PVP) in nanoparticle synthesis. Dalton Trans 44:17883–17905. https://doi.org/10.1039/c5dt02964c
56. Scurti S, Monti E, Rodríguez-Aguado E et al (2021) Effect of polyvinyl alcohol ligands on supported gold nano-catalysts: morphological and kinetics studies. Nanomaterials 11:879. https://doi.org/10.3390/nano11040879

57. Denicourt-Nowicki A, Roucoux A (2014) Ammonium surfactant-capped Rh(0) nanoparticles for biphasic hydrogenation. In: Tao F (ed) Metal nanoparticles for catalysis: advances and applications, pp 99–111
58. Ott LS, Finke RG (2007) Transition-metal nanocluster stabilization for catalysis: a critical review of ranking methods and putative stabilizers. Coord Chem Rev 251:1075–1100. https://doi.org/10.1016/j.ccr.2006.08.016
59. Denicourt-Nowicki A, Mordvinova N, Roucoux A (2021) Metal nanoparticles in water: a relevant toolbox for green catalysis. In: Philippot K, Roucoux A (eds) Nanoparticles in catalysis: advances in synthesis and applications1st edn. Wiley, pp 45–71
60. Amiens C, Ciuculescu-Pradines D, Philippot K (2016) Controlled metal nanostructures: fertile ground for coordination chemists. Coord Chem Rev 308:409–432. https://doi.org/10.1016/j.ccr.2015.07.013
61. Axet MR, Philippot K (2021) Organometallic metal nanoparticles for catalysis. In: Nanoparticles in catalysis: advances in synthesis and applications. Wiley-VCH, Weinheim, pp 73–97
62. Drouet S, Creus J, Collière V et al (2017) A porous Ru nanomaterial as an efficient electrocatalyst for the hydrogen evolution reaction under acidic and neutral conditions. Chem Commun 53:11713–11716. https://doi.org/10.1039/c7cc05615j
63. Pelzer K, Philippot K, Chaudret B (2003) Synthesis of monodisperse heptanol stabilized ruthenium nanoparticles. Evidence for the presence of surface hydrogens. Z Phys Chem 217:1539–1547. https://doi.org/10.1524/zpch.217.12.1539.20474
64. De Tovar J, Romero N, Denisov SA et al (2018) Light-driven water oxidation using hybrid photosensitizer-decorated Co3O4 nanoparticles. Mater Today Energy 9:506–515. https://doi.org/10.1016/j.mtener.2018.07.008
65. Muñoz J, Álvarez-Prada I, Lopez-Lopez E et al (2020) Synthesis of 0D to 3D hybrid-carbon nanomaterials carrying platinum(0) nanoparticles: towards the electrocatalytic determination of methylparabens at ultra-trace levels. Sens Actuators B Chem 305:127467. https://doi.org/10.1016/j.snb.2019.127467
66. Álvarez-Prada I, Peral D, Song M et al (2021) Ruthenium nanoparticles supported on carbon-based nanoallotropes as co-catalyst to enhance the photocatalytic hydrogen evolution activity of carbon nitride. Renew Energy 168:668–675. https://doi.org/10.1016/j.renene.2020.12.070
67. Tasbihi M, Fresno F, Álvarez-Prada I et al (2021) A molecular approach to the synthesis of platinum-decorated mesoporous graphitic carbon nitride as selective CO_2 reduction photocatalyst. J CO2 Util 50:101574. https://doi.org/10.1016/j.jcou.2021.101574
68. Jeong S, Kim GM, Kang GS et al (2019) Selectivity modulated by surface ligands on Cu_2O/TiO_2 catalysts for gas-phase photocatalytic reduction of carbon dioxide. J Phys Chem C 123:29184–29191. https://doi.org/10.1021/acs.jpcc.9b05780
69. Álvarez-Prada I, Nguyen AD, Romero N et al (2022) Insights into the light-driven hydrogen evolution reaction of mesoporous graphitic carbon nitride decorated with Pt or Ru nanoparticles. Dalton Trans 51:731–740. https://doi.org/10.1039/d1dt03006j
70. Romero N, Bofill R, Francàs L et al (2021) Light-driven hydrogen evolution assisted by covalent organic frameworks. Catalysts 11:754. https://doi.org/10.3390/catal11060754
71. Xu Y, Xie M, Li X et al (2022) Regulating the electronic structure of Fe-based metal organic frameworks by electrodeposition of Au nanoparticles for electrochemical overall water splitting. J Colloid Interface Sci 626:426–434. https://doi.org/10.1016/j.jcis.2022.06.163
72. Li X, Dong Q, Tian Q et al (2022) Recent advance in metal- and covalent-organic framework-based photocatalysis for hydrogen evolution. Mater Today Chem 26:101073. https://doi.org/10.1016/j.mtchem.2022.101037
73. Rasal AS, Yadav S, Yadav A et al (2021) Carbon quantum dots for energy applications: a review. ACS Appl Nano Mater 4:6515–6541. https://doi.org/10.1021/acsanm.1c01372
74. Humayun M, Wang C, Luo W (2022) Recent progress in the synthesis and applications of composite photocatalysts: a critical review. Small Methods 6:2101395. https://doi.org/10.1002/smtd.202101395

75. Guo DJ, Jing ZH (2011) Electrocatalytic properties of platinum nanoparticles supported on fluorine tin dioxide/multi-walled carbon nanotube composites for methanol electrooxidation in acidic medium. J Colloid Interface Sci 359:257–260. https://doi.org/10.1016/j.jcis.2011.03.019
76. Huo L, Jin C, Jiang K et al (2022) Applications of nickel-based electrocatalysts for hydrogen evolution reaction. Adv Energy Sustain Res 3:2100189. https://doi.org/10.1002/aesr.202100189
77. Nunez-Bajo E, Blanco-López MC, Costa-García A, Fernández-Abedul MT (2017) Electrogeneration of gold nanoparticles on porous-carbon paper-based electrodes and application to inorganic arsenic analysis in white wines by chronoamperometric stripping. Anal Chem 89:6415–6423. https://doi.org/10.1021/acs.analchem.7b00144
78. Romero N, Guerra RB, Gil L et al (2020) TiO_2-mediated visible-light-driven hydrogen evolution by ligand-capped Ru nanoparticles. Sustain Energy Fuels 4:4170–4178. https://doi.org/10.1039/d0se00446d
79. Creus J, Mallón L, Romero N et al (2019) Ruthenium nanoparticles supported on carbon microfibers for hydrogen evolution electrocatalysis. Eur J Inorg Chem 2019:2071–2077. https://doi.org/10.1002/ejic.201801438
80. Dubouis N, Grimaud A (2019) The hydrogen evolution reaction: from material to interfacial descriptors. Chem Sci 10:9165–9181. https://doi.org/10.1039/c9sc03831k
81. Hansen JN, Prats H, Toudahl KK et al (2021) Is there anything better than Pt for HER? ACS Energy Lett 6:1175–1180. https://doi.org/10.1021/acsenergylett.1c00246
82. Sabatier P (1911) Hydrogénations et déshydrogénations par catalyse. Ber Dtsch Chem Ges 44: 1984–2001. https://doi.org/10.1002/cber.19110440303
83. Ooka H, Huang J, Exner KS (2021) The Sabatier principle in electrocatalysis: basics, limitations, and extensions. Front Energy Res 9. https://doi.org/10.3389/fenrg.2021.654460
84. Trasatti S (1972) Work function, electronegativity, and electrochemical behaviour of metals: III. Electrolytic hydrogen evolution in acid solutions. J Electroanal Chem Interfacial Electrochem 39:163–184. https://doi.org/10.1016/S0022-0728(72)80485-6
85. Rossmeisl J, Logadottir A, Nørskov JK (2005) Electrolysis of water on (oxidized) metal surfaces. Chem Phys 319:178–184. https://doi.org/10.1016/j.chemphys.2005.05.038
86. Nørskov JK, Bligaard T, Rossmeisl J, Christensen CH (2009) Towards the computational design of solid catalysts. Nat Chem 1:37–46. https://doi.org/10.1038/nchem.121
87. Rossmeisl J, Qu ZW, Zhu H et al (2007) Electrolysis of water on oxide surfaces. J Electroanal Chem 607:83–89. https://doi.org/10.1016/j.jelechem.2006.11.008
88. Nørskov JK, Rossmeisl J, Logadottir A et al (2004) Origin of the overpotential for oxygen reduction at a fuel-cell cathode. J Phys Chem B 108:17886–17892. https://doi.org/10.1021/jp047349j
89. Zadick A, Dubau L, Sergent N et al (2015) Huge instability of Pt/C catalysts in alkaline medium. ACS Catal 5:4819–4824. https://doi.org/10.1021/acscatal.5b01037
90. Creus J, De Tovar J, Romero N et al (2019) Ruthenium nanoparticles for catalytic water splitting. ChemSusChem 12:2493–2514. https://doi.org/10.1002/cssc.201900393
91. Romero N, Creus J, García-Antón J et al (2021) Rebirth of ruthenium-based nanomaterials for the hydrogen evolution reaction. In: Nanoparticles catalysis, pp 257–277
92. Nørskov JK, Bligaard T, Logadottir A et al (2005) Trends in the exchange current for hydrogen evolution. J Electrochem Soc 152:J23–J26. https://doi.org/10.1149/1.1856988
93. Jaramillo TF, Jørgensen KP, Bonde J, Nielsen JH, Horch S (1979) Chorkendorff I (2007) Identification of active edge sites for electrochemical H_2 evolution from MoS_2 nanocatalysts. Science 317:100–102
94. del Rosal I, Poteau R (2021) Sabatier principle and surface properties of small ruthenium nanoparticles and clusters: case studies. In: Nanoparticles in catalysis. Wiley, pp 331–351
95. Cusinato L, Martínez-Prieto LM, Chaudret B et al (2016) Theoretical characterization of the surface composition of ruthenium nanoparticles in equilibrium with syngas. Nanoscale 8: 10974–10992. https://doi.org/10.1039/c6nr01191h

96. García-Antón J, Axet MR, Jansat S et al (2008) Reactions of olefins with ruthenium hydride nanoparticles: NMR characterization, hydride titration, and room-temperature C-C bond activation. Angew Chem Int Ed 47:2074–2078. https://doi.org/10.1002/anie.200704763
97. Atrei A, Lesiak-Orlowska B, Tóth J (2022) Magnetite nanoparticles functionalized with citrate: a surface science study by XPS and ToF-SIMS. Appl Surf Sci 602:154366. https://doi.org/10.1016/j.apsusc.2022.154366
98. Johnson D, Qiao Z, Djire A (2021) Progress and challenges of carbon dioxide reduction reaction on transition metal based electrocatalysts. ACS Appl Energy Mater 4:8661–8684
99. Xie C, Niu Z, Kim D et al (2020) Surface and interface control in nanoparticle catalysis. Chem Rev 120:1184–1249
100. Lu Q, Jiao F (2016) Electrochemical CO_2 reduction: Electrocatalyst, reaction mechanism, and process engineering. Nano Energy 29:439–456. https://doi.org/10.1016/j.nanoen.2016.04.009
101. Narouz MR, Osten KM, Unsworth PJ et al (2019) N-heterocyclic carbene-functionalized magic-number gold nanoclusters. Nat Chem 11:419–425. https://doi.org/10.1038/s41557-019-0246-5
102. Riscoe AR, Wrasman CJ, Herzing AA et al (2019) Transition state and product diffusion control by polymer–nanocrystal hybrid catalysts. Nat Catal 2:852–863. https://doi.org/10.1038/s41929-019-0322-7
103. Han X, Liu L, Yuan J et al (2021) Polyacrylamide-mediated silver nanoparticles for selectively enhancing electroreduction of CO_2 towards CO in water. ChemSusChem 14:721–729. https://doi.org/10.1002/cssc.202002481
104. Ma L, Hu W, Pan Q et al (2019) Polyvinyl alcohol-modified gold nanoparticles with record-high activity for electrochemical reduction of CO_2 to CO. J CO2 Util 34:108–114. https://doi.org/10.1016/j.jcou.2019.06.002
105. Kim D, Yu S, Zheng F et al (2020) Selective CO_2 electrocatalysis at the pseudocapacitive nanoparticle/ordered-ligand interlayer. Nat Energy 5:1032–1042. https://doi.org/10.1038/s41560-020-00730-4
106. Chou TC, Chang CC, Yu HL et al (2020) Controlling the oxidation state of the Cu electrode and reaction intermediates for electrochemical CO_2 reduction to ethylene. J Am Chem Soc 142:2857–2867. https://doi.org/10.1021/jacs.9b11126
107. Zhang W, Huang C, Xiao Q et al (2020) Atypical oxygen-bearing copper boosts ethylene selectivity toward electrocatalytic CO_2 reduction. J Am Chem Soc 142:11417–11427. https://doi.org/10.1021/jacs.0c01562
108. Zhou Y, Che F, Liu M et al (2018) Dopant-induced electron localization drives CO_2 reduction to C_2 hydrocarbons. Nat Chem 10:974–980. https://doi.org/10.1038/s41557-018-0092-x
109. Pankhurst JR, Iyengar P, Okatenko V, Buonsanti R (2021) Copper nanocrystal morphology determines the viability of molecular surface functionalization in tuning electrocatalytic behavior in CO_2 reduction. Inorg Chem 60:6939–6945. https://doi.org/10.1021/acs.inorgchem.1c00287
110. Motagamwala AH, Dumesic JA (2021) Microkinetic modeling: a tool for rational catalyst design. Chem Rev 121:1049–1076. https://doi.org/10.1021/acs.chemrev.0c00394
111. Kovačič Ž, Likozar B, Huš M (2020) Photocatalytic CO_2 reduction: a review of ab initio mechanism, kinetics, and multiscale modeling simulations. ACS Catal:14984–15007. https://doi.org/10.1021/acscatal.0c02557
112. Kopač D, Likozar B, Huš M (2020) How size matters: electronic, cooperative, and geometric effect in perovskite-supported copper catalysts for CO_2 reduction. ACS Catal 10:4092–4102. https://doi.org/10.1021/acscatal.9b05303
113. Ješić D, Lašič Jurković D, Pohar A et al (2021) Engineering photocatalytic and photoelectrocatalytic CO_2 reduction reactions: mechanisms, intrinsic kinetics, mass transfer resistances, reactors and multi-scale modelling simulations. Chem Eng J 407:126799. https://doi.org/10.1016/j.cej.2020.126799
114. An H, Wu L, Mandemaker LDB et al (2021) Sub-second time-resolved surface-enhanced Raman spectroscopy reveals dynamic CO intermediates during electrochemical CO_2 reduction on copper. Angew Chem Int Ed 60:16576–16584. https://doi.org/10.1002/anie.202104114

115. Zhao S, Christensen O, Sun Z et al (2023) Steering carbon dioxide reduction toward C-C coupling using copper electrodes modified with porous molecular films. Nat Commun 14:844. https://doi.org/10.1038/s41467-023-36530-z
116. de Ruiter J, An H, Wu L et al (2022) Probing the dynamics of low-overpotential CO_2-to-CO activation on copper electrodes with time-resolved Raman spectroscopy. J Am Chem Soc 144: 15047–15058. https://doi.org/10.1021/jacs.2c03172
117. Sa YJ, Lee CW, Lee SY et al (2020) Catalyst-electrolyte interface chemistry for electrochemical CO_2 reduction. Chem Soc Rev 49:6632–6665. https://doi.org/10.1039/d0cs00030b
118. Nguyen QT, Rousset E, Nguyen VTH et al (2021) Covalent grafting of ruthenium complexes on iron oxide nanoparticles: hybrid materials for photocatalytic water oxidation. ACS Appl Mater Interfaces 13:53829–53840. https://doi.org/10.1021/acsami.1c15051
119. Hoertz PG, Il KY, Youngblood WJ, Mallouk TE (2007) Bidentate dicarboxylate capping groups and photosensitizers control the size of IrO_2 nanoparticle catalysts for water oxidation. J Phys Chem B 111:6845–6856. https://doi.org/10.1021/jp070735r
120. Youngblood WJ, Lee SA, Kobayashi Y et al (2009) Photoassisted overall water splitting in a visible light-absorbing dye-sensitized photoelectrochemical cell. J Am Chem Soc 131:926–927
121. Corby S, Rao RR, Steier L, Durrant JR (2021) The kinetics of metal oxide photoanodes from charge generation to catalysis. Nat Rev Mater 6:1136–1155
122. Martín Morales E, Coppel Y, Lecante P et al (2020) When organophosphorus ruthenium complexes covalently bind to ruthenium nanoparticles to form nanoscale hybrid materials. Chem Commun 56:4059–4062. https://doi.org/10.1039/d0cc00442a
123. Reynal A, Willkomm J, Muresan NM et al (2014) Distance dependent charge separation and recombination in semiconductor/molecular catalyst systems for water splitting. Chem Commun 50:12768–12771. https://doi.org/10.1039/c4cc05143b
124. Kawawaki T, Akinaga Y, Yazaki D et al (2023) Promoting photocatalytic carbon dioxide reduction by tuning the properties of cocatalysts. Chem A Eur J 29
125. Liu L, Zhang X, Yang L et al (2017) Metal nanoparticles induced photocatalysis. Natl Sci Rev 4:761–780
126. Nasrallah H, Douma F, Hamoud HI, El-Roz M (2021) Metal nanoparticles in photocatalysis: advances and challenges. In: Nanostructured photocatalysts: from fundamental to practical applications. Elsevier, pp 119–143
127. Melchionna M, Fornasiero P (2020) Updates on the roadmap for photocatalysis. ACS Catal 10:5493–5501. https://doi.org/10.1021/acscatal.0c01204
128. Kamat PV, Jin S (2018) Semiconductor photocatalysis: tell us the complete story! ACS Energy Lett 3:622–623. https://doi.org/10.1021/acsenergylett.8b00196
129. Zhang X, Zhou Z (2022) Perspective on theoretical models for CO_2 electrochemical reduction. J Phys Chem C 126:3820–3829. https://doi.org/10.1021/acs.jpcc.1c10870

Top Organomet Chem (2024) 75: 133–198
https://doi.org/10.1007/3418_2024_119
© The Author(s), under exclusive license to Springer Nature Switzerland AG 2024
Published online: 4 October 2024

N-Heterocyclic Carbene Modified Metal Nanocatalysts

Patricia Lara and Andrés Suárez

Contents

Abstract The scientific community has widely studied the synthesis of metal nanoparticles (MNPs) and their applications in a plethora of catalytic transformations for the last 25 years. In these nanometric structures, the catalytic activity and selectivity can be modulated by the controlled tuning of the surface properties of the particles through the use of coordinating donor ligands bound to the metal surface. Among the ligands employed for MNP stabilization, N-heterocyclic carbene (NHC) derivatives have emerged in the last 15 years as a prominent class of compounds due to their strong electron-donor properties and structural diversity. Herein, we discuss

P. Lara (✉)
Departamento de Química Inorgánica, and Centro de Innovación en Química Avanzada (ORFEO-CINQA), Instituto de Investigaciones Químicas (IIQ), Universidad de Sevilla, Sevilla, Spain
e-mail: patricia@iiq.csic.es

A. Suárez
Instituto de Investigaciones Químicas (IIQ) and Centro de Innovación en Química Avanzada (ORFEO-CINQA), Consejo Superior de Investigaciones Científicas (CSIC), Sevilla, Spain

the unique ability of NHC ligands to control the reactivity of nanosized metal catalytic systems by describing the synthetic procedures employed for the preparation of NHC-stabilized nanoparticles, their characterization, and the applications of these nanosystems in different catalytic reactions, such as hydrogenation, C-C and C-X bond forming reactions, H/D exchange, oxidation, and electrocatalytic processes.

Keywords Catalysis · Metal · Nanoparticles · N-Heterocyclic carbenes · Surface chemistry

1 Introduction

The increasing interest on the use of metal nanoparticles (MNPs) as catalysts stems from their particular electronic configurations along with their larger surface area in comparison to bulk metals, which provide numerous catalytically active sites [1]. Among the different modes of stabilizing MNPs, the use of electron-donating ligands, in addition to preventing metal aggregation, allows to easily modify the surface properties, and therefore the catalytic performance of the particles, by simply performing variations on the chemical structure of the ligand and adjusting the metal/ligand ratio [2]. In this context, different families of organic ligands, also frequently employed in molecular catalysis, have been utilized for the stabilization of metal nanoparticles. Ubiquitous examples of these derivatives include thiols, thioethers, alcohols, amines, and phosphines [1, 2].

Since the first isolation of a N-heterocyclic carbene (NHC) in 1991 by Arduengo et al. [3], these derivatives have been widely employed as ligands for the modification of the reactivity of metal complexes with catalytic applications [4] and the functionalization of metal surfaces [5, 6]. The straightforward synthesis of the precursors of the NHC ligands, along with their large structural and steric variability and strong electron-donor properties, [7] somewhat resembling that of the ubiquitous phosphines, makes them ideal molecules to be used as stabilizing agents for the synthesis of MNPs [1, 2]. As a consequence of the high strength of the NHC-metal linkages formed, in addition to determine the size, shape, and surface properties, enhanced stability of the nanoparticles is achieved by the use of these ligands. Moreover, since these ligands are chemically bonded to specific metal atoms located at the surface, the catalytic behavior of the MNPs can be affected either by enhancing the electron density of the metal surface or by blocking highly active, but unselective, catalytic sites. Finally, the NHC ligands, which are readily amenable to structural variations through the appropriate introduction of substituents, can impart the MNPs with desired solubility properties, i.e., solubility in water or in apolar reaction media, and facilitate NP immobilization through attractive interactions of the NHC appended groups with the support.

While the presence of carbenes as transient species in MNPs formed in imidazolium-based ionic liquids was proposed as early as 2005 and 2007 by the groups of Finke [8] and Dupont, [9] respectively, it was not until a few years later than Vignolle and Don Tilley reported the first synthesis of NHC-capped nanoparticles by the chemical reduction of preformed Au(I)-NHC complexes [10]. In 2010, the group of Glorius made use of Pd/Fe_3O_4 nanoparticles capped with chiral NHC ligands to catalyze the enantioselective α-arylation of ketones, being this the first report of the application of NHC ligands in the modification of the catalytic behavior of supported metal nanoparticles [11]. This seminal work, soon followed by the rational synthesis of colloidal Ru NPs stabilized by NHC ligands performed by Chaudret and co-workers, [12] clearly demonstrated the capability of NHCs to stabilize and modify the reactivity of metal nanoparticles, leading to an emergence of the synthesis of well-defined NHC-stabilized nanosized systems and their catalytic applications, which are the topic of this book chapter.

2 Synthetic Approaches to the Preparation of NHC-Modified Metal Nanocatalysts

Different methodologies, which do not differ significantly from those used for the synthesis of MNPs incorporating other ligand classes, have been employed in the preparation of both colloidal and supported metal nanoparticles containing NHC ligands. Thus, a survey covering common methods reported in the literature reveals four main procedures utilized for preparing MNPs based on late transition metals and an array of NHC ligand families (Figs. 1 and 2). First, NHC-containing MNPs are accessible through decomposition of well-defined metal complexes incorporating NHC-based ligands by thermal treatment or using suitable reducing agents (Fig. 1a). This methodology, which has been employed for the preparation of nanoparticles based on noble metals, such as Au, [13–16] Ru, [17], Ir, [18] Pd, [19, 20] or Pt [21],

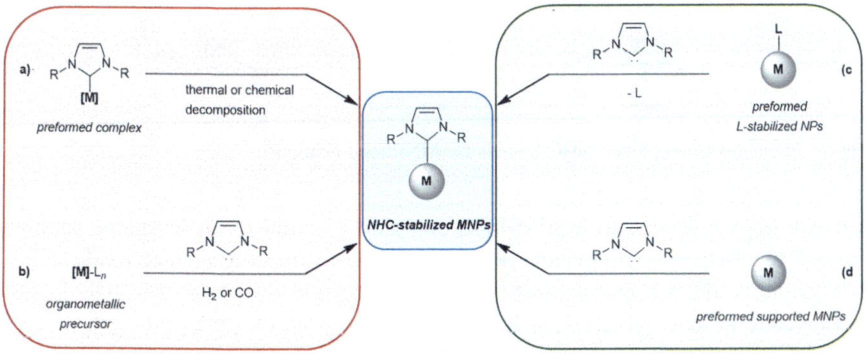

Fig. 1 Common synthetic procedures for the synthesis of colloidal and supported NHC-stabilized MNPs

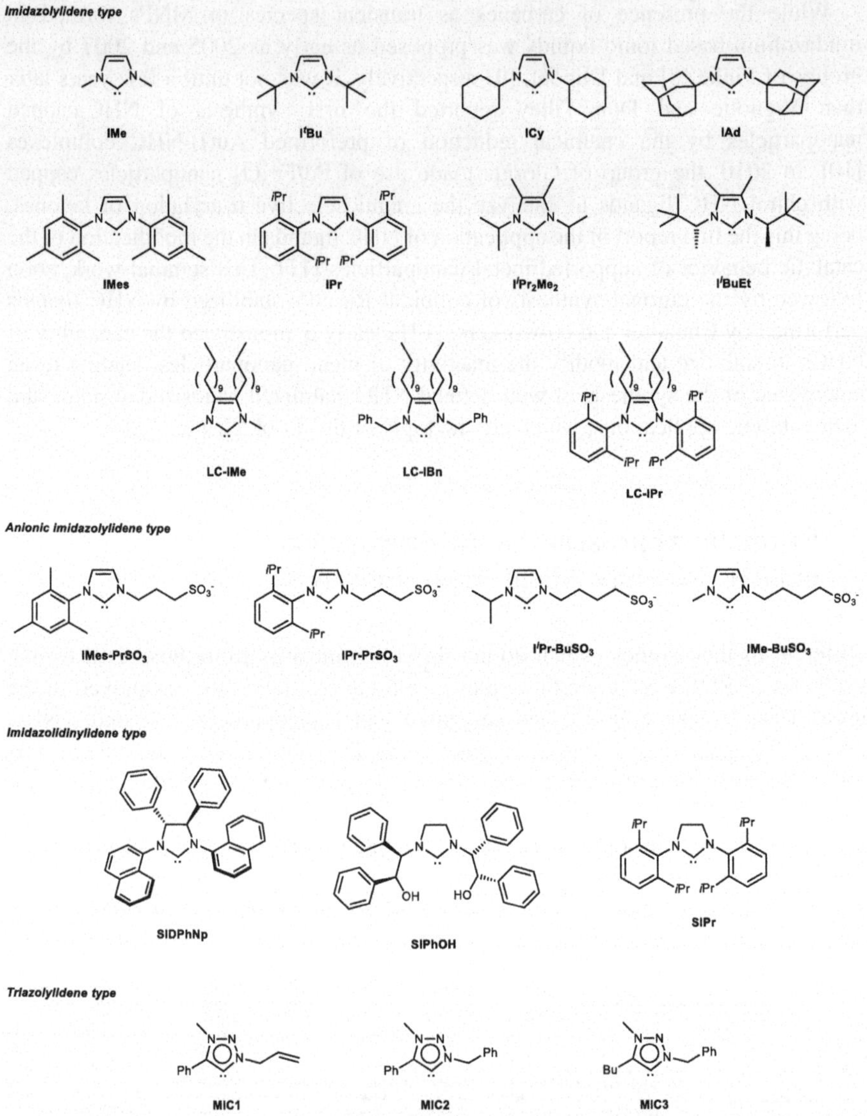

Fig. 2 Examples of employed NHC ligands for MNP modification

has also been extended to immobilized complexes bound to an inorganic support through the formation of covalent bonds [22] and to reduced graphene oxide (rGO) through attractive π-π interactions between pyrene substituents at the NHC ligand and the carbon material [23–25]. In addition to the need of a previous synthesis of well-defined complexes incorporating NHC ligands, the preparation of nanoparticles by decomposition of these derivatives is usually not amenable to easy modifications of ligands, ligand/metal ratios, and decomposition experimental procedures.

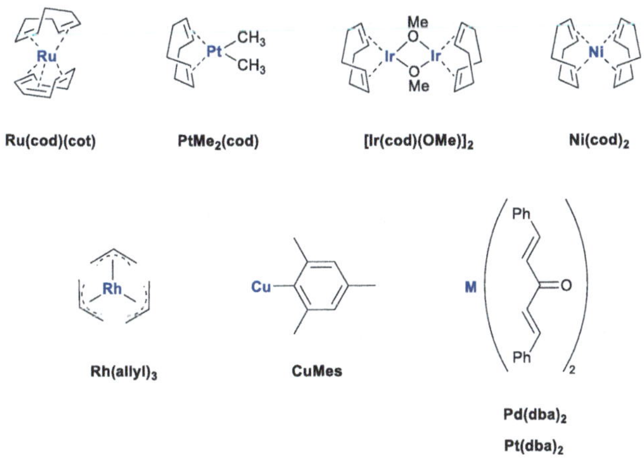

Fig. 3 Examples of organometallic precursors used for the preparation of M·NHC NPs following the "organometallic approach"

Moreover, the MNP surface thus obtained is usually contaminated with ligands from the preformed complexes other than the NHCs, potentially limiting the access of substrates to reactive metal sites.

A versatile procedure that allows a high control on the size and properties of colloidal nanoparticles involves the decomposition of a suitable organometallic precursor with a gaseous reductant (H_2 or CO), in the presence of a polymer (for steric stabilization) or a sub-stoichiometric amount of a stabilizing ligand, such as amines or phosphines (Fig. 1b). The organometallic precursor normally coordinates ligands that upon complex reduction lead to species with poor binding affinity for the metal surface, providing nanoparticles which ideally only contains the stabilizing ligands, and hydrides or CO coming from the reducing agent (Fig. 3). This procedure, first developed by Philippot, Chaudret and coworkers, [26] is widely known in the literature as the "organometallic approach," and was initially applied to the synthesis of Ru·NHC NPs in 2011 by decomposition of [Ru(cod)(cot)] (cod = 1,5-cyclooctadiene; cot = 1,3,5-cyclooctatriene) with H_2 in the presence of 0.2 or 0.5 equiv. of a NHC ligand [12]. Later, this methodology was extended to the preparation of NHC-stabilized Pt nanoparticles by Lara et al., [27] and to other metals (Ni, [28] Cu, [29] Pd, [29] Ir, [30] Rh [31]) as reported by different groups [32]. Some interesting features of this synthetic procedure reside in the efficient control and reproducibility of the properties, particularly size and surface state, of the synthetized MNPs. At the same time, it also allows for tailoring the reactivity of the MNPs by simple modification of the structure of the stabilizing ligand and the metal/ligand ratio. Following this "organometallic approach," MNPs incorporating different classes of N-heterocyclic ligands, including imidazolylidene, imidazolidinylidene [33] and triazolylidene, [34] have been reported. While initial studies involving this preparation method made use of isolated carbene ligands, several procedures have been developed that are also tolerant to the "in situ" generation of non-isolable NHC ligands from the corresponding azolium salts and a strong base, [33] or using zwitterionic NHC-CO_2 derivatives [28]. Furthermore, another interesting feature of this synthetic method is that NHC-modified bimetallic nanoparticles are easily accessed through proper choice of two different metal precursors [29, 35–37].

The third procedure for carbene-modified MNP preparation involves the exchange of labile ligands, such as thioethers, stabilizing preformed MNPs by strong donating NHC derivatives (Fig. 1c) [38–44]. While this methodology usually does not produce significant changes in the size of the MNPs, a recurrent disadvantage of using a ligand exchange process resides in that an excess of NHCs might be needed to displace the precursor ligand, leading to extensive work-up to purify the particles and a limited control of the final ligand/metal ratio.

Finally, as shown by Glorius and co-workers in 2010 with the synthesis of heterogeneous Pd nanoparticles capped with NHCs, [11] supported NHC-modified MNPs can be obtained by simple addition of isolated or "in situ" generated carbene ligands to immobilized nanoparticles (Fig. 1d). This approach has been employed with nanoparticles supported on both inorganic [45–49] or organic, such as polyvinylpyrrolidone (PVP), [50] supports. In addition to easily assemble NHC-capped nanocatalysts, this protocol allows to modify the reactivity of the nanomaterials by using different NHC ligands and ligand/metal ratios.

3 Determination of NHC Coordination Mode and Sites in Metal Nanocatalysts

Determining the coordination mode and surface locations of ligands modifying metal nanoparticles is crucial to explain their observed properties, particularly stability and catalytic reactivity. To this end, different spectroscopic techniques can be employed to establish the location, dynamics, and coordination modes of species present at the surface of metal nanosystems. In striking contrast to the study of NHC-coordinated molecular species, where coordination properties of the NHC ligands is determined from a few readily available instrumental techniques, particularly solution NMR and IR spectroscopies, a more complicated scenario is given in the case of NHC-ligated nanoparticles. In fact, the study of the coordination of NHC ligands to nanoparticle surfaces has been less explored, although significant knowledge has been acquired through, mainly, the use of XPS, FT-IR and solid-state magic angle spinning (MAS) NMR spectroscopies.

In this context, solid state NMR spectroscopy is one of the most persuasive techniques to gain information regarding the coordination mode of NHC to the metal surfaces of nanoparticles, providing a direct evidence of the formation of NHC-metal bonds involving the C-2 carbenic carbon. These analysis, however, can be hampered by the presence of Knight shift effects, which are originated by the free electrons on the surface of large MNPs, causing the displacement and broadening of

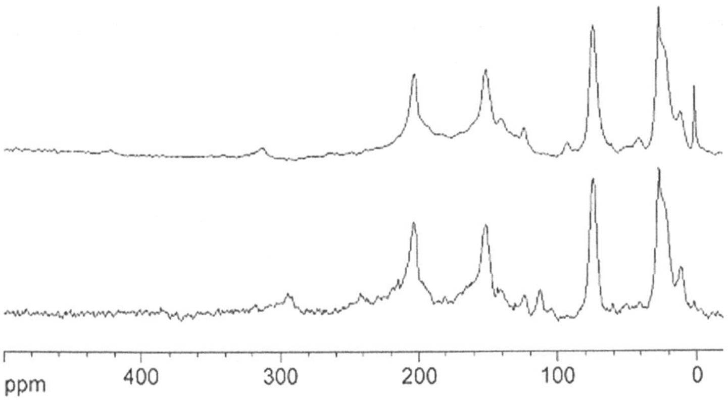

Fig. 4 ^{13}C MAS (bottom) and CP-MAS (top) NMR spectra of Ru·IPr*$^{0.2}$ nanoparticles. Adapted with permission from [12] Copyright 2011 Wiley-VCH Verlag GmbH&Co. KGaA, Weinheim

the signals associated to the molecules located at the surface [51]. The first example involving the study of the surface of NHC-stabilized nanoparticles by solid state NMR was reported by Chaudret and co-workers in 2011 [12]. Solution ^{1}H NMR spectra of colloids using 0.2 or 0.5 equiv. of the I^{t}Bu or IPr carbenes, sized 1.5–1.7 nm, did not allow to observed the resonances of the CH groups of the imidazole backbones and the phenyl groups of IPr, probably as a consequence of their close proximity to the metal surface. Further evidence of the coordination of the ligands as carbene species was obtained by registering the solid state ^{13}C NMR spectra of the particles prepared using ^{13}C labeled NHCs at the C-2 position (Fig. 4). While Ru·I^{t}Bu*$^{0.5}$ and Ru·IPr*$^{0.2}$ NPs produced resonances at 190 and 205 ppm, respectively, attributable to the carbene atom, in the colloid containing a higher amount of the ^{13}C-labeled IPr* ligand (Ru·IPr*$^{0.5}$) another resonance, besides the one observed at 205 ppm, was detected at 195 ppm, which constitutes an indication of two different environments for the carbene at the MNP surface. The observed chemical shifts for the coordinated NHCs were significantly different to those of the corresponding free carbenes values, being this a clear demonstration of the coordination of the carbenes to the Ru surfaces.

Another illustrative example of the use of solid state NMR spectroscopy to confirm carbene coordination was reported with colloidal water-soluble NHC-stabilized Pt nanoparticles (mean size 1.3–2.0 nm) (Fig. 5) [21]. The ^{13}C CP MAS-NMR spectrum of the Pt NPs showed a broad resonance at ca. 177 ppm that was attributed to the carbene carbon (Fig. 5). Moreover, since ^{195}Pt nucleus has nuclear spin $I = 1/2$, it was possible to observe the coupling constant ^{13}C-^{195}Pt (940 ± 20 Hz) for the first time in a nanometric system, and compare it with that of a molecular complex. In the case of the MNP surface, and taking into account the value of the coupling constant, the authors concluded that the coordination of the NHC ligand to the Pt surface was weaker than in a Pt(0)-NHC complex (J_{CPt} = 1,365 Hz), although stronger than in the case of Pt(II) derivatives, where

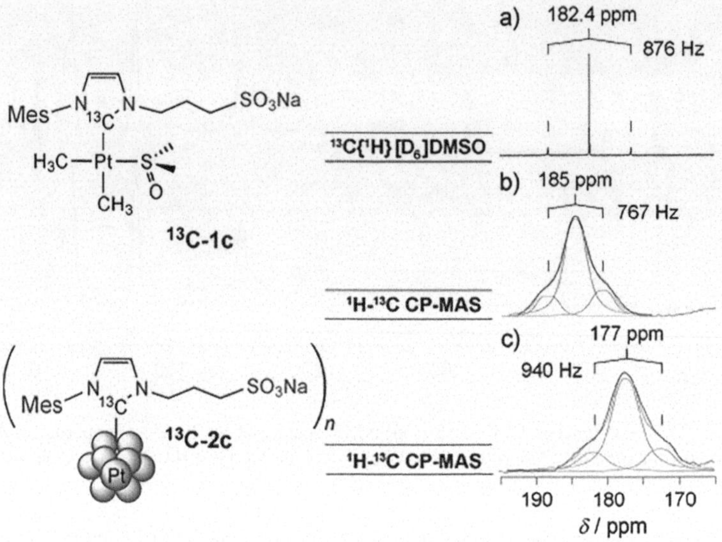

Fig. 5 Resonance signals, and deconvolution curves, associated to the Pt-bonded carbenic carbons in a) the $^{13}C\{^{1}H\}$ NMR spectrum of a NHC-stabilized Pt(II) complex, and b) and c) ^{1}H-^{13}C CP-MAS NMR spectra of the Pt(II) complex and NHC-stabilized Pt NPs, respectively. Adapted with permission from [21] Copyright 2014 Wiley-VCH Verlag GmbH&Co. KGaA, Weinheim

the +2 oxidation state of the metal atom decreases the coupling constant ($J_{CPt} = 767$ Hz).

The CO molecule has been widely employed as a probe to determine ligand location, as its coordination to a metal nanoparticle surface provides valuable information of the types of available catalytically active sites [52]. CO might coordinate to a metal surface in a bridging mode (CO_b), usually on the faces of the MNPs, or in a terminal mode (CO_t) at exposed atoms (edges, apexes and defects), which lead to the observation of different signals. The ratio between terminal and bridging coordinated CO molecules is dependent on the particle size since it increases as the NP diameter decreases [53]. As a consequence of the characteristic and strong absorptions caused by CO bound to a metal surface, FT-IR spectroscopy is useful to identify coordination sites and dynamic processes of ligands by detecting the different CO coordination modes to the metallic surfaces. In electron-rich metal surfaces ligated by NHC ligands, strong back-donation to CO might produce a red shift of the CO absorption modes.

As an example of this, Glorius, Chaudret, van Leuween et al. investigated the coordination of CO to the surface of Pt nanoparticles capped by NHCs containing long alkyl chains at the 4 and 5 positions of the ligands (LC-IMe and LC-IPr of Fig. 2) [54]. Depending on the structure of the NHC attached to the surface of the particles, terminal CO strong absorptions were detected at 2035 cm^{-1} for Pt·LC-IPr (1.2 ± 0.2 nm), and at 2,030 cm^{-1} in the case of Pt·LC-IMe (1.9 ± 0.2 nm), while weaker bridging CO bands appeared at 1,820 cm^{-1} and 1,800 cm^{-1} for Pt·LC-IPr

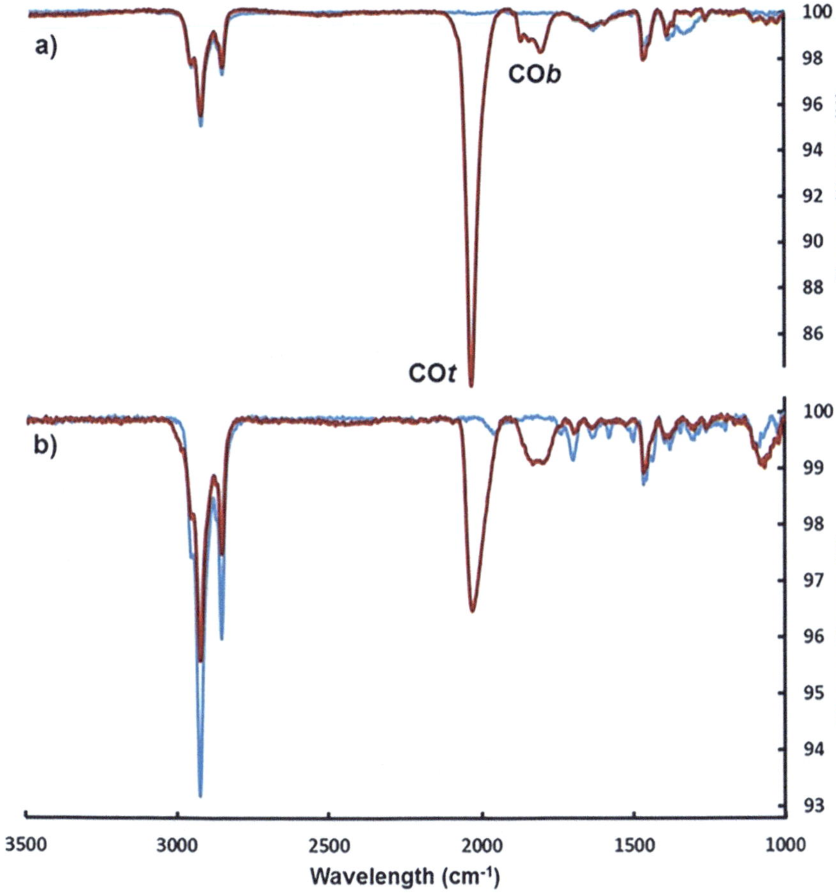

Fig. 6 FT-IR spectra of Pt·LC-IPr (**a**) and Pt·LC-IMe (**b**) before (blue) and after (red) CO exposure. Adapted with permission from [54] Copyright 2017 Wiley-VCH Verlag GmbH&Co. KGaA, Weinheim

and Pt·LC-IMe, respectively (Fig. 6). Interestingly, the CO_b/CO_t intensity ratio was found to be higher in the case of Pt·LC-IMe, which was found by TEM to exhibit larger sizes in comparison to Pt·LC-IPr, suggesting the existence of more accessible faces for the coordination of CO in a bridging mode.

In addition, MAS-NMR spectroscopy also provide information on the surface state of NHC-capped MNPs through the coordination of ^{13}C-enriched CO. [55, 56] For example, the location of the active sites of Ru NPs stabilized with LC-NHC ligands (Ru·LC-IPr and Ru·LC-IMe) was examined by ^{13}C MAS-NMR spectroscopy [55]. Both colloidal Ru NPs exhibited resonances attributable to bridging CO_b as broad peaks centered at ca. 230 ppm, and to terminal CO_t as sharp signals at ca. 198 ppm (Fig. 7). However, in agreement with the expected higher bulkiness of the LC-IPr N-substituents, the CO_b resonance for Ru·LC-IMe was found to be

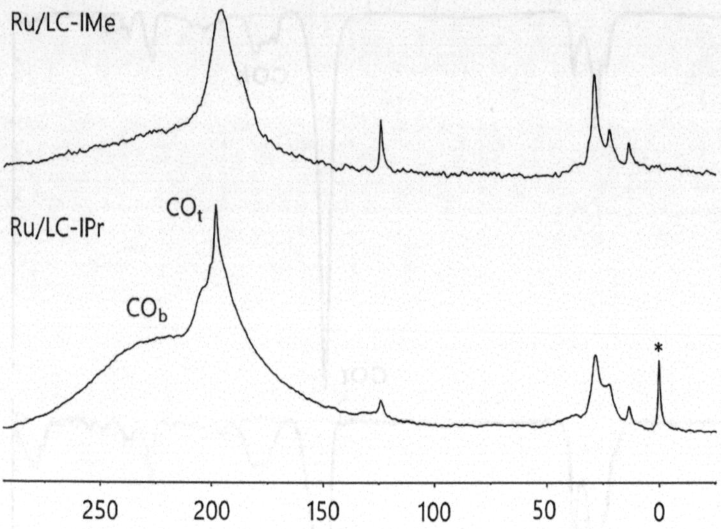

Fig. 7 ^{13}C-MAS NMR spectra of Ru·LC-IMe (top) and Ru·LC-IPr (bottom) after exposure to ^{13}CO. The peak with asterisk corresponds to grease. Adapted with permission from [55] Copyright 2016 Royal Society of Chemistry

weaker than in the case of Ru·LC-IPr, suggesting that more face sites are available for Ru·LC-IPr as LC-IPr coordination occurs preferentially at edges and apexes. Moreover, in the ^{13}C CP MAS-NMR spectra of both Ru·LC-NHC colloids, the CO_b resonance disappears and the CO_t signal become weaker, suggesting that bridging CO molecules do not coordinate in close proximity to the NHCs, while terminal CO ligands partially do. This information was correlated to the catalytic activity of the Ru NPs in hydrogenation reactions, as discussed in Sect. 4.1.

The observation of Knight shift is also useful to demonstrate the coordination of NHC ligands to MNPs by MAS-NMR spectroscopy, as shown by Chaudret, de Jesús and co-workers [57]. These researchers synthetized Pd nanoparticles stabilized by a sulfonated NHC ligand having sizes ranging from 1.3 to 4.8 nm. In the particular case of Pd NPs stabilized with ^{13}C-labeled NHCs of mean size 3.7 nm, the ^{13}C CPMG echo MAS-NMR spectrum exhibited a Knight-shifted resonance of the carbenic carbon appearing at ca. 600 ppm (Fig. 8). Moreover, an interesting influence of the MNP size on the chemical shift of the ^{13}C NMR signals of CO adsorbed molecules was observed. For small particles of mean size below 2 nm, typical chemical shifts between 150 and 250 ppm for adsorbed ^{13}CO were detected, while for larger particles (3.8 and 4.8 nm) this signal was Knight-shifted to 750–800 ppm.

X-ray photoelectron spectroscopy (XPS) is a useful technique that provides information about the surface of the particles, particularly regarding the metal oxidation states and the coordination of capping ligands to the metallic surface. An illustrative example of the use of XPS to determine the ligation of carbene ligands to the MNP surface was reported by Glorius and coworkers already in

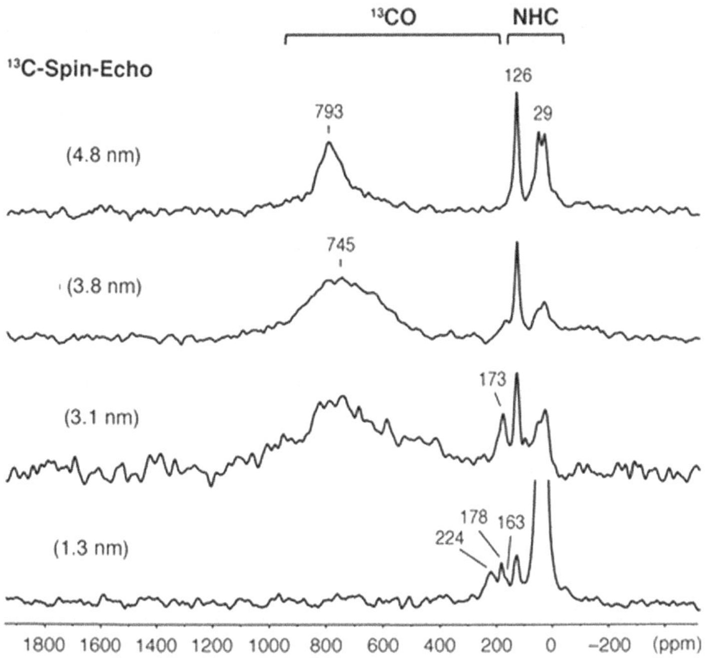

Fig. 8 Effect of nanoparticle size in the solid-state ^{13}C spin-echo NMR spectra of sulfonated NHC-capped Pd NPs after ^{13}CO adsorption. Adapted with permission from [57] Copyright 2019 Wiley-VCH Verlag GmbH&Co. KGaA, Weinheim

2010 [11]. In this case, supported Pd nanoparticles on magnetite were decorated with quiral NHC ligands. XPS analysis of the nanomaterials confirmed the zero oxidation state for the Pd surface atoms, and the presence of signals associated to the N 1s and C 1s energies at 400 and 285 eV in binding energy (BE), respectively, assignable to the coordinated carbene ligands. In 2016, the same group made use of XPS to investigate the coordination of a bidentate NHC-thioether ligand to the surface of Pd NPs (Fig. 9) [39]. The value of the N 1s energy signal exhibiting a shift to lower BEs in comparison to the imidazolium salt used as precursor of the NHC ligand (401.6–401.8 eV) and to higher energies with respect to the free ligand (399.5 eV), which can be attributed to a loss of electron density on the N atoms adjacent to the carbenic carbon upon NHC coordination to the metal surface. Meanwhile, the S 2p signal exhibited no shift compared to the imidazolium salt, suggesting a weak interaction of the thiother group with nanoparticle surface. Moreover, signal integration confirmed the N/S ratio 2 to 1 expected for the stabilizing ligand located at the surface of the particles. These experiments allowed to establish XPS as an adequate technique to confirm the coordination of NHC ligands to the surfaces of nanoparticles, thus avoiding the time-demanding synthesis of ^{13}C-enriched carbene molecules to be used to detect NHC ligation by NMR spectroscopy.

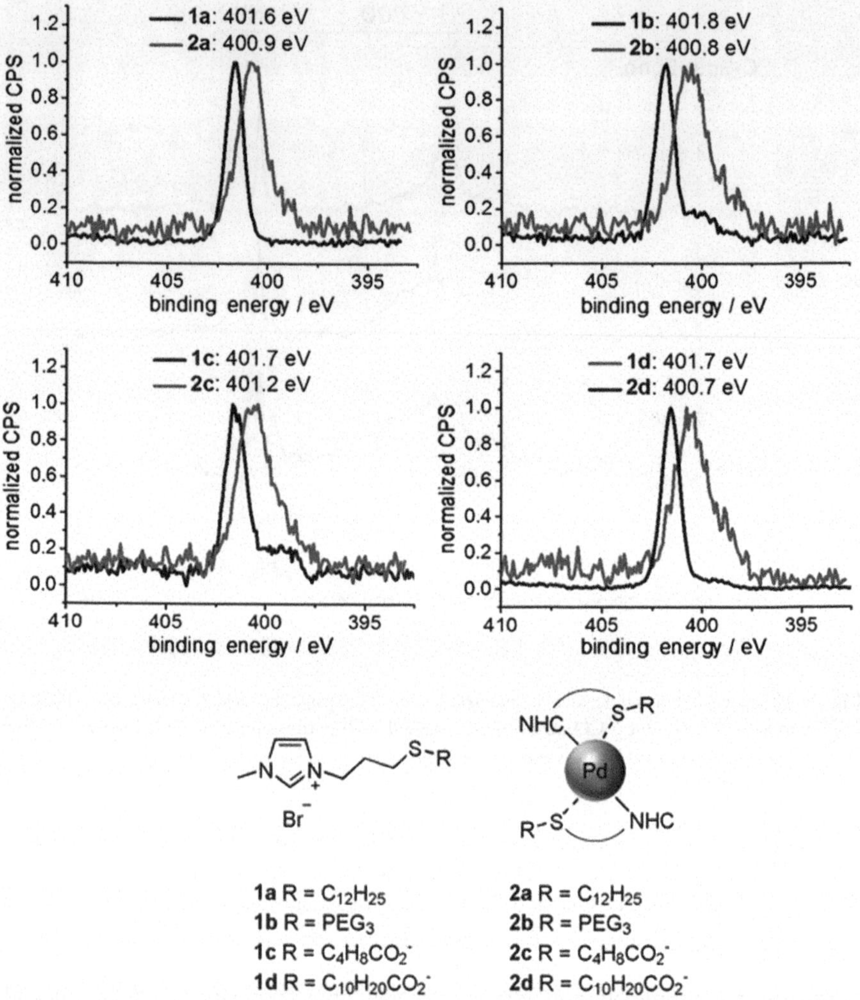

Fig. 9 X-ray photoelectron spectroscopy (XPS) of the N 1s signal of Pd NPs capped with bidentate NHC-S ligands and the corresponding NHC-S imidazolium salt precursors. Adapted with permission from [39] Copyright 2016 Wiley-VCH Verlag GmbH&Co. KGaA, Weinheim

Recently, Rendón, Lara, Suárez et al. proved the coordination of mesoionic carbene ligands at the surface of Ru nanoparticles, Ru·MIC [34]. XPS spectrum analysis allowed to fit the Ru 3p photoemission peak at 461.9 eV binding energy into two different oxidation states, Ru(0) and Ru(IV) (Fig. 10), which demonstrated a slight oxidation of the metal surface (around 8%). In addition, the linkage of the carbene to the surface of the particles was confirmed by the presence of a signal for N 1s at binding energy of 401.2 eV, which was deconvoluted into two different components of relative ratio 1:2, one located at 339.4 eV and a second one at 401.4 eV. These signals are associated to the N atom adjacent to the carbene carbon,

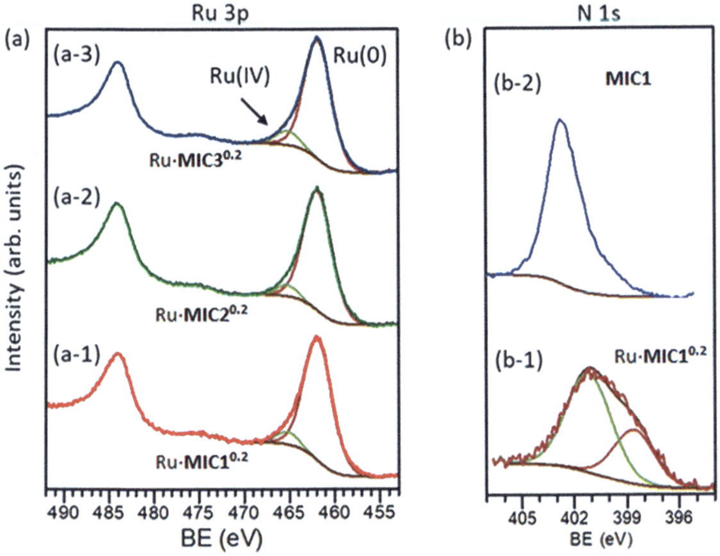

Fig. 10 X-ray photoelectron spectroscopy (XPS) peaks of: (**a**) Ru 3p regions of the Ru·MIC nanoparticles (experimental and fitted spectra), and (**b**) N 1s regions of Ru·MIC1$^{0.2}$ (b-1) (experimental and fitted spectra) and the triazolium salt precursor of MIC1 (b-2) (experimental spectrum). Reproduced with permission from [34] Copyright 2023 Royal Society of Chemistry

and to the two N atoms present at the ligand backbone, respectively. While the position of the N 1s peak of the non-isolable, free triazolylidene ligand could not be determined, the observed BE value is lower than that of the parent triazolium salt, as expected from the loss of charge density on the N atom adjacent to the carbenic carbon upon NHC coordination to the Ru surface. As a result of all these analyses, both the presence and the integrity of the carbene at the surface of these Ru nanoparticles were demonstrated.

4 Catalytic Applications of NHC-Modified Metal Nanocatalysts

4.1 Catalytic Hydrogenations

Hydrogenation reactions of unsaturated organic molecules are important transformations both in industrial settings and laboratories. Taking this fact with the well-known ability of metal nanoparticles to activate the H_2 molecule and transfer it to unsaturated derivatives, [58] it is not surprising that hydrogenation reactions represent by far the largest applications of reported NHC-modified MNPs.

4.1.1 Hydrogenation of Olefins

Supported and colloidal Pd NPs catalyze the hydrogenation of olefins, although efficient control of the reaction chemoselectivity is usually difficult to achieve. In this respect, modification of Pd NPs using NHC ligands has been found to provide improved selectivities. Moreover, through the use of appropriate NHC ligands, it has been possible to synthesize colloidal nanoparticles with adequate solubility in organic and aqueous reaction media.

For instance, Glorious, Ravoo et al. reported the synthesis of palladium nanoparticles via a ligand exchange protocol using tioether-stabilized Pd NPs and NHC ligands bearing long alkyl chains at their backbone (Pd·LC-IMe and Pd·LC-IBn; Fig. 11) [38]. It was expected that NHCs substituted at positions 4 and 5 with long aliphatic chains would form a lipophilic protective monolayer around the metal surface, thus providing steric repulsion between the MNPs and giving the nanomaterials good solubility in nonpolar media. At the same time, the presence of small (methyl) or flexible (benzyl) substituents at the nitrogen atoms of the NHC modifier would reduce the steric repulsion between the metal surface and the ligand. These Pd nanosystems were tested in the chemoselective hydrogenation of a series of olefins that differ in the degree of substitution and in their conjugation to other unsaturated groups. Competition experiments revealed that Pd·LC-IMe and Pd·LC-IBn could discriminate between substrates with different number of substituents. For example, terminal olefins, such as styrene and 1-decene, were selectively hydrogenated at the C=C bond, although partial isomerization of the double C=C bond in the case of 1-decene was observed leading to the formation of internal decenes. Meanwhile, hydrogenation of trisubstituted olefinic derivatives, such as 3-methyl-2-cyclohexenone and citronellol, was negligible under the same reaction conditions. Based on these observations, hydrogenation of 3,7-dimethyloct-6-en-1-yl acrylate was tested, leading chemoselectivily to the corresponding ethyl ester 3,7-dimethyloct-6-en-1-yl propionate in very high yield (Fig. 11, bottom). The Hg test showed that the catalyst mode of action is heterogeneous, ruling out the

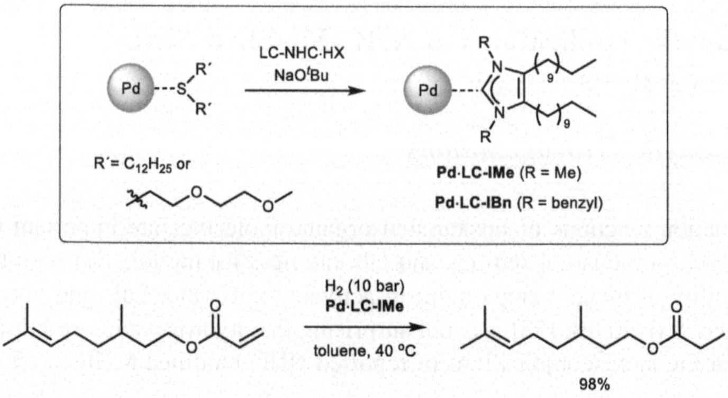

Fig. 11 Synthesis of Pd·LC NHC NPs, and application in the hydrogenation of olefins

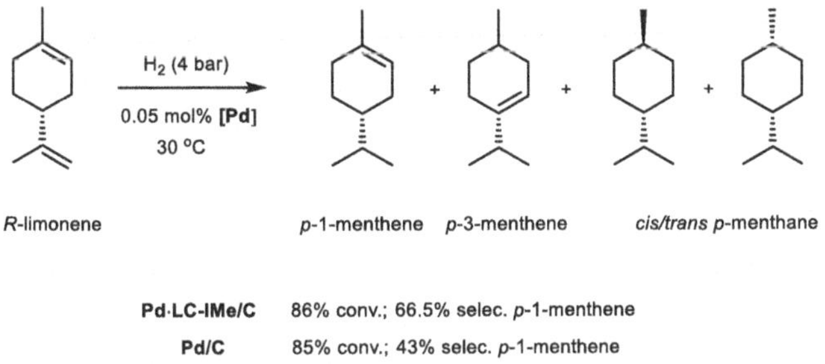

Fig. 12 Hydrogenation of *R*-limonene mediated by Pd·LC-IMe/C and unmodified Pd/C catalysts

formation of active homogeneous palladium species. Finally, by comparison of the performance of the LC-NHC modified materials with that of the thioether-stabilized parent MNPs and commercial Pd catalysts, the observed activity and selectivity obtained with the Pd·LC-NHC nanoparticles was attributed to the modification of the metal surface by the presence of the NHC ligands.

The LC-IMe ligand was also employed in the functionalization of nanomaterials prepared from ligand-free Pd NPs formed by an unusual metal vapor synthesis [59]. The bare and NHC-capped Pd NPs were impregnated on a carbon support (Vulcan® XC72), and their activity examined in the hydrogenation of *R*-limonene (Fig. 12). While both Pd/C catalysts hydrogenated limonene with similar conversions, a significant increase on the selectivity towards the hydrogenation of the exocyclic olefin was observed in the case of the NHC-modified nanoparticles, limiting the reduction of the endocyclic C=C bond to yield *cis/trans p*-menthane or the isomerization to *p*-3-menthene. Moreover, the stability of the Pd·LC-IMe/C nanoparticles was evidenced by the recycling of the catalyst up to five times, providing a maintained selectivity towards the formation of *p*-1-menthene with a slight decrease of the catalytic activity.

Similarly, a series of Pd NPs (3.6–5.1 nm) capped with bidentate hybrid NHC-thioether ligands (NHC-S) was prepared through ligand exchange methodology, and tested in alkene hydrogenation (Fig. 13) [39]. Coordination of the stabilizing ligands to the metal surface was demonstrated to occur primary through the stronger donating NHC moiety, with the thioether group binding through a weaker dative interaction. The thioether fragment incorporates substituents with different expected polarity, ranging from nonpolar (NHC-S-1) to polar (NHC-S-3), with the PEG-containing (NHC-S-2) and long-chain carboxylic acid (NHC-S-4) groups having an intermediate polarity. Depending on the ligands polarity, the corresponding nanoparticles could exhibit solubility in non-polar to polar reaction media. Therefore, as expected, the catalytic activity of the Pd·NHC-S NPs in the hydrogenation of styrene and isophorone was found to be solvent dependent, with Pd NPs stabilized with polar NHCs showing better reactivity in polar media and vice versa. At variance with commercial Pd/C catalyst, the Pd·NHC-S NPs showed a high chemoselectivity in the addition of H_2 to terminal C=C bonds over di- and trisubstituted ones in all the solvents tested.

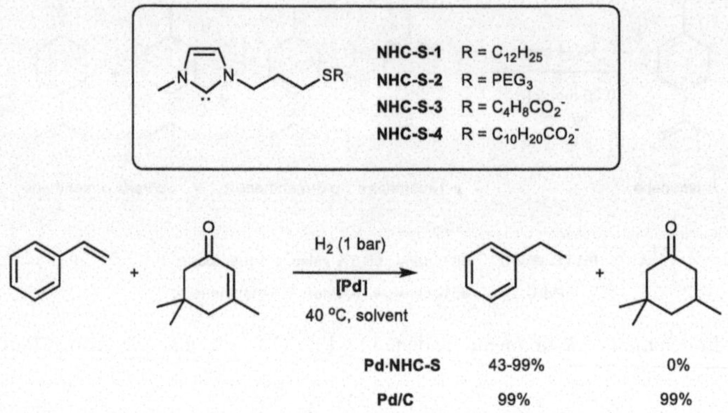

Fig. 13 Hydrogenation of olefins catalyzed by Pd·NHC-S nanoparticles

Water-soluble Pd NPs functionalized with NHC ligands containing carboxylate and sulfonate groups were synthetized from Pd NPs of ~4 nm size stabilized with didodecylsulphide following a ligand exchange procedure (Fig. 14) [40]. The presence of charged groups on the NHC ligands allowed for electrostatic stabilization and solubilization in water of the nanoparticles. By the performance of competition experiments in aqueous media, the hydrogenation of olefins with different degrees of substitution (styrene, 1-decene, 3-methyl-2-cyclohexenone, and citronellol) was achieved with good yields using low pressures (2 bar H_2) and short reaction times (1 h), with TON values of up to 2,500. Under more demanding conditions, i.e., lower catalyst loadings and shorter reaction times, the charged NHC-stabilized Pd NPs exhibited different catalytic behavior. Thus, Pd·NHC-CO_2-2 NPs were found to hydrogenate styrene, 1-decene, 3-methyl-2-cyclohexenone quantitatively, while Pd·I^iPr-$BuSO_3$ and Pd·NHC-CO_2-1 were only efficient for two olefins, styrene and 1-decene, and Pd·IMe-$BuSO_3$ only reduced styrene. The hydrogenation of other olefins was also tested. For example, the Pd·I^iPr-$BuSO_3$ catalyst was employed in the hydrogenation of the tetrasubstituted alkene pulegone, leading to a 1:1 mixture of two diastereoisomers. Since Pd·IMe-$BuSO_3$ was only effective in the reduction of styrene, the chemoselective hydrogenation of the γ,δ-C=C double bond of sorbinaldehyde could be accomplished with these nanoparticles. Moreover, both Pd·NHC-CO_2-2 and Pd·I^iPr-$BuSO_3$ nanoparticles catalyzed the reduction of (+)-carvone with low diastereoselectivity, suggesting that the chiral substituents of the NHC-CO_2-2 ligand exerts a poor stereochemical induction in the reaction. Finally, catalyst separation and reuse experiments were carried out with Pd·NHC-CO_2-1. Following an initial catalytic reaction without noting significant changes in the size distributions of the MNPs, the catalyst was precipitated from the reaction media by lowering the pH of the solution. Next, the Pd NPs were redispersed in water by changing the solution to slightly basic conditions and tested again in the hydrogenation of the substrates. The recycled MNPs were found to remain active in the hydrogenation of styrene, 1-decene and 3-methyl-2-cyclohexenone, although inactive in the reduction of citronellol.

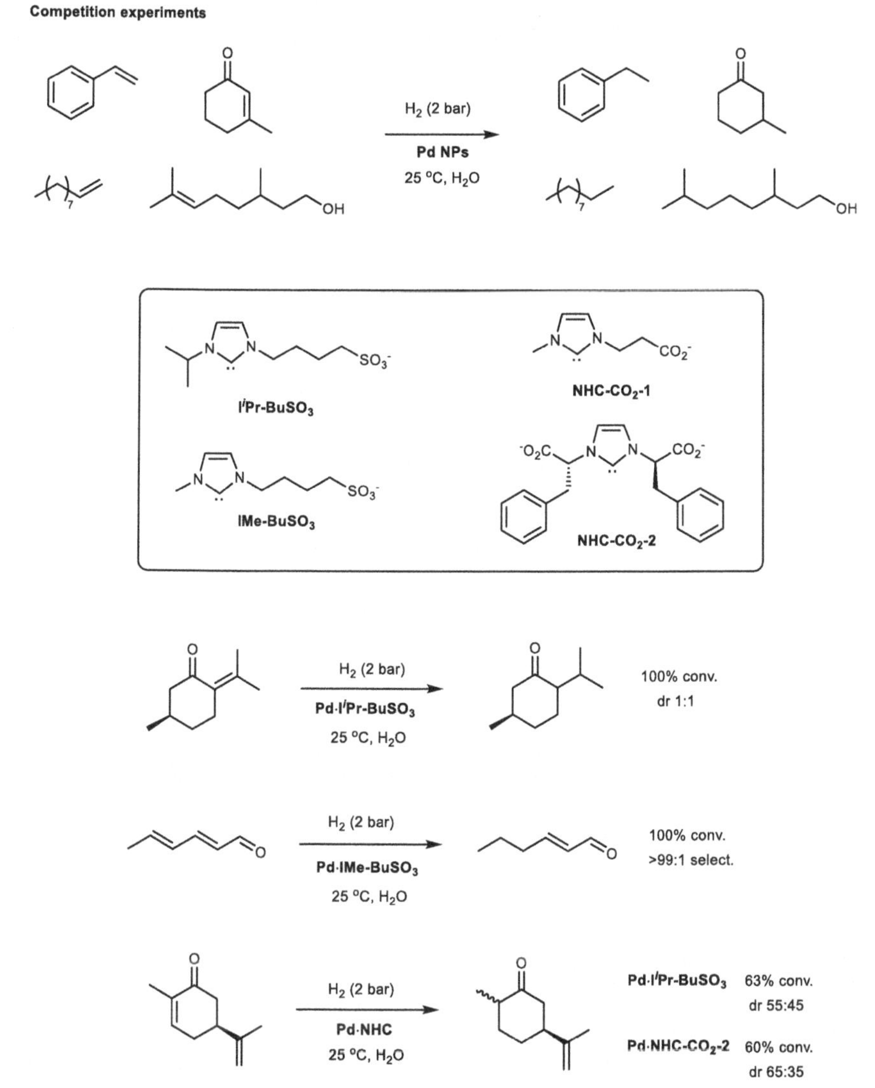

Fig. 14 Examples of hydrogenation of olefins in H_2O mediated by Pd·NHC-SO_3 and Pd·NHC-CO_2 nanoparticles

Chaudret, de Jesús and co-workers examined the synthesis of Pd NPs stabilized with sulfonated N-heterocyclic carbene ligands using three different synthetic approaches involving the decomposition of a dimethyl Pd(II) complex by i) thermal treatment, ii) CO reduction, and iii) reaction with H_2 (Fig. 15) [19]. The nanoparticles were soluble in water, and depending on the employed synthetic methodology and experimental conditions, they showed spherical (thermal decomposition and CO reduction) or worm-like structures (H_2 reduction), with average

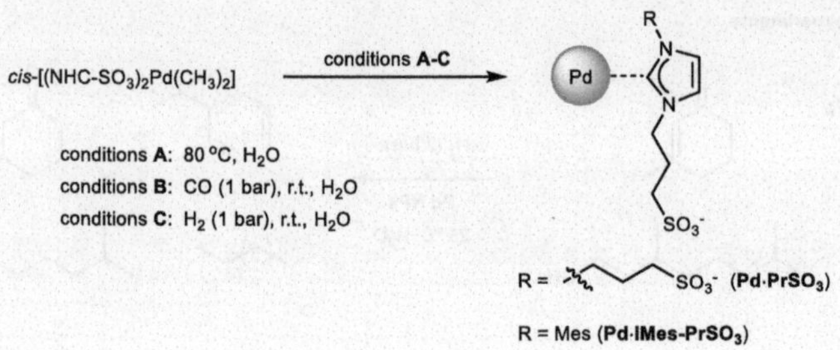

Fig. 15 Synthesis of sulfonated NHC-capped Pd nanoparticles reported by Chaudret, de Jesús and co-workers

Fig. 16 Synthesis of Pt·NHC-SO_3 nanoparticles used in hydrogenation reactions in water

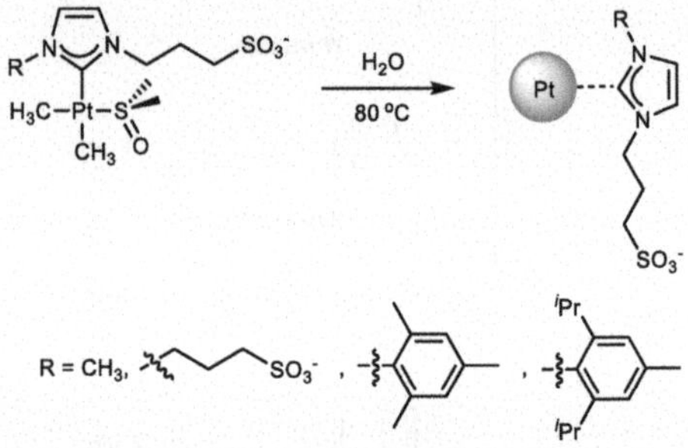

sizes ranging between 1.5 and 7 nm. The Pd·IMes-PrSO_3 nanoparticles prepared by thermal decomposition were active in the hydrogenation of styrene, and could be reused up to ten times with substrate conversions to ethylbenzene higher than 80%. After the tenth reaction, significant nanoparticle coalescence forming worm-like structures similar to those obtained by Pd(II) complex reduction with H_2 was observed.

The same groups also prepared Pt NPs stabilized with similar sulfonated N-heterocyclic carbene ligands by decomposition of the corresponding dimethyl Pt complexes in water (average diameters 1.3–2.0 nm) (Fig. 16) [21, 60]. These nanomaterials were tested in the hydrogenation in water of styrene and benzene derivatives substituted with olefin, acetylene, aldehyde, and nitro groups. Chemoselective reduction of the groups attached to the aromatic ring was observed, with a reactivity trend following the order: unsaturated C–C bond > nitro > carbonyl group. With the exception of the reactions with acetophenone, the nanoparticles could be reused up to 11 times without significant erosion of the catalytic activity and only a slight increase of the MNP size. While relevant differences in activity or selectivity were not observed with the nature of the NHC ligand, biphasic separation and recycling studies pointed out to a higher durability of the catalysts based on NHC ligands N-substituted with aryl rings.

Finally, diverse Ru·NHC NPs prepared by decomposition of [Ru(cod)(cot)] with H_2 in the presence of suitable NHC ligands have also been tested in the hydrogenation of olefins [33, 55, 61, 62]. Examples of these reactions include the reduction of styrene, 3-phenyl-2-cyclohexenone, isophorone, methyl (*Z*)-2-acetamidocinnamate, dimethyl itaconate, and geraniol. A common observation with all these catalytic systems is that the reduction of the C=C bonds takes place at higher reaction rates than that of other reducible functional groups, such as arenes and carbonyls. Moreover, chiral induction has not been achieved in the hydrogenation of pro-chiral olefins using Ru NPs capped with optically pure N-heterocyclic carbenes [33].

4.1.2 Alkyne Semihydrogenation

Selective hydrogenation of alkynes to alkenes (i.e., alkyne semihydrogenation) has received an important attention from both academia and industry contexts [63]. Efficient catalytic systems for this transformation should not only prevent the formation of over-reduced alkane products but also be able to control the *Z*/*E* selectivity in the reduction of disubstituted derivatives.

For instance, non-supported and supported on reduced graphene oxide (rGO) Pd NPs (<4 nm) were prepared through decomposition with H_2 of pyrene-substituted NHC-stabilized Pd complexes (Fig. 17) [23]. In the semihydrogenation of alkynes, the supported Pd NPs exhibited higher reaction rates and similar *Z*-selectivity in comparison to the colloidal catalyst, and was recycled up to 10 times with negligible loss of the catalytic activity and minimal leaching (10 wt% of the initial Pd amount after the tenth recycling experiment). The improved efficiency of the rGO-supported Pd material was attributed to support effects increasing the MNP stability. Analysis of the Pd NPs by XPS and TEM after each catalyst re-use showed an increase in the mean diameters (3.5–6.6 nm) and size distributions of the MNPs, and a decrease of Pd^{2+}/Pd^0 ratios. These results were interpreted considering that the Pd NPs are composed of Pd^{2+} at the metal surface and Pd^0 at the core, and upon size increase of the nanoparticles lower relative amounts of Pd^{2+} at the surface could be expected.

Díaz de los Bernardos, Godard et al. reported the synthesis of Ni nanoparticles supported on multiwall carbon nanotubes (CNTs) through an "organometallic approach" involving the decarboxylation of the zwitterionic IMe-CO_2 adduct (Fig. 18) [28]. The heterogeneous nanosystems Ni·$IMe^{0.2}$/CNTs, Ni·$IMe^{0.5}$/CNTs and Ni·$IMe^{1.0}$/CNTs were tested in the semihydrogenation of diphenylacetylene under relatively mild reaction conditions (3 mol% Ni, 50°C and 5 bar H_2). Full conversions to the alkene derivatives (*Z*/*E* ratio > 94:6) were achieved for Ni·$IMe^{0.2}$/CNTs and Ni·$IMe^{0.5}$/CNTs, while the Ni·$IMe^{1.0}$/CNTs system provided lower TON values likely as a consequence of the presence of a larger amount of the ligand on the surface hindering substrate coordination. The catalyst Ni·$IMe^{0.5}$/CNTs was also examined in the semihydrogenation of other disubstituted alkynes, in all cases exhibiting high conversions to the *Z*-alkene with little to none over-reduced product formation. Finally, recycling and metal leaching control experiments were performed with Ni·$IMe^{0.5}$/CNTs, which showed that the catalyst could be re-used over three runs with only a slight decrease in the catalytic activity.

In a subsequent study carried out by the same group, the synthetic methodology based on the decarboxylation of the IMe-CO_2 adduct for the preparation of Ni·NHC

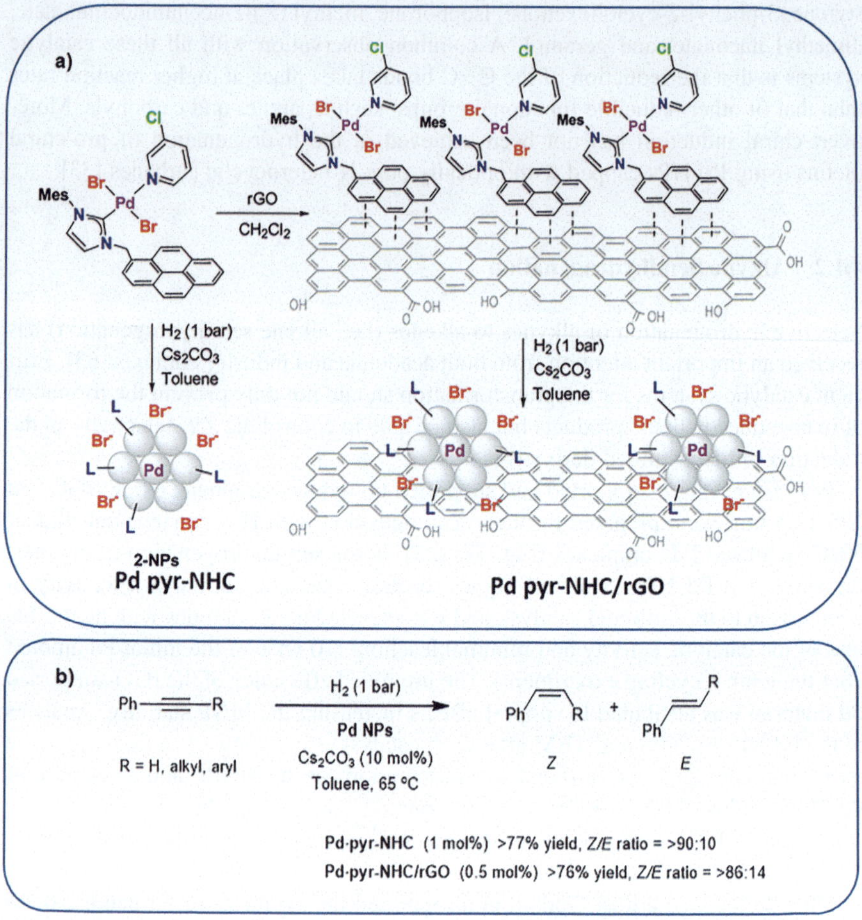

Fig. 17 (**a**) Synthesis of non-supported and supported on rGO Pd·NHC NPs, and (**b**) application of the Pd NPs in the hydrogenation of alkynes. Adapted with permission from [23] Copyright 2018 American Chemical Society

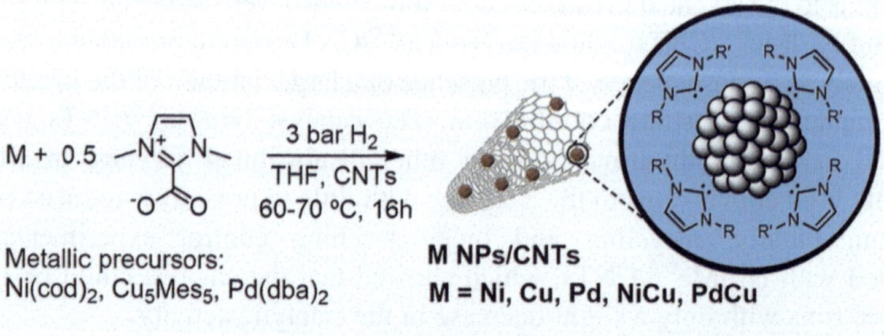

Fig. 18 Synthesis of mono- and bimetallic M·IMe/CNTs nanoparticles. Adapted with permission from [29] Copyright 2019 Wiley VCH Verlag GmbH&Co. KGaA, Weinheim

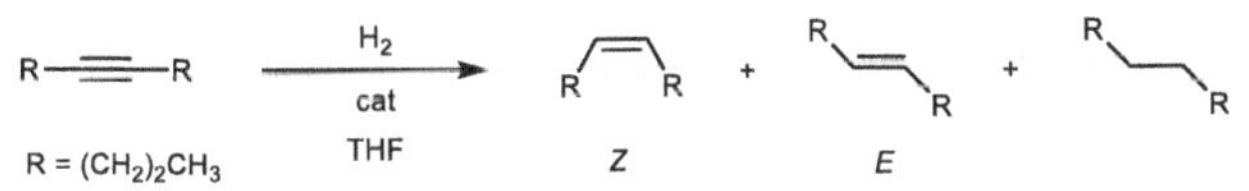

Ni·IMe$^{0.5}$/CNTs (3 mol% Ni, 5 bar H_2, 50 °C, 16 h): >99% conv.; *Z*:*E*:alkane = 99:0:1

Cu·IMe$^{0.5}$/CNTs (3 mol% Cu, 50 bar H_2, 50 °C, 16 h): 0% conv.

Pd·IMe$^{0.5}$/CNTs (0.01 mol% Pd, 3 bar H_2, 30 °C, 3 h): >99% conv.; *Z*:*E*:alkane = 66:31:3

Ni$_{0.5}$Cu$_{0.5}$·IMe$^{0.5}$/CNTs (3 mol% Ni, 50 bar H_2, 50 °C, 16 h): 93% conv.; *Z*:*E*:alkane = 93:0:7

Pd$_{0.5}$Cu$_{0.5}$·IMe$^{0.5}$/CNTs (0.05 mol% Pd, 3 bar H_2, 30 °C, 0.5 h): >99% conv.; *Z*:*E*:alkane = 95:4:1

Fig. 19 Hydrogenation of 4-octyne catalyzed by mono- and bimetallic M·IMe$^{0.5}$/CNTs nanoparticles

NPs supported on CNTs was extended to the synthesis of other carbon-supported mono- (Cu and Pd) and bimetallic (NiCu and PdCu) catalysts (Fig. 19) [29]. Comparison of the activity and selectivity of the M·IMe$^{0.5}$/CNTs (M = Ni, Cu and Pd) nanoparticles in the hydrogenation of 4-octyne showed that, while the CNT-supported Ni NPs provided excellent conversions and alkene selectivity, the Cu·IMe$^{0.5}$/CNTs catalyst was inactive even under elevated H_2 pressures (50 bar H_2) (Fig. 19). This observation was attributed to the low activity of copper surfaces in the activation of hydrogen. Meanwhile, the Pd·IMe$^{0.5}$/CNTs catalyst showed a high activity and moderate selectivity under very mild reaction conditions (0.1 mol% Pd, 3 bar H_2, 30°C), although also catalyzing the isomerization of 4-octyne. On the other hand, the bimetallic Ni$_{0.5}$Cu$_{0.5}$·IMe$^{0.5}$/CNTs and Pd$_{0.5}$Cu$_{0.5}$·IMe$^{0.5}$/CNTs nanoparticles were found to be less active than their monometallic Ni and Pd counterparts, although a promoting effect of the presence of copper on the selectivity was noted.

Based on the promising catalytic performance of Pd$_{0.5}$Cu$_{0.5}$·IMe$^{0.5}$/CNTs, the effect of the Pd:Cu composition was further examined using Pd$_{0.25}$Cu$_{0.75}$·IMe$^{0.5}$/CNTs and Pd$_{0.75}$Cu$_{0.25}$·IMe$^{0.5}$/CNTs. In the hydrogenation of 1- and 4-octyne, the catalyst with a higher Pd content (>50% Pd) led to full substrate conversions, although significant levels of over-reduction and isomerization products were observed. Meanwhile, an excellent olefin *Z*-selectivity was achieved with Pd$_{0.25}$Cu$_{0.75}$·IMe$^{0.5}$/CNTs, although at the expenses of a moderate activity. The reduction of disubstituted alkynes containing aliphatic, aromatic, hydroxy, and carbonyl groups using Pd$_{0.5}$Cu$_{0.5}$·IMe$^{0.5}$/CNTs exhibited a substrate-dependent activity and selectivity. Finally, the bimetallic Pd$_{0.5}$Cu$_{0.5}$·IMe$^{0.5}$/CNTs catalyst was examined in the gas phase semihydrogenation of acetylene in ethylene-rich model gas feeds using a micro fixed-bed reactor at 70°C and a total pressure of 1 bar. Reaction monitoring for 40 h showed a high stable reaction rate with moderate to good selectivity towards ethylene that decreases slightly over time.

A particularly interesting catalytic system for alkyne semihydrogenation based on copper was reported by Fedorov, Copéret and co-workers [22]. This material was prepared by deposition of the IMes carbene ligand on copper nanoparticles (2.0 nm) supported on passivated silica prepared through H_2 reduction of grafted [$Cu_4(HMDS)_4$] (HMDS = hexamethyldisilazido) complexes (Fig. 20). The IMes-

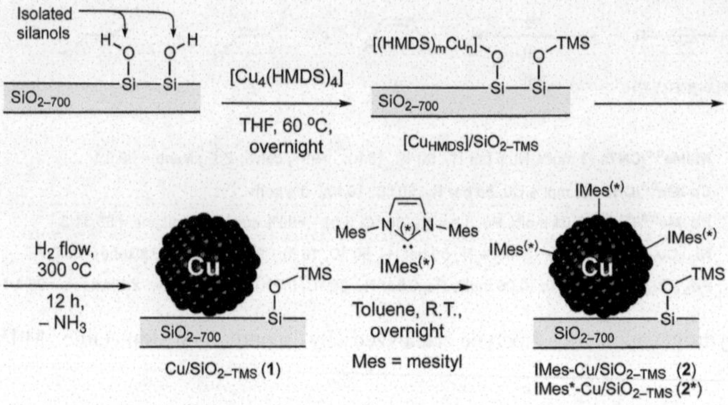

Fig. 20 Synthesis of IMes-modified Cu NPs supported on passivated silica. Reproduced with permission from [22] Copyright 2018 Royal Society of Chemistry

containing copper nanocatalyst was found to be active in the catalytic partial hydrogenation of disubstituted alkynes (20 bar H_2, 60°C, 1.8 mol% Cu), leading to the corresponding *Z*-olefins with high levels of chemo- and stereoselectivity. In striking contrast, NHC-unfunctionalized Cu NPs were found to fully hydrogenate the alkynes to alkanes. The high chemoselectivity towards alkene formation of the NHC-modified nanoparticles was attributed to the selective binding of the IMes ligand to Cu surface sites that were catalytically unselective, as revealed by solid-state NMR spectroscopy [64].

4.1.3 Hydrogenation of (Hetero)arenes

The hydrogenation of aromatic substrates has received considerable attention from both research laboratories and industry [65]. In this context, (hetero)arene hydrogenation reactions catalyzed by MNPs, particularly colloidal NHC-stabilized Ru NPs, have been frequently studied. The first use of NHC-stabilized nanoparticles in the hydrogenation of arenes was described by van Leeuwen et al., and consisted on the application of previously reported Ru·I^{t}Bu$^{0.5}$, Ru·IPr$^{0.2}$ and Ru·IPr$^{0.5}$ NPs in the reduction of various arenes [66]. Striking ligand effects on the catalytic activity of the Ru NPs were observed, with nanoparticles based on the IPr carbene generally leading to higher rates than those containing the I^{t}Bu ligand. For example, catalytic activity in the hydrogenation of *o*-methylanisol was found to follow the trend: Ru·I^{t}Bu$^{0.5}$ < < Ru·IPr$^{0.5}$ < Ru·IPr$^{0.2}$, leading to complete and partial arene hydrogenation products. In the case of colloids based on the IPr ligand with low ligand/metal ratios, as shown by CO coordination studies, the NHC ligand binds preferentially at the corners, edges, and defects of the nanoparticles, leaving faces available for arene coordination. However, at high ligand/metal ratios, IPr may also be coordinated at the faces of the particles. The reduced reactivity of the Ru·I^{t}Bu$^{0.5}$ was explained as a result of several factors, such as the ligand higher surface coverage that hinders substrate coordination or the more basic properties of I^{t}Bu in

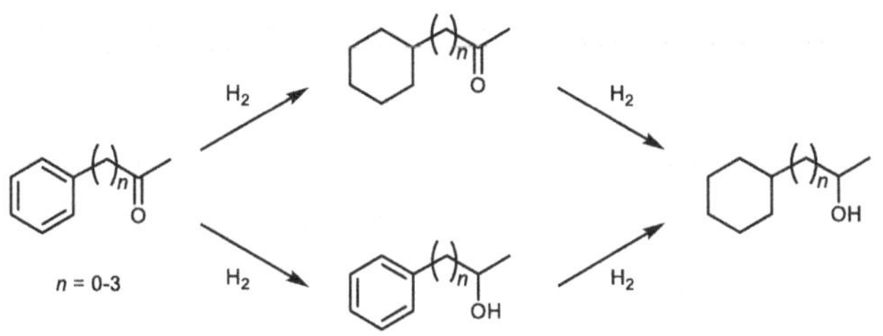

Fig. 21 Hydrogenation of aryl ketones by colloidal Ru·NHC nanoparticles

comparison to IPr that could lead to more electron-rich surfaces that will disfavor arene coordination.

The hydrogenation of benzene and arenes substituted with electron donating and withdrawing groups was also tested with these nanomaterials. An analogous catalyst reactivity trend was observed, with arenes containing electron donating groups being reduced much faster than those with electron withdrawing substituents. However, some substrates such as bromobenzene and benzonitrile could not be reduced, while only reduction of the nitro group to yield aniline was achieved in the hydrogenation of nitrobenzene.

Acetophenone, an arene functionalized with a reducible carbonyl group, has been used as a benchmark substrate in order to compare the activity and selectivity of catalysts. Hydrogenation of acetophenone can lead to the formation of three possible reaction products as result of the reduction of the aromatic ring, the carbonyl group or both (Fig. 21). Ru·IPr$^{0.2}$ nanoparticles catalyzed the hydrogenation of acetophenone leading to product distributions that depends on the solvent, substrate/catalyst ratio and hydrogen pressure, although reduction of the arene moiety was found to occur faster than the hydrogenation of the C=O group. Moreover, in the hydrogenation of arenes tethered to carbonyl groups through an alkyl chain of variable length, the selectivity toward the reduction of the hydrocarbon ring increases as it does the number of CH_2 units of the linker. These results show a high preference of the nanoparticles to hydrogenate the arene groups in aryl ketones probably due to a more favorable arene coordination over ketone coordination to the metal surfaces. In the hydrogenation of acetophenone, coordination of the aromatic moiety facilitates the reduction of the nearby carbonyl group yielding products with both hydrogenated fragments.

Long-chain NHC-stabilized Ru NPs based on LC-IPr and LC-IMe carbenes were prepared by Philippot, Chaudret, Glorius et al. (LC-IMe and LC-IPr of Fig. 2) [55]. It was expected that, as shown with Pd·LC-NHC nanoparticles, [38] the presence of long aliphatic chains at the backbone of the ligand would form a lipophilic monolayer around the metallic surface leading to micelle-type structures and increasing the nanoparticles stability. A significant influence of the N-substituents of the ligands on the catalytic activity of Ru·LC-IPr$^{0.1}$, Ru·LC-IPr$^{0.3}$, Ru·LC-IMe$^{0.1}$ and Ru·LC-IMe$^{0.3}$ NPs in the hydrogenation of arenes was evidenced. Extensive nanoparticle characterization by IR and NMR spectroscopies supported that the bulkier

LC-IPr ligands were mainly located in apexes and edges, whereas smaller LC-IMe molecules could also be found at the faces of the nanoparticles, thus potentially limiting substrate coordination. Ru·LC-IMe$^{0.3}$ nanoparticles were ineffective in the hydrogenation of acetophenone, whereas complete reduction of the substrate, i.e., arene and carbonyl moieties, was observed in the reactions with Ru·LC-IPr$^{0.3}$. In line with these observations, using the Ru·LC-IMe$^{0.1}$ NPs having a lower ligand content than Ru·LC-IMe$^{0.3}$ led to a moderate catalytic activity (19% versus 0%). This fact was attributed to a higher accessibility of the substrate to the metallic surface when employing less amount of stabilizing ligand. In marked contrast to this observation, the use of Ru·LC-IPr$^{0.1}$ led to a lower conversion in comparison to those particles stabilized with a higher amount of IPr ligand in the hydrogenation of acetophenone (89% versus 100%), probably due to the larger size and polydispersity of the latter catalyst in comparison to those of Ru·LC-IPr$^{0.3}$. Similar catalyst reactivity trends were found in the hydrogenation of other substrates, including an acetamide, anisol and an α,β-unsaturated ketone. Finally, the Ru·LC-IPr and Ru·LC-IMe NPs were also applied to the oxidation and tandem hydrogenation-oxidation processes (see Sect. 4.6).

Chaudret, Glorius et al. reported the synthesis of Ru NPs stabilized with rigid lipophilic NHC ligands derived from cholesterol (Fig. 22) [61]. The spherical Ru·chol1 and Ru·chol2 nanoparticles (mean diameters: 1.4–1.5 nm), having experimental Ru/ligand ratios of 4.8:1 and 4.0:1, respectively, were employed in the hydrogenation of various aromatic substrates (naphthalene, biphenyl, anisol, phenol and acetophenone). The Ru·chol1 NPs provided complete conversions to the fully hydrogenated derivatives, while a lower catalytic activity was observed in the case of Ru·chol2. Differences in the activity and selectivity between both catalysts were proposed to arise from the ligand structures and, particularly, from an influence of the N-substituents of the ligands. In comparison to the smaller substituents of chol2, the higher steric demand of the bulkier N-substituents of chol1 should limit its coordination to the nanoparticle faces, thus facilitating arene coordination and subsequent hydrogenation.

Chiral NHCs have been employed as ligands in a plethora of metal-catalyzed enantioselective reactions [67]. However, use of chiral carbene ligands in nanoparticles stabilization has only been briefly explored. For example, the synthesis of Ru NPs stabilized with the optically pure imidazolidinylidene SIDPhNp and SIPhOH ligands was reported by Chaudret, Glorius and co-workers (Fig. 2) [33]. In addition to the coordination of the ligands through the carbene carbon, the SIDPhNp ligand might additionally interact with the metal surfaces by the aromatic rings through π-interactions; meanwhile, it was expected that the hydroxyl groups of the SIPhOH ligand could lead to favorable ligand–substrate interactions. IR and MAS NMR spectroscopies of these Ru·NHC NPs exposed to carbon monoxide showed significant CO coordination in the terminal mode and limited bridging coordination, suggesting a preferential coordination of both NHC ligands to the nanoparticles faces. Both catalysts were tested in the hydrogenation of substrates containing arene moieties, C=C bonds and carbonyl groups, and their performance was compared with that of Ru·ICy nanoparticles. While no enantiomeric induction

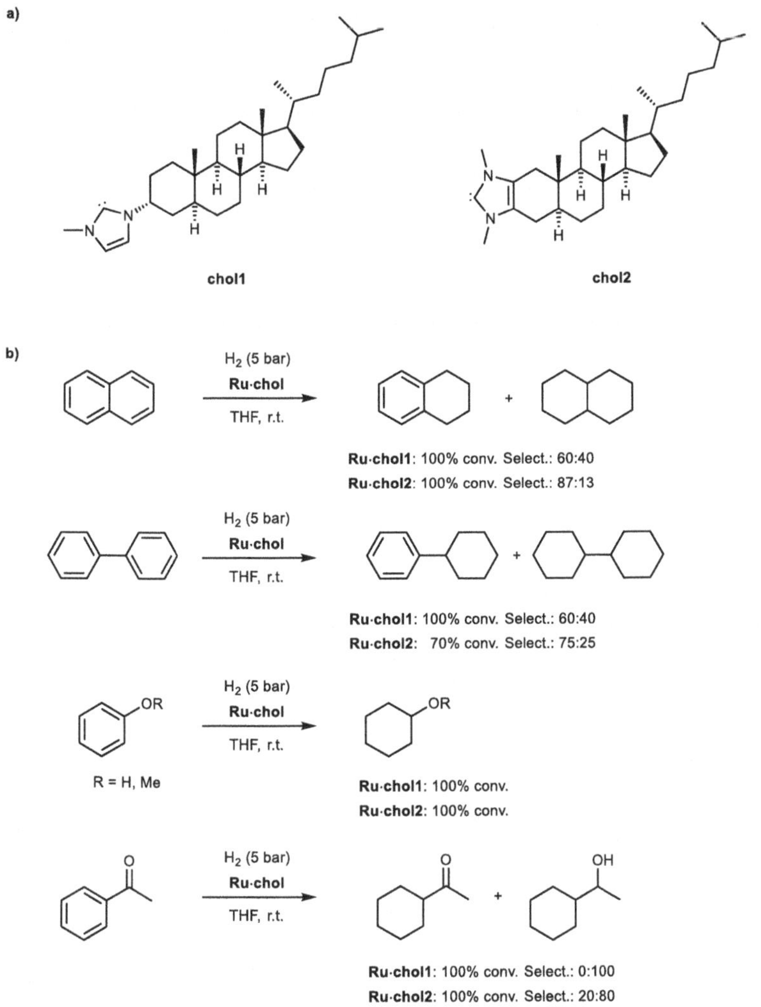

Fig. 22 (**a**) Cholesterol-derived NHC ligands, and (**b**) hydrogenation of arenes catalyzed by Ru·chol nanoparticles

was observed in the hydrogenation of pro-chiral substrates, the ligand structure influenced the catalytic activity, being Ru·ICy$^{0.5}$ (1.2 nm) more active than Ru·SIDPhNp$^{0.5}$ (1.3 nm), and the Ru·SIPhOH$^{0.5}$ (1.5 nm) nanoparticles the less active. Thus, hydrogenation of acetophenone led to complete hydrogenation of both arene and carbonyl functions with Ru·ICy$^{0.5}$ and Ru·SIDPhNp$^{0.5}$, and a mixture of the different reduction products with Ru·SIPhOH$^{0.5}$ (3 days, 10 bar H_2, 25°C). The same catalytic activity trend was observed in the hydrogenation of anisol and the arene moiety of quinoxaline. The higher activity observed with Ru·ICy$^{0.5}$ was mainly attributed to the presence of more free active sites due to the lower bulkiness of the ligand N-substituents and the presence of alkyl groups that do not interact with the metallic surface through π-interactions, as in the case of SIDPhNp, or OH coordination, as for the SIPhOH ligand.

Similarly, Lara, Martínez-Prieto et al. studied the catalytic performance of chiral Ru·I^{t}BuEt$^{0.5}$ nanoparticles (Fig. 2) [56]. These catalysts were able to perform the hydrogenation of a series of (hetero)arenes such as 2-acetonaphtone, naphtol, quinoxalines, and 2-methylbenzofuran with high chemoselectivity although negligible enantiomeric induction.

Rhodium nanoparticles stabilized with the IPr carbene ligand prepared using different IPr/Rh ratios (Rh·IPr$^{0.2}$, Rh·IPr$^{0.4}$ and Rh·IPr$^{0.6}$) were examined in the reduction with H_2 of various aromatic substrates by Godard and co-workers (Fig. 23) [31]. Nanoparticles prepared using a ligand/Rh ratio of 0.2 were the most active catalytic system in the hydrogenation of phenol, leading to partially hydrogenated cyclohexanone with high selectivity or to completely reduced cyclohexanol depending on the reaction conditions. However, this colloid showed much lower catalytic activities in the hydrogenation of substituted phenols. The Rh·IPr$^{0.2}$ NPs were also found to catalyze the reduction of benzene and pyridine, used as a N-heteroarene model, under relatively mild conditions (30°C, 20 bar of H_2). Based on the latter result, the effect of substituents in *ortho* and *para* positions of the pyridine ring was examined. 2-Methylpyridine was reduced at a similar rate than pyridine, whereas hydrogenation of 2,6-dimethyl, 4-methyl and 4-trifluoromethyl pyridines took place at significantly lower rates. By employing the same Rh NPs, the selectivity in the hydrogenation of ketoarenes and pyridine derivatives containing keto groups was also examined. Under high H_2 pressures, the fully hydrogenated (arene and keto positions) derivatives were the major products, whereas mixtures of species resulting from partial hydrogenation were observed at lower pressures. Finally, by adjusting the reaction temperature, quinoline partial reduction to 1,2,3,4-tetrahydroquinoline or full hydrogenation to decahydroquinoline was observed.

In 2016, Muratsugu, Glorius et al. reported the modification using NHC ligands of Ru NPs supported on potassium-doped alumina (Ru/K-Al_2O_3) employing different Ru/NHC ratios (Fig. 24) [45]. Detailed characterization of the NHC-modified materials by ^{13}C CP-MAS NMR spectroscopy and XPS evidenced the coordination of the NHC ligands to the metal surface, while TEM and EXAFS analysis indicated that no significant changes of the nanoparticles initial size (1.5 nm) in the presence of increasing amounts of NHCs took place. In the hydrogenation of a series of arenes, such as acetophenone, *trans*-stilbene and phenyl acetylene, use of higher amounts of ligand modifier was found to inhibit the substrate conversion, being this effect more pronounced for the bulkier IMes ligand in comparison to ICy. These observations were accounted on the basis of the expected shielding of active sites upon using bulkier N-substituents at the NHCs and higher amounts of ligand. Interestingly, by adjusting the NHC/Ru ratio, selective hydrogenation of the C-C triple bond of phenyl acetylene could be performed, since the use of a larger amount of NHC for the stabilization of the particles only allows the coordination of the substrate through the smaller alkyne moiety.

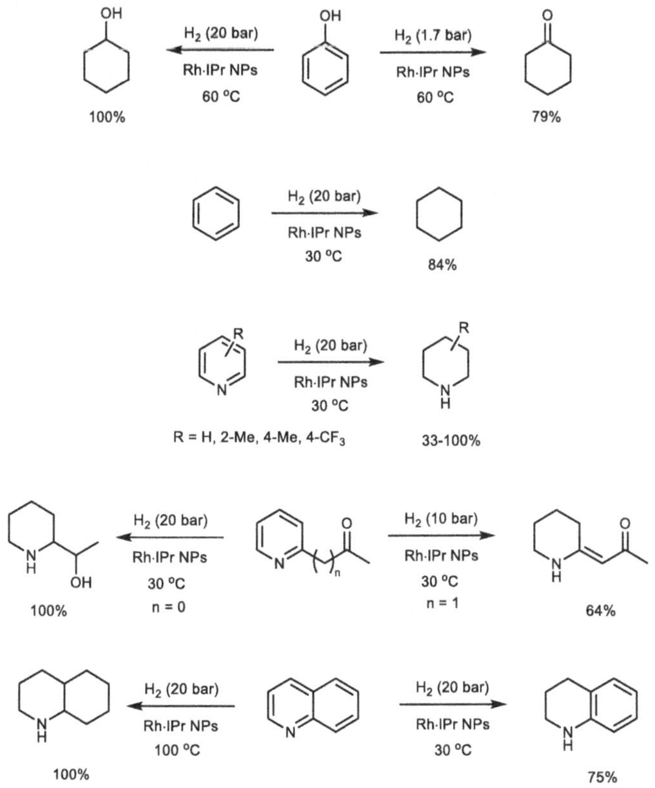

Fig. 23 Hydrogenation reactions catalyzed by colloidal Rh·IPr$^{0.2}$ nanoparticles

Martínez-Prieto et al. synthetized Ru NPs (1.5 nm) supported on reduced graphene oxide (Ru/rGO) that were subsequently modified by the coordination of different amounts of a pyrene substituted NHC ligand, pyr-IMes (Fig. 25) [62]. Extensive spectroscopic characterization of the Ru·pyr-IMes^{n}/rGO (n = 0.2, 0.5, 0.8, 1) nanomaterials showed that the pyr-IMes ligands are found in the coordination sphere of the Ru NPs interacting with the rGO support through non-covalent π-interactions or as protonated carbenes coordinated to the Ru surface. The Ru·pyr-IMes/rGO catalysts were initially evaluated in the hydrogenation of acetophenone after preactivation of the materials with H_2. As expected, the activity and selectivity of the Ru·pyr-IMes/rGO nanoparticles were influenced by the Ru/pyr-IMes ratio, as this is an indication of the ligand surface coverage degree. In addition to the hydrogenated products resulting from the reduction of the carbonyl and arene fragments, formation of ethylbenzene and ethylcyclohexene, produced from the hydrodeoxygenation of the hydroxyl group, was observed. Catalysts having higher amounts of surface NHC ligands gave rise to improved selectivity towards 1-phenylethanol, although at expense of catalytic activity. Moreover, the Ru·pyr-IMes$^{0.5}$/rGO catalyst was also tested in the hydrogenation of different (hetero)arenes

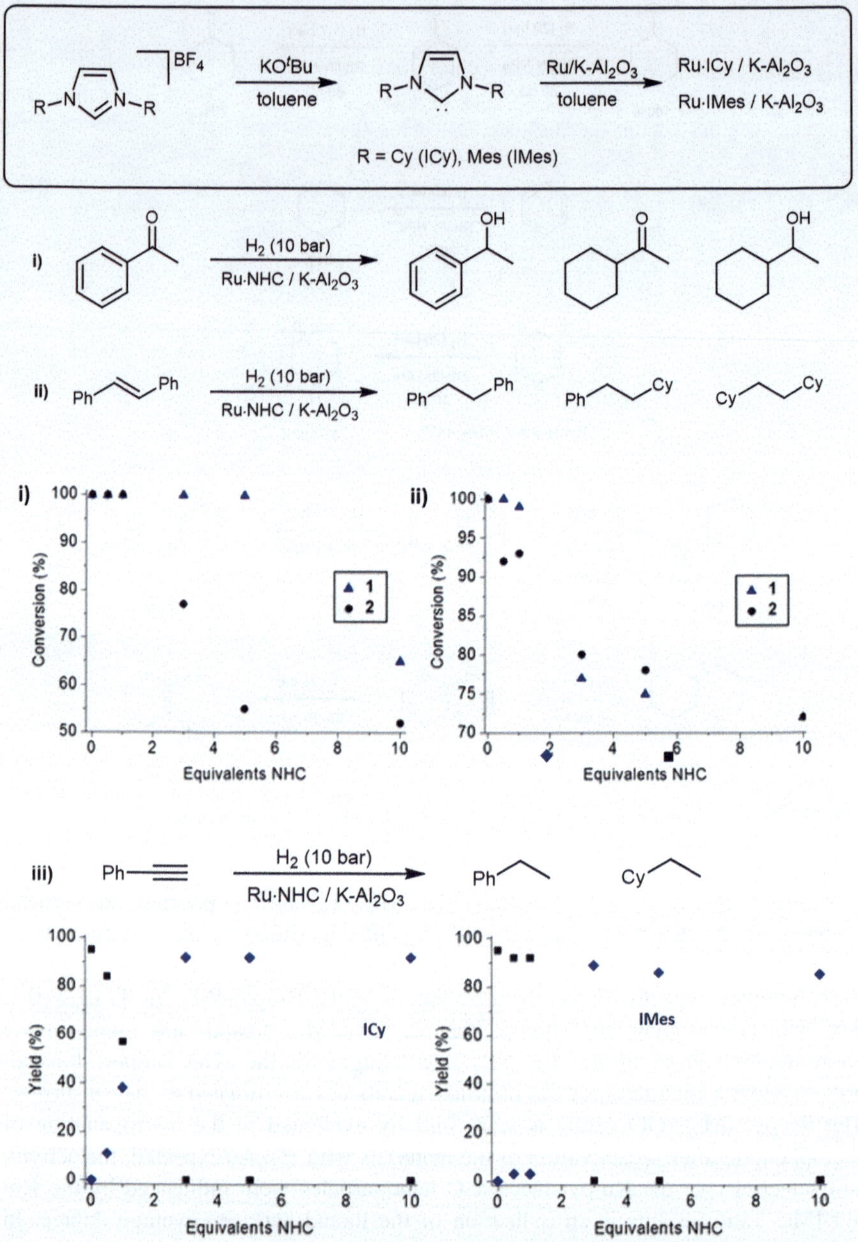

Fig. 24 Synthesis and catalytic applications of NHC-modified Ru/K-Al_2O_3 nanoparticles (1 = ICy; 2 = IMes). Adapted with permission from [45] Copyright 2016 American Chemical Society

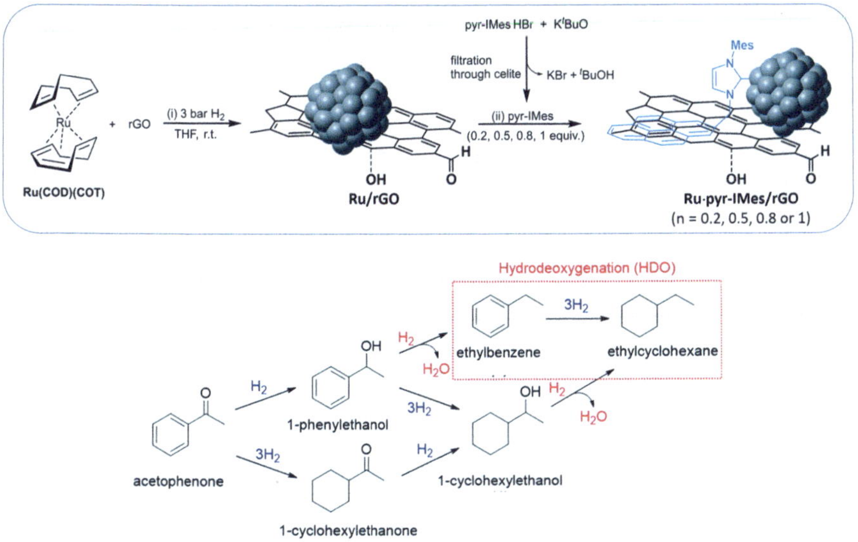

Fig. 25 Synthesis of NHC-modified Ru·pyr-IMes/rGO nanoparticles and hydrogenation of acetophenone. Adapted with permission from [62] Copyright 2022 Royal Society of Chemistry

containing other reducible functions, i.e., nitrobenzene, styrene, quinoline, 1-methylindole, and 5-hydroxymethylfurfural (Fig. 26). Most substrates were fully hydrogenated to their completely reduced products with high selectivity, with the exception of styrene that provided ethylcyclohexane in 43% yield and 5-hydroxymethylfurfural that was preferentially reduced at the carbonyl group. Finally, the recyclability of Ru·pyr-IMes$^{0.5}$/rGO, which showed a good trade-off between activity and selectivity, was investigated by performing seven successive additions of acetophenone.

Hydrogenation of lignocellulosic biomass-derived furfural provides a renewable access to important value-added products [68]. In this context, Axet et al. examined the catalytic behavior of polyvinylpyrrolidone stabilized Ru NPs (Ru/PVP) modified with organic ligands, such as amines, NHCs and phosphines, in the hydrogenation of furfural to 1,2-pentanediol [50]. Reduction of furfural with H_2 can lead to different product distributions resulting from hydrogenation and C-O hydrogenolysis processes (Fig. 27). Previous results from the same group showed that the parent Ru/PVP NPs catalyze the hydrogenation of furfuryl alcohol to tetrahydrofurfuryl alcohol and 1,2-pentanediol in 73% and 27% yield, respectively, in 1-propanol as solvent (125°C, 20 bar H_2). However, hydrogenation of furfural with the same MNPs produced tetrahydrofurfuryl alcohol and 1,2-pentanediol in up to 42% and 24%, along with the formation of the acetal hydrogenation product, with an initial TOF of 55 h^{-1}. In an attempt to increase the electron density on the metal surface, and thus to improve the selectivity towards the formation of 1,2-pentanediol and prevent the acetalization process resulting from the reaction of the alcoholic solvent

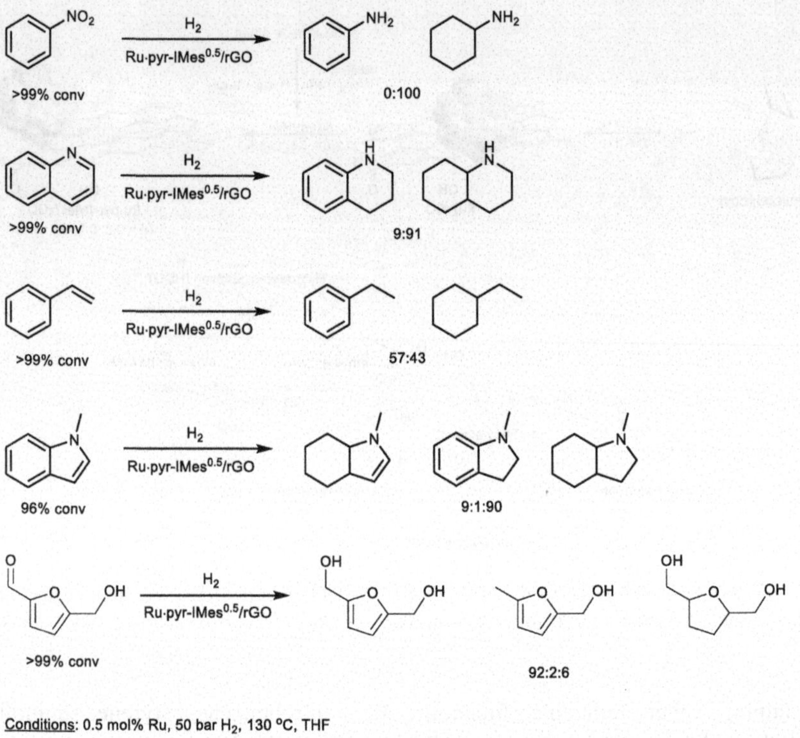

Fig. 26 Examples of hydrogenation of (hetero)arenes catalyzed by the Ru·pyr-IMes$^{0.5}$/rGO nanoparticles

with the starting aldehyde, MNP modification with electron-donating ligands was examined. Addition of amine ligands (hexadecylamine (HDA), 2,2,6,6-tetramethylpiperidine (TMP), and tert-butylamine (TBA)) increased the catalytic activity of the MNPs (TOF = 66–99 h^{-1}), and suppressed significantly the formation of acetal, providing tetrahydrofurfuryl alcohol and 1,2-pentanediol with up to 64% and 36%, respectively, in the case of the less bulky HDA amine. In contrast, the use of a NHC ligand (IPr) produced furfuryl alcohol with high selectivity (>80%) with an initial TOF of 51 h^{-1}.

4.1.4 Hydrogenation of Nitroarenes

Reduction of nitroarenes to anilines is an important process in the production of dyes, agrochemicals, and pharmaceuticals [69, 70]. Using the organometallic approach, Lara et al. reported the synthesis of Pt NPs stabilized with diverse NHC ligands (IPr and I^{i}Pr$_2$Me$_2$ of Fig. 2; ligand/metal ratios = 0.2 and 0.5) [27]. The nanoparticles were found to be active in the hydrogenation of nitroarenes under

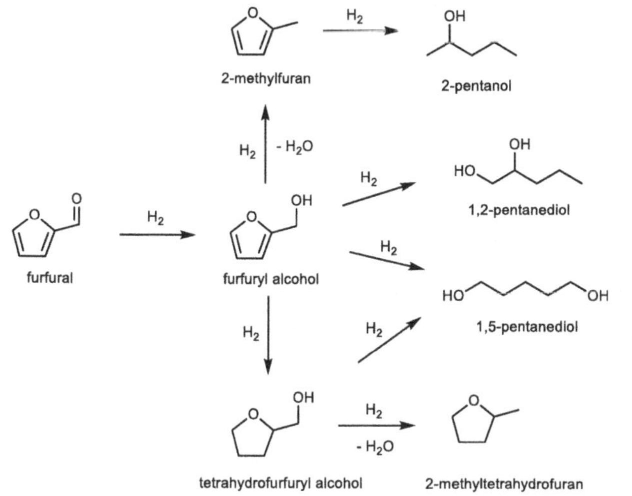

Fig. 27 Potential products formation in the hydrogenation of furfural

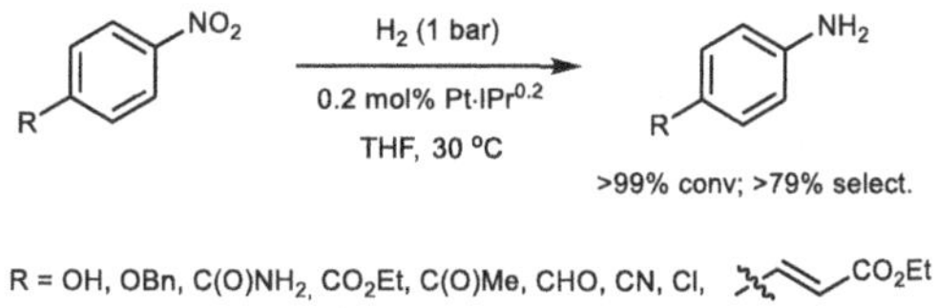

Fig. 28 Hydrogenation of nitroarenes to anilines catalyzed by Pt·IPr$^{0.2}$ nanoparticles

relatively mild conditions in different solvents with TOF values of up to 4.9×10^3 h^{-1} (1 bar of H_2, 30°C) (Fig. 28). More interestingly, the most active catalyst, Pt·IPr$^{0.2}$, was examined in the reduction of nitroarenes incorporating a range of functional groups, such as hydroxyl, benzyloxy, carbonyl, olefin, and halogens, providing the corresponding anilines with selectivities higher than 79%.

Axet and co-workers investigated the influence of the stabilizing ligands and supports in the hydrogenation of nitrobenzene catalyzed by Ru NPs (Fig. 29) [71]. To this end, Ru NPs with comparable sizes (0.8–1.6 nm) and structures stabilized by fullerene C_{60} (Ru·C_{60}), poly(vinylpyrrolidone) (Ru·PVP20 and Ru·PVP4), hexadecylamine (Ru·HAD), and 1,3-bis(2,6-diisopropylphenyl) imidazol-2-ylidene (Ru·IPr) were chosen on the basis of their different electronic properties. Thus, while HDA and IPr serve as donor ligands, fullerene C_{60} attracts electron density from the Ru surfaces and polymeric PVP only interacts weakly with the nanoparticles. The electron rich or deficient character of the surface metals was confirmed by analyzing the CO absorption band in the IR spectrum of the nanoparticles exposed to CO and by DFT calculations. The examination of the stepwise hydrogenation of nitrobenzene to aniline and cyclohexylamine in EtOH (30 bar H_2, 80°C) showed a catalytic activity trend as follows: Ru·C_{60} (136.9 h^{-1}) > Ru·HAD (129.2 h^{-1}) > Ru·PVP20 (82.8 h^{-1}) > Ru·PVP4 (71.4 h^{-1}) > Ru·IPr (64.8 h^{-1}), with the more basic IPr ligand being the less active and selective catalyst.

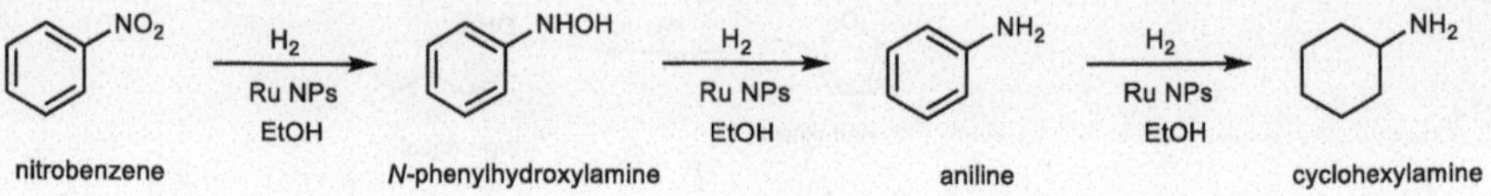

Fig. 29 Sequential hydrogenation of nitrobenzene using Ru·C_{60}, Ru·PVP20, Ru·PVP4, Ru·HAD, and Ru·IPr nanoparticles

Reactions with the latter catalyst also produced significant accumulation of the *N*-phenylhydroxylamine intermediate, suggesting that strong donor ligands disfavor its hydrogenation, which is thought to be the rate-determining step in the hydrogenation of nitrobenzene to aniline. In the hydrogenation of aniline to cyclohexylamine, a ligand effect was also evidenced, with Ru·IPr being one of the less active catalysts, although exhibiting the highest selectivity towards the formation of cyclohexylamine.

Since Corma and Serna reported seminal work on the use of gold catalysts for the hydrogenation of nitroarenes, [72] Au NPs have been widely explored in the reduction of nitroaromatics. For example, by employing a ligand exchange procedure, Glorius, Ravoo et al. prepared water-soluble NHC-capped monometallic (Pd and Au) and bimetallic (AuPd) nanoparticles (mean diameter ca. 4.2 nm) from oleylamine-stabilized MNPs supported on titania. The alloyed nanoparticles, which were stable over a period time of years, were applied as catalysts in the hydrogenation of nitroaromatics (Fig. 30) [41]. While nitroarenes substituted at ortho positions were not tolerated, full conversion of para- and meta-substituted substrates with electron-donating and electron-withdrawing groups was achieved using relatively low catalyst loadings under mild conditions (5 bar H_2, room temperature).

4.2 *Reduction Reactions Using Reductants Other than H_2*

While less convenient in terms of atom economy, the use of reducing agents other than H_2 eliminates the need of using gases and pressurization equipment. In addition to the use of NHC-modified MNPs in hydrogenation reactions, various nanosystems have been examined in the reduction of organic molecules using different reductants. For example, the aforementioned water-soluble, N-heterocyclic carbene-stabilized Au, Pd, and AuPd nanoparticles supported on Ti_2O developed by Glorius, Ravoo et al. were tested in the transfer hydrogenation of 1,2-diphenylacetylene using formic acid (see Sect. 4.1.4; Fig. 31) [41]. In this transformation, monometallic Au·I^{i}Pr-$BuSO_3/TiO_2$ nanoparticles were ineffective, whereas the analogous Pd·I^{i}Pr-$BuSO_3$/ TiO_2 and $Au_{90}Pd_{10}$·I^{i}Pr-$BuSO_3/TiO_2$ nanomaterials catalyzed the hydrogenation of the alkyne to mixtures of (*E*)- and (*Z*)-stilbene. Interestingly, in the case of the bimetallic $Au_{90}Pd_{10}$·I^{i}Pr-$BuSO_3/TiO_2$ catalyst, the product *E*/*Z* selectivity was found to depend on the reaction solvent. Thus, when the reaction was performed in polar protic solvents such as isopropanol a reverse selectivity (*Z*/*E* ratio = 12:88 vs. 90:10

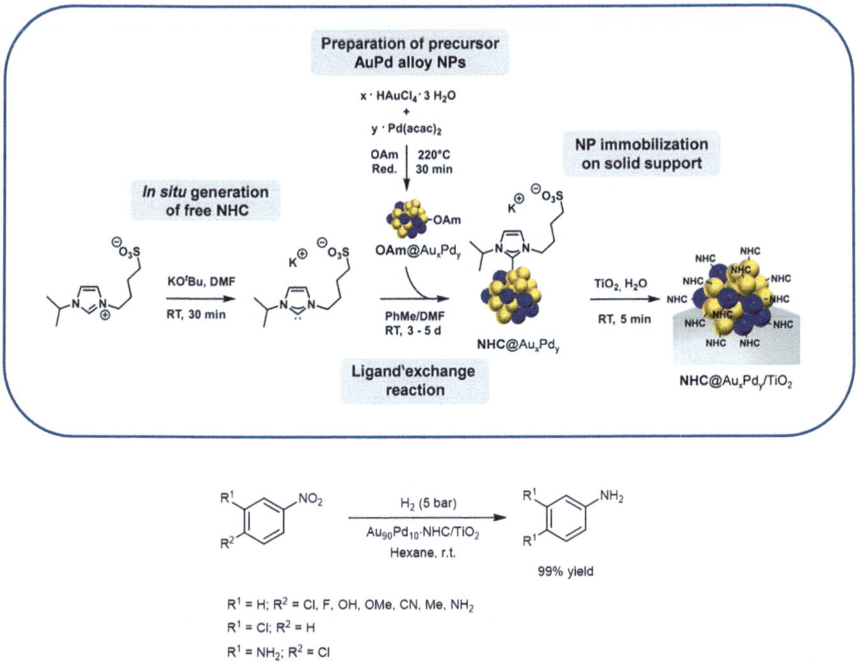

Fig. 30 Synthesis of water-soluble, NHC-stabilized Pd, Au and AuPd nanoparticles, and hydrogenation of nitroarenes catalyzed by Ti_2O-supported NHC-modified PdAu nanoparticles. Adapted with permission from [41] Copyright 2018 Wiley-VCH Verlag GmbH&Co. KGaA, Weinheim

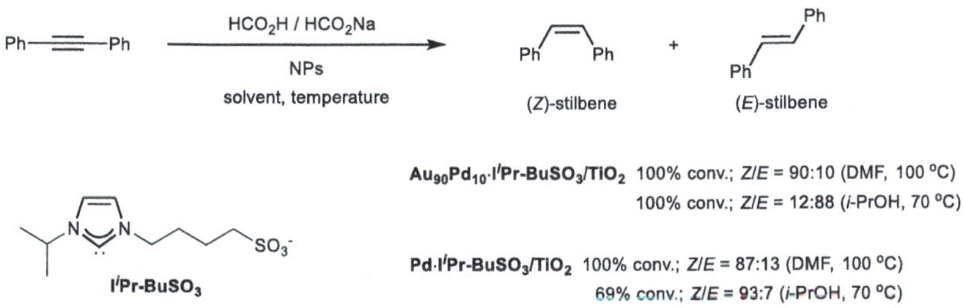

Fig. 31 Transfer hydrogenation of diphenylacetylene with formic acid catalyzed by $Au_{90}Pd_{10}$·NHC/TiO_2 nanoparticles

in DMF) was observed, which was attributed to nanoparticles catalyzed *Z* to *E* isomerization of stilbene.

Samuelson et al. detected the formation of NHC-capped Ru NPs (average diameter: 2–3 nm) under catalytic conditions when using tethered and untethered ruthenium half-sandwich complexes in the transfer hydrogenation of ketones with 2-propanol in basic media (Fig. 32) [17]. Moreover, catalytically active Ru NPs were also formed from isopropanol solutions of a mixture of $[RuCl_2(p\text{-cymene})]_2$, the NHC ligand precursor and KOH under reflux. The Ru catalysts were successful

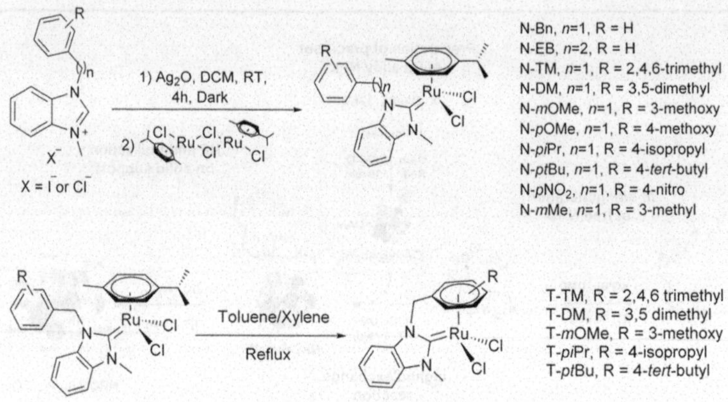

Fig. 32 Synthesis of tethered and untethered ruthenium half-sandwich complexes used in ketone transfer hydrogenation. Adapted with permission from [17] Copyright 2020 Wiley-VCH Verlag GmbH&Co. KGaA, Weinheim

at reducing aryl and heteroaromatic ketones under basic conditions (ketone/Ru/KOH = 100:1:25, reflux). The heterogeneous nature of the catalytically active species was thoughtfully demonstrated by the performance of different control experiments including poisoning, hot filtration and three-phase tests.

The application of Au materials to the reduction of nitroarenes to anilines with sodium borohydride in aqueous media has been profusely studied [73]. One first example of the application of NHC-stabilized Au NPs in these reactions involves the use of nanomaterials synthetized through thermal decomposition of pre-formed pentafluorophenyl Au(I) complexes in the absence of solvent (Fig. 33) [13]. From the $[Au(C_6F_5)_2][Imid]$ complex, MNPs of 1.5 nm together with nanocrystals of polyhedral shape (>50 nm) were formed. Meanwhile, thermolysis of the Au$(NHC)(C_6F_5)$ complex produced Au NPs of 1.0 to 1.3 nm depending on the decomposition temperature. The 1.0 nm Au NPs were loaded on aminopropyl-functionalized SiO_2 leading to a nanocomposite material that was employed in the reduction of 4-nitrophenol with $NaBH_4$ in H_2O. The reaction proceeded with a rate constant of $k = 2.3 \times 10^{-3}\ s^{-1}$, following a pseudo first-order kinetics.

Similarly, hybrid nanocomposites of NHC-functionalized π-conjugated conducting polymers with gold nanoparticles of 3.6 nm, Au·NHC-CP NPs, were examined in the reduction of 4-nitrophenol with $NaBH_4$ [14]. This material was prepared from bithiophene-functionalized NHC-Au(I) complexes by disproportionation of Au(I) to Au(III) and Au(0), and concomitant oxidative polymerization of the thiophene monomers mediated by Au(III) species (Fig. 34). The Au·NHC-CP NPs were found to catalyze the reduction of 4-nitrophenol with a high reaction rate of $6.29\ s^{-1}$.

Dimitratos, Casini and co-workers disclosed the synthesis of hydrophilic Au NPs having narrow particle size distributions and average diameters of 2.3–2.6 nm via the reduction of $[(NHC)_2Au]$ complexes with $NaBH_4$ (Fig. 35) [15]. As determined by XPS, the use of a complex based on a sulfonated bidentate NHC ligand yielded Au

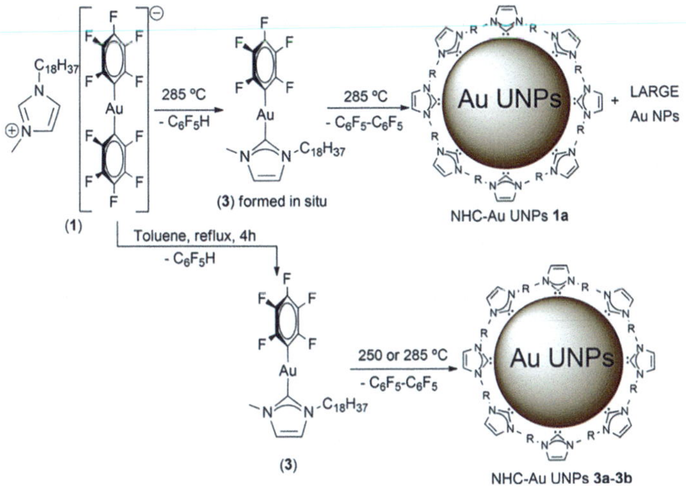

Fig. 33 Synthesis of Au·NHC NPs through thermolysis of pentafluorophenyl Au(I) complexes. Adapted with permission from [13] Copyright 2014 Royal Society of Chemistry

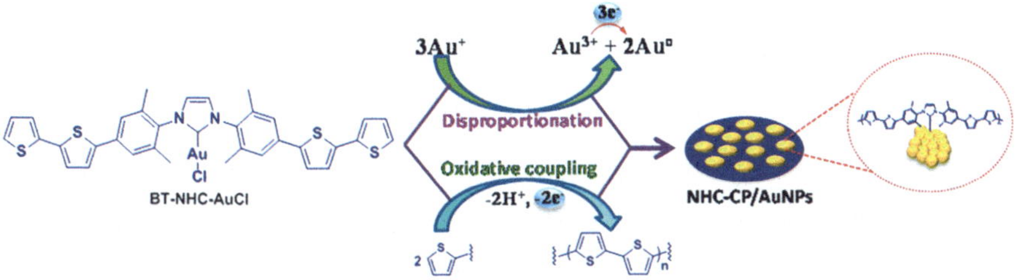

Fig. 34 Synthesis of nanocomposites of NHC-functionalized conducting polymers with gold nanoparticles, Au·NHC-CP NPs. Reproduced with permission from [14] Copyright 2014 American Chemical Society

NPs (AuNP-2) containing a higher amount of surface Au(I) and having higher ligand surface coverage than those derived from an Au complex with monodentate carbenes (AuNP-1). In the reduction of 4-nitrophenol, a lower catalytic activity was observed in the case of AuNP-2 than for AuNP-1, which was attributed to the presence of lower amounts of Au(0) and a limited accessibility of the substrate to the metal surface. Reduction of 2- and 3-nitrophenol was also investigated, observing lower conversions but similar rates for both catalysts. Finally, the same catalytic activity trend was also observed for the reduction of resazurin to resorufin using NH_2OH in water (Fig. 36). The immobilization of AuNP-1 and AuNP-2 colloids on titania (TiO_2) was attempted with limited MNP incorporation and significant changes in their sizes. The MNP size increased in the case of AuNP-1/TiO_2 (6.7 $\pm$ 2.0 nm) but became smaller for AuNP-2/TiO_2 (1.7 $\pm$ 0.4 nm). Moreover, XPS data showed the exclusive presence of surface Au(0) in AuNP-1/TiO_2, and some amount of Au(I) for AuNP-2/TiO_2. In the reduction of 4-nitrophenol, both heterogeneous systems

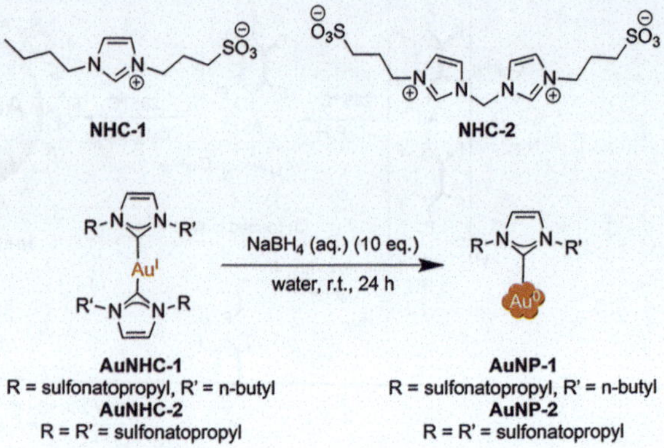

Fig. 35 Synthesis of hydrophilic Au NPs from preformed Au-NHC complexes. Adapted with permission from [15] Copyright 2022 Wiley-VCH Verlag GmbH&Co. KGaA, Weinheim

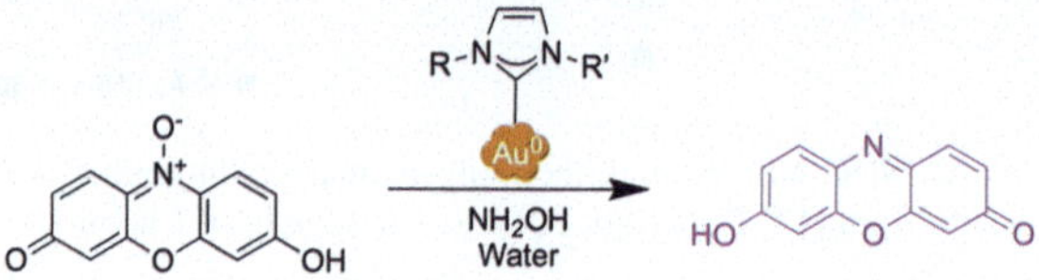

Fig. 36 Reduction of resazurin to resorufin by NH_2OH in water catalyzed by hydrophilic Au NPs. Adapted with permission from [15] Copyright 2022 Wiley-VCH Verlag GmbH&Co. KGaA, Weinheim

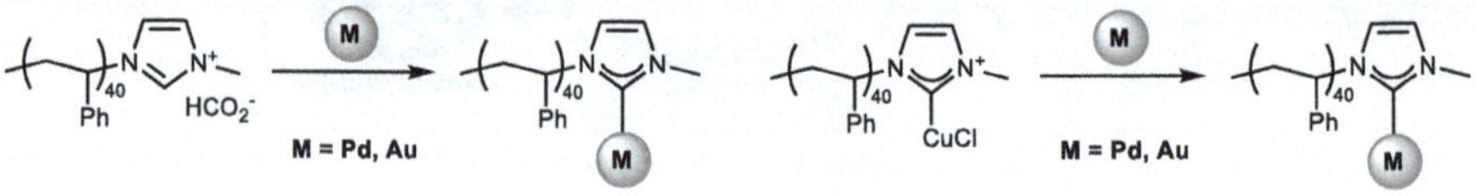

Fig. 37 Synthesis of polystyrene-NHC stabilized Au and Pd NPs

exhibited similar reaction rates ($k = 0.08–0.09\ min^{-1}$), which were slower than those provided by the colloidal AuNP-1 catalyst but similar to those of AuNP-2.

Gold and palladium nanomaterials were prepared by ligand exchange from citrate-stabilized Au and Pd NPs using methylimidazolium bicarbonate and NHC-copper end-group functionalized polystyrenes (PS) (Fig. 37) [42]. The obtained materials, which showed no changes in the average size of the MNPs with respect to the parent particles, were tested in the reduction of 4-nitrophenol with $NaBH_4$. Reactions catalyzed by both catalysts exhibited two distinct kinetic regimes composed of an initial fast reaction (k_1) that was followed by a slower process (k_2). The reaction rate slowdown was attributed to the blockade of active sites of the catalysts by the 4-aminophenol product. Kinetic rate constants k_1 of 0.03–0.04 s^{-1} were determined for the Au NPs stabilized by the PS-NHC ligands, which were higher than those provided by the citrate-Au NPs ($k_1 = 0.0077\ s^{-1}$) and the analogous polystyrene-NHC stabilized Pd catalysts ($k_1 = 0.012–0.014\ s^{-1}$).

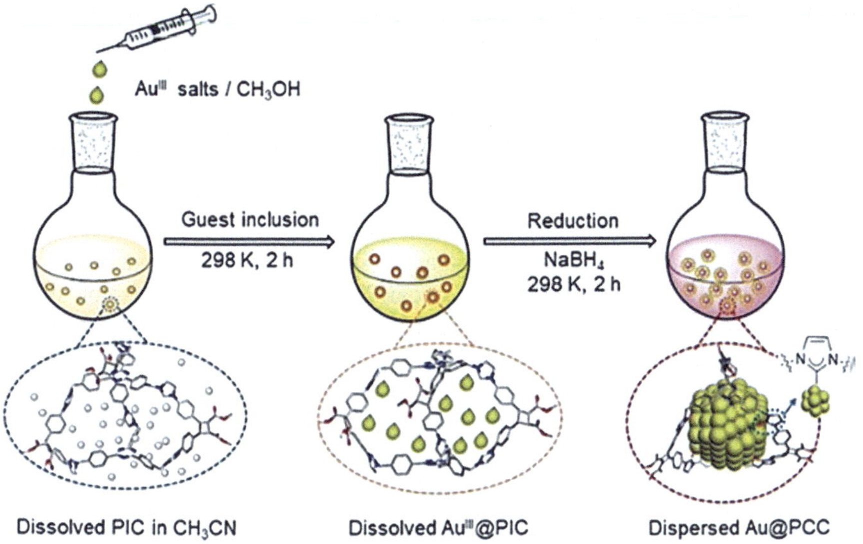

Fig. 38 Synthesis of polycarbene cage-stabilized gold nanoparticles Au@PCCs. Adapted with permission from [74] Copyright 2020 Wiley-VCH Verlag GmbH&Co. KGaA

Finally, Han et al. reported the synthesis of polycarbene cage-stabilized gold nanoparticles (Au@PCCs; PCC = polycarbene cage) using three-dimensional polyimidazolium cages (PICs) of different sizes as nanoparticle forming templates (Fig. 38) [74]. The catalytic activity of the Au@PCCs materials (2.0–3-3 nm) was examined in the reduction of 2- and 4-nitroaniline and 2- and 4-nitrophenol with $NaBH_4$. The reactions proceeded with high apparent kinetic rate constants of up to 3.12 min^{-1}. Moreover, Au@PCCs were also active in the reduction of organic dyes, such as methyl orange and congo red. Separation and reuse of the most active catalytic system could be accomplished up to 8 times without observing a decline in the reaction rates or changes in the size and structure of the nanoparticles.

4.3 *Catalytic H/D Exchange*

Organic molecules labeled with hydrogen isotopes have found applications in many important Chemistry related fields, including drug development, clinical analysis, and reaction mechanistic studies. A convenient method for the labeling of organic derivatives consists on the catalytic exchange of hydrogen by deuterium or tritium using a hydrogen isotope gas, D_2 or T_2 [75]. Seminal observations from the Finke group, [8] who reported deuterium incorporation at imidazolium cations of nanosystems stabilized with ionic liquids, led other research groups to examine the use of mono- and bimetallic nanoparticles in hydrogen isotope exchange (HIE) processes [76]. This reactivity is directly attributable to the fact that metal nanoparticles readily activate C-H and H-H bonds under mild conditions, therefore lending suitable catalytic systems for H/D exchange [58].

In this context, a series of reported studies have focused on the HIE in amino acids using water-soluble, colloidal MNPs prepared by the organometallic approach. For example, Martínez-Prieto, Lippens, Chaudret et al. employed Ru NPs stabilized with sulfonated NHC ligands, Ru·IPr-$PrSO_3^{0.2}$ and Ru·IMes-$PrSO_3^{0.2}$, in the deuteration of L-lysine with D_2 (Fig. 39) [77]. The nanoparticles (1.4–1.5 nm) were synthetized through the controlled decomposition of [Ru(cod)(cot)] with H_2 in the presence of the carbene ligands generated "in situ" by deprotonation of the corresponding imidazolium salts. As a consequence of the incorporation of an anionic group on the NHC ligand, these nanoparticles were soluble in water, and the interaction of the L-lysine substrate with the Ru NPs could be measured by solution NMR spectroscopy using chemical shift perturbation (CSP) experiments. CSP titration allows to determine the strength and binding site of the substrate to a receptor molecule [78]. At pH = 10.4, nearly complete deuteration of the α and ε positions, and limited H/D exchange at the γ position (12.5%), of L-lysine was observed after 42 h (2 bar D_2, 55°C), in good agreement with results obtained in CSP experiments carried out under similar conditions to those employed in the catalytic reaction. Upon pH modification, deuterium incorporation degree at the different positions varies due to the existence of L-lysine in different forms. Thus, at pH values of 13.2, the MNP catalytic activity increases and a significant level of H/D

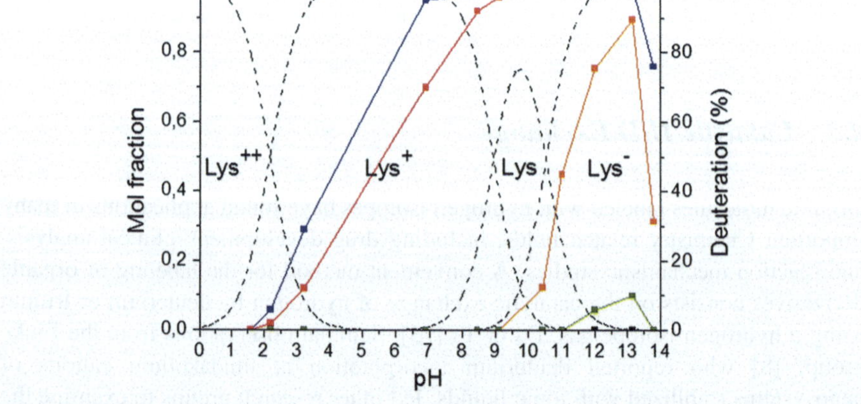

Fig. 39 Deuteration of L-lysine using Ru NPs stabilized with sulfonated NHC ligands at different pH values. Adapted with permission from [77] Copyright 2017 Royal Chemical Society

exchange at the γ position (89.5%), accompanied of a moderate deuteration of the δ site (10%), was achieved. At this pH value, L-lysine is deprotonated at the carboxylic group and both NH_2 groups can coordinate to the nanoparticle metal surface facilitating deuteration of the adjacent C-H positions. Moreover, the simultaneous coordination of the amino moieties also facilitates the deuteration of the γ and δ positions. Conversely, at highly acidic pH values, deuteration is suppressed since the amino groups of the L-lysine are protonated as NH_3^+ and substrate binding to the Ru surface does not take place.

The sulfonated N-heterocyclic carbene ligand IPr-$PrSO_3$ was also employed in the stabilization of bimetallic RuPt NPs following the organometallic approach [35]. A series of nanomaterials were prepared from [Ru(cod)(cot)] and platinum precursors having various decomposition rates (Pt$(nbe)_3$, nbe = norbornene; Pt $(CH_3)_2$(cod), cod = 1,5-cyclooctadiene; $Pt_2(dba)_3$, dba = dibenzylideneacetone), providing nanoparticles of similar sizes (1.3–1.5 nm), although with different structures and surface compositions: RuPt·IPr-$PrSO_3$(nbe), RuPt·IPr-$PrSO_3$(dmc), RuPt·IPr-$PrSO_3$(dba) and $RuPt_2$·IPr-$PrSO_3$(dba). Extensive nanoparticles characterization showed that the materials are alloys with anisotropic (RuPt·IPr-$PrSO_3$(nbe) and RuPt·IPr-$PrSO_3$(dmc)) or homogeneous metal distributions (RuPt·IPr-$PrSO_3$(dba) and $RuPt_2$·IPr-$PrSO_3$(dba)). Also, Pt-enrichment of the surface was found to follow the order: RuPt·IPr-$PrSO_3$(nbe) = RuPt·IPr-$PrSO_3$(dba) < < RuPt·IPr-$PrSO_3$(dmc) = $RuPt_2$·IPr-$PrSO_3$(dba). These nanoparticles, along with their Ru and Pt NPs monometallic analogues, were used in the isotopic H/D exchange of L-lysine in water (6–7 mol% catalyst, D_2O, 2 bar D_2, 55°C) (Fig. 40). The α and ε positions of L-lysine were almost completely deuterated with the bimetallic MNPs; however, deuterium incorporation was somewhat lower than that observed for Ru NPs and deuteration of the γ-CH_2 was not achieved. Reaction rates in the case of $RuPt_2$·IPr-$PrSO_3$(dba) were particularly low, as expected from their higher Pt content and the fact that Pt NPs were ineffective for L-lysine deuteration. Chemical shift perturbation (CSP) experiments suggested that the lack of H/D exchange in the reactions with Pt-enriched nanoparticles could be attributed to strong Pt–carboxylate interactions.

Similarly, water soluble monometallic Ir and Ru, and bimetallic RuIr NPs (1.1–1.3 nm) stabilized with the IMes-$PrSO_3$ ligand (0.2 equiv) were examined in the deuteration of L-lysine at pH = 9 in D_2O (Fig. 40) [36]. Complete deuteration of the α and ε positions and moderate labelling of the γ position was achieved using monometallic Ru·IMes-$PrSO_3$ and bimetallic $RuIr_2$·IMes-$PrSO_3$ and Ir_2Ru·IMes-$PrSO_3$ nanocatalysts, while Ir·IMes-$PrSO_3$ NPs were significantly less active.

Bimetallic PdNi NPs stabilized by the IMes-$PrSO_3$ ligand were also found to catalyze the deuteration of amino acids using D_2 gas (4–5 mol% cat, 2 bar D_2, 55°C, D_2O) (Fig. 41) [37]. These nanoparticles have a spherical shape with mean diameters of 1.8–2.9 nm, and an alloy composition with the presence of both metals on the surface. The deuteration of L-phenylalanine occurs at the β position referred to the acid group, with the first deuteration of the CH_2-β moiety taking place in an enantiospecific manner. Interestingly, unlike in the deuteration of aromatic amino acids with other noble metal-based catalysts, no reduction of the aromatic system

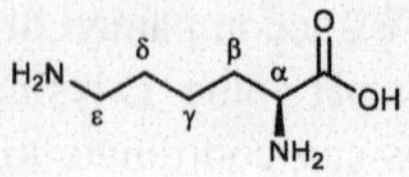

Catalyst	α (%)	γ (%)	ε (%)
Ru·IPr-PrSO$_3$	99	12	98
RuPt·IPr-PrSO$_3$(nbe)	94	0	96
RuPt·IPr-PrSO$_3$(dmc)	97	0	97
RuPt·IPr-PrSO$_3$(dba)	91	0	97
RuPt$_2$·IPr-PrSO$_3$(dba)	69	0	67
Pt·IPr-PrSO$_3$	6	3	0

Conditions: 6-7 mol% cat, 2 bar D_2, 55 °C, D_2O, 42 h

Catalyst	α (%)	γ (%)	ε (%)
Ru·IMes-PrSO$_3$	>98	14	>98
Ru$_2$Ir·IMes-PrSO$_3$	>98	28	>98
RuIr$_2$·IMes-PrSO$_3$	>98	27	>98
Ir·IMes-PrSO$_3$	34	0	35

Conditions: 5 mol% cat, 2 bar D_2, 55 °C, D_2O, 48 h

Fig. 40 Deuteration of L-lysine using RuPt and RuIr NPs stabilized with sulfonated NHC ligands

Fig. 41 Deuteration of L-phenylalanine using PdNi NPs stabilized with sulfonated NHC ligands

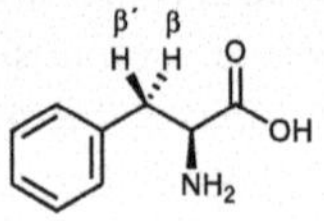

Catalyst	CHD (β′) (%)	CD$_2$ (β′+β) (%)
Pd·IMes-PrSO$_3$	2.5	3
Pd$_7$Ni$_3$·IMes-PrSO$_3$	1.5	97
Pd$_5$Ni$_5$·IMes-PrSO$_3$	24.5	34
Pd$_3$Ni$_7$·IMes-PrSO$_3$	6.5	8
Ni·IMes-PrSO$_3$	7	12

Conditions: 4-5 mol% cat, 2 bar D_2, 55 °C, D_2O, pH 0 8-9, 48 h

was observed. The achieved selectivity contrasts with that provided by Ru NPs, which promote the deuteration at the CH-α position in L-lysine. Moreover, a synergistic effect was observed since analogous monometallic Pd and Ni nanoparticles are less active than the bimetallic systems containing high amounts of Pd (50 or 70%). In the deuteration of L-tyrosine, however, low deuterium incorporation under the same reaction conditions was observed, probably as a consequence of the presence of a hydroxyl group in the substrate.

In 2019, Feuillastre, Pieters et al. examined the use of Ru NPs as catalysts in the HIE of a series of purine derivatives using D_2 or T_2 as isotopic sources [79]. Deuteration at both aromatic C-H positions with high levels of selectivity was achieved using Ru·PVP and Ru·ICy NPs under mild conditions (5–20 mol% Ru, 2 bar D_2, 55° C) (Fig. 42). The NHC-modified Ru NPs consistently led to higher deuterium enrichments than the Ru·PVP system and were also amenable to perform tritium labelling of Didanosine and Idelalisib, two pharmaceutical molecules, under modified reaction conditions (25–50 mol% cat, 0.9 bar T_2).

Monometallic Ru and Ir and bimetallic RuIr NPs (1.1–1.3 nm) stabilized with the IMes ligand (0.2 equiv) were tested in the deuteration of N-heteroarenes (Fig. 43) [36]. Initial tests performed with 2-phenylpyridine (5 mol% cat, 2 bar D_2, 55°C) showed that the monometallic Ru·IMes and Ru enriched Ru$_2$Ir·IMes NPs catalyzed

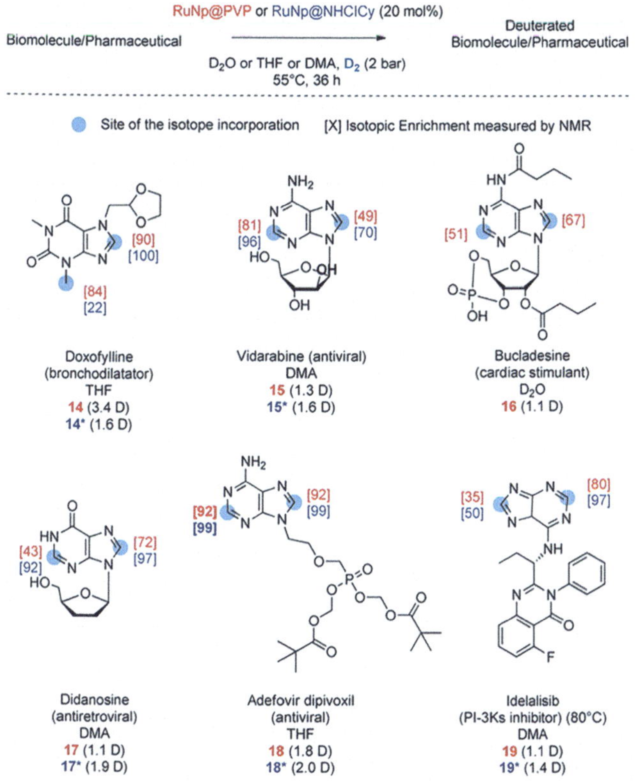

Fig. 42 Deuteration of purine derivatives catalyzed by Ru·PVP and Ru·ICy nanoparticles. Reproduced with permission from [79] Copyright 2019 Wiley-VCH Verlag GmbH&Co. KGaA

the hydrogenation of the aromatic ring, whereas no reduction was observed in the case of Ir·IMes and Ir_2Ru·IMes catalysts. Moreover, the Ir and $RuIr_2$ NPs promote H/D exchange at remote sites in addition to positions close to the N atom. Comparison of the IMes-modified nanoparticles with those stabilized by PVP showed an increased reactivity of the former catalysts suggesting that the presence of the NHC ligand promotes the catalytic activity probably by enhancing the electron density at the metal surface. Deuteration of 2-methyl-naphthylamine and 5,6,7,8-tetrahydronaphthtylamine using the IMes-capped nanoparticles showed similar results, with significant formation of hydrogenated product in the reactions catalyzed by Ru·IMes and Ru_2Ir·IMes NPs. Overall, these results suggest the existence of a synergetic effect of both metals, which is enhanced by the presence of NHC ligands on the surface.

As shown by Asensio, Tricard, Chaudret and co-workers, Ni NPs stabilized by the ICy and IMes NHCs (0.25 equiv.; mean size 1.8–2.0 nm) catalyze the H/D exchange of nitrogen-containing heterocycles of pharmaceutical interest with high chemoselectivity, avoiding the formation of undesired reduction side products that

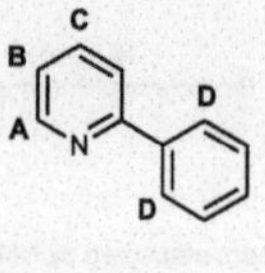

Catalyst	red	A (%)	B (%)	C (%)	D (%)
Ru·IMes	yes	–	–	–	–
Ru_2Ir·IMes	yes	–	–	–	–
$RuIr_2$·IMes	no	88	81	70	85
Ir·IMes	no	15	25	20	4

Conditions: 5 mol% cat, 2 bar D_2, 55 °C, THF, 24 h

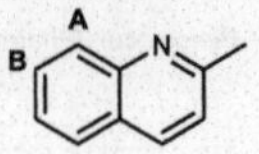

Catalyst	red (%)	A (%)	B (%)
Ru·IMes	84	91	78
Ru_2Ir·IMes	48	87	16
$RuIr_2$·IMes	<5	83	8
Ir·IMes	<5	99	8

Conditions: 5 mol% cat, 2 bar D_2, 55 °C, THF, 24 h

Fig. 43 Deuteration of N-heterocycles catalyzed by M·IMes (M = Ru, Ir) and RuIr·IMes nanoparticles

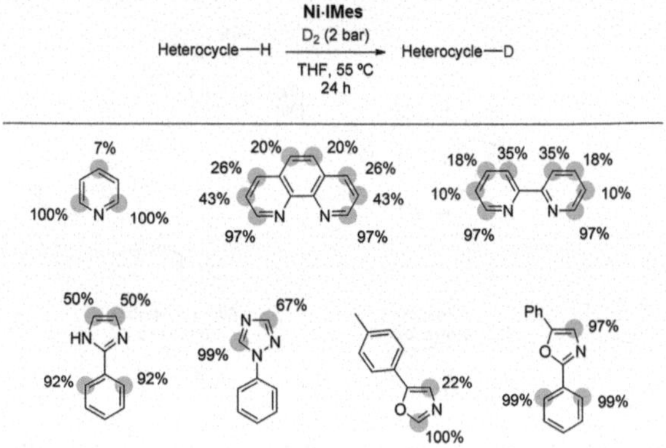

Fig. 44 Deuteration of N-heterocycles catalyzed by Ni·NHC nanoparticles. Adapted with permission from [80] Copyright 2020 Royal Society of Chemistry

are normally formed using Ru·ICy NPs (Fig. 44) [80]. For example, initial catalyst screening using 2-phenylpyridine as a model substrate (12 mol% Ni, 2 bar D_2, 55°C, 24 h) showed that while Ru·ICy nanoparticles lead to hydrogenated products, Ni NPs perform H/D exchange at the α position of the nitrogen atom (95% D incorporation with both catalysts) and at the ortho positions of the phenyl group (12% and 95% H/D exchange with Ru·ICy and Ru·IMes, respectively). Moreover, the Ni·IMes catalyst was found to be more active than Ni·ICy nanoparticles, likely because of the lower steric hindrance of the N-substituents of the ligand, leading also to deuterium incorporation at other positions of the pyridine ring (<14%). The most active Ni·IMes NPs were further applied in the deuteration of other nitrogen-containing heterocycles, with deuterium incorporation taking place preferentially at the α position of the nitrogen atom.

Pieters et al. were successful at modifying the activity and selectivity in the deuteration of alcohols, aldehydes and N-heterocycles of a commercial Ru on carbon (Ru/C) catalyst by functionalizing the Ru surface with the IAd ligand (Fig. 2) [46]. Solid-state NMR spectroscopy and XPS analysis of the IAd-modified (0.1–1 equiv) catalysts showed the presence of both NHC and imidazolium species

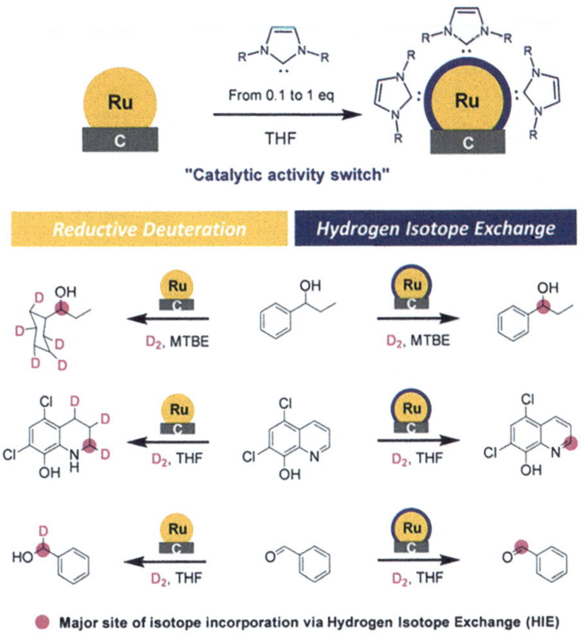

Fig. 45 Examples of deuteration of alcohols, aldehydes and N-heterocycles catalyzed by Ru/C and Ru·IAd/C catalysts. Reproduced with permission from [46] Copyright 2020 Wiley-VCH Verlag GmbH&Co. KGaA

resulting from NHC protonation at the metal surfaces. While no significant differences were detected in the HIE of 1-dodecanol and $OH(CH_2)_5OH$, higher regioselectivity towards the deuteration of the α-position was observed using the IAd ligand (0.1 equiv) in the labeling of other alcohols (Fig. 45). Moreover, alcohols with aromatic groups, which are readily hydrogenated with the unmodified catalyst, could be deuterated with low levels of phenyl reduction using the Ru·$IAd^{0.1}$/C catalyst. On the other hand, higher reaction rates and deuterium enrichments were observed in the deuteration of pharmaceutically relevant heterocycles with Ru·$IAd^{0.1}$/C in comparison to the unmodified catalyst. In addition, through adjustment of the NHC/Ru ratio, heterocycle hydrogenation of biologically relevant substrates could be suppressed. Finally, while Ru/C hydrogenated the aldehyde moiety leading to the corresponding α-deuterated alcohols, selective HIE at the carbonyl position of aldehydes could be performed using Ru·$IAd^{1.0}$/C as catalyst. These changes in regioselectivity were attributed to structural effects due to NHC coordination and the bulkiness of the adamantyl substituents. The steric hindrance exerted by the adamantyl group impedes the π-coordination of the arene moieties of the substrates needed for the hydrogenation process, while it allows to leave enough active sites for the approach to the metal surface of the substrate nitrogen or oxygen atoms to facilitate the C-H activation reaction.

Pieters, Chaudret, Derdau et al. used air-stable ICy-capped Ir NPs (mean size: 1.1–1.3 nm) in the selective HIE of anilines with D_2 (Fig. 46) [30]. High isotopic enrichments were achieved for a series of aniline derivatives with both electron-donating and -withdrawing groups, leading selectively to H/D exchange at the *ortho*

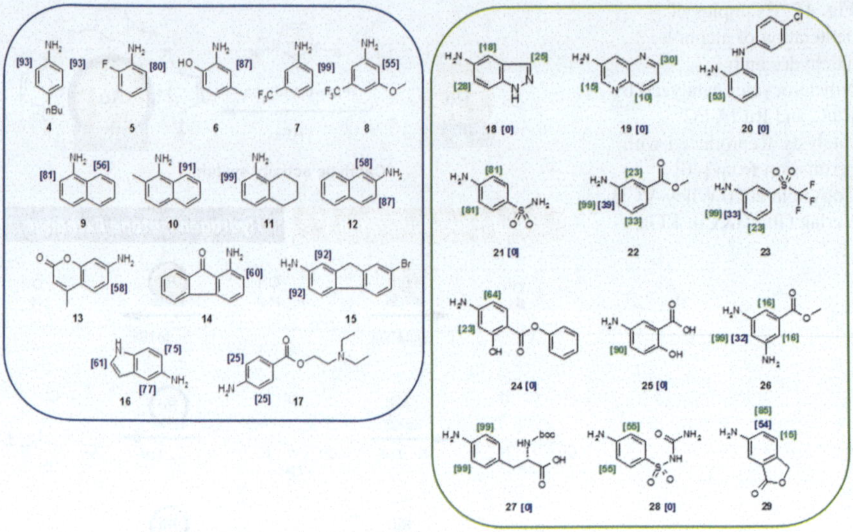

Fig. 46 Deuteration of anilines catalyzed by Ir·ICy nanoparticles (4 mol% cat, 1 bar D_2, 55–80°C). Adapted with permission from [30] Copyright 2020 Wiley-VCH Verlag GmbH&Co. KGaA

positions. The catalyst was tolerant to the presence of diverse functional groups, including halogens, phenols, indoles, esters, and ketones. Deuteration of anilines bearing polar groups, including relevant drug-like complex molecules, was performed under slightly modified conditions using 80°C and THF/D_2O mixtures. Finally, the developed HIE procedure was found to be amenable to the use of T_2 as the isotopic source, as shown by the selective tritiation of a Volixibat pharmacophore.

Recently, the group of Rendón, Lara and Suárez reported the synthesis of a series of Ru NPs stabilized with 0.1–0.2 equiv of mesoionic 1,2,3-triazolylidene ligands, MIC1-MIC3 (Fig. 2), through decomposition of [Ru(cod)(cot)] with H_2 in the presence of the stabilizing ligands generated "in situ" [34]. The most active Ru·MIC1$^{0.2}$ NPs were employed in the labelling of the E-H (E = B, Si, Ge, Sn) bond in hydrosilanes, hydroboranes, hydrogermananes, and a hydrostannane using D_2 under relatively mild conditions (1.0 mol% Ru, 1 bar D_2, 55°C) with high yields and complete selectivity (Fig. 47).

4.4 Catalytic C-C Bond Formation

In 2010, Glorius and co-workers disclosed the synthesis via wet impregnation of magnetically recoverable Pd NPs supported on magnetite (Fe_3O_4). The nanoparticles surface was subsequently modified with chiral carbene ligands, NHC*-1 to NHC*-5, formed "in situ" from the corresponding imidazolinium salts under basic conditions

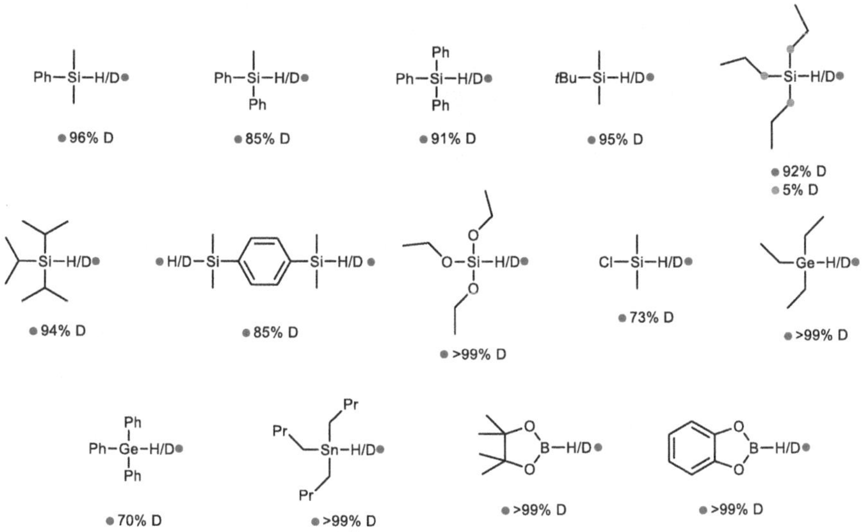

Fig. 47 Deuteration of substrates containing E-H (E = B, Si, Ge, Sn) bonds catalyzed by Ru·MIC1$^{0.2}$ nanoparticles (1.0 mol% Ru, 1 bar D_2, 55°C, THF)

(Fig. 48) [11]. These nanosystems were evaluated in the α-arylation of 2-methyl-1-tetralone with bromobenzene (toluene, 80°C, NaO^{t}Bu), showing that nanoparticles modified with 2.5 mol% of NHC*-1 were the most efficient system providing the C-C coupling product in 74% yield and 53% enantiomeric excess. Catalyst recycling could be performed using magnetic separation up to five times without a significant decrease of activity and enantioselectivity. Moreover, leaching studies of the NHC-modified and non-modified catalysts showed a stabilizing effect attributed to the coordination of the NHC ligand to the Pd surface. Finally, the Pd·NHC*-1/Fe_3O_4 nanocatalyst was further tested in the coupling of 2-methyl-1-tetralone and 2-methyl-1-indanone with different aryl halides, and in intramolecular C-C coupling reactions, yielding the corresponding products with moderate to good yields and optical purities between 33 and 85%*ee*.

Pleixats, Shafir et al. prepared Pd NPs (2.9–4.2 nm) stabilized by tris-imidazolium salts (trimid) via decomposition of $Pd(dba)_2$ with 3 bar of H_2 using Pd/L ratios 1:1 (Fig. 49) [81]. ^{13}C CP-MAS NMR spectroscopy showed in the case of Pd·trimid-1, having a more basic I^- counteranion, a low intense and broad resonance for C-2 carbon of the ligand at ca. 167 ppm, suggesting the formation of a NHC-metal interaction. This nanomaterial was determined to be an efficient catalyst for the Suzuki C-C coupling of aryl bromides and activated aryl chlorides using catalyst loadings of 0.2 mol% Pd, even in the presence of diverse functional groups at the aryl halides and boronic acids. The Pd·trimid-2 nanoparticles were found to be less active in the reactions with bromides, whereas, nanoparticles prepared from trimid-3 exhibited similar catalytic activities to those of Pd·trimid-1 in the reactions of bromides, although were less efficient in the coupling of more

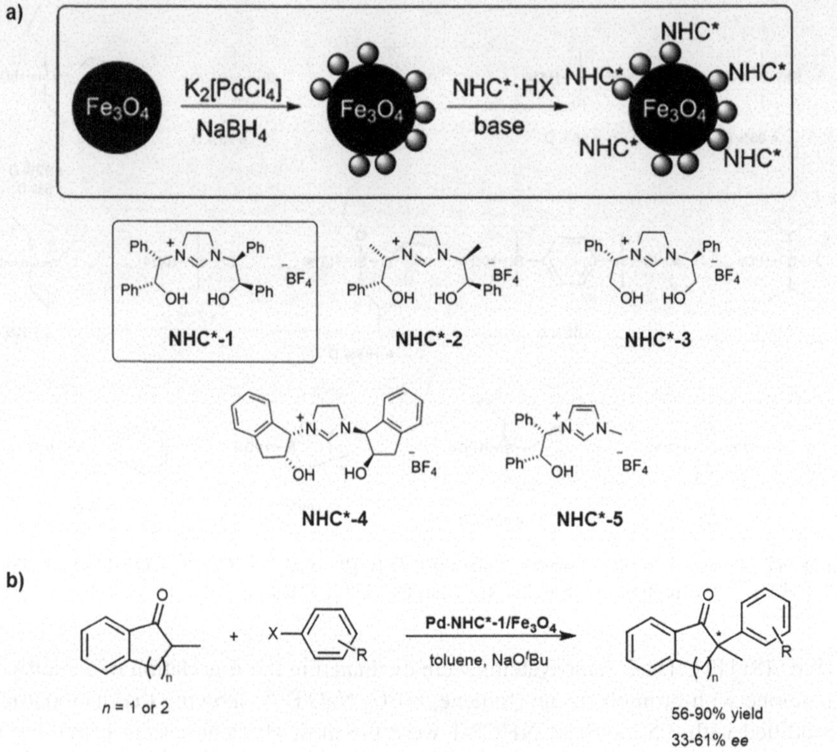

Fig. 48 (**a**) Synthesis of Pd·NHC*/Fe_3O_4 nanocatalysts, and (**b**) their catalytic applications in α-arylation reactions. Adapted with permission from [11] Copyright 2010 Wiley-VCH Verlag GmbH&Co. KGaA

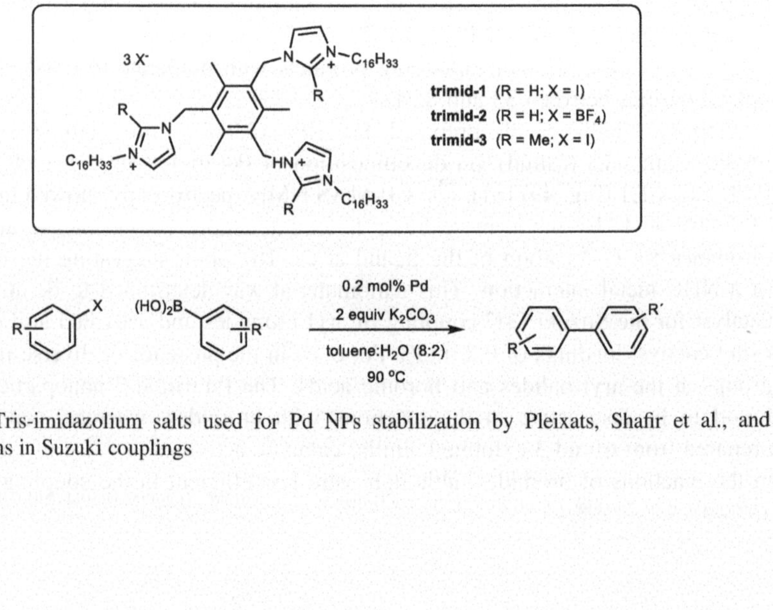

Fig. 49 Tris-imidazolium salts used for Pd NPs stabilization by Pleixats, Shafir et al., and applications in Suzuki couplings

challenging activated chlorides. The latter observation was attributed to the presence of the Me group at the C-2 position that impedes the formation of NHC-metal linkages.

Philippot, García-Antón and co-workers examined the use of hybrid pyrazole-imidazolium and pyrazole-carbene scaffolds in the stabilization of Pd NPs prepared following the organometallic approach (Fig. 50) [20]. While the use of the salts as stabilizers led to the formation of aggregates, the NHC-based ligand provided efficient nanoparticles stabilization, allowing the formation of isolated spherical Pd NPs (2.1–2.4 nm) with NHC/Pd ratios of 0.3, 0.5 and 1.0. Conversely, a lower ligand/metal ratio of 0.1 produced aggregated nanoparticles of 3.6 nm forming superstructures with sizes varying from 50 to 250 nm. The catalytic behavior of these nanocatalysts was compared to that of Pd(II) complexes incorporating the same ligands in the Suzuki–Miyaura reaction. Significantly different results were observed, with the molecular catalysts producing C-C coupling derivatives as the major products and the Pd NPs leading to the preferential formation of C-C products resulting from bromoarene homocoupling.

Muratsugu, Tada et al. synthesized NHC-coordinated bimetallic Cr/Rh catalysts supported on ceria (ICy-*r*-$Cr_{0.19}Rh_{0.06}CeO_z$) through ICy coordination to H_2-reduced $Cr_{0.19}Rh_{0.06}CeO_z$ (*r*-$Cr_{0.19}Rh_{0.06}CeO_z$), a material composed of small Rh nanoclusters and $Cr(OH)_3$ species on the ceria surface. Detailed studies by FT-IR spectroscopy revealed the presence of the NHC ligand in the catalyst, while XPS analysis showed its coordination to the Rh atoms [47]. In striking contrast to the unmodified nanomaterial, the NHC-containing catalysts were found to be active in the 1,4-arylation of cyclohexanone with phenylboronic acid (Fig. 51). After optimization of the amount of NHC (0.8 wt%), and comparison with related catalytic systems, it was determined that ICy coordination to the Rh nanoclusters was key for achieving catalytic turnover. Further catalytic studies demonstrated that catalysts based on other NHCs (IMe, ItBu, IMes and IBz) were less efficient in catalyzing this process, and that ICy-*r*-$Cr_{0.19}Rh_{0.06}CeO_z$ was able to promote the reaction of other α,β-unsaturated ketones with different boronic acids. Moreover, control experiments demonstrated that the catalytic reactions were heterogeneous in nature. Through the combination of different techniques, including XAFS (Rh *K*-edge), XPS and photoluminescence spectroscopy, the electron donation from ICy to the Rh nanoclusters was found to be negligible, ruling out this as the key factor for triggering the catalytic activity. However, DFT calculations pointed out to coordination of the ICy as responsible for controlling the proximity of adsorption sites of the phenyl group and cyclohexanone on the Rh surface and promote the formation of the C–C bond between the reaction partners.

Caballero, Mata, Pérez et al. developed the synthesis of a series of colloidal and rGO-supported Au NPs stabilized with NHC ligands. The catalytic behavior of the Au NPs based on NHC ligands containing pyrene substituents was examined in the cyclopropanation of styrene (Fig. 52) [82]. As previously reported by the same group in the hydration and hydroamination of alkynes (see Sects. 4.5 and 4.6), [24, 25] nanoparticles immobilization and ligand design were found to be crucial parameters for achieving catalytic turnovers. Thus, the supported Au·pyrIPr/rGO NPs provided

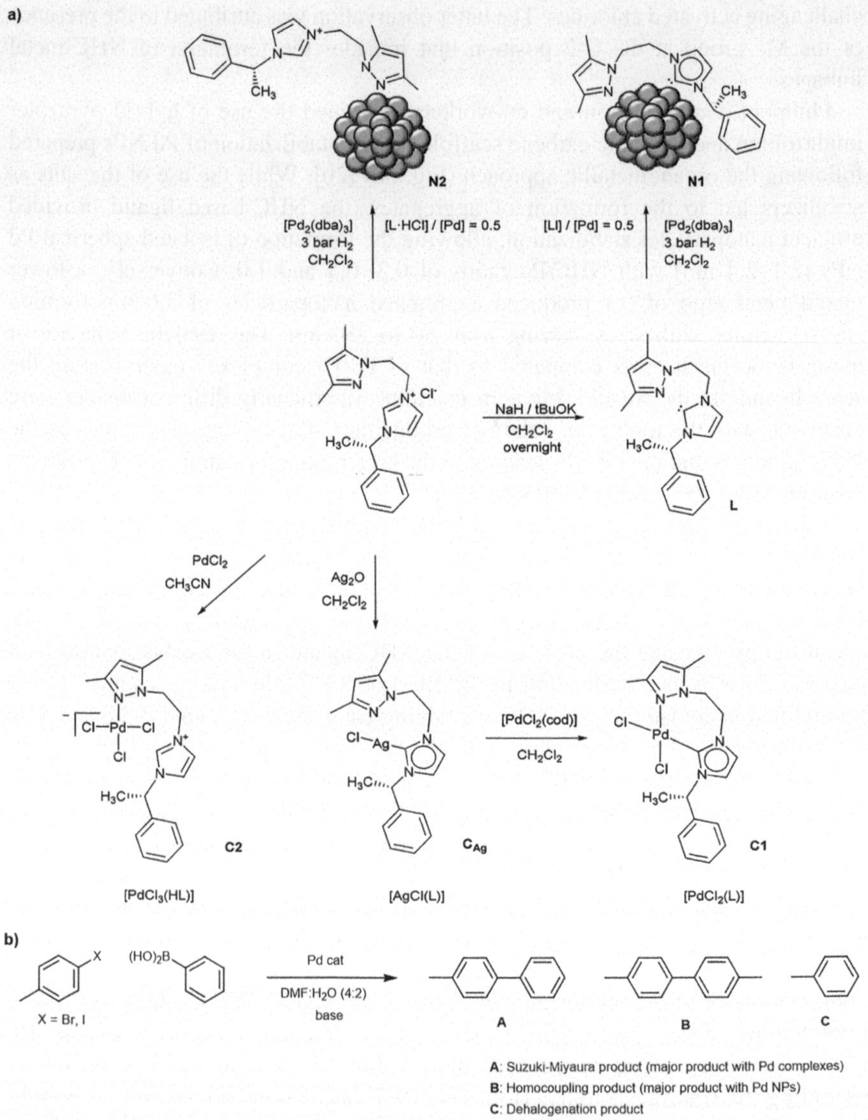

Fig. 50 Pd NPs and molecular complexes employed by Philippot, García-Antón et al. in the Suzuki-Miyaura reaction. Adapted with permission from [20] Copyright 2017 Royal Society of Chemistry

better catalytic performances than the Au·pyrIPr colloidal NPs. Since carbene insertion into C(sp^3)-H and C(sp^2)-H bonds provides an interesting method for C-C bond formation, the Au·pyrIPr/rGO catalyst was further tested in the reaction of ethyldiazoacetate with cyclohexane and benzene, respectively. A 5- to 10-fold

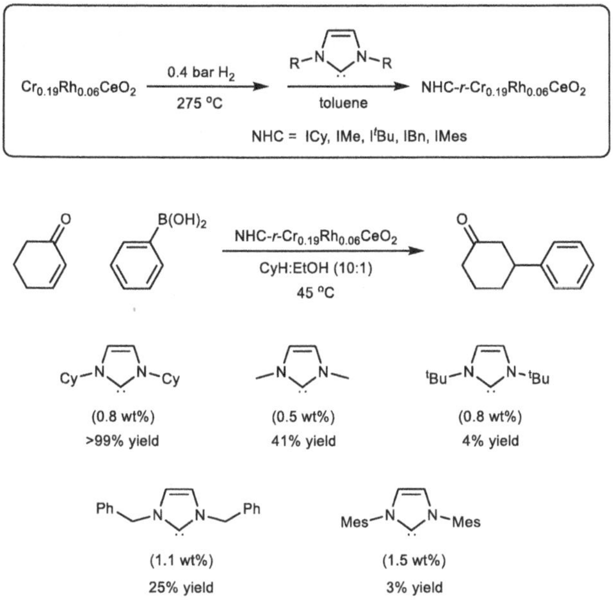

Fig. 51 Preparation of NHC-coordinated Cr/Rh catalysts supported on ceria (NHC-*r*-$Cr_{0.19}Rh_{0.06}CeO_z$) and their application in the 1,4-arylation of α,β-unsaturated ketones with boronic acids

increase of the TON values with respect to those provided by previously reported molecular gold catalysts was observed, with catalyst recycling allowing to reach a total TON of 1,400 after six consecutive uses.

4.5 Catalytic C-X (X = Heteroatom) Bond Formation

The influence of NHCs as ligands on Pd NPs supported on alumina (Pd/Al_2O_3) was investigated by Muratsugu, Doltsinis, Glorius, and co-workers [48]. NHC coordination to the Pd metal surface was confirmed by NMR spectroscopy and XPS, whereas size modifications of the MNPs was not observed according to TEM and STEM-EDS analysis. Initially, the Pd·NHC/Al_2O_3 catalysts were tested in the hydrogenolysis of bromobenzene, a process that should be favored by the presence of electron-rich metal surfaces (Fig. 53). Catalyst modified with NHC ligands with aromatic N-substituents (IPr, IMes) provided faster reactions, while those functionalized with NHCs having N-alkyl groups (ICy, IMe) were less efficient although still more active than the unmodified catalyst. DFT calculations performed on Pd_{13}-nanoclusters showed that binding of the NHC ligands to the palladium surface takes place through the carbene carbon atom, and in the case of N-aryl substituted NHCs also via the delocalized π-systems. Later, the Pd·NHC/Al_2O_3

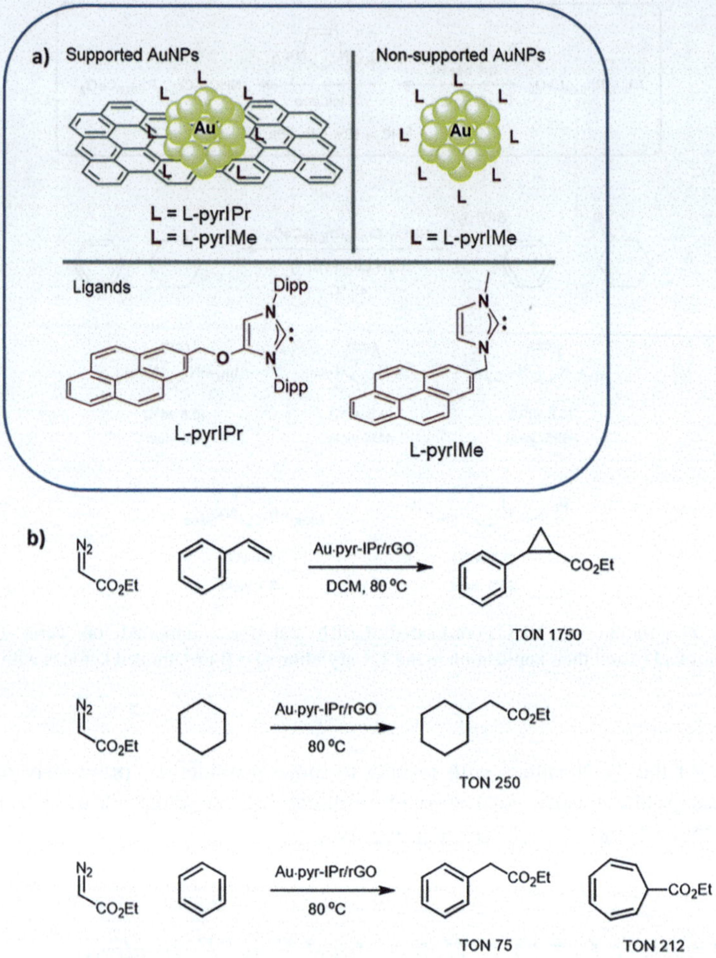

Fig. 52 (**a**) Non-supported and rGO-supported Au NPs functionalized with pyrNHC ligands reported by Caballero, Mata, Pérez et al., and (**b**) applications in cyclopropanation and carbene insertion into C(sp^3)-H and C(sp^2)-H bonds reactions. Adapted with permission from [82] Copyright 2022 Royal Society of Chemistry

catalysts formed "in situ" were tested in Buchwald-Hartwig amination reactions. While product formation was not observed with the unmodified Pd/Al_2O_3 catalyst, a strong dependence of the reactivity with the NHC structure and catalyst loading was evidenced as only analogues of the IPr ligand (IPr* and SIPr) showed efficiency. Mercury, hot filtration, and three-phase tests demonstrated the heterogeneous nature of the catalytic process.

Chiral NHC-based Au NPs embedded in dendrimers and supported on SBA-15 have been shown to catalyze the stereoselective lactonization of allene-carboxylic acid derivatives (Fig. 54) [16]. The gold nanoparticles (ca. 4.7 nm) were prepared by

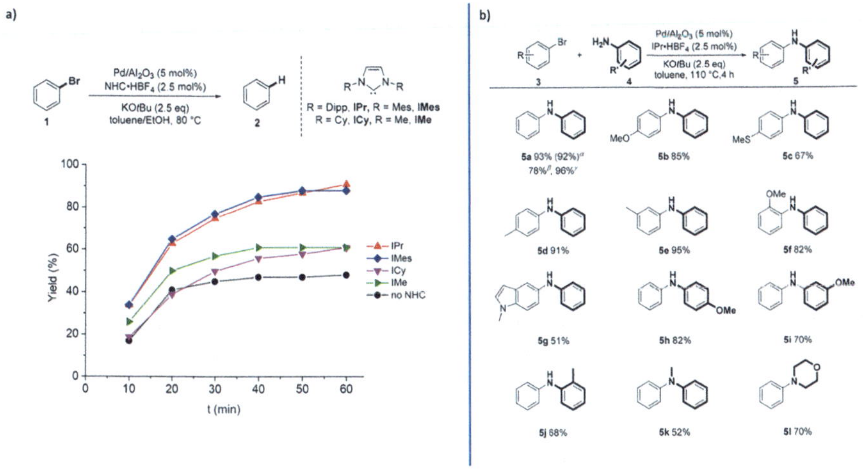

Fig. 53 (**a**) Hydrogenolysis of bromobenzene, and (**b**) Buchwald-Hartwig amination reactions catalyzed by NHC-modified Pd/Al_2O_3 catalysts. Adapted with permission from [48] Copyright 2017 American Chemical Society

reduction of chiral NHC-Au(I) complexes with $^{t}BuNH_2{\cdot}BH_3$, subsequently stabilized with the dendrimers and loaded onto the mesoporous silica. Analysis of sample catalysts by TEM, XPS, XANES and ^{1}H NMR spectroscopy, inter alia, suggested that the nanoparticles were composed of a core of Au(0) coated by a monolayer of NHC-coordinated Au(I) atoms, probably forming potentially catalytic $(NHC)Au^+$ and inactive $(NHC)_2Au^+$ species. While the NHC-ligated Au NPs were shown to catalyze allene lactonization at 20°C, albeit with low enantiomeric induction (<16% *ee*), dendrimer-stabilized Au NPs (ca. 3 nm) synthetized by reduction of $HAuCl_4$ with $NaBH_4$ in the absence of NHC ligands were inactive even at 80°C. Based on catalytic studies, and particularly Eyring analysis, protodeauration was proposed to be the turnover-limiting reaction step, as previously found in analogous homogeneous Au(I) catalyzed processes. Therefore, the catalytic activity of the NHC-capped Au NPs was attributed to the high electron-donating properties of the NHC ligands that reduces the energy barrier of the protodeauration step. Moreover, the presence of hydroxyl groups in the ligands, which could facilitate the protodeauration step via H-bonding interactions, was determinant for achieving catalytic turnover. Finally, the reaction heterogeneity was determined by catalyst filtration and ICP-OES analysis of the filtrate, and the catalyst was employed in successive runs up to five times providing lower catalytic activities after each reuse.

Supported Au NPs capped with NHC ligands (3.3–4.8 nm) were obtained by reduction of Au(I)-NHC triflate complexes with reduced graphene oxide (rGO), and tested in the intramolecular hydroamination of alkynes to yield indols under mild conditions (0.05 Au mol%) (Fig. 55) [25]. A direct influence on the catalytic activity and recyclability of the NHC stabilizing ligands was observed, with the Au NPs functionalized with pyrenyl substituted ligand (γ-NPs-rGO NPs) exhibiting better

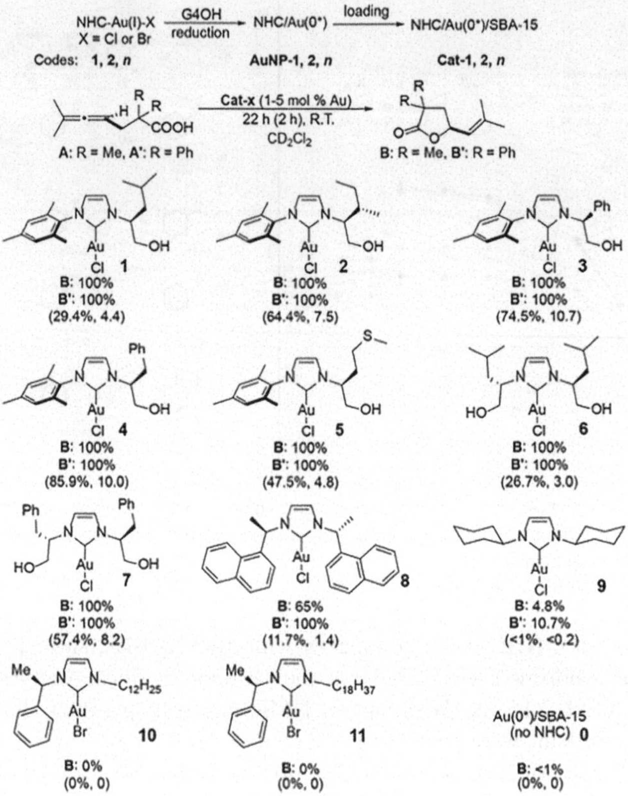

Fig. 54 Lactonization of allenes catalyzed by chiral supported Au·NHC nanoparticles. The yields after 22 h reactions of **A** and **A′**. In parentheses, yield after 2 h reactions of **A**, and the corresponding TOF (h^{-1}). Reproduced with permission from [16] Copyright 2018 American Chemical Society

catalytic performance attributable to potential π-π interactions between the pyrene group and the support. The materials based on NHC ligands lacking pyrene wingtips (α-NPs-rGO and β-NPs-rGO) could be reused four times with only a minor decrease in activity, while a pronounced deactivation was observed in the fifth run. Conversely, in the case of the γ-NPs-rGO NPs catalyst, gradual deactivation was only observed after six reactions, although full conversion was still achieved in the tenth run. ICP/MS and XPS analysis of the recovered catalysts suggested metal leaching as the reason for the poorer stability and recyclability of the nanoparticles without the pyrene substituent.

Finally, two different Pt NPs (1.2–1.9 nm) capped with long-chain substituted N-heterocyclic carbenes, LC-IPr and LC-IMe, obtained through the organometallic approach were tested in the hydroboration of phenylacetylene (Fig. 56) [54]. The presence of long alkyl chains at the 4 and 5 positions of the NHC ligands provided improved stability and solubility to the nanoparticles in apolar media. Moreover, the size of the N substituents of the ligands was found to play a key role on the catalytic

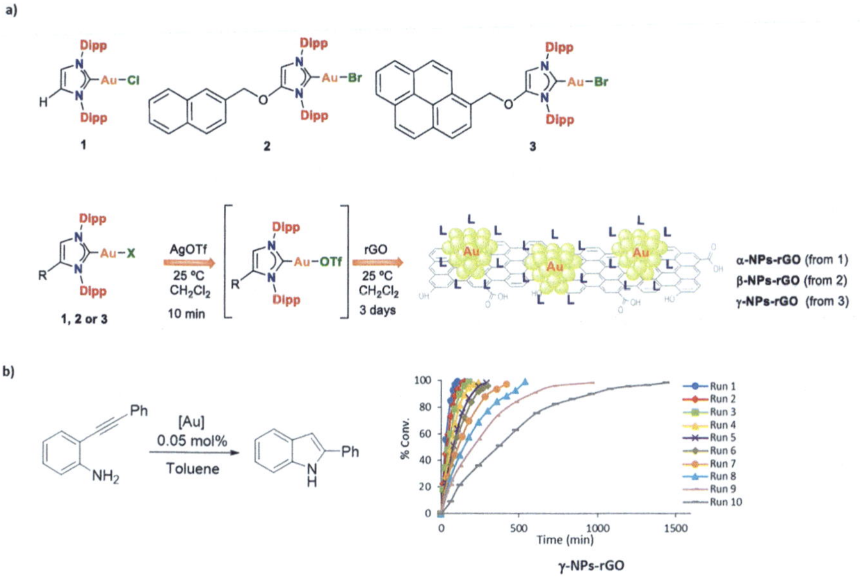

Fig. 55 (**a**) Synthesis of rGO-supported Au·NHC nanoparticles, and (**b**) intramolecular hydroamination catalyzed by γ-NPs-rGO NPs. Adapted with permission from [25] Copyright 2021 Elsevier

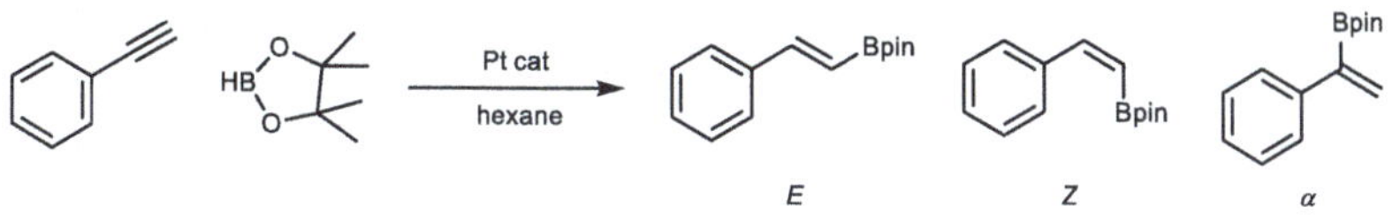

Pt·LC-IPr (3 mol%)	>99% conv.	*E*/*Z*/*α* = 81:7:8
Pt·LC-IMe (3 mol%)	15% conv.	ethylbenzene + styrene
Pt/C (3 mol%)	28% conv.	ethylbenzene + styrene
[Pt(IPr)$_2$] (1 mol%)	85% conv.	*E*/*Z*/*α* = 64:3:5

Fig. 56 Hydroboration of phenylacetylene catalyzed by Pt·LC-NHC nanoparticles

activity. Thus, while larger Pt·LC-IMe colloids (1.9 ± 0.3 nm) converted phenylacetylene to mixtures of the hydrogenated derivatives ethylbenzene and styrene, smaller Pt·LC-IPr nanoparticles (1.2 ± 0.2 nm) catalyzed the hydroboration reaction with good selectivity towards the formation of the trans-isomer. The observed catalytic behavior of the LC-IMe capped nanoparticles resembles that of common heterogeneous Pt catalysts such as Pt/C, while the selectivity achieved with the Pd·LC-IPr catalyst suggested the formation of homogeneous species under catalytic conditions. ESI-MS analysis of a colloidal solution of Pt·LC-IPr NPs evidenced the presence of the platinum complexes Pt(LC-Pr)$_2$ and Pt$_2$(LC-Pr)$_2$.

However, post-catalysis TEM analysis of the nanoparticles showed no significant differences in the particles mean size and distribution. Based on these observations, it was proposed that either metal leaching occurs in very limited amounts or the homogeneous active species were re-absorbed into the nanoparticles. The Pd·LC-IPr colloids also catalyzed the hydroboration of other substrates with electron withdrawing and donating groups at the *para* position with moderate to good conversions and *trans* selectivity.

4.6 Oxidation Catalysis

The aforementioned Ru NPs capped with LC-NHCs employed in hydrogenation reactions (see Sect. 4.1.3) were also tested in the oxidation of primary alcohols to aldehydes and benzylamine into benzonitrile, using oxygen as oxidant (Fig. 57) [55]. The Ru·LC-IPr$^{0.3}$ NPs demonstrated a higher activity than Ru·LC-IMes$^{0.3}$ colloids, whereas the catalysts incorporating lower amounts of NHC ligands were poorly active, likely because of their lower stability under oxidative conditions. Moreover, since post-catalysis analysis of the Ru·LC-IPr$^{0.3}$ and Ru·LC-IMes$^{0.3}$ showed no significant changes in the mean sizes and distributions, tandem oxidation/hydrogenation reactions were run with selected substrates.

In addition, the previously mentioned AuPd NPs stabilized with sulfonated NHC ligands, reported by Glorius, Ravoo et al., were tested as biomimetic catalysts in the aerobic oxidation of D-glucose to gluconic acid and H_2O_2 (see Sects. 4.1.4 and 4.2; Figs. 30 and 58) [41]. The resulting H_2O_2 can be titrated using tetramethylbenzidine (TMB) as chromogenic agent in the presence of horseradish peroxidase, providing a

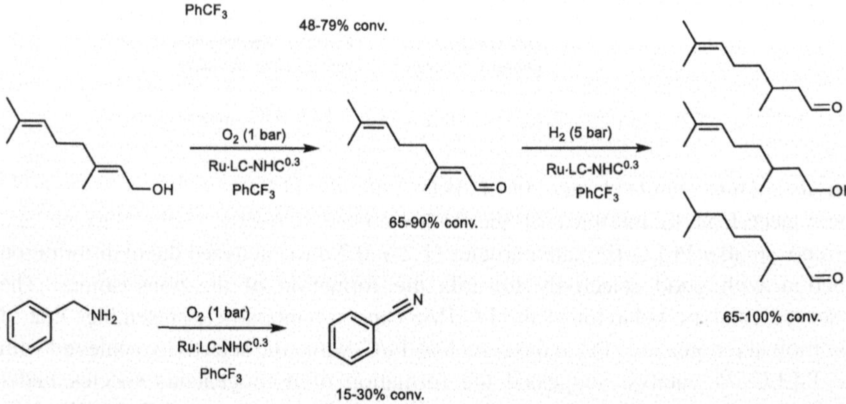

Fig. 57 Oxidation and tandem oxidation/hydrogenation reactions catalyzed by Ru·LC-NHC nanoparticles

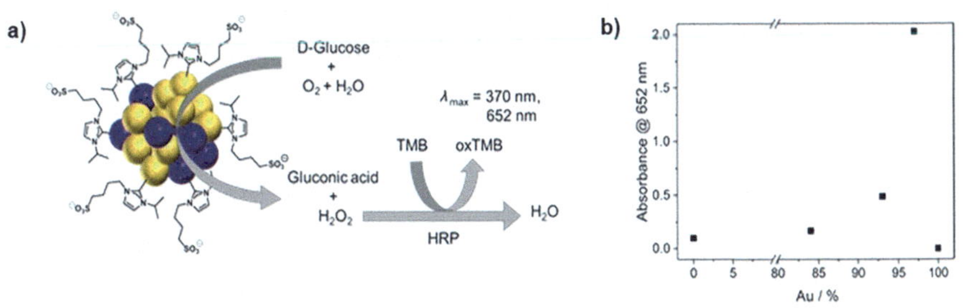

Fig. 58 (**a**) D-Glucose oxidation catalyzed by AuPd·NHC-SO_2 nanoparticles, and (**b**) catalytic performance of AuPd alloy NPs as a function of Au surface fraction. Adapted with permission from [41] Copyright 2018 Wiley-VCH Verlag GmbH&Co. KGaA

potential method for clinical analysis of D-glucose. Although examples of Au nanoparticles catalyzing the D-glucose oxidation to gluconic acid have been reported, both monometallic, sulfonated NHC-modified Au and Pd NPs have limited stability under the catalytic alkaline and high ionic strength conditions giving rise to the formation of aggregates that are catalytically inactive. On the contrary, the AuPd alloys serve as biomimetic catalysts as they are not unstable under catalytic conditions. These results suggest that the presence of Pd favors the colloidal stability of the alloy MNPs.

As described above, the group of Mata reported the synthesis of immobilized Au NPs (3.2 nm) capped with a pyrene-substituted NHC ligand via reduction of an Au-NHC triflate complex with reduced graphene oxide (rGO) (Figs. 55 and 59) [24]. The presence of the NHC ligand and triflate on the nanoparticle Au(I) surface was confirmed by XPS. This nanomaterial was tested in the hydration of alkynes, being the catalytic behavior of the MNPs superior to that of the parent gold complex. Thus, the supported nanoparticles exhibited good catalytic activities even at low catalyst loadings (0.02 mol%), while a 10-fold loading increase was required in the case of the complex to achieved similar catalytic performance. Moreover, the nanocatalyst could be separated and reused up to ten times providing full conversions in each run, although erosion of the catalytic activity attributable to metal leaching was observed after each re-use. To evaluate the importance of the presence of the pyrene substituent on the nanoparticles stability, Au NPs stabilized with NHC ligands lacking the pyrene tag or having a naphthyl substituent were also synthetized (Fig. 55a) [25]. These nanoparticles exhibited a lower stability than those incorporating the pyrNHC ligand as reflected in both the catalytic activity in the hydration of 4-octyne and in the catalyst recyclability.

Large global demand of H_2O_2 as oxidant and bleaching agent made direct synthesis from H_2 oxidation with O_2 an attractive process. At this respect, NHC-modified Pd/TiO_2 catalysts have been examined in the thermal production of H_2O_2 from H_2 and O_2 in the absence of promoters, such as halides or acids, which are typically needed in this process (Fig. 60) [49]. The synthesis of the Pd catalysts was performed using 1% Pd/TiO_2 and a range of different NHC ligands, including

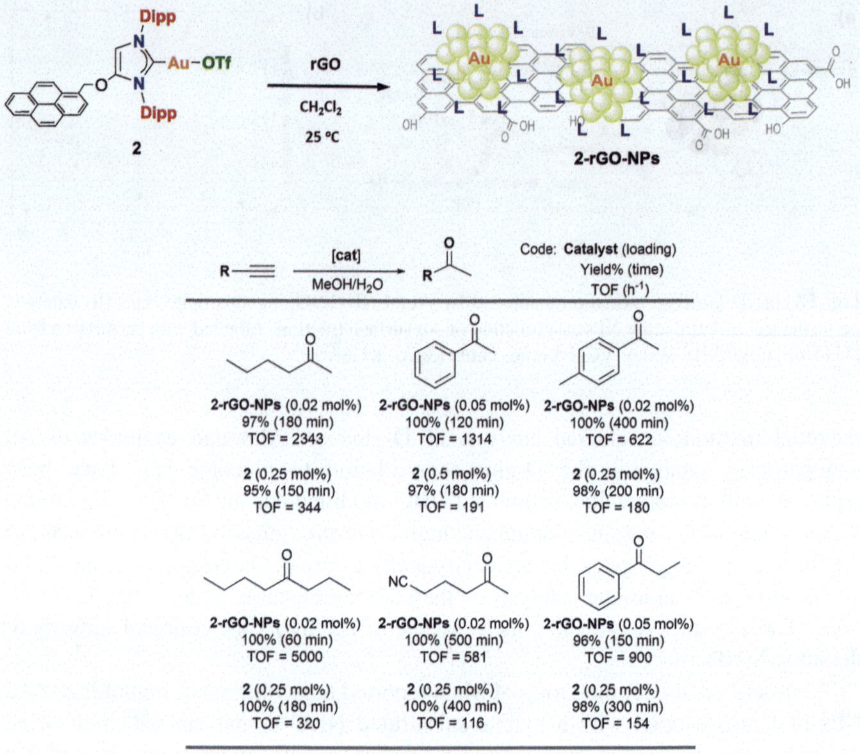

Fig. 59 Synthesis of rGO-supported Au·NHC nanoparticles and application in the hydration of alkynes. Adapted with permission from [24] Copyright 2019 Elsevier

ICy, IMes, IPr, and IPr derivatives, via deprotonation of the corresponding imidazolium salts and subsequent immobilization on the supported metal. TEM analysis of the most efficient IPr-containing material revealed no variations on the nanoparticles sizes upon carbene ligand modification of the catalyst, while XPS showed a shift in the Pd oxidation state towards Pd^0. The IPr-decorated catalyst gave rise to a notable catalytic efficiency with higher TOFs for the H_2O_2 synthesis (160 $mmol_{H2O2}$ $kg_{cat}{}^{-1}$ h^{-1}) and lower degradation rates (184 $mmol_{H2O2}$ $kg_{cat}{}^{-1}$ h^{-1}) than those provided by the unmodified Pd/TiO_2 (80 and 221 $mmol_{H2O2}$ $kg_{cat}{}^{-1}$ h^{-1}, respectively), and rivalling those of the current state-of-the-art catalysts. Analysis using CO-DRIFTS of the Pd·IPr/TiO_2 catalysts with different Pd/IPr molar ratios exhibited a shift in the band attributed to terminal coordinated CO to lower wavenumbers as the IPr:Pd ratio increased. Such an observation suggests that introduction of the NHC to the catalyst surface increases backdonation to CO, what allow to ascribe the enhanced activity to the presence of highly electrodonating NHC ligands on the metal surface. Finally, catalyst recyclability was demonstrated. Through five sequential reactions using the same catalyst, relatively high concentrations of 0.98 wt% of H_2O_2 were achieved.

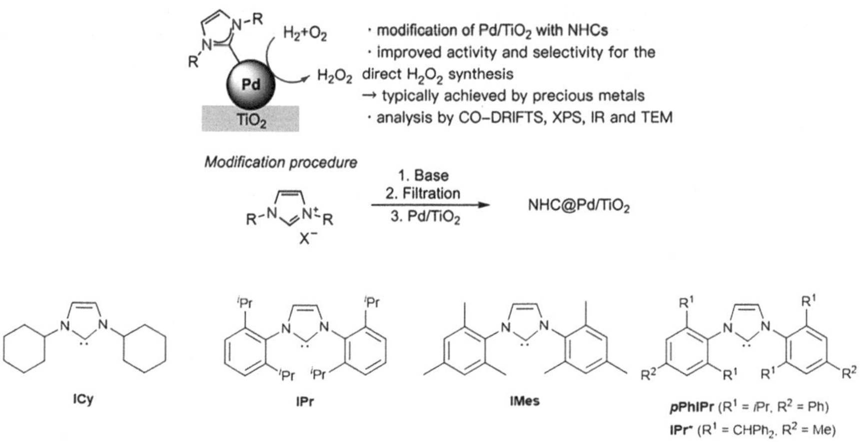

Fig. 60 Synthesis of Pd·NHC/Ti$_2$O catalysts examined in H_2O_2 synthesis and degradation. Adapted with permission from [49] Copyright 2022 American Chemical Society

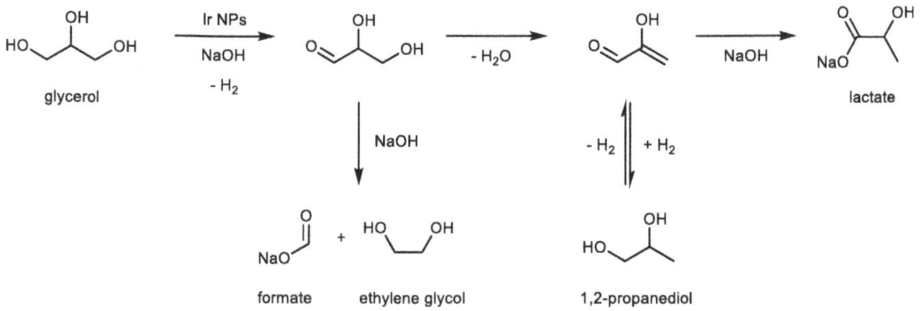

Fig. 61 Aceptorless dehydrogenation of glycerol catalyzed by colloidal Ir·IPr and supported Ir·IPr/C^k NPs

Synthesis of IPr-containing Ir colloids (1.8 nm), Ir·IPr NPs, through reaction with hydrogen of the [Ir(Cl)(cod)(IPr)] complex was performed by Oberhauser and co-workers (Fig. 61) [18]. Similarly, analogous supported Ir NPs (1.4 nm), Ir·IPr/C^k NPs, were also prepared after complex impregnation on Ketjenblack (C^k) followed by exposure to H_2. Using the same methodologies, decomposition of the [Ir(Cl)(cod)]$_2$ complex led to Ir black precipitate and a supported C^K iridium nanomaterial (1.2 nm). All these Ir-based materials were employed for the conversion of glycerol to lactic acid under basic conditions. Non-supported Ir·IPr NPs provided in dioxane maximum glycerol conversion TOF values of up to 9,857 h^{-1}, yielding lactic acid with a selectivity of 93% along with minor amounts of 1,2-propanediol, ethylene glycol, and formic acid. The activity of the catalytic system was essentially maintained during three reactions after catalyst recovery via decantation under air. On the other hand, reactions with Ir·IPr/C^K and Ir/C^K were performed in aqueous media and led to much lower TOF values of 4,003 and 2,413 h^{-1}, respectively, and reduced selectivities for lactic acid production. The

Ir·IPr/C^K nanosystem could be re-used with a lower loss of catalytic activity than Ir/C^K, probably as a result of NHC enhanced stabilization that prevents nanoparticle aggregation. ICP-OES analysis of post-catalytic solutions showed leaching of 2.6–2.8% of Ir for both supported catalysts, whereas it was <0.3% in the case of colloidal Ir·IPr NPs.

4.7 *Electrocatalytic Reduction of CO_2*

Efficient conversion of the greenhouse CO_2 gas into valuable feedstocks such as CO, methane, formate, and alcohols can be achieved in a sustainable manner by the use electrochemical methods [83]. In 2016, Chang, Yang and co-workers reported the first use of Au NPs in the electroreduction of CO_2. The catalytic nanomaterial (5–6 nm size) was prepared by ligand exchange from oleylamine-stabilized Au NPs with IMes, and was subsequently supported on carbon black and deposited on carbon paper as working electrode (Fig. 62) [43]. The Au·NHC NPs exhibited enhanced reduction of CO_2 to CO in water under neutral conditions in comparison to unmodified MNPs, with Faradaic efficiencies of up to 83% at an overpotential of 0.46 V. Interestingly, NHC ligands having larger steric bulk led to lower efficiencies, as determined by testing the I^{t}Bu, IPr, and IMes containing materials. The better performance of the NHC-modified catalysts was attributed to the formation of electron-rich surfaces due to the attachment of the strong σ-donating carbene ligands.

NHC-stabilized Au and Pd NPs, prepared from polymeric imidazolium salts with bicarbonate counteranions or NHC-CO_2 adducts through a ligand exchange methodology, were found to be more efficient than related materials based on other ligands, such as thiols and amines, in CO_2 electroreduction (Fig. 63) [44]. The Au nanocatalysts exhibited an activity retention of 86% after reduction of CO_2 at −0.9 V for 11 h; meanwhile, Pd NPs showed a Faradaic efficiency to CO of approximately 65%, being superior to that provided by commercial Pd/C. The polymeric NHC-based stabilizer was regarded to provide stabilization against nanoclustering, while at the same time to form a hydrophobic layer preventing potentially competing proton access to the metal surface and subsequent reduction to H_2.

5 Conclusions and Future Perspectives

Although an impressive development involving the synthesis and catalytic uses of NHC-capped MNPs has been achieved in the last decade, there are still many areas that have been only briefly, if any, explored and deserve further attention. First, regarding MNP metal compositions, with the notable exception of some NHC-containing mono- and bimetallic Ni and Cu NPs, most examples reviewed in

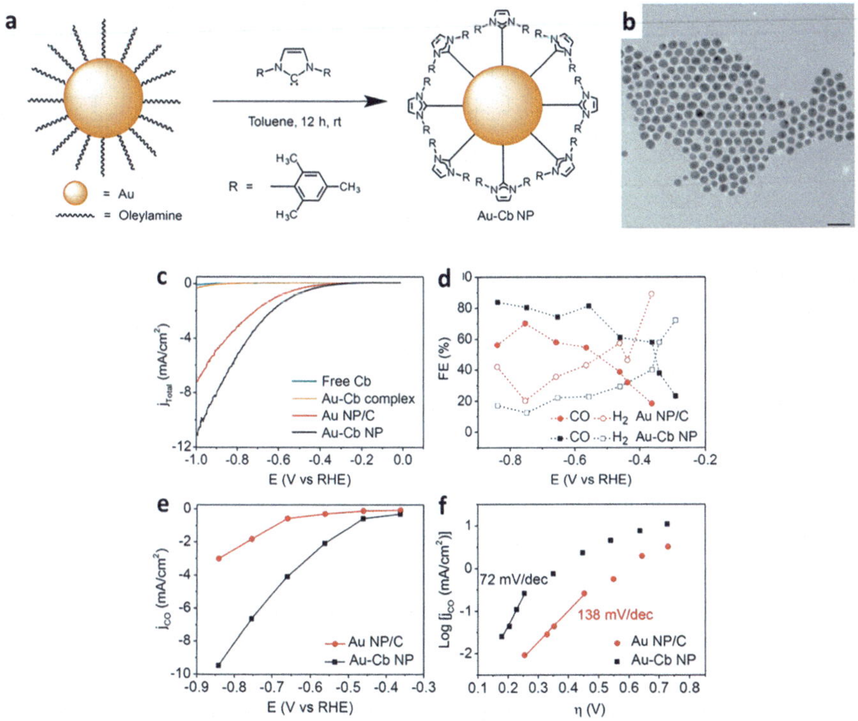

Fig. 62 Electrochemical reduction of CO_2 catalyzed by supported Au·IMes nanoparticles: (**a**) Ligand exchange synthesis of IMes-containing Au NPs. (**b**) TEM image of Au·IMes NPs, scale bar: 20 nm. (**c**) LSV scans of Au·IMes NPs, Au/C NPs, free carbene, and molecular Au·NHC complex under CO_2-saturated 0.1 M $KHCO_3$ at pH 6.8. (**d**) Faradaic efficiencies of products formed from Au·NHC NPs and Au/C NPs. (**e**) Specific CO current density (based on electrochemically active surface area) plots for Au·NHC NPs and Au/C NPs. (**f**) Tafel plots of Au·NHC NPs and Au/C NPs. Adapted with permission from [43] Copyright 2016 American Chemical Society

this chapter consist on nanoparticles based on second- and third-row late transition metals. As a consequence of the scarcity of these elements in the Earth's crust, it is desirable to develop active catalysts based on Earth-abundant metals, such as Mn, Co, or Fe, due to their much lower economic and environmental costs. At this respect, while monometallic nanoparticles based on first-row transition metals may exhibit a reduced catalytic activity in comparison to their noble metal counterparts, the use of bimetallic nanoparticles combining a precious metal and an Earth-abundant element might be interesting [84]. Dilution of a noble metal with a Earth-abundant element might give rise to synergistic effects in catalysis improving catalytic activity and selectivity with respect to the monometallic analogues. Further developments in this direction should deal with an increased control of the metal compositions and distributions (alloy, core-shell, cluster-in-cluster,...) in the nanomaterials, through the use of systematic approaches for the rational synthesis of nanosystems with desirable catalytic properties.

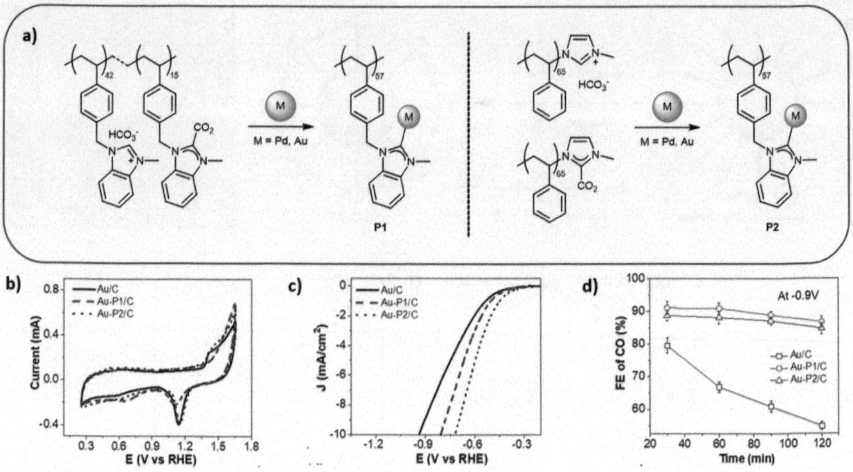

Fig. 63 Synthesis and applications in the electrochemical reduction of CO_2 of polymeric Au·NHC and Pd·NHC nanoparticles: (**a**) MNP synthesis; (**b**) cyclic voltagrams measured in 0.5 M H_2SO_4 saturated by N_2 at a scan rate of 100 mV s^{-1}; (**c**) linear sweep voltametric curves measured in 0.1 M $KHCO_3$ at a scan rate of 10 mV s^{-1}, and (**d**) Faradaic efficiencies of CO at −0.9 V for different electrolysis times. Adapted with permission from [44] Copyright 2019 Wiley-VCH Verlag GmbH&Co. KGaA

In addition, if the large NHC ligands diversity is considered, most examples of NHC-containing MNPs are based on commonly employed imidazolylidene and imidazolidinylidene derivatives. Conversely, use of other NHC ligand families, which are easily synthetized and are regarded as stronger electron-donors, has received little attention. One example of such NHC ligands are triazolylidenes, [85] which have only been recently examined in the stabilization of MNPs by the preparation of Ru and Au nanosystems [34, 86]. Similarly, cyclic (alkyl)(amino) carbenes (CAAC) [87] have been found to coordinate to Au surfaces, [88] but have not been employed in the stabilization of nanoparticles for catalytic purposes. Although these classes of ligands are difficult to isolate, the lack of examples is somewhat surprising since the preparation of many NHC-containing nanosized materials have been carried out through "in situ" NHC generation from the corresponding azolium salts or CO_2 adducts. Besides, in comparison to molecular catalysis, where high levels of enantiocontrol has been achieved by the use of metal catalysts modified with chiral NHC ligands in a plethora of catalytic processes, [67] MNPs modified with chiral carbene ligands have received a limited attention, usually leading to reaction products with negligible or low optical purity [11, 16, 33, 52]. At this respect, design of novel chiral NHC ligands for nanoparticle modification might be necessary.

Therefore, in spite of the enormous interest that NHC-modified metal nanoparticles with catalytic applications has stirred up in the last years, studies covering the previously mentioned and other important fields are foreseen in the near future.

Acknowledgments Financial support from MCIN/AEI/10.13039/501100011033/FEDER,UE (TED2021-129181B-I00, PID2022-136570OB-I00) is gratefully acknowledged.

References

1. Kobayashi S (ed) (2020) Nanoparticles in catalysis. Topics in organometallic chemistry, vol 66. Springer. https://doi.org/10.1007/978-3-030-56630-2
2. Lu L, Zou S, Fang B (2021) The critical impacts of ligands on heterogeneous nanocatalysis: a review. ACS Catal 11:6020–6058. https://doi.org/10.1021/acscatal.1c00903
3. Arduengo AJ, Harlow RL, Kline M (1991) A stable crystalline carbene. J Am Chem Soc 113: 361–363. https://doi.org/10.1021/ja00001a054
4. Bellotti P, Koy M, Hopkinson MN, Glorius F (2021) Recent advances in the chemistry and applications of N-heterocyclic carbenes. Nat Rev Chem 5:711–725. https://doi.org/10.1038/s41570-021-00321-1
5. Koy M, Bellotti P, Das M, Glorius F (2021) N-heterocyclic carbenes as tunable ligands for catalytic metal surfaces. Nat Catal 4:352–363. https://doi.org/10.1038/s41929-021-00607-z
6. Smith CA, Narouz MR, Lummis PA, Singh I, Nazemi A, Li C-H, Crudden CM (2019) N-heterocyclic carbenes in materials chemistry. Chem Rev 119:4986–5056. https://doi.org/10.1021/acs.chemrev.8b00514
7. Huynh HV (2018) Electronic properties of N-heterocyclic carbenes and their experimental determination. Chem Rev 118:9457–9492. https://doi.org/10.1021/acs.chemrev.8b00067
8. Ott LS, Cline ML, Deetlefs M, Seddon KR, Finke RG (2005) Nanoclusters in ionic liquids: evidence for *N*-heterocyclic carbene formation from imidazolium-based ionic liquids detected by ^{2}H NMR. J Am Chem Soc 127:5758–5759. https://doi.org/10.1021/ja0423320
9. Scholten JD, Ebeling G, Dupont J (2007) On the involvement of NHC carbenes in catalytic reactions by iridium complexes, nanoparticle and bulk metal dispersed in imidazolium ionic liquids. Dalton Trans 5554–5560. https://doi.org/10.1039/B707888A
10. Vignolle J, Don Tilley T (2009) N-heterocyclic carbene-stabilized gold nanoparticles and their assembly into 3D superlattices. Chem Commun:7230–7232. https://doi.org/10.1039/B913884F
11. Ranganath KVS, Kloesges J, Schäfer AH, Glorius F (2010) Asymmetric nanocatalysis: N-heterocyclic carbenes as chiral modifiers of Fe_3O_4/Pd nanoparticles. Angew Chem Int Ed 49:7786–7789. https://doi.org/10.1002/anie.201002782
12. Lara P, Rivada-Wheelaghan O, Conejero S, Poteau R, Philippot K, Chaudret B (2011) Ruthenium nanoparticles stabilized by N-heterocyclic carbenes: ligand location and influence on reactivity. Angew Chem Int Ed 50:12080–12084. https://doi.org/10.1002/anie.201106348
13. Crespo J, Guari Y, Ibarra A, Larionova J, Lasanta T, Laurencin D, López-de-Luzuriaga JM, Monge M, Olmos ME, Richeter S (2014) Ultrasmall NHC-coated gold nanoparticles obtained through solvent free thermolysis of organometallic Au(I) complexes. Dalton Trans 43:15713–15718. https://doi.org/10.1039/c4dt02160f
14. Song SG, Satheeshkumar C, Park J, Ahn J, Premkumar T, Lee Y, Song C (2014) N-heterocyclic carbene-based conducting polymer–gold nanoparticle hybrids and their catalytic application. Macromolecules 47:6566–6571. https://doi.org/10.1021/ma501817n
15. Thomas SR, Yang W, Morgan DJ, Davies TE, Li JJ, Fischer RA, Huang J, Dimitratos N, Casini A (2022) Bottom-up synthesis of water-soluble gold nanoparticles stabilized by N-heterocyclic carbenes: from structural characterization to applications. Chem A Eur J 28:e202201575. https://doi.org/10.1002/chem.202201575
16. Ye R, Zhukhovitskiy AV, Kazantsev RV, Fakra SC, Wickemeyer BB, Toste FD, Somorjai GA (2018) Supported Au nanoparticles with N-heterocyclic carbene ligands as active and stable heterogeneous catalysts for lactonization. J Am Chem Soc 140:4144–4149. https://doi.org/10.1021/jacs.8b01017

17. Kathuria L, Din Reshi NU, Samuelson AG (2020) N-heterocyclic carbene (NHC)-stabilized Ru^0 nanoparticles: in situ generation of an efficient transfer hydrogenation catalyst. Chem A Eur J 26:7622–7630. https://doi.org/10.1002/chem.202000142
18. Oberhauser W, Evangelisti C, Liscio A, Kovtun A, Cao Y, Vizza F (2018) Glycerol to lactic acid conversion by NHC-stabilized iridium nanoparticles. J Catal 368:298–305. https://doi.org/10.1016/j.jcat.2018.10.024
19. Asensio JM, Tricard S, Coppel Y, Andrés R, Chaudret B, de Jesús E (2017) Synthesis of water-soluble palladium nanoparticles stabilized by sulfonated N-heterocyclic carbenes. Chem A Eur J 23:13435–13444. https://doi.org/10.1002/chem.201702204
20. Gómez-Villarraga F, De Tovar J, Guerrero M, Nolis P, Parella T, Lecante P, Romero N, Escriche L, Bofill R, Ros J, Sala X, Philippot K, García-Antón J (2017) Dissimilar catalytic behavior of molecular or colloidal palladium systems with a new NHC ligand. Dalton Trans 46: 11768–11778. https://doi.org/10.1039/c7dt02729j
21. Baquero EA, Tricard S, Flores JC, de Jesús E, Chaudret B (2014) Highly stable water-soluble platinum nanoparticles stabilized by hydrophilic N-heterocyclic carbenes. Angew Chem Int Ed 53:13220–13224. https://doi.org/10.1002/anie.201407758
22. Kaeffer N, Liu HJ, Lo HK, Fedorov A, Copéret C (2018) An N-heterocyclic carbene ligand promotes highly selective alkyne semihydrogenation with copper nanoparticles supported on passivated silica. Chem Sci 9:5366–5371. https://doi.org/10.1039/C8SC01924J
23. Mollar-Cuni A, Ventura-Espinosa D, Martín S, Mayoral A, Borja P, Mata JA (2018) Stabilization of nanoparticles produced by hydrogenation of palladium–N-heterocyclic carbene complexes on the surface of graphene and implications in catalysis. ACS Omega 3:15217–15228. https://doi.org/10.1021/acsomega.8b02193
24. Ventura-Espinosa D, Martín S, Mata JA (2019) The non-innocent role of graphene in the formation/immobilization of ultra-small gold nanoparticles functionalized with N-heterocyclic carbene ligands. J Catal 375:419–426. https://doi.org/10.1016/j.jcat.2019.06.009
25. Ventura-Espinosa D, Martín S, García H, Mata JA (2021) Ligand effects in the stabilization of gold nanoparticles anchored on the surface of graphene: implications in catalysis. J Catal 394: 113–120. https://doi.org/10.1016/j.jcat.2020.12.027
26. Chaudret B, Philippot K (2007) Organometallic derived metals, colloids, and nanoparticles. Comprehensive organometallic chemistry III, vol 12, pp 71–99. https://doi.org/10.1016/B0-08-045047-4/00168-0
27. Lara P, Suárez A, Collière V, Philippot K, Chaudret B (2014) Platinum N-heterocyclic carbene nanoparticles as new and effective catalysts for the selective hydrogenation of nitroaromatics. ChemCatChem 6:87–90. https://doi.org/10.1002/cctc.201300821
28. Díaz de los Bernardos M, Pérez-Rodríguez S, Gual A, Claver C, Godard C (2017) Facile synthesis of NHC-stabilized Ni nanoparticles and their catalytic application in the Z-selective hydrogenation of alkynes. Chem Commun 53:7894–7897. https://doi.org/10.1039/c7cc01779k
29. Lomelí-Rosales DA, Delgado JA, Díaz de los Bernardos M, Pérez-Rodríguez S, Gual A, Claver C, Godard C (2019) A general one-pot methodology for the preparation of mono and bimetallic nanoparticles supported on carbon nanotubes: application in the semi-hydrogenation of alkynes and acetylene. Chem A Eur J 25:8321–8331. https://doi.org/10.1002/chem.201901041
30. Valero M, Bouzouita D, Palazzolo A, Atzrodt J, Dugave C, Tricard S, Feuillastre S, Pieters G, Chaudret B, Derdau V (2020) NHC-stabilized iridium nanoparticles as catalysts in hydrogen isotope exchange reactions of anilines. Angew Chem Int Ed 59:3517–3522. https://doi.org/10.1002/anie.201914369
31. Martinez-Espinar F, Blondeau P, Nolis P, Chaudret B, Claver C, Castillón S, Godard C (2017) NHC-stabilised Rh nanoparticles: surface study and application in the catalytic hydrogenation of aromatic substrates. J Catal 354:113–127. https://doi.org/10.1016/j.jcat.2017.08.010
32. Cerezo-Navarrete C, Lara P, Martínez-Prieto LM (2020) Organometallic nanoparticles ligated by NHCs: synthesis, surface chemistry and ligand effects. Catalysts 10:1144. https://doi.org/10.3390/catal10101144

33. Martínez-Prieto LM, Ferry A, Lara P, Richter C, Philippot K, Glorius F, Chaudret B (2015) New route to stabilize ruthenium nanoparticles with non-isolable chiral N-heterocyclic carbenes. Chem A Eur J 21:17495–17502. https://doi.org/10.1002/chem.201502601
34. Molinilllo P, Puyo M, Vattier F, Lacroix B, Rendón N, Lara P, Suárez A (2023) Ruthenium nanoparticles stabilized by 1,2,3-triazolylidene ligands in the hydrogen isotope exchange of E–H bonds (E = B, Si, Ge, Sn) using deuterium gas. Nanoscale 15:14488–14495. https://doi.org/10.1039/d3nr02637j
35. Bouzouita D, Lippens G, Baquero EA, Fazzini PF, Pieters G, Coppel Y, Lecante P, Tricard S, Martínez-Prieto LM, Chaudret B (2019) Tuning the catalytic activity and selectivity of water-soluble bimetallic RuPt nanoparticles by modifying their surface metal distribution. Nanoscale 11:16544–16552. https://doi.org/10.1039/c9nr04149d
36. Zuluaga-Villamil A, Mencia G, Asensio JM, Fazzini PF, Baquero EA, Chaudret B (2022) N-heterocyclic carbene-based iridium and ruthenium/iridium nanoparticles for the hydrogen isotope exchange reaction through C–H bond activations. Organometallics 41:3313–3319. https://doi.org/10.1021/acs.organomet.2c00288
37. Suárez-Riaño O, Mencia G, Tricard S, Esvan J, Fazzini PF, Chaudret B, Baquero EA (2023) Water-soluble NHC Pd/Ni bimetallic nanoparticles for H/D exchange in aromatic amino-acids. Chem Commun 59:1062–1065. https://doi.org/10.1039/d2cc06019a
38. Richter C, Schaepe K, Glorius F, Ravoo BJ (2014) Tailor-made N-heterocyclic carbenes for nanoparticle stabilization. Chem Commun 50:3204–3207. https://doi.org/10.1039/C4CC00654B
39. Rühling A, Schaepe K, Rakers L, Vonhören B, Tegeder P, Ravoo BJ, Glorius F (2016) Modular bidentate hybrid NHC-thioether ligands for the stabilization of palladium nanoparticles in various solvents. Angew Chem Int Ed 55:5856–5860. https://doi.org/10.1002/anie.201508933
40. Ferry A, Schaepe K, Tegeder P, Richter C, Chepiga KM, Ravoo BJ, Glorius F (2015) Negatively charged N-heterocyclic carbene-stabilized Pd and Au nanoparticles and efficient catalysis in water. ACS Catal 5:5414–5420. https://doi.org/10.1021/acscatal.5b01160
41. Tegeder P, Freitag M, Chepiga KM, Muratsugu S, Möller N, Lamping S, Tada M, Glorius F, Ravoo BJ (2018) N-heterocyclic carbene-modified Au–Pd alloy nanoparticles and their application as biomimetic and heterogeneous catalysts. Chem A Eur J 24:18682–18688. https://doi.org/10.1002/chem.201803274
42. Zhang L, Wei Z, Meng M, Ung G, He J (2020) Do polymer ligands block the catalysis of metal nanoparticles? Unexpected importance of binding motifs in improving catalytic activity. J Mater Chem A 8:15900–15908. https://doi.org/10.1039/d0ta03906c
43. Cao Z, Kim D, Hong D, Yu Y, Xu J, Lin S, Wen X, Nichols EM, Jeong K, Reimer JA, Yang P, Chang CJ (2016) A molecular surface functionalization approach to tuning nanoparticle electrocatalysts for carbon dioxide reduction. J Am Chem Soc 138:8120–8125. https://doi.org/10.1021/jacs.6b02878
44. Zhang L, Wei Z, Thanneeru S, Meng M, Kruzyk M, Ung G, Liu B, He J (2019) A polymer solution to prevent nanoclustering and improve the selectivity of metal nanoparticles for electrocatalytic CO_2 reduction. Angew Chem Int Ed 58:15834–15840. https://doi.org/10.1002/anie.201909069
45. Ernst JB, Muratsugu S, Wang F, Tada M, Glorius F (2016) Tunable heterogeneous catalysis: N-heterocyclic carbenes as ligands for supported heterogeneous Ru/K-Al_2O_3 catalysts to tune reactivity and selectivity. J Am Chem Soc 34:10718–10721. https://doi.org/10.1021/jacs.6b03821
46. Palazzolo A, Naret T, Daniel-Bertrand M, Buisson DA, Tricard S, Lesot P, Coppel Y, Chaudret B, Feuillastre S, Pieters G (2020) Tuning the reactivity of a heterogeneous catalyst using N-heterocyclic carbene ligands for C–H activation reactions. Angew Chem Int Ed 59:20879–20884. https://doi.org/10.1002/anie.202009258
47. Ikemoto S, Muratsugu S, Koitaya T, Tsuji Y, Das M, Yoshizawa K, Glorius F, Tada M (2023) Coordination-induced trigger for activity: N-heterocyclic carbene-decorated ceria catalysts

incorporating Cr and Rh with activity induction by surface adsorption site control. J Am Chem Soc 145:1497–1504. https://doi.org/10.1021/jacs.2c07290
48. Ernst JB, Schwermann C, Yokota G, Tada M, Muratsugu S, Doltsinis NL, Glorius F (2017) Molecular adsorbates switch on heterogeneous catalysis: induction of reactivity by N-heterocyclic carbenes. J Am Chem Soc 139:9144–9147. https://doi.org/10.1021/jacs.7b05112
49. Lewis RJ, Koy M, Macino M, Mowpriya Das M, Carter JH, Morgan DJ, Davies TE, Ernst JB, Freakley SJ, Glorius F, Hutchings GJ (2022) N-heterocyclic carbene modified palladium catalysts for the direct synthesis of hydrogen peroxide. J Am Chem Soc 144:15431–15436. https://doi.org/10.1021/jacs.2c04828
50. Bruna L, Cardona-Farreny M, Colliere V, Philippot K, Axet MR (2022) In situ ruthenium catalyst modification for the conversion of furfural to 1,2-pentanediol. Nanomaterials 12:328. https://doi.org/10.3390/nano12030328
51. Van der Klink JJ, Brom HB (2000) NMR in metals, metal particles and metal cluster compounds. Prog Nucl Magn Reson Spectrosc 36:89–201. https://doi.org/10.1016/S0079-6565(99)00020-5
52. Novio F, Philippot K, Chaudret B (2010) Location and dynamics of CO co-ordination on Ru nanoparticles: a solid state NMR study. Catal Lett 140:1–7. https://doi.org/10.1007/s10562-010-0428-7
53. Bradley JS, Hill EW, Behal S, Klein C, Duteil A, Chaudret B (1992) Preparation and characterization of organosols of monodispersed nanoscale palladium. Particle size effects in the binding geometry of adsorbed carbon monoxide. Chem Mater 4:1234–1239. https://doi.org/10.1021/cm00024a023
54. Martínez-Prieto LM, Rakers L, López-Vinasco AM, Cano I, Coppel Y, Philippot K, Glorius F, Chaudret B, van Leeuwen PWNM (2017) Soluble platinum nanoparticles ligated by long-chain N-heterocyclic carbenes as catalysts. Chem A Eur J 23:12779–12786. https://doi.org/10.1002/chem.201702288
55. Martínez-Prieto LM, Ferry A, Rakers L, Richter C, Lecante P, Philippot K, Chaudret B, Glorius F (2016) Long-chain NHC-stabilized RuNPs as versatile catalysts for one-pot oxidation/hydrogenation reactions. Chem Commun 52:4768–4771. https://doi.org/10.1039/C6CC01130F
56. Lara P, Martínez-Prieto LM, Roselló-Merino M, Richter C, Glorius F, Conejero S, Philippot K, Chaudret B (2016) NHC-stabilized Ru nanoparticles: synthesis and surface studies. Nano-Struct Nano-Objects 6:39–45. https://doi.org/10.1016/j.nanoso.2016.03.003
57. Asensio JM, Tricard S, Coppel Y, Andrés R, Chaudret B, de Jesús E (2016) Knight shift in ^{13}C NMR resonances confirms the coordination of N-heterocyclic carbene ligands to water-soluble palladium nanoparticles. Angew Chem Int Ed 56:865–869. https://doi.org/10.1002/anie.201610251
58. Asensio JM, Bouzouita C, Piet WNM, van Leeuwen PWNM, Chaudret B (2020) σ-H–H, σ-C–H, and σ-Si–H bond activation catalyzed by metal nanoparticles. Chem Rev 120:1042–1084. https://doi.org/10.1021/acs.chemrev.9b00368
59. Tegeder P, Marelli M, Freitag M, Polito L, Lamping S, Psaro R, Glorius F, Ravoo BJ, Evangelisti C (2018) Metal vapor synthesis of ultrasmall Pd nanoparticles functionalized with N-heterocyclic carbenes. Dalton Trans 47:12647–12651. https://doi.org/10.1039/C8DT02535E
60. Ruiz-Varilla AM, Baquero EA, Chaudret B, de Jesús E, Gonzalez-Arellano C, Flores JC (2020) Water-soluble NHC-stabilized platinum nanoparticles as recoverable catalysts for hydrogenation in water. Cat Sci Technol 10:2874–2881. https://doi.org/10.1039/d0cy00481b
61. Rakers L, Martínez-Prieto LM, López-Vinasco AM, Philippot K, van Leeuwen PWNM, Chaudret B, Glorius F (2018) Ruthenium nanoparticles ligated by cholesterol-derived NHCs and their application in the hydrogenation of arenes. Chem Commun 54:7070–7073. https://doi.org/10.1039/c8cc02833h
62. García-Zaragoza A, Cerezo-Navarrete C, Mollar-Cuni A, Oña-Burgos P, Mata JA, Corma A, Martínez-Prieto LM (2022) Tailoring graphene-supported Ru nanoparticles by

functionalization with pyrene-tagged N-heterocyclic carbenes. Cat Sci Technol 12:1257–1270. https://doi.org/10.1039/d1cy02063c
63. Delgado JA, Benkirane O, Claver C, Curulla-Ferré D, Godard C (2017) Advances in the preparation of highly selective nanocatalysts for the semi-hydrogenation of alkynes using colloidal approaches. Dalton Trans 46:12381–12403. https://doi.org/10.1039/c7dt01607g
64. Kaeffer N, Mance D, Copéret C (2020) N-heterocyclic carbene coordination to surface copper sites in selective semihydrogenation catalysts from solid-state NMR spectroscopy. Angew Chem Int Ed 59:19999–20007. https://doi.org/10.1002/anie.202006209
65. Wiesenfeldt MP, Nairoukh Z, Dalton T, Glorius F (2019) Selective arene hydrogenation for direct access to saturated carbo- and heterocycles. Angew Chem Int Ed 58:10460–10476. https://doi.org/10.1002/anie.201814471
66. Gonzalez-Galvez D, Lara P, Rivada-Wheelaghan O, Conejero S, Chaudret B, Philippot K, van Leeuwen PWNM (2013) NHC-stabilized ruthenium nanoparticles as new catalysts for the hydrogenation of aromatics. Cat Sci Technol 3:99–105. https://doi.org/10.1039/C2CY20561K
67. Jayaraj A, Raveedran AV, Latha AT, Priyadarshini D, Swamy PCA (2023) Coordination versatility of NHC-metal topologies in asymmetric catalysis: synthetic insights and recent trends. Coord Chem Rev 478:214922. https://doi.org/10.1016/j.ccr.2022.214922
68. Racha A, Samanta C, Sreekantan S, Marimuthu B (2023) Review on catalytic hydrogenation of biomass-derived furfural to furfuryl alcohol: recent advances and future trends. Energy Fuel 37: 11475–11496. https://doi.org/10.1021/acs.energyfuels.3c01174
69. Haydar Goksu H, Sert H, Kilbas B, Sen F (2017) Recent advances in the reduction of nitro compounds by heterogenous catalysts. Curr Org Chem 21:794–820. https://doi.org/10.2174/1385272820666160525123907
70. Lara P, Philippot K (2014) The hydrogenation of nitroarenes mediated by platinum nanoparticles: an overview. Cat Sci Technol 4:2445–2465. https://doi.org/10.1039/C4CY00111G
71. Axet MR, Conejero S, Gerber IC (2018) Ligand effects on the selective hydrogenation of nitrobenzene to cyclohexylamine using ruthenium nanoparticles as catalysts. ACS Appl Nano Mater 1:5885–5894. https://doi.org/10.1021/acsanm.8b01549
72. Corma A, Serna P (2006) Chemoselective hydrogenation of nitro compounds with supported gold catalysts. Science 313:332–334. https://doi.org/10.1126/science.1128383
73. Zhang K, Suh JM, Choi J-W, Jang HW, Shokouhimehr M, Varma RS (2019) Recent advances in the nanocatalyst-assisted $NaBH_4$ reduction of nitroaromatics in water. ACS Omega 4:483–495. https://doi.org/10.1021/acsomega.8b03051
74. Gou XX, Liu T, Wang YY, Han YF (2020) Ultrastable and highly catalytically active N-heterocyclic-carbene-stabilized gold nanoparticles in confined spaces. Angew Chem Int Ed 59: 16683–16689. https://doi.org/10.1002/anie.202006569
75. Kopf S, Bourriquen F, Li W, Neumann H, Junge K, Beller M (2022) Recent developments for the deuterium and tritium labeling of organic molecules. Chem Rev 122:6634–6718. https://doi.org/10.1021/acs.chemrev.1c00795
76. Lepron M, Daniel-Bertrand M, Mencia G, Chaudret B, Feuillastre S, Pieters G (2021) Nanocatalyzed hydrogen isotope exchange. Acc Chem Res 54:1465–1480. https://doi.org/10.1021/acs.accounts.0c00721
77. Martínez-Prieto LM, Baquero EA, Pieters G, Flores JC, de Jesús E, Nayral C, Delpech F, van Leeuwen PWNM, Lippens G, Chaudret B (2017) Monitoring of nanoparticle reactivity in solution: interaction of L-lysine and Ru nanoparticles probed by chemical shift perturbation parallels regioselective H/D exchange. Chem Commun 53:5850–5853. https://doi.org/10.1039/c7cc02445b
78. Zuiderweg ERP (2002) Mapping protein–protein interactions in solution by NMR spectroscopy. Biochemistry 41:1–7. https://doi.org/10.1021/bi011870b
79. Palazzolo A, Feuillastre S, Pfeifer V, Garcia-Argote S, Bouzouita D, Tricard S, Chollet C, Marcon E, Buisson DA, Cholet S, Fenaille F, Lippens G, Chaudret B, Pieters G (2019) Efficient access to deuterated and tritiated nucleobase pharmaceuticals and oligonucleotides using

hydrogen-isotope exchange. Angew Chem Int Ed 58:4891–4895. https://doi.org/10.1002/anie.201813946
80. Bouzouita D, Asensio JM, Pfeifer V, Palazzolo A, Lecante P, Pieters G, Feuillastre S, Tricard S, Chaudret B (2020) Chemoselective H/D exchange catalyzed by nickel nanoparticles stabilized by N-heterocyclic carbene ligands. Nanoscale 12:15736–15742. https://doi.org/10.1039/d0nr04384b
81. Planellas M, Pleixats R, Shafir A (2012) Palladium nanoparticles in Suzuki cross-couplings: tapping into the potential of tris-imidazolium salts for nanoparticle stabilization. Adv Synth Catal 354:651–662. https://doi.org/10.1002/adsc.201100574
82. Martínez-Laguna J, Mollar-Cuni A, Ventura-Espinosa D, Martín S, Caballero A, Mata JA, Pérez PJ (2022) Gold nanoparticle-catalysed functionalization of carbon–hydrogen bonds by carbene transfer reactions. Dalton Trans 51:5250–5256. https://doi.org/10.1039/d1dt04351j
83. Lu Q, Jiao F (2016) Electrochemical CO_2 reduction: Electrocatalyst, reaction mechanism, and process engineering. Nano Energy 29:439–456. https://doi.org/10.1016/j.nanoen.2016.04.009
84. Mustieles Marin I, Asensio JM, Chaudret B (2021) Bimetallic nanoparticles associating noble metals and first-row transition metals in catalysis. ACS Nano 15:3550–3556. https://doi.org/10.1021/acsnano.0c09744
85. Vivancos A, Segarra C, Albrecht M (2018) Mesoionic and related less heteroatom-stabilized N-heterocyclic carbene complexes: synthesis, catalysis, and other applications. Chem Rev 118:9493–9586. https://doi.org/10.1021/acs.chemrev.8b00148
86. Cui L, Du M, Guo S (2022) Preparation, characterization and stability studies of gold nanoparticles capped by 1,2,3-triazole-based mesoionic carbenes. ChemistrySelect 7:e20220199. https://doi.org/10.1002/slct.202201999
87. Sau SC, Hota PK, Mandal SK, Soleilhavoup M, Bertrand G (2020) Stable abnormal N-heterocyclic carbenes and their applications. Chem Soc Rev 49:1233–1252. https://doi.org/10.1039/c9cs00866g
88. Ren J, Freitag M, Gao Y, Bellotti P, Das M, Schulze Lammers B, Mönig H, Zhang Y, Daniliuc CG, Du S, Fuchs H, Glorius F (2022) Reversible self-assembly of an N-heterocyclic carbene on metal surfaces. Angew Chem Int Ed 61:e202115104. https://doi.org/10.1002/anie.202115104

Top Organomet Chem (2024) 75: 199–242
https://doi.org/10.1007/3418_2024_118
© The Author(s), under exclusive license to Springer Nature Switzerland AG 2024
Published online: 8 October 2024

Metal Nanoparticles Functionalized with Hydrosoluble Ligands

Oscar Suárez-Riaño and Edwin A. Baquero

Contents

Abstract Metal nanoparticles (MNPs) functionalized with hydrosoluble ligands have emerged as versatile tools in various fields, including catalysis and biomedical applications, among others. This chapter explores recent advances in the synthesis, characterization, and applications of water-soluble MNPs stabilized by diverse ligands. The significance of working with water is emphasized, driven by the imperative for environmentally friendly and sustainable chemistry. To address the challenge of achieving solubility and stability in aqueous environments for MNPs, a range of hydrosoluble ligands have been explored. These ligands, including carboxylic acids, polymers, surfactants, thiols, phosphines, N-heterocyclic carbene (NHC) ligands, proteins, peptides, and bio-extracts, play crucial roles in conferring water solubility, mitigating aggregation and enhancing the catalytic properties of MNPs. The chapter provides a comprehensive overview of the synthesis methods, structural characterization techniques, and applications of water-soluble MNPs across various metal species, including Au, Ru, Pd, Pt, Ag, Rh, Fe, and Cu. Applications of water-

O. Suárez-Riaño and E. A. Baquero (✉)
Estado Sólido y Catálisis Ambiental (ESCA), Departamento de Química, Facultad de Ciencias, Universidad Nacional de Colombia, Bogota, Colombia
e-mail: eabaquerov@unal.edu.co

soluble MNPs discussed in this chapter encompass mainly catalytic processes and biomedical applications including imaging. The unique properties of MNPs functionalized with hydrosoluble ligands offer opportunities for tailored applications with enhanced selectivity, activity, and biocompatibility.

Keywords Catalysis · Chemistry of materials · Hydrosoluble ligands · Metallic nanoparticles · Surface chemistry

1 Introduction

In the contemporary era, the significance of working with water cannot be overstated, primarily attributed to the imperative for a more environmentally friendly and sustainable approach to chemistry. Water offers numerous advantages, including its abundant availability, non-toxic nature, environmental compatibility, and affordability. Additionally, its inherent properties such as high heat capacity and polarity further enhance its appeal [1, 2]. Hence, there exists a critical need to develop catalytic systems capable of operating in aqueous environments, exhibiting superior performance in terms of both selectivity and activity.

One effective strategy to accomplish this objective is through the utilization of metallic nanoparticles (MNPs), which have garnered considerable attention in recent years due to their exceptional properties compared to traditional homogeneous and heterogeneous catalysts. These innovative systems promise to combine the high catalytic activity, encompassing conversion and selectivity, characteristic of homogeneous catalysts, with the advantageous feature of separability and reusability associated with heterogeneous systems [3].

However, harnessing the potential of MNPs in aqueous environments necessitates the attainment of requisite solubility and stability for these metals. To address this challenge, a diverse array of hydrosoluble ligands have been explored. These ligands not only serve to confer water solubility through the presence of polar moieties but also to mitigate aggregation and exploit additional properties conferred by the ligands in MNP applications, in addition to enhancing hydrophilicity and even selectivity and activity in catalytic processes when they are used as nanocatalysts. Among the myriad of stabilizing agents utilized the most employed are carboxylic acids (both aliphatic and aromatic), polymers, surfactants, thiols, phosphines, N-heterocyclic carbene (NHC) ligands, proteins, peptides, and their bio-extracts.

In this chapter, we present the recent advances of water-soluble MNPs elucidating their functionalization with diverse ligands (presented along the text) and corresponding applications. The manuscript is organized according to different metals, with a focus on Au, Ru, Pd, Pt, Ag, Rh, Fe, and Cu-based NPs discussed throughout. It is important to note that this chapter does not aim to comprehensively cover all reported examples. Therefore, we encourage readers to consult various reviews and book chapters that delve into the intricacies of water-soluble MNPs, where such nanoparticles are of primary interest [1, 4, 5].

2 Gold-Based Nanoparticles

Gold nanoparticles (Au NPs) have been extensively studied among various metals due to their diverse applications in catalysis, sensing, and biotechnology. For instance, Guo et al. synthesized Au NPs stabilized by a nitrogen-enriched PEG polymer through the reduction of $HAuCl_4$ in the presence of $NaBH_4$ (Scheme 1) [6]. Various $\mathbf{L_1}$:Au ratios were investigated, revealing that as the ligand concentration relative to the metal increases, the NPs (**Au-1@**$\mathbf{L_1}$) decrease in size. Sizes as small as 1.1 nm were achieved with a ligand/metal ratio of 2 equivalents, whereas NPs of 4.4 nm were obtained with 0.04 equivalents. This phenomenon occurs because higher stabilizer concentrations favor the nucleation process. Additionally, the localized surface plasmon resonance (LSPR) of the Au NPs was influenced by the ligand concentration, exhibiting sharper peaks as the NP size increased and the ligand amount decreased. These Au NPs were employed in the selective reduction of nitroarenes to their corresponding aniline derivatives, achieving high conversion rates of up to 99% and yields ranging from 82% to 99% in shorter reaction times, typically less than 1.5 h. Remarkably, the catalytic activity of these NPs was preserved even after four consecutive runs.

It is well-known that NHC ligands endow MNPs with exceptional stability and enriched electron density. With this in mind, various examples have been developed in Au-based systems. For instance, MacLeod et al. synthesized **Au-2@**$\mathbf{L_2}$ NPs through the reduction of an Au-NHC-PEG complex, prepared via the transmetalation of an Ag-NHC complex in the presence of an Au(III) salt (Scheme 2) [7]. The incorporation of the PEG moiety aims to provide neutral and bulkier hydrophilic steric stabilization facilitated by the NHC ligand. The Au NPs were synthesized by the addition of borane *tert*-butylamine, yielding Au NPs with average sizes around 4.2 nm as determined by TEM and 16.2 nm as determined by dynamic light scattering (DLS), a common technique for studying colloidal systems and their sizes in solution. The discrepancy in sizes is attributed to the presence of the long-chain PEG-NHC ligand on the surface. FTIR and ^1H NMR studies confirmed the presence of $\mathbf{L_2}$ in the Au NPs. However, direct evidence of NHC-Au coordination

Scheme 1 Synthesis of **Au-1@**$\mathbf{L_1}$ NPs (Adapted with permission from Ref. [6] Copyright 2015 Wiley)

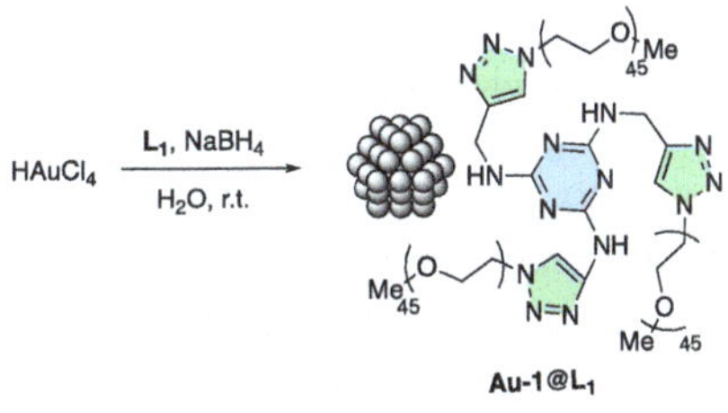

Scheme 2 Synthesis of (**a**) L_2-Au^I-NHC precursor and (**b**) **Au-2@L$_2$** NPs (Adapted with permission from Ref. [7] Copyright 2015 ACS Publications)

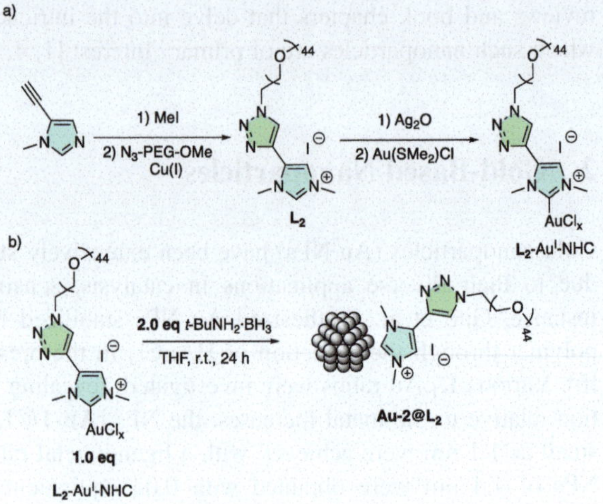

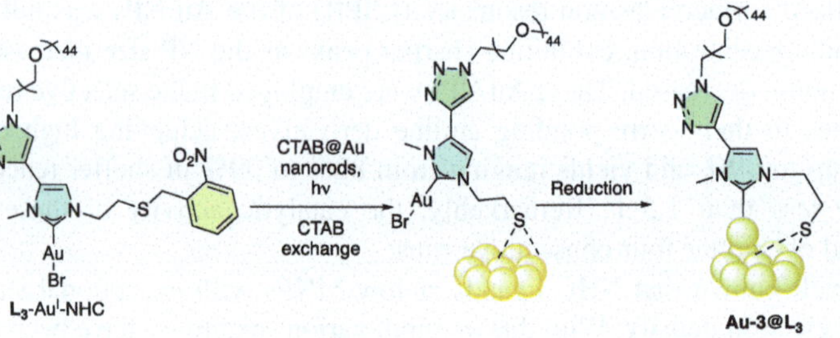

Scheme 3 Synthesis of **Au-3@L$_3$** NPs (Adapted with permission from Ref. [8] Copyright 2018 Nature)

within the NP system was not presented. The stability of these MNPs was evaluated in aqueous solutions under various conditions. They exhibited good stability at both low and high temperatures (−78°C to 95°C), across a wide range of pH values (3–14), in cell culture media, and in aqueous H_2O_2 solutions. These results suggest that **Au-2@L$_2$** NPs could find utility in various biological applications.

In a subsequent investigation, MacLeod et al. employed a sulfur-modified NHC-PEG ligand (L_3) as stabilizer to obtain Au NPs via ligand exchange methodology, utilizing Au@CTAB nanorods as precursors and assisted by UV-Vis to induce thiolate formation, followed by the reduction of the Au^I-NHC complex (Scheme 3) [8]. For the reduction of the Au^I-NHC complex, they utilized potassium bis(trimethylsilyl)amide and borane *tert*-butylamine, observing no discernible difference in the final **Au-3@L$_3$** nanorods, thus preserving their shape and size (ca. ~30–40 nm in length and ~8–10 nm in width, as determined by TEM). Significantly, through various DFT studies, the authors elucidated the high stability of these systems, attributing it to an Au adatom that impedes structural reorganization for proper NHC coordination. Consistent with previous research, these Au

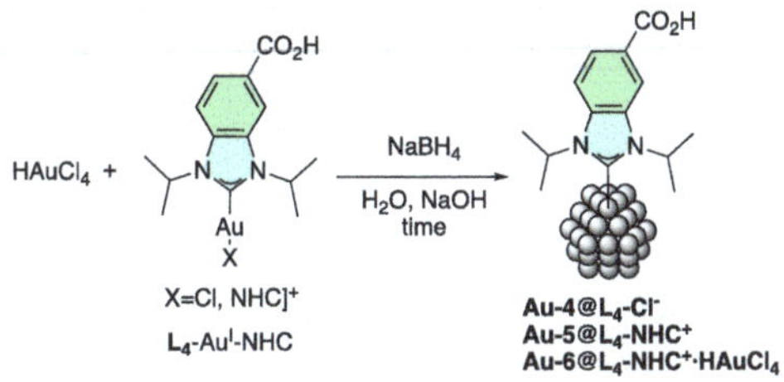

Scheme 4 Synthesis of Au@NHC NPs (Adapted with permission from Ref. [9] Copyright 2017 Wiley)

nanorods were examined under various aqueous solution conditions, demonstrating robust stability at both low and high temperatures, in the presence of high salt concentrations, across a wide pH range and in biological media and cell cultures.

In another study employing Au-NHC complexes as precursors, Salorinne et al. obtained Au NPs by reducing a monocarbene-Au complex, a biscarbene-Au complex, and a mixture of the biscarbene-Au complex with $HAuCl_4$ salt in the presence of $NaBH_4$ (Scheme 4) [9]. NP formation was monitored via polyacrylamide gel electrophoresis, representing a novel method to track synthesis. Utilizing different starting materials resulted in varying rates of MNP formation; MNPs formed more rapidly with the monocarbene-Au complex and the mixture of the biscarbene complex and $HAuCl_4$ compared to the slower formation observed with the biscarbene complex alone. However, after a certain time (5 h), similar MNP sizes were achieved (2–2.4 nm, as determined by TEM). The size of the Au NPs depended on reaction time, with smaller systems observed at 5 h (around 2 nm) and slightly larger NPs at 24 h (around 3.3 nm). Additionally, these Au NPs demonstrated high stability over time and resistance to electrolytes and biological nucleophiles, such as glutathione, when tested under various aqueous conditions. Finally, the NHC-stabilized Au NPs (Au@NHC NPs) were evaluated as photoacoustic probes, yielding promising results even with the smaller systems.

Following the concept of employing NHC as stabilizers for Au NPs, Nosratbad et al. utilized various polymer-modified imidazolium salts (**L_5** and **L_6**) as precursors for the in situ generation of NHC ligands to stabilize Au NPs via ligand exchange using previously oleylamine (OLY) stabilized Au NPs (Scheme 5) [10]. These Au@NHC polymer NPs were characterized by ^{1}H NMR to determine NHC coordination, where the corresponding signal of the acidic proton was not observed, unlike in the imidazolium salt. Optical properties of **Au@OLY**, **Au-7@L_5**, and **Au-8@L_6** NPs showed no significant differences. On the other hand, the NH-stabilized Au NPs exhibited high colloidal stability under various aqueous conditions, encompassing a wide pH range (3–12), and in the presence of NaCl and dithiothreitol. Subsequently, characterization via DLS revealed polydisperse systems with the mono-NHC polymer (**Au-7@L_5**), ranging in size from 10 to 100 nm, significantly larger compared to **Au@OLY** (10 nm), while the multi-NHC polymer (**Au-8@L_6**) yielded monodisperse systems with sizes closer to those of **Au@OLY**, ranging from 10 to 20 nm.

In comparison to previous studies where PEGylated NHC ligands were used to stabilize Au NPs, Nguyen et al. employed a different type of NHC ligand: a polymeric mesoionic carbene with a 1,2,3-triazole backbone as novel stabilizers,

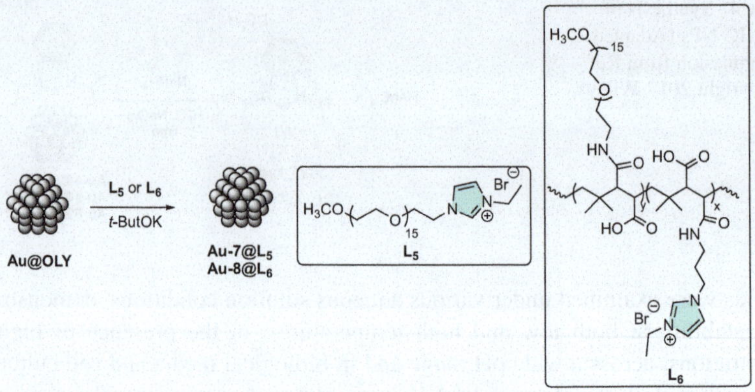

Scheme 5 Synthesis of **Au-7@L$_5$** and **Au-8@L$_6$** NPs (Adapted with permission from Ref. [10] Copyright 2023 Springer)

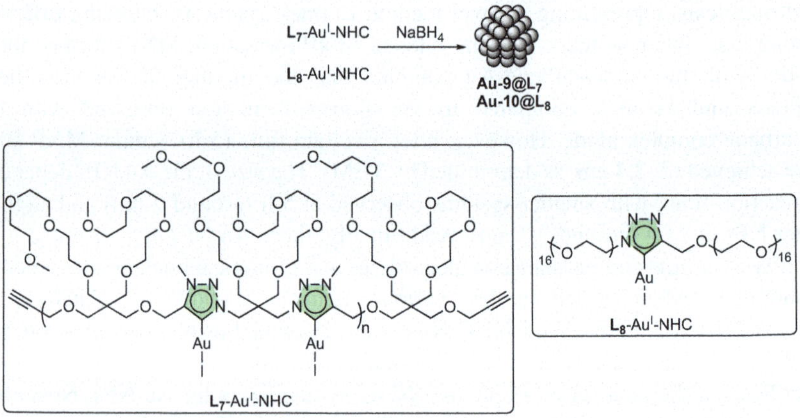

Scheme 6 Synthesis of Au@polyNHC NPs (Adapted with permission from Ref. [11] Copyright 2021 ACS Publications)

due to their higher sigma-donor character compared to conventional NHC ligands (**L$_7$** and **L$_8$**) [11]. The synthesis of the Au NPs was carried out using either polymeric or monomeric mesoionic Au^{I}-NHC complexes (Scheme 6) as precursors in the presence of $NaBH_4$, requiring more time and a higher equivalent of reducing agent for the polymeric Au complex due to the steric confinement. Au NPs generated from the polymeric complex (**Au-9@L$_7$**) exhibited a mean size around 11.8 nm, whereas Au NPs from the monomeric complexes (**Au-10@L$_8$**) showed MNP sizes around 4.1 nm, as determined by TEM, illustrating the effect of the metal precursor. Characterization via XPS revealed both Au metallic and Au^{I} species present in the NPs. Additionally, by ^{1}H NMR, broadening of signals close to the Au surface was observed in the Au NPs suggesting coordination of the NHCs to the surface. These Au NPs were evaluated under various conditions, demonstrating high stability across

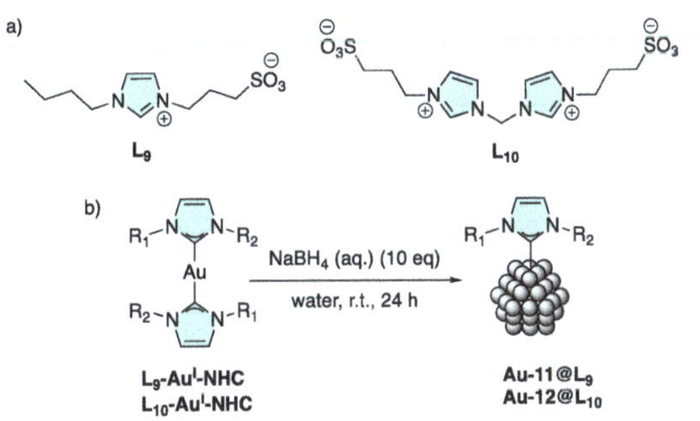

Scheme 7 (**a**) Chemical structure of ligands L_9 and L_{10} and (**b**) synthesis of **Au-11@L$_9$** and **Au-12@L$_{10}$** NPs (Adapted with permission from Ref. [12] Copyright 2022 Wiley)

a wide pH range (2–12), with higher stability at acidic pH for **Au-9@L$_7$** NPs. Notably, they exhibited high stability against high concentrations of glutathione (up to 6 mM for **Au-9@L$_7$**), representing the highest concentration reported using Au NPs. Finally, the Au@polyNHCs NPs were tested in the catalytic hydrogenation of 4-nitrophenol, yielding better results for **Au-10@L$_8$** NPs due to the higher steric hindrance of **Au-9@L$_7$** NPs.

Thomas et al. synthesized two distinct Au@NHC NPs (**Au-11@L$_9$** and **Au-12@L$_{10}$**) by reducing Au^I-NHC complexes containing mono- and bidentate NHC ligands in the presence of an excess of $NaBH_4$ (Scheme 7) [12]. These MNPs underwent characterization via ^{1}H NMR and ATR-FTIR to verify the presence or absence of the acidic proton of the carbenic carbon. As anticipated, no signals were detected using both techniques, suggesting NHC coordination through the carbenic carbon to the surface. Regarding size, no differences were observed between the two NHC ligands used as stabilizers, yielding Au NPs with 2.3–2.6 nm in diameter, as determined by TEM. Furthermore, XPS results indicated the presence of both Au metallic and Au-oxidized species in these systems, likely attributable to a monolayer of Au^I-NHC complex enveloping the Au^0 core of the MNPs. Finally, these Au NPs were evaluated in the reduction of 4-nitrophenol to 4-aminophenol, demonstrating high conversion rates after 1.5 min (81%) for the monodentate Au@NHC NPs (**Au-11@L$_9$**), whereas the bidentate system (**Au-12@L$_{10}$**) required more time (6 min) to achieve similar results, possibly due to greater coverage of the MNP surface.

NHCs have been presented as Au NP stabilizers; however, it is not the only way how these heterocycles can be used as capping agents. In contrast, these can be used as imidazolium salts and stabilize MNPs not by a direct coordination bond but through electrostatic interactions, as presented by Monti et al. Several imidazolium salts were used in the presence of $HAuCl_4$ and hydrazine (Scheme 8) to obtain the corresponding Au NPs (**Au-13@L$_{11}$**, **Au-14@L$_{12}$**, and **Au-15@L$_{13}$**), with sizes between 15 and 9 nm, showing the smallest diameter of the bulkiest ligand-stabilized MNPs [13–15]. These systems were characterized by FTIR and ^{1}H NMR, finding significant changes in both intensity and wavenumber and chemical shift values, respectively, for different functional groups of the stabilizing imidazolium salt. It confirmed the stabilization of the Au NPs by the imidazolium salts, and that

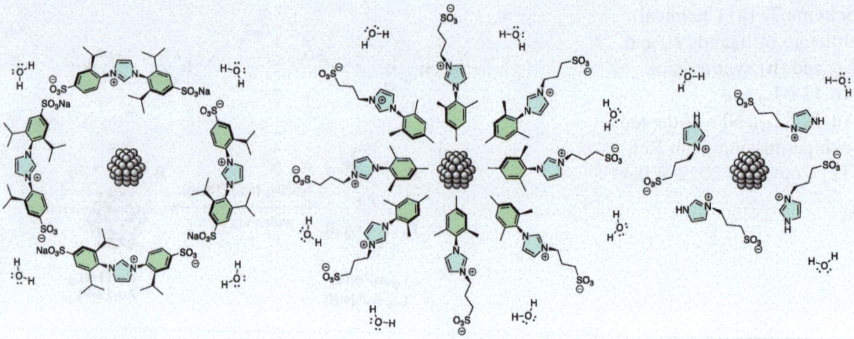

Scheme 8 Synthesis of Au@imidazolium salt NPs (Adapted with permission from Ref. [14] Copyright 2020 The Royal Society of Chemistry)

depending on the ligand, different ligand arrangements were found. Finally, these Au NPs were employed in the selective reduction of aromatic nitro compounds in water, finding better catalytic activities compared to previous studies (k: 46.83×10^{-3} vs 1×10^{-4}) but low recyclability, preserving the catalytic activity only for three cycles.

In a separate investigation, Enciso et al. utilized different stabilizers than the aforementioned NHC ligands, employing various tetrachloroaurate salts of triazine dendrons as stabilizers and precursors in the presence of $NaBH_4$ to synthesize Au NPs (Scheme 9) [16]. The inclusion of 4,7,10-trioxa-1,14-tridecanediamine groups conferred the desired hydrophilicity to the system. It is noteworthy that in this instance, the same Au salt served as both the stabilizing agent and precursor, presenting similarities with the methodologies detailed above where Au complexes were utilized as MNP precursors. The resultant Au NPs exhibited varying sizes within the range of 7 to 10 nm, with no discernible trend observed based on the bulkiness of the stabilizer. Lastly, the authors attempted to characterize Au-C bond formation via XPS analysis. However, due to the thickness of the dendrimer shell, successful probing was achievable for the first two dendrons (L_{14} and L_{15}) but not for the largest one (L_{16}).

Further, Ahmad et al. also explored a methodology wherein the stabilizing agent was intrinsic to the Au salt. Specifically, they utilized a water-soluble aryldiazonium Au^{III} salt to synthesize Au NPs, with benzoic acid acting as stabilizing agent [17]. The synthesis of Au NPs was facilitated by the addition of 9-borabicyclo [3.3.1]nonane (9-BBN) as a reducing agent under mild conditions (Scheme 10). Unlike conventional methods, the size of the nanoparticles was determined using UV-Vis spectroscopy, showing a diameter of 70 nm, a value consistent with that obtained from DLS (i.e., 68.2 nm). The stability of the Au NPs was rigorously assessed, demonstrating robust resistance to varying NaCl concentrations (ranging from 0.01 to 1.00 M), a broad pH range (1–13), and diverse temperatures (20–90°C). Furthermore, through DFT calculations, the authors determined the Au-C bond

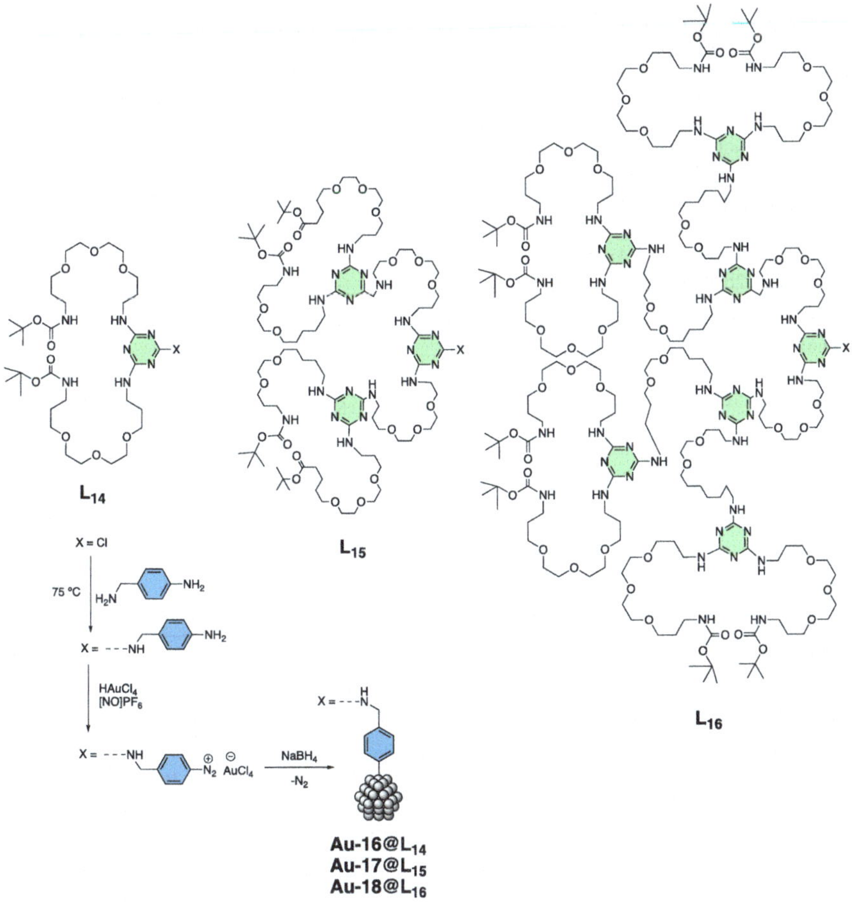

Scheme 9 Synthesis of Au@dendron NPs (Adapted with permission from Ref. [16] Copyright 2017 The Royal Society of Chemistry)

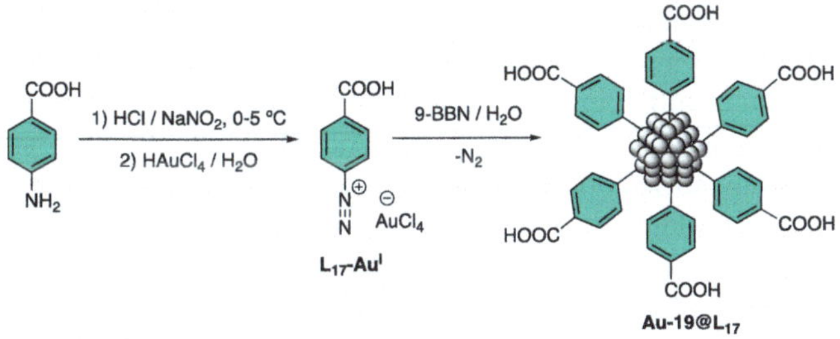

Scheme 10 Synthesis **Au-19@L_{17}** NPs (Adapted with permission from Ref. [17] Copyright 2017 The Royal Society of Chemistry)

Scheme 11 Synthesis of Au@denaturated-protein NPs (Adapted with permission from Ref. [18] Copyright 2020 SpringerOpen)

Scheme 12 Copolymer L_{18} used in the stabilization of **Au-20@L18** NPs (Adapted with permission from Ref. [19] Copyright 2020 Elsevier)

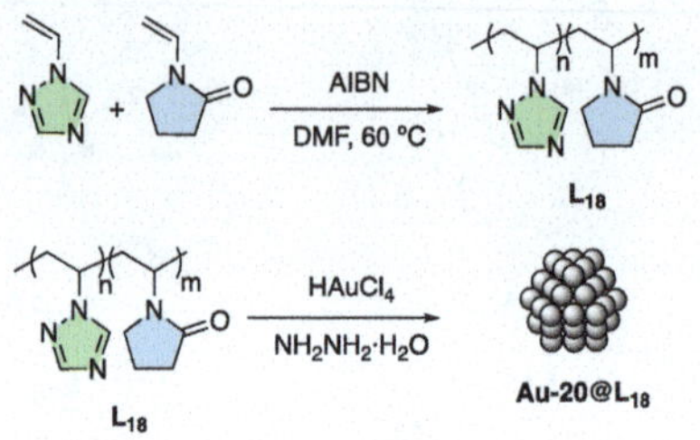

distance to be 2.04 Å. Lastly, the catalytic performance of the nanoparticles was evaluated in the reduction of 4-nitrophenol, achieving complete conversion in less than 5 min. However, the nanoparticles exhibited limited recyclability, with effective use observed only for three catalytic cycles.

Au NPs were synthesized using denatured proteins with a high number of amino groups as stabilizers. The rationale behind employing proteins lies in the potential utilization of biopolymers resulting from their denaturation process. In this context, Chen et al. employed denatured human serum albumin (HSA) modified with PEG to prevent protein aggregation. Au NPs were synthesized in the presence of $HAuCl_4$ using either $NaBH_4$ or ascorbic acid as reducing agents (Scheme 11) [18]. When $NaBH_4$ was employed as reducing agent, and the proportion between the metal precursor and biopolymer was varied, spherical Au NPs with sizes ranging between 2 and 5 nm were obtained, with smaller sizes observed as the proportion of the polymer increased. As a control experiment, Au NPs were prepared in the absence of denatured protein, resulting in a significantly reduced stability. Alternatively, when ascorbic acid was used as reducing agent, Au NPs with different shapes were obtained, including Au nanoflower-like structures with diameters ranging between 30 and 70 nm. Interestingly, increasing the amount of stabilizer led to an increase in the size of the Au nanoflowers, unlike the trend observed with $NaBH_4$. Finally, these Au spherical and nanoflower-shaped NPs were evaluated in *p*-nitrophenol reduction, with superior performance observed for the Au NPs synthesized using $NaBH_4$, particularly for the smaller-sized particles.

Pozdnyajov et al. synthesized Au NPs stabilized by a copolymer with a triazole and vinylpyrrolidone backbone, aiming to enhance the properties of Au NPs by leveraging the unique characteristics of the two different monomers (Scheme 12)

[19]. The nanoparticles were synthesized using $HAuCl_4$ in the presence of the copolymer and hydrazine hydrate to obtain the corresponding **Au-20@L$_{18}$** NPs. Various ratios of stabilizing agent to metal were investigated, resulting in larger sizes as the copolymer-to-metal ratio decreased, ranging from 2 to 10 nm. This change in size also had an impact on the SPR band, exhibiting a redshift with increasing size, shifting from 511 nm to 554 nm. Notably, these Au NPs exhibited properties akin to organic high-resistance semiconductors, demonstrating electrical conductivity values between 10^{-9} and 10^{-10} S/cm, significantly higher compared to the copolymer alone (4.5×10^{-14} S/cm).

In another study employing copolymers as stabilizing agents, Traynor et al. explored copolymers composed of methacrylic acid (MAA), oligo(ethylene glycol) methyl ether methacrylate (OEGMA), ethylene glycol dimethylacrylate (EGDMA), 2-dodecyl 2-bromoisobutyrate (DBIB), and dodecane thiol (DDT) (Scheme 13) [20]. The synthesis of the Au NPs involved the addition of $NaBH_4$ to a solution containing the copolymer and $HAuCl_3 \cdot H_2O$. Various copolymers were synthesized by adjusting the number of units per monomer from the aforementioned constituents, resulting in Au NPs typically ranging between 2 and 5 nm, contingent upon the polymer content and the monomer distribution. Through the exploration of different combinations, the authors identified the essential role of the thioether group in achieving stable MNPs and preventing aggregation. Intriguingly, the presence of MAA influenced the MNP size relative to the polymer-to-metal ratio, resulting in smaller sizes even at lower polymer-to-metal proportions when the MAA content ranged between 15% and 20%. This finding contrasts with prior reports where lower polymer content typically yielded larger systems. These Au NPs were assessed in the catalytic reduction of 4-nitrophenol to 4-aminophenol, revealing a considerable range of outcomes, approximately a 45-fold variability in reaction rates and a 2000-fold variation in induction times, depending on each Au NP and its polymeric composition, among other factors.

Different bio-based stabilizers have been utilized in the synthesis of monometallic and bimetallic Au-based NPs. In this context, eggshell membranes (ESM) have emerged as a promising stabilizer and reducing agent for the production of monometallic Au and Ag NPs, as well as bimetallic AuAg NPs, owing to its low cost and abundant availability (Scheme 14) [21]. ESM was introduced into the respective metal precursor solution and allowed to react at room temperature until the MNPs were formed. Analysis via UV-Vis spectroscopy revealed the formation of MNPs, evidenced by the appearance of SPR bands. For monometallic Au and Ag systems, SPR bands with maxima at 535 nm and 436 nm, respectively, were observed, while a single band at 546 nm indicated alloy formation in the bimetallic systems. Concerning size and morphology, monometallic Au NPs exhibited sizes ranging from 5 to 50 nm, with a mean size of 18.12 nm and displaying triangular and spherical shapes. Conversely, Ag NPs displayed sizes ranging from 5 to 20 nm, with a mean size of 12.4 nm and spherical-like shapes. Bimetallic AuAg NPs showcased sizes ranging from 30 to 80 nm, with a mean size of 42.21 nm. Additionally, Raman studies confirmed MNP formation and the involvement of functional groups in the metal ion reduction process, as evidenced by changes observed in the signals corresponding to amino, thiol, carbonyl, and hydroxy groups. Finally,

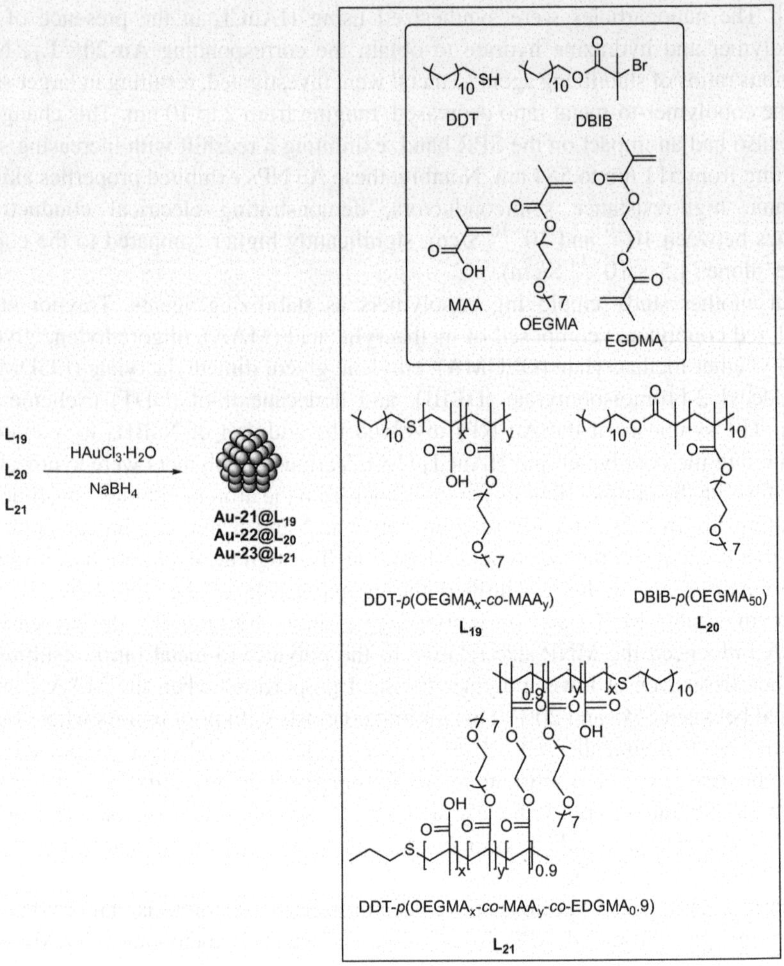

Scheme 13 Synthesis of copolymer-stabilized Au NPs (Adapted with permission from Ref. [20] Copyright The Royal Society of Chemistry 2022)

Scheme 14 Synthesis of Au, Ag, and AuAg NPs stabilized by ESM (Adapted with permission from Ref. [21] Copyright 2015 The Royal Society of Chemistry)

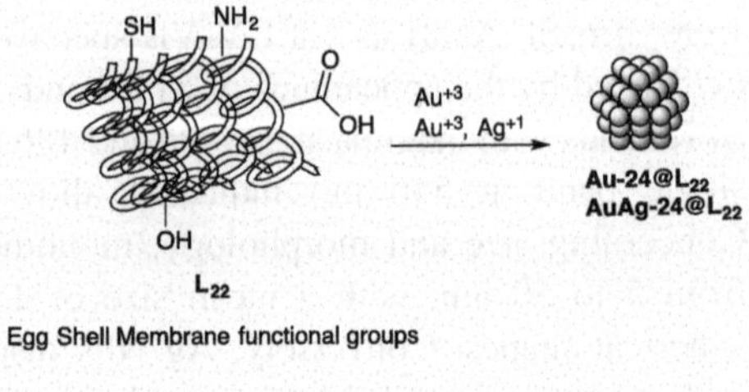

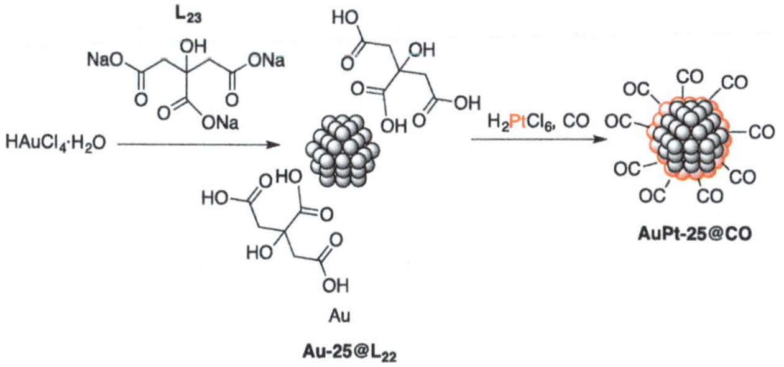

Scheme 15 Synthesis of Au-Pt core-shell NPs stabilized by CO (Adapted with permission from Ref. [22] Copyright 2022 Nature)

Scheme 16 Synthesis of **Au-26@L$_{23}$** NPs (Adapted with permission from Ref. [23] Copyright 20217 ACS Publications)

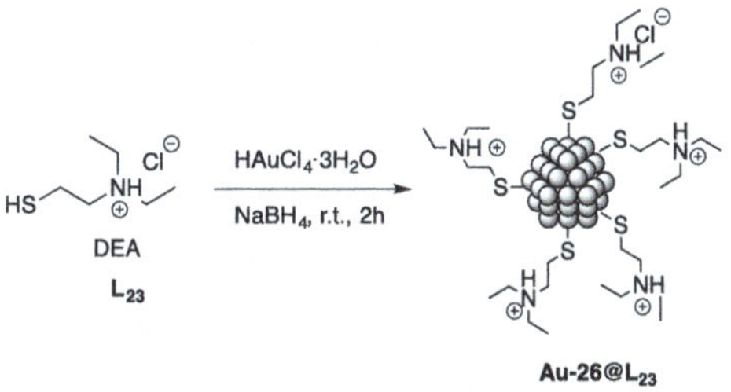

the MNPs were evaluated for dye removal applications, with monometallic Au NPs demonstrating superior performance in the removal of methyl orange compared to AuAg and Ag NPs.

Another example of bimetallic Au-based NPs was done by Ledendecker et al. where Au NPs were used as templates for Pt growing in the surface, in order to obtain core-shell Au-Pt NPs, and taking into account the importance of core-shell NPs with thin noble metal shell in different catalytic, biomedical, and pharmaceutical applications [22]. The Au core was synthesized by using $HAuCl_4 \cdot 3H_2O$ in the presence of trisodium citrate dihydrate to get Au NPs (Scheme 15) with a mean size around 25 nm. To obtain the Pt shell, ascorbic acid was used as a reducing agent in the presence of H_2PtCl_6 as a metal precursor and CO as a stabilizing agent, owing to its robust coordination with Pt. The resulting Au-Pt core-shell NPs exhibited similar sizes than Au NPs, attributed to the thick Pt shell with a thickness of 0.76 nm. Analysis via ATR-FTIR revealed the Pt deposition through the emergence of two new signals corresponding to CO bridging and terminal coordination modes to Pt, observed at approximately 1900 cm^{-1} and 2065 cm^{-1}, respectively. This synthetic strategy was extended to other types of Au NPs, including Au nanorods and Au nanocubes, resulting in the formation of thick Pt shells in all cases.

In addition to exploring various applications and synthesis methods of Au NPs, it is crucial to investigate the interface between the ligand and the nanoparticle. To address this, Venditti et al. utilized Au NPs stabilized by 2-diethylaminoethanethiol hydrochloride (DEA) and examined their formation and characteristics through NMR and XPS analyses (Scheme 16) [23]. The synthesis involved the use of

$HAuCl_4 \cdot 3H_2O$ as the metal precursor in the presence of DEA and $NaBH_4$ to produce the Au NPs. The size of the Au NPs exhibited a pH dependence, with a hydrodynamic radius of 5 nm at pH 2, whereas at pH 7, aggregates ranging from 50 to 200 nm were identified by DLS. Various solution NMR and XPS experiments enabled the determination of the coordination through the Au-S bond, revealing that DEA not only coordinated but also physisorbed to the MNP surface, a finding of particular significance suggesting that DEA can be easily exchanged.

3 Ruthenium-Based Nanoparticles

In pursuit of more sustainable chemistry and recognizing the utility of Ru-based systems in numerous chemical transformations, it is imperative to delve into studies involving water-soluble Ru NPs. Guerrero et al. embarked on this exploration by synthesizing Ru NPs through the organometallic approach method, employing Ru (COD)(COT) as the organometallic precursor in the presence of H_2, with alkyl-sulfonated diphosphines (PR_3SO_3Na) serving as stabilizers (Ru@PR_3SO_3Na; Scheme 17) [24]. Throughout their investigation, the authors synthesized diverse Ru NPs by manipulating the ratio between the ligands and the metal, as well as varying the length of the alkyl chain. The results unveiled Ru NPs with sizes ranging from 1.2 to 1.5 nm. Notably, altering the concentration of ligands resulted in smaller MNPs when a higher quantity of ligand was used (e.g., 0.2 eq. vs 0.1 eq.), yielding sizes of 1.51 nm vs 1.27 nm for $\mathbf{L_{24}}$, 1.48 nm vs 1.26 nm for $\mathbf{L_{25}}$, and 1.50 nm vs 1.23 nm for $\mathbf{L_{26}}$. Conversely, no discernible change in MNP size was observed upon varying the chain length. These nanocatalysts were tested in hydrogenation reactions using different substrates (styrene, tetradecane, and acetophenone). The authors observed differences in selectivity and conversion with styrene at shorter reaction times (1 h) depending on the length of the alkyl chain, showing better results for the Ru NPs with longer alkyl chain ligands (**Ru-1@**$\mathbf{L_{24}}$), presumably due to the high flexibility of the ligand, that allow the substrate coordination to the nanoparticle surface. On the other hand, by varying the amount of ligand, the conversion from styrene to ethylbenzene was reduced as more ligand was present in the system, due to a higher surface coverage. Regarding to tetradecene, due to its insolubility in water,

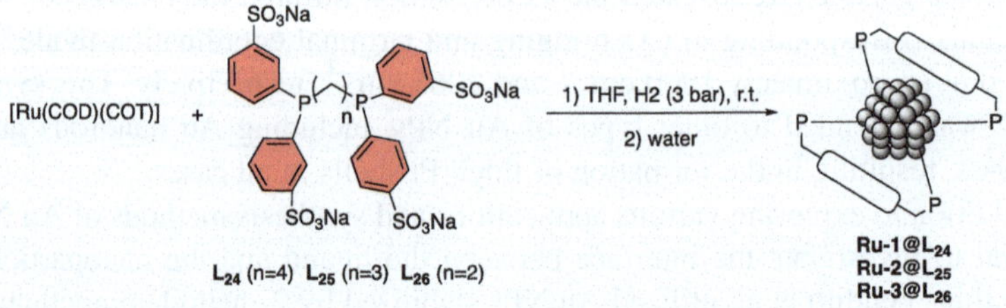

Scheme 17 Synthesis of Ru@PR_3SO_3Na NPs (Adapted with permission from Ref. [24] Copyright 2012 Elsevier)

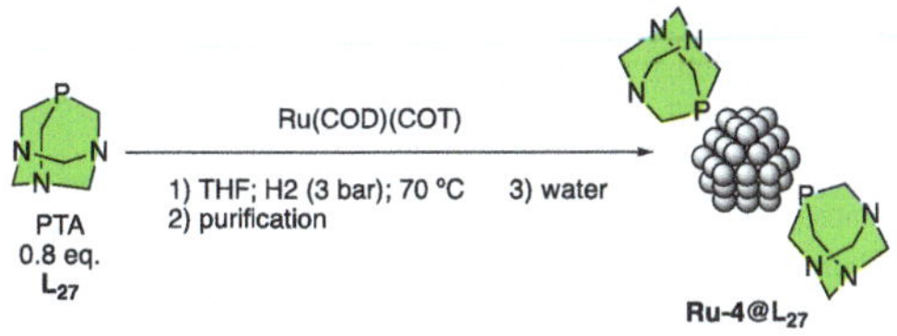

Scheme 18 Synthesis of **Ru-4@L$_{27}$** NPs (Adapted with permission from Ref. [25] Copyright 2012 Wiley)

lower conversion values were obtained. Finally, with acetophenone, the main product was obtained after the hydrogenation of the carbonyl bond (i.e., 1-phenylethanol).

Debouttière et al. conducted another study involving phosphorus-based ligands in Ru NPs. In this investigation, the authors synthesized different nanoparticles utilizing 1,3,5-thiaza-7-phosphaadamantane (PTA) as stabilizer. They also followed the organometallic approach method, once again employing Ru(COD)(COT) as Ru precursor under H_2 (Scheme 18). However, this time the reaction was conducted at 70°C due to the limited solubility of PTA in THF [25]. The authors explored different ratios between PTA and Ru, with the most favorable outcome observed at 0.8 eq., resulting in MNPs with a mean size of 1.3 nm. As previously mentioned, surface analysis plays a crucial role in identifying the species present on the MNP surface and elucidating the coordination of ligands to the surface. To achieve this, the authors conducted various liquid- and solid-state magic-angle spinning NMR (MAS-NMR) studies. Through ^{31}P NMR spectroscopy, they confirmed the coordination of PTA through the P atom, as evidenced by the absence of free PTA. Additionally, employing DOSY ^{1}H NMR spectroscopy, they observed a strong coordination of PTA to the surface, as no ligand exchange was observed when dodecanethiol was introduced. Regarding catalytic tests, dodecane, toluene, and m-methylanisole were utilized as substrates. The results demonstrated complete hydrogenation for dodecene and toluene, while m-methylanisole underwent partial hydrogenation.

NHC-stabilized Ru NPs have emerged as a promising platform, owing to the exceptional stereoelectronic properties of the coordinated NHC ligands to the metallic surface. Among the array of examples, Martínez-Prieto et al. showcased the synthesis of water-soluble **Ru-5@L$_{28}$** and **Ru-6@L$_{29}$** NPs employing sulfonated NHC ligands. The introduction of the sulfonate group as a substituent in the imidazole ring endowed colloidal stability to the MNPs in water [26]. The synthesis process followed the organometallic approach method, wherein the NHC-free carbene, formed in situ, was added to the solution containing the Ru precursor (Scheme 19). This method yielded Ru NPs with sizes around 1.4 and 1.5 nm, as determined by TEM, with the specific size depending on the NHC used. Remarkably, the authors conducted an in-depth exploration of the surface state of the Ru NPs utilizing FTIR, solid-state MAS-NMR, and XPS techniques. Their investigation revealed that the surfaces of the MNPs were amenable to substrate coordination. Furthermore, they confirmed NHC coordination through the carbenic carbon to the nanoparticle surface, shedding light on the intricate interplay between the ligands and the MNPs. Regarding the catalytic activity, Ru NPs were employed in hydrogen isotope exchange (HIE) reactions, a field gaining recent attention due to the growing

Scheme 19 Synthesis of **Ru-5@L_{28}** and **Ru-6@L_{29}** NPs (Adapted with permission from Ref. [26] Copyright 2017 The Royal Society of Chemistry)

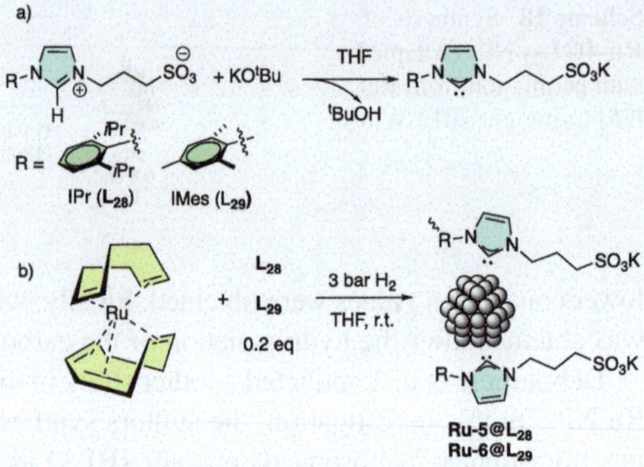

interest in deuterated molecules and their diverse applications [27]. L-lysine served as model substrate, yielding high deuteration percentages (>96%) at the α and ε positions in regard to the amino acid moiety within a pH range of 10.4 to 13 and at the γ position (89.5%) for pH values exceeding 13. This observed phenomenon was primarily attributed to the accessibility of amino groups for coordination to the MNP surface. This hypothesis was substantiated by chemical shift perturbation (CPS) analysis, revealing that both α and ε protons were oriented toward the nanoparticle surface. Moreover, intriguing findings emerged when examining reactions at lower pH levels (pH 6.9), where less than 1% of amino groups remained unbound. Despite this, significant deuteration occurred at the α and ε positions (95% and 70%, respectively). This phenomenon was attributed to proton exchange between the -NH_3^+ groups and deuterides at the nanoparticle surface, leading to the formation of -NH_2 and subsequent coordination of the substrate.

In a separate investigation involving Ru NPs stabilized by water-soluble NHC ligands (**L_{28}**), Palazzolo et al. achieved the deuteration of complex oligonucleotides [28], preserving their pyrimidine bases, which are pivotal as fundamental components in various biochemical processes. Notably, while reactions conducted with Ru@PVP NPs in D_2O led to the deuteration of diverse purine derivatives, these polymer-based systems did not yield analogous outcomes to the NHC-based MNPs due to their limited solubility and stability in water.

Expanding on the concept of HIE reactions using water-soluble MNPs, Bouzouita et al. explored bimetallic RuPt NPs stabilized with water-soluble NHC ligands via the organometallic approach (Scheme 20) [29]. In this study, the authors investigated the impact of employing various Pt organometallic precursors and the metal ratio between Ru and Pt on both the surface composition and catalytic performance of the nanoparticles. The resulting RuPt NPs exhibited sizes ranging between 1.3 and 1.5 nm, with no discernible trend depending on the Pt precursor used. Furthermore, by employing different Pt precursors and adjusting the Ru-Pt ratios in the RuPt NPs, the surface properties of the MNPs could be tailored. This was evidenced by FTIR analysis using carbon monoxide as a probe molecule, as well as MAS-NMR using ^{13}CO. Both techniques revealed distinct wavenumber values and chemical shifts for the CO_b (bridge) and CO_t (terminal) signals,

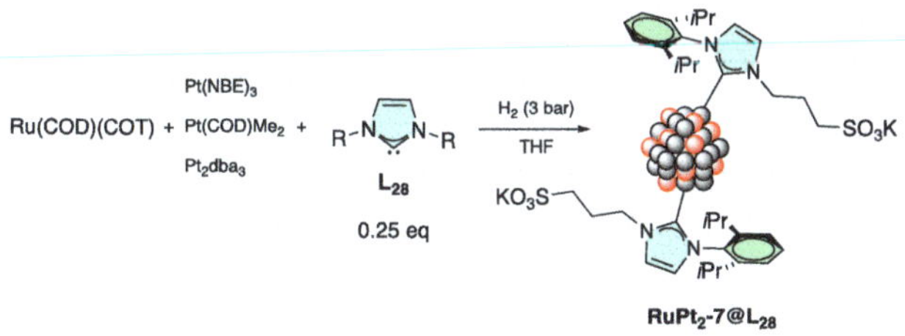

Scheme 20 Synthesis of RuPt NPs stabilized by NHC ligands (Adapted with permission from Ref. [29] Copyright 2019 The Royal Society of Chemistry)

Scheme 21 Synthesis of RuIr NPs (Adapted with permission from Ref. [30] Copyright 2022 ACS Publications)

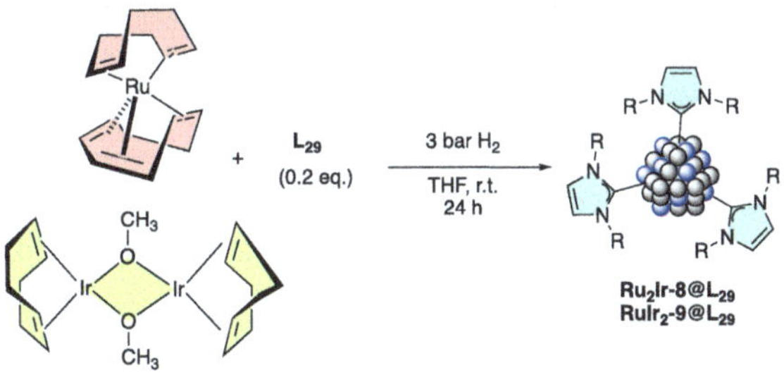

respectively, compared to pure Ru and Pt systems, demonstrating the modification of the NP surface composition. In terms of catalysis, the RuPt NPs were employed as nanocatalysts in the deuteration of L-lysine. Interestingly, the introduction of Pt into the system resulted in a reduction in catalytic activity. Specifically, deuteration at the γ position was suppressed compared to the pure Ru system. Moreover, as the amount of Pt was increased, the deuteration percentages at the α and ε positions decreased further relative to the pure Ru system. Through NMR analysis, the authors discovered that with a $RuPt_2$ composition, the proton at the α position exhibited greater susceptibility to alteration compared to other positions. This was attributed to the affinity of Pt for carboxylate groups, resulting in strong coordination of both the acid and amino groups of the substrate to the nanoparticle surface. This strong coordination hindered desorption of the substrate from the surface, thereby leading to diminished catalytic activity according to Sabatier's principle.

On the other hand, Zuluaga-Villamil et al. conducted a study on Ru-based bimetallic NPs, wherein they synthesized various RuIr NPs by manipulating the metal ratio and utilizing water-soluble NHC ligands as stabilizers through the organometallic approach (Scheme 21) [30]. Characterization via TEM revealed that the MNPs had sizes ranging between 1.1 and 1.7 nm. Surface studies employing solid-state MAS-NMR suggested that the coordination of the NHC ligand occurred via the carbenic carbon. Additionally, using FTIR coupled with carbon monoxide, the authors confirmed the bimetallic nature of the synthesized NPs, with CO_t signals exhibiting intermediate wavenumber values compared to those of the pure Ru and Ir systems. The motivation behind synthesizing these bimetallic systems was to modulate catalytic activity in HIE reactions. Remarkably, when employing L-lysine as a

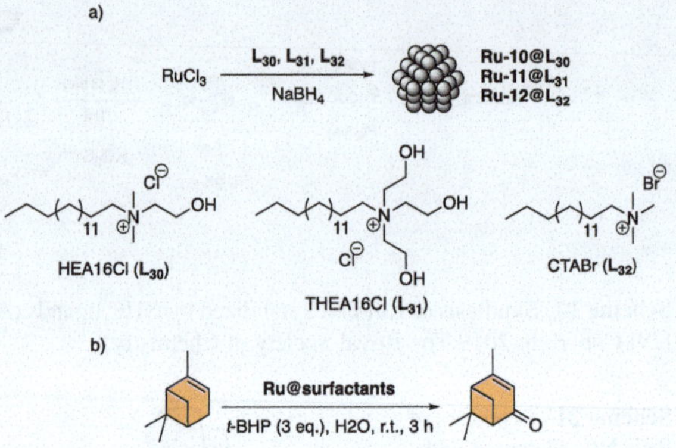

Scheme 22 (**a**) Synthesis and (**b**) catalytic activity of Ru@Surfactants NPs (Adapted with permission from Ref. [31] Copyright 2018 Elsevier)

substrate, a synergistic effect was observed with RuIr NPs, promoting deuteration at the γ position (27–28%) at a pH value of 9.2. This finding is particularly noteworthy as it enables HIE reactions to occur under conditions more akin to biological environments.

Some Ru NPs have been stabilized also with surfactants, as Rauchdi et al. reported in 2018, using different modified hydroxymethylamonium salts. The synthesis followed the salt reduction route using $RuCl_3$ as a metal precursor in the presence of sodium borohydride (Scheme 22) [31]. As a result, they obtained Ru NPs with sizes between 2.0 and 2.5 nm. These Ru NPs (Ru@Surfactants) were used as catalysts in the oxidation of α-pinene to verbenone in the presence of *tert*-butyl hydroperoxide in water. By comparing the different surfactants, they found that the most active catalyst was the ones with Cl^- as a counter ion, with conversion values of 100% and yield of 35%. Interestingly, by comparing their Ru NPs with commercial surfactants, they found better results in terms of conversion and yield, showing the beneficial effect of the modified salts. In addition, they evaluate the effect of the Ru/L ratio, obtaining better results for 0.2 eq. molar ratio compared to 1.0 eq. due to more stability in the aqueous phase. Finally, recyclability tests showed no change in MNP size after the reaction.

4 Palladium-Based Nanoparticles

In the realm of catalysis, Pd has emerged as a widely utilized noble metal for numerous chemical transformations. Therefore, it is imperative to examine the diverse array of studies conducted with water-soluble Pd NPs. For instance, Bouriazos et al. investigated Pd NPs stabilized with various nitrogen-containing ligands as catalysts for the hydrogenation of renewable polyunsaturated methyl esters derived from soybean, achieving turnover frequency (TOF) values of 71,000 h^{-1} and 68,000 h^{-1} [32]. In this study, the authors highlighted several advantages associated with the nitrogen-containing ligands (Scheme 23), including

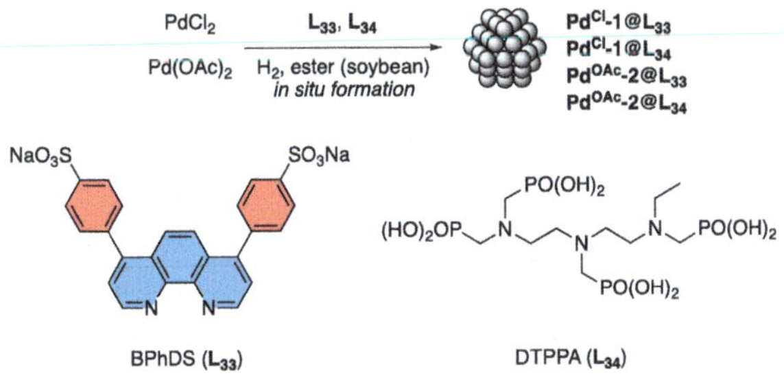

Scheme 23 Stabilizing agents for the in situ Pd NPs obtained in the selective hydrogenation of polyunsaturated methyl esters (Adapted with permission from Ref. [32] Copyright 2014 Elsevier)

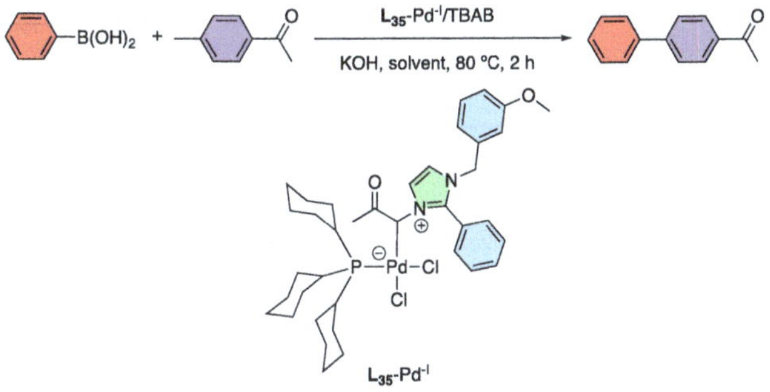

Scheme 24 Suzuki-Miyaura reaction catalyzed by the in situ formation of Pd NPs stabilized by the addition of TBAB (Adapted with permission from Ref. [33] Copyright 2017 Elsevier)

their heightened hydrophilicity compared to phosphorus-containing counterparts, as well as their lower cost and superior water stability relative to phosphine ligands. However, despite the primary objective of working with Pd complexes in situ, the reaction conditions led to the formation of Pd NPs exhibiting high catalytic activity as black precipitates. DLS measurements were employed to characterize the MNPs in solution. Notably, distinct MNP sizes were obtained depending on the Pd precursor used, with MNPs of approximately 57 nm and 36 nm when using $PdCl_2$ and $Pd(OAc)_2$ as precursors, respectively. To ascertain whether the catalysis was indeed attributed to the Pd NPs, the authors removed all excess Pd complex, achieving significant conversion to the monounsaturated methyl esters. Furthermore, the addition of Hg resulted in a reduction in catalytic activity, further confirming the pivotal role of the Pd NPs in catalysis.

Another example of in situ generated Pd NPs was reported by Lee et al. They conducted a study involving zwitterionic Pd complexes containing imidazolium and phosphite ligands. In their approach, *n*-tetrabutylammonium bromide (TBAB) was added in the Suzuki-Miyaura reaction (Scheme 24) [33]. Upon analyzing the TBAB phase during recyclability tests, Pd NPs were observed to have sizes ranging from 4 to 11 nm in the first cycle, with MNPs measuring around 18 to 24 nm in the subsequent cycles. In this context, TBAB served a dual role as both a stabilizing agent and a phase transfer agent for the substrates, facilitating the reaction in water. Moreover, the formation of in situ Pd NPs required the presence of Pd^{2+} complexes.

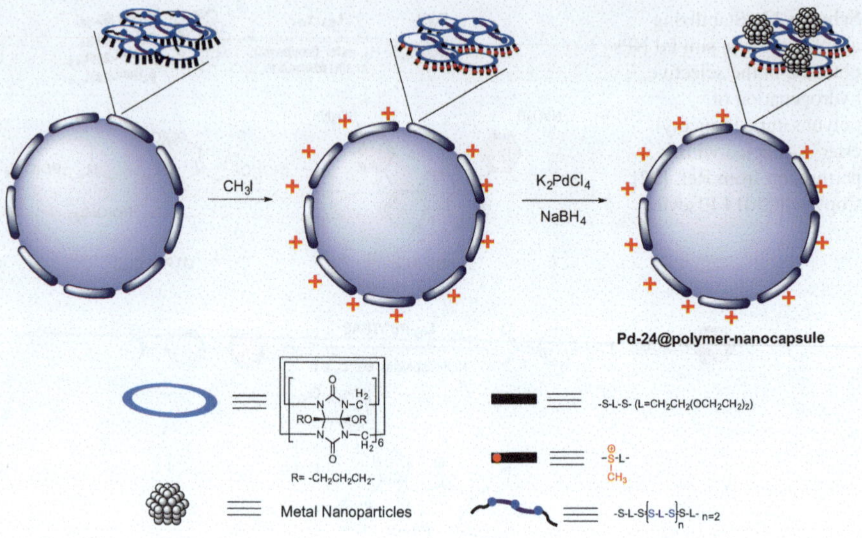

Scheme 25 Synthesis of Pd@polymer nanocapsule NPs (Adapted with permission from Ref. [34] Copyright 2014 Elsevier)

Interestingly, when $PdCl_2$ or Pd^0-NHC complex was used, no formation of NPs was observed.

As previously mentioned, polymers serve as effective stabilizers for numerous MNPs. In this sense, Yun et al. utilized polymer nanocapsules based on cucurbit[6]uril to stabilize Pd NPs (Pd@polymer nanocapsule NPs), employing K_2PdCl_4 as metal precursor in the presence of $NaBH_4$ via the metal salt reduction method (Scheme 25) [34]. To render the polymer nanocapsules water-soluble, they underwent treatment with MeI to introduce sulfonium groups, without altering their size and morphology. Remarkably, besides its primary role as a reducing agent for Pd^{2+} to Pd^0, $NaBH_4$ also facilitated the generation of thiols at the surface of the nanocapsules by cleaving disulfide bonds. These thiols then formed bonds with the Pd atoms. This hypothesis was substantiated by XPS studies revealing the appearance of a new sulfur peak. Furthermore, varying proportions of Pd and stabilizer were investigated, resulting in larger systems with lower amounts of nanocapsules. The obtained MNPs exhibited an average size of approximately 1.9 nm and displayed a face-centered cubic (*fcc*) crystalline structure, as confirmed by HR-TEM. In terms of catalytic activity, these Pd NPs were evaluated in both Suzuki-Miyaura and Buchwald-Hartwig reactions. In Suzuki-Miyaura reactions, regardless of the aryl halide and phenylboronic acid employed, high conversion rates (100%) were achieved with short reaction times (1–2 h) at room temperature. Recyclability tests demonstrated no loss of catalytic activity even after five cycles, and no leaching of the catalyst was observed. For the Buchwald-Hartwig reaction, aryl iodide and *p*-anisidine were employed, resulting in complete conversion to biaryl amines in a H_2O/THF reaction mixture. This finding is particularly

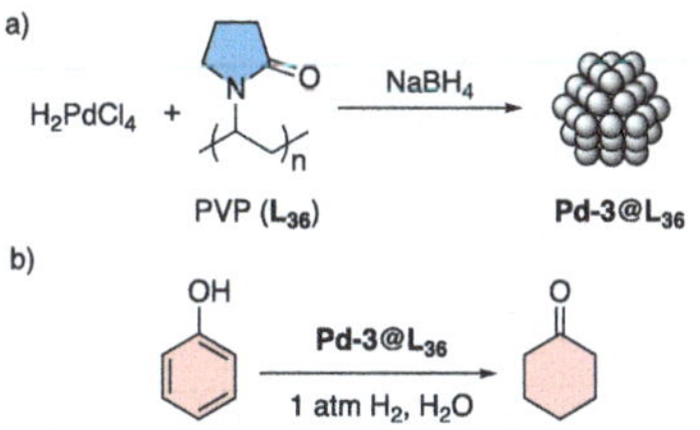

Scheme 26 (**a**) Synthesis of **Pd-3@L$_{36}$** NPs and (**b**) their application in the selective hydrogenation of phenol to cyclohexanone (Adapted with permission from Ref. [35] Copyright 2014 The Royal Society of Chemistry)

Scheme 27 Biocompatible phosphonic acids used as stabilizing agents of different Pd NPs (Adapted with permission from Ref. [36] Copyright 2014 The Royal Society of Chemistry)

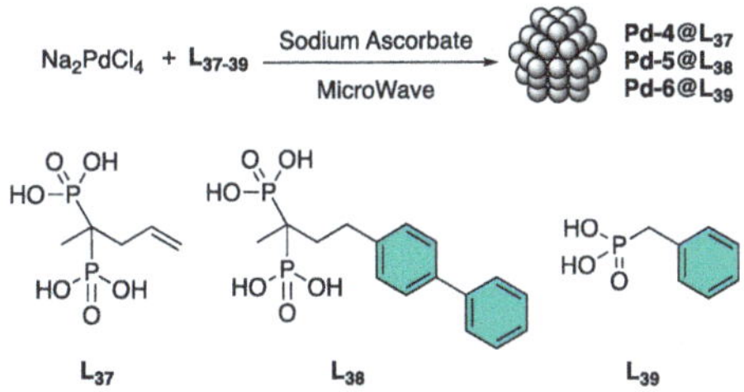

noteworthy as it underscores the novelty of employing water-soluble Pd NPs in the Buchwald-Hartwig reaction, presenting promising opportunities for further exploration in this area.

Continuing the exploration of Pd NPs stabilized by polymers, Zhu et al. utilized PVP as a stabilizing agent to synthesize Pd@PVP NPs (**Pd-3@L$_{36}$** NPs) for the selective hydrogenation of phenol to cyclohexanone in water (Scheme 26) [35]. The synthesis of **Pd-3@L$_{36}$** NPs involved the reduction of H_2PdCl_4 with $NaBH_4$, resulting in MNPs with a mean size of 2.9 nm, irrespective of the molar ratio between Pd and PVP. XPS analysis confirmed the presence of Pd^0 species without signals indicative of oxidized Pd species, while XRD revealed the *fcc* crystalline structure of Pd. In terms of catalytic activity, **Pd-3@L$_{36}$** NPs exhibited high efficiency in the selective hydrogenation of phenol to cyclohexanone, achieving conversion and selectivity values of up to 99% at 90°C after 16 h at 1 atm of H_2. This selectivity can be attributed to the strong stability of cyclohexanone on the Pd surface, facilitating its decoordination. Interestingly, varying the PVP/Pd molar ratio resulted in decreased catalytic activity when a higher amount of PVP was utilized as a stabilizer, likely due to the coverage of the active Pd sites by the polymer. Furthermore, recycling tests demonstrated preserved full conversion and selectivity over six cycles, with no observed MNP aggregation. Finally, a comparison with commercial Pd/C catalysts revealed lower conversion values (80.9%).

Several environmentally friendly methods have emerged for synthesizing Pd NPs, with one notable approach demonstrated by Ayad et al. involving the reduction of Na_2PdCl_2 with sodium ascorbate in the presence of biocompatible phosphonic acids as stabilizing agents, assisted by microwave heating (Scheme 27) [36]. The resulting MNPs exhibited pH-dependent characteristics, with smaller MNPs formed

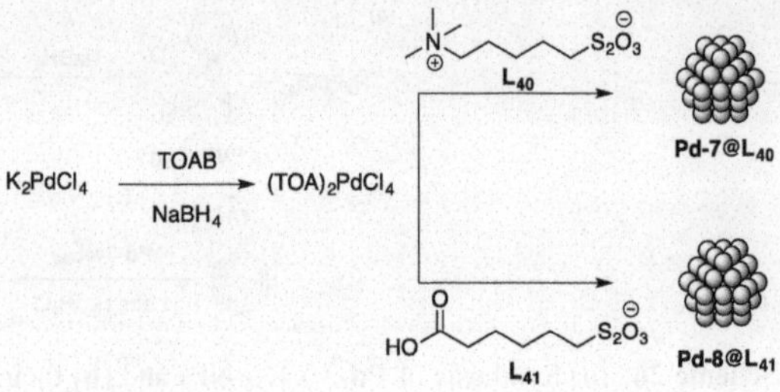

Scheme 28 Synthesis of **Pd-7@L$_{40}$** and **Pd-8@L$_{41}$** NPs (Adapted with permission from Ref. [37] Copyright 2023 MDPI)

at basic pH 10, measuring approximately 2 to 7 nm, while at acidic and neutral pH conditions (pH 3 and 7, respectively), polydisperse systems were observed, with sizes ranging from less than 5 nm to about 50 nm for the larger particles. HR-TEM analysis revealed the expected *fcc* crystalline structure. Additionally, XPS studies elucidated that phosphate coordination played a crucial role in stabilizing the MNPs, as evidenced by the shift in the binding energy of the P $2p_{3/2}$ peak. Further, catalytic assessments were conducted in typical C-C cross-coupling reactions catalyzed by these Pd NPs, including Suzuki-Miyaura, Sonogashira, and Mizoroki-Heck reactions. The smaller MNPs synthesized at basic pH exhibited superior catalytic performance, irrespective of the stabilizing agents used, with aromatic moieties yielding the best results due to enhanced contact between the substrates and the nanoparticle surface.

Alternative ligands have been explored for the synthesis of Pd NPs, with thiolate ligands being one such example. Avila et al. successfully achieved the regioselective transformation of allylbenzene to 3-phenylpropanal using Pd NPs stabilized by thiolate ligands featuring carboxylate and ammonium tails (Scheme 28) [37]. The Pd NPs were synthesized through the reduction of K_2PdCl_4 by $NaBH_4$ in the presence of an ammonium quaternary salt (TOAB) serving as a phase transfer agent. Different Pd NPs were obtained depending on the type of stabilizing agent employed, with ammonium tail thiolate ligands (**Pd-7@L$_{40}$** NPs with **L$_{40}$** = 5-(trimethylammonium)pentanethiolate) yielding 3.2 nm MNPs and carboxylate tail thiolate ligands (6-(carboxylate)hexanethiolate = **L$_{41}$**) (**Pd-8@L$_{41}$** NPs with **L$_{41}$** = 6-(carboxylate)hexanethiolate) resulting in 1.7 nm MNPs. Characterization of these systems was performed using ^{1}H NMR in solution, revealing broader signals for groups close to the sulfur binding atom and the complete absence of the two methylene groups adjacent to the sulfur atom. These results suggest the coordination of the ligands to the MNP surface. In terms of catalytic activity, the selective oxidation of allylbenzene to 3-phenylpropanal was accomplished, with superior results observed for Pd NPs stabilized by the ammonium tail thiolate ligand. Furthermore, the addition of excess H_2 led to increased selectivity toward the anti-Markovnikov addition of H_2O.

As previously mentioned, NHC ligands have emerged as a promising alternative for stabilizing MNPs. Various water-soluble Pd@NHC NPs have been synthesized using different approaches. For example, Ferry et al. obtained a range of Pd@NHC NPs via a ligand exchange process (**Pd-9@L$_{42}$**, **Pd-10@L$_{43}$**, **Pd-11@L$_{44}$**,

Scheme 29 Synthesis by ligand exchange methodology of Pd@NHC NPs (Adapted with permission from Ref. [38] Copyright 2015 ACS Publications)

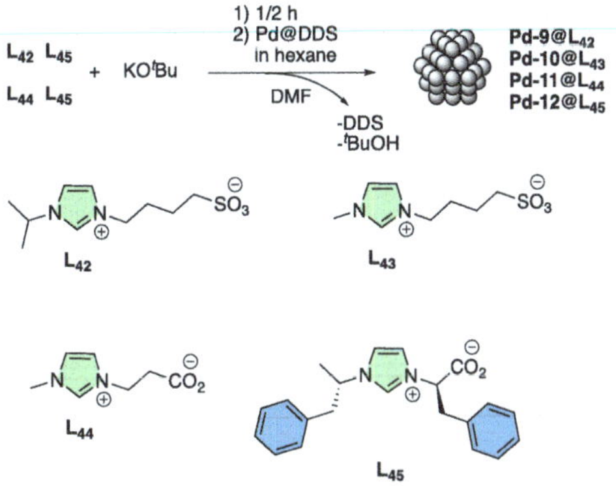

Pd-12@L$_{45}$), starting from Pd NPs stabilized by didodecyl disulfide (dds) ligands [38]. The synthesis involved introducing Pd@dds NPs in hexane into an in situ NHC-formed solution in DMF (Scheme 29). These MNPs were stabilized by various NHCs featuring sulfonate or carboxylate groups. Through various NMR studies in solution, they confirmed the displacement of dds from the MNP surface and the coordination of NHC via Pd-C bond formation, evidenced by a broad peak around 166.7 ppm. Additionally, TEM analysis revealed no significant changes in the mean size between Pd@dds and Pd@NHC NPs, with sizes around 4 nm. Catalytic evaluations of these Pd@NHC NPs were conducted in the hydrogenation of several olefins, including styrene, 1-decene, 3-methyl-2-cyclohexanone, and citronellol. The reductions were performed in water, at 2 bar of H_2 pressure, for 1 h, yielding values ranging from 88 to 100%. By comparing the performance of **Pd-12@L$_{45}$** with commercial Pd/C catalyst, the high catalytic activity of these Pd NPs was demonstrated, showcasing their advantageous effects. Moreover, to ascertain the responsibility of the MNPs for the catalytic activity, the authors utilized a PEPPSI-type Pd complex. However, no hydrogenation was observed, further affirming the significant catalytic activity of the Pd NPs.

In an alternative approach to synthesizing Pd@NHC NPs, Asensio et al. employed thermal decomposition of Pd-NHC complexes under different conditions, including temperatures of 80°C and a temperature ramp from room temperature to 60°C and a CO atmosphere (1 bar) (Scheme 30) [39]. Notably, employing these distinct methods resulted in varying mean sizes of the Pd NPs: larger NPs, ranging between 3.0 and 4.6 nm, were obtained via thermal decomposition, while using CO as a reducing agent yielded MNPs with sizes around 1.3 nm. Furthermore, the authors introduced an unconventional method to confirm NHC coordination to the Pd surface by observing Knight shift (KS) signals in MAS-NMR spectra. By utilizing ^{13}C-labeled Pd complexes and employing the thermal decomposition method, they observed a broad signal around 600 ppm, confirming the coordination of NHC to the Pd surface through the Pd-C bond. However, it is worth noting that this KS is size-dependent and can also be examined using adsorbed molecules on the surface. By employing ^{13}CO as a probe molecule, highly shifted signals of CO around 700–800 ppm were observed in the ^{13}C NMR spin-echo spectra for the

Scheme 30 Synthesis of Pd@NHC NPs through different methods (Adapted with permission from Ref. [39] Copyright 2017 Wiley)

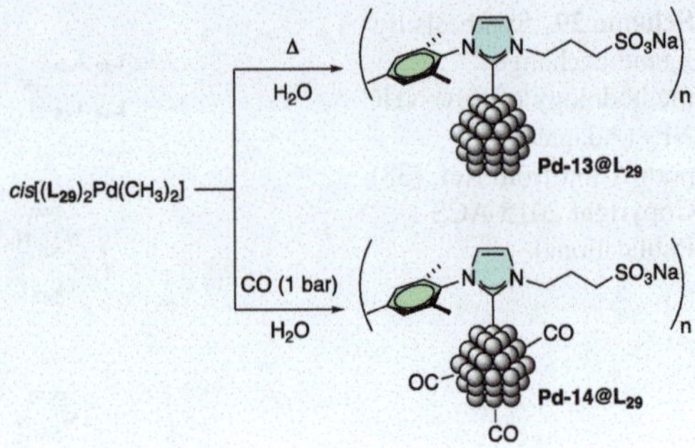

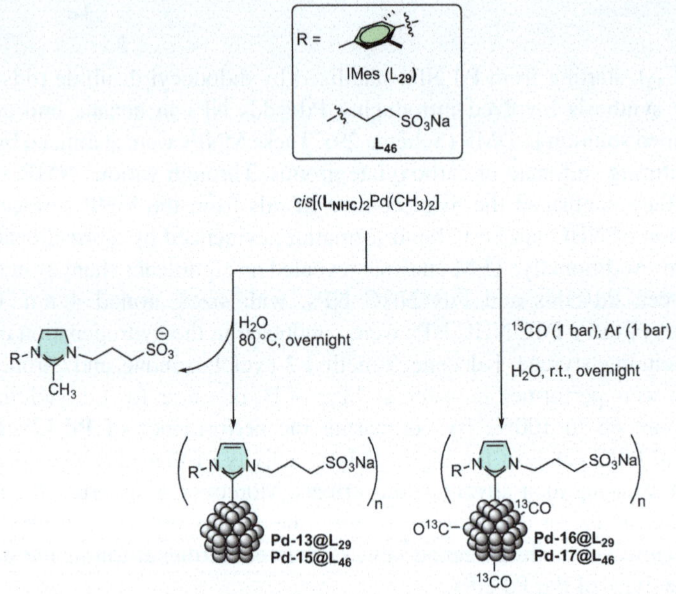

Scheme 31 Different synthetic procedures to obtain Pd NPs by using Pd-NHC complexes as precursors (Adapted with permission from Ref. [40] Copyright 2017 Wiley)

thermally decomposed MNPs with sizes around 3.8 and 4.8 nm. Conversely, for MNPs with a size of 3.1 nm, results fell between KS and non-KS CO signals due to their proximity to the critical size where KS emerges (around 2 nm). Interestingly, for the CO-decomposed MNPs, no KS signals were observed due to their small size.

In a subsequent study, Asensio et al. explored distinct pathways for synthesizing water-soluble Pd@NHC NPs using Pd-NHC complexes as precursors (Scheme 31) [40]. Through thermal decomposition, they obtained Pd NPs ranging from 3 to 4.6 nm in size. Notably, while there was no disparity in the hydrodynamic diameter determined by DLS and DOSY, TEM provided additional insights, revealing challenges in identifying the ligand and solvation layers with the latter technique. For characterization via MAS-NMR, an enriched ^{13}C carbenic carbon Pd complex

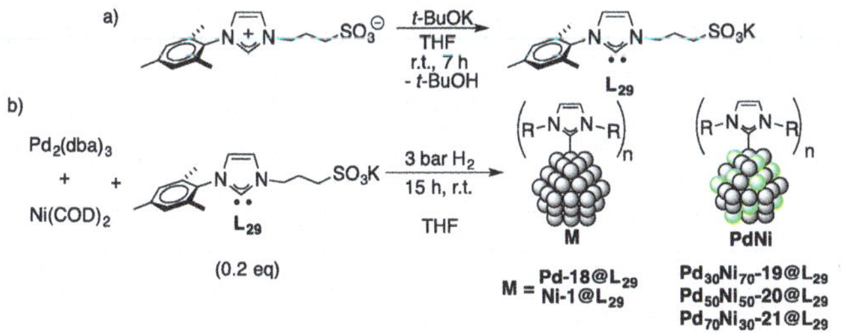

Scheme 32 Synthesis of (**a**) L_{29} and (**b**) Pd, Ni, and PdNi NPs stabilized by NHC ligands (Adapted with permission from Ref. [41] Copyright 2023 The Royal Society of Chemistry)

was employed. The resulting MNPs exhibited a broad KS peak, as previously mentioned, affirming NHC coordination. Furthermore, investigation into the thermal decomposition mechanism revealed a sequence where methane is first released, yielding a cationic Pd complex. Subsequent reductive coupling between methyl and NHC ligand yielded the Pd@NHC NPs. Conversely, employing ^{13}CO reduction yielded smaller MNPs, with sizes ranging from 1.6 to 2.4 nm after purification via dialysis. In MAS-NMR spectra, signals around 800 ppm indicative of KS were observed for the 2.4 nm Pd NPs. The proposed mechanism suggests CO insertion into the Pd-Me bond, forming an acetyl complex that decomposes into acetate, followed by reductive coupling between methyl and NHC, akin to the thermal decomposition method. Furthermore, attempts using H_2 as a reducing agent did not yield stable nano-objects, possibly due to the instability of hydride Pd-NHC complexes. Finally, the thermally decomposed NPs were utilized in the selective hydrogenation of styrene to ethylbenzene, achieving full conversion and 80% yield after ten catalytic cycles, albeit with variations in size and shape of the MNPs observed over successive cycles.

Suárez-Riaño et al. investigated the selective deuteration of aromatic amino acids using bimetallic PdNi systems stabilized by water-soluble NHC ligands, presenting a notable advancement in this area [41]. Employing the organometallic approach, they synthesized these monometallic and bimetallic NPs by employing Pd_2dba_3 and $NiCOD_2$ as metal precursors in the presence of an in situ generated NHC under a H_2 atmosphere (Scheme 32). Characterization via FTIR spectroscopy revealed shifts in the wavenumber of the CO_b band, confirming the bimetallic nature of the PdNi systems. Further validation through STEM-EDX analysis and XRD studies, employing Vegard's law, supported the FTIR findings. Surface analyses using MAS-NMR and XPS provided insights into the coordination and interaction of the NHC ligands with the NP surface. Solid-state MAS-NMR studies using ^{13}CO as a probe molecule confirmed the accessibility of the MNP surface for substrate coordination. XPS analysis revealed alterations in the binding energy of the N1s signal, indicating NHC coordination through the carbenic carbon to the MNP surface, as evidenced by shifts relative to the imidazolium salt. Regarding catalytic activity, the bimetallic PdNi systems demonstrated remarkable performance in achieving selective deuteration without affecting the aromatic ring of L-phenylalanine.

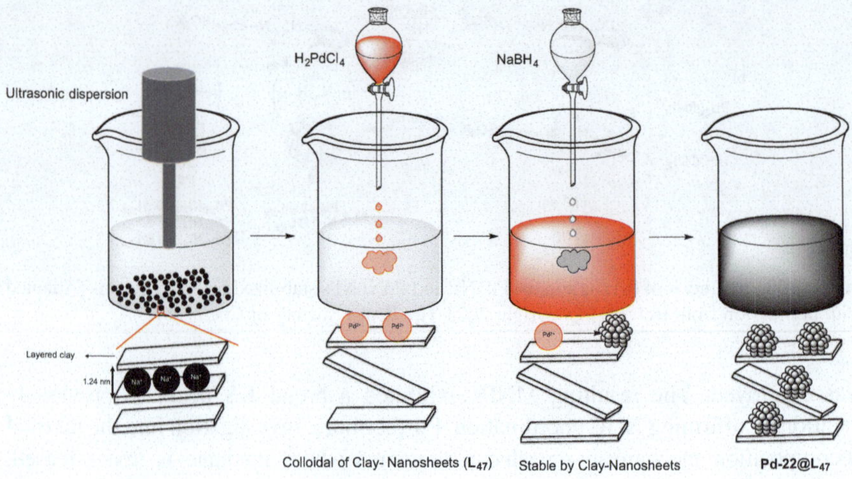

Scheme 33 Synthesis of the supported/stabilized Pd NPs with clay nanosheets (Adapted with permission from Ref. [42] Copyright 2022 Elsevier)

Impressively, deuteration percentages of up to 98.5% were attained at the β position relative to the amino acid group. Notably, the deuteration occurred in an enantiospecific manner, involving a CHD intermediate leading to the formation of the CD_2 molecule, observed across all Pd, Ni, and PdNi systems.

Despite the numerous examples showcasing different stabilizers for MNPs, there is a concern that excessive ligand usage could potentially diminish catalytic activity by obstructing active sites. However, sometimes ligands provide the MNPs with unique electronic and steric properties which might enhance the catalytic activity. Thus, achieving a clean surface while harnessing the catalytic benefits stemming from stabilizer-MNPs interactions necessitates a solid material capable of serving as both a support and a stabilizer. In pursuit of superior systems featuring pristine surfaces and enhanced catalytic activity, Li et al. devised and synthesized Pd NPs supported on clay nanosheets [42]. The clay nanosheets were synthesized through the ultrasonication of montmorillonite in water. Subsequently, the addition of H_2PdCl_4 and $NaBH_4$ yielded water-dispersible supported Pd NPs (Scheme 33). TEM analysis revealed that the resulting MNPs had sizes around 2.3 nm, while SEM imaging detected the 2D structure of the clay nanosheet as a single layer, thereby affirming its dual role as both stabilizer and support. By modulating the stirring rate during synthesis, Pd NPs of varying sizes were produced. Elevating the stirring rate corresponded to a reduction in MNP size, with diameters starting at 8.3 nm at 400–800 rpm, decreasing to 5.4 nm at 1,200–1,600 rpm, and ultimately reaching 2.3 nm at 2,000–2,300 rpm. This phenomenon is attributed to the diminished growth rate of Pd nuclei, resulting in smaller nanoparticles. Moreover, these MNPs exhibited superior performance in the hydrogenation of phenol to cyclohexanone, with improved results observed as the mean MNP size decreased. A

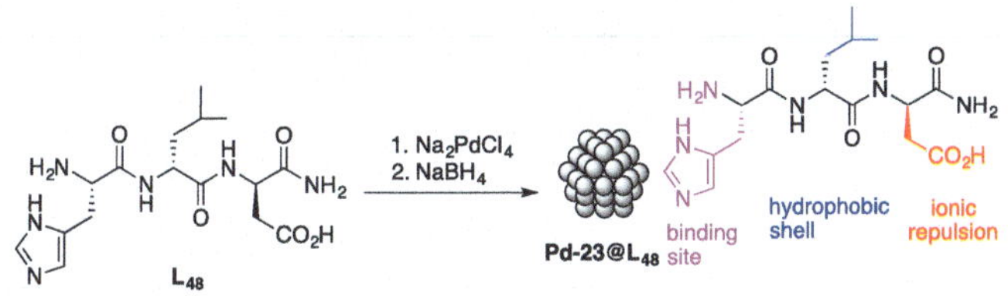

Scheme 34 Synthesis of **Pd-23@L$_{48}$** NPs (Adapted with permission from Ref. [43] Copyright 2016 Wiley)

comparison with Pd@PVP NPs revealed lower conversion values, indicative of blocked Pd active sites, thereby underscoring the advantage of attaining clean surfaces.

In addition to the aforementioned methods, an alternative approach to synthesizing water-soluble Pd NPs was explored by Corra et al., who employed peptides as stabilizing agents to introduce novel biocompatible systems (Scheme 34) [43]. The synthesis involved utilizing Na_2PdCl_4 as metal precursor in the presence of the peptide at a 1:10 M ratio, with tetrabutylammonium borohydride as reducing agent in a mixture of H_2O and ACN (3:1). Characterization via TEM revealed an average MNP size of approximately 3 nm, while TGA analysis enabled the calculation of peptide coverage per nm, indicating an average of six peptide units per MNP. Notably, by employing different peptide sequences, the distinct roles of individual amino acids in MNP stabilization were elucidated. Histidine was identified as the primary amino acid responsible for coordination to the MNP surface in a bidentate manner. Leucine played a crucial role in forming a hydrophobic shell around the MNPs, thereby preventing solvation. Additionally, aspartic acid contributed to stabilizing the MNPs in water by preventing their agglomeration through its carboxylate moiety.

5 Platinum-Based Nanoparticles

As observed with Ru and Pd, investigations into water-soluble NHC ligands have extended to Pt NPs. Baquero et al. conducted a comprehensive study on the formation of Pt@NHC NPs through three distinct synthetic routes: thermal decomposition, reduction with CO, and reduction with H_2 (Scheme 35) [44, 45]. TEM characterization revealed mean sizes ranging from 1 to 4 nm, exhibiting a discernible trend linked to the ligand structure, with smaller MNPs using bulkier carbenes (L_{28} and L_{29}). Notably, the thermal decomposition and CO reduction methods yielded smaller MNPs (1–2 nm) compared to reduction with H_2 (1.7–4 nm). Employing ^{13}C-labeled carbenic carbon NHC ligands in conjunction with MAS-NMR, the authors not only identified the Pt-NHC bond signal around 160–180 ppm but also detected coupling between ^{13}C and ^{195}Pt, characterized by coupling constants of approximately 1,000–1,100 Hz. This result unequivocally demonstrates that NHC

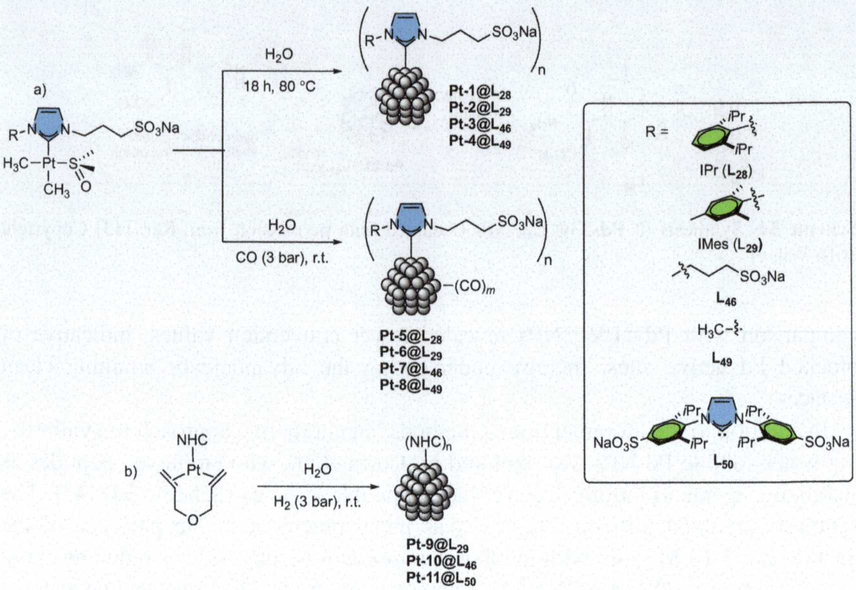

Scheme 35 Synthesis of Pt@NHC NPs through (**a**) thermal and CO reduction and (**b**) reduction with H_2 (Adapted with permission from Refs. [44, 45] Copyright 2014 and 2018 Elsevier and The Royal Society of Chemistry)

Scheme 36 Nano-object with Pt@NHC and Pt-NHC complexes observed by Baquero et al. (Adapted with permission from Ref. [44] Copyright 2014 Elsevier)

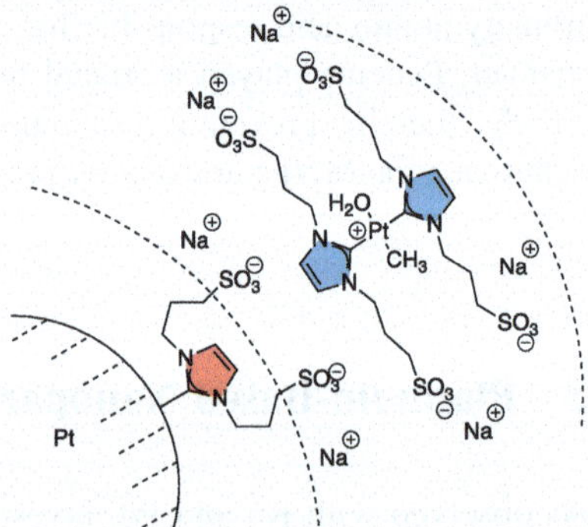

ligand is coordinated to one Pt atom. However, the authors did not study the oxidation state of the Pt surface. Hence, XPS analysis would be helpful for this study. Intriguingly, employing an NHC ligand featuring propyl sulfonate chains as substituents on both nitrogen atoms (**L$_{46}$**) within the imidazole ring resulted in the formation of a distinct nanostructure. This nanostructure exhibited a secondary coordination sphere (Scheme 36), validated through DLS and NMR analyses, indicating larger MNP sizes determined via DLS due to the presence of a second coordination sphere of Pt complex. Furthermore, different diffusion coefficients determined by DOSY were observed compared to the complex and the imidazolium salt, with these discrepancies being more pronounced for MNPs synthesized via thermal treatment. Lastly, the Pt NPs synthesized via thermal decomposition were utilized as catalysts in the selective reduction of styrene to ethylbenzene, demonstrating full conversion and sustained high catalytic activity over nine cycles without significant alterations in MNP size or morphology.

In a subsequent investigation building upon the previously mentioned Pt@NHC NPs (Scheme 35a top), Ruiz-Varilla et al. explored the utility of these nanocatalysts in the selective hydrogenation of various aromatic unsaturated substrates under biphasic conditions (water/organic solvent), including phenylacetylene, benzaldehyde, acetophenone, and nitrobenzene [46]. TEM characterization revealed MNPs with mean sizes ranging from 1.2 to 2 nm, with smaller sizes observed when employing NHC ligands with larger steric hindrance. In the hydrogenation of phenylacetylene, notable results were achieved, with high conversion and moderate selectivity toward styrene (selectivity, 70–75%) over ethylbenzene (selectivity, 25–30%). The catalysts exhibited remarkable recyclability, retaining catalytic activity even after ten successive runs. When benzaldehyde served as substrate, the Pt NPs achieved full conversion and selective production of benzyl alcohol without aromatic ring reduction, demonstrating sustained high catalytic activity over ten cycles. However, in the case of acetophenone, although the MNPs exhibited selectivity toward 1-phenylethanol, their robustness was not as pronounced as observed with other substrates. Pt black precipitates were observed after multiple runs, likely due to the presence of nucleophilic alcohols, the primary product. Finally, similar to benzaldehyde and phenylacetylene, nitrobenzene hydrogenation yielded high conversion rates and selective formation of aniline, alongside sustained high catalytic activity over 11 cycles.

As an innovative approach to synthesizing Pt NPs with unconventional ligands, Mao et al. proposed utilizing polyphenols derived from Black Wattle Tannin (BWT = $\mathbf{L_{51}}$), a plant source, for their ability to confer hydrophilicity to the MNPs via hydroxyl groups while preventing aggregation through their bulky hydrophobic backbones (Scheme 37) [47]. The synthesis involved initially forming chelate Pt^{+4} complexes through interaction between polyphenols and K_2PtCl_4, followed by reduction with $NaBH_4$ to yield water-soluble Pt NPs (**Pt-12@$\mathbf{L_{51}}$** NPs). Notably, successful MNP formation required a high concentration of BWT (15 mg), with lower amounts resulting in aggregation. TEM analysis revealed a mean MNP size of approximately 1.8 nm. XPS analysis indicated shifts in the O1s spectrum, suggesting coordination of hydroxyl groups with the Pt surface. In catalytic applications, these Pt NPs were employed in the oxidation of both aromatic and aliphatic alcohols. High conversion rates were observed for aromatic alcohols (e.g., benzyl alcohol, 89%) compared to aliphatic ones (e.g., hexanol, 36%), likely attributed to the strong adsorption of aromatic moieties onto the MNP surface. Recyclability tests demonstrated the robustness and stability of the Pt NPs, maintaining high conversion rates over seven cycles without leaching or aggregation. Comparison with Pt@PVP and Pt@PEG NPs revealed that **Pt-12@$\mathbf{L_{51}}$** NPs retained catalytic activity over four cycles without leaching or aggregation, while the

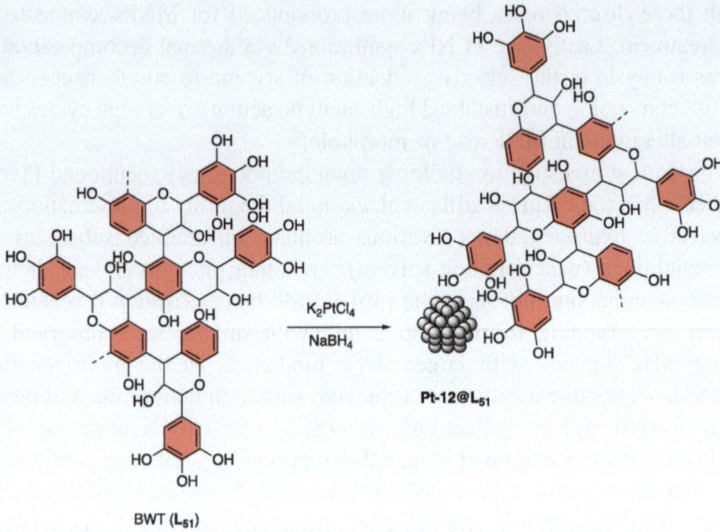

Scheme 37 Synthesis of **Pt-12@L_{51}** NPs stabilized by polyphenols (Adapted with permission from Ref. [47] Copyright 2015 The Royal Society of Chemistry)

polymer-stabilized Pt NPs exhibited significant decrease in catalytic activity due to weak NP stabilization, leading to leaching and aggregation problems.

6 Silver-Based Nanoparticles

Ag NPs have gained prominence in various antibacterial and antiviral applications, necessitating the development of water-soluble formulations. Numerous studies have explored this avenue, often employing biocompatible stabilizing agents. For example, Agasti et al. synthesized Ag NPs using $AgNO_3$ as metal precursor in the presence of glycine and $NaBH_4$ serving as stabilizer and reducing agent, respectively [48]. The impact of glycine concentration on MNP properties was investigated, revealing no discernible relationship between glycine concentration and MNP size or optical properties. Both higher (e.g., 200 mM) and lower (e.g., 50 mM) concentrations yielded MNPs with sizes around 60 nm. XRD analysis indicated a preference for the (111) plane in ligand coordination, evidenced by variations observed with changing glycine concentration. Additionally, FTIR analysis confirmed glycine coordination via the carboxylic group, as evidenced by the shift in the C=O signal, rather than the amine moiety.

Regarding the antibacterial and antiviral applications, Cao et al. devised a method for synthesizing Ag NPs using cellulose acetate (**L_{52}**), which served both as a stabilizer and a reducing agent, due to its abundance of highly reactive hydroxyl groups (Scheme 38) [49]. Notably, these Ag NPs were not produced in the

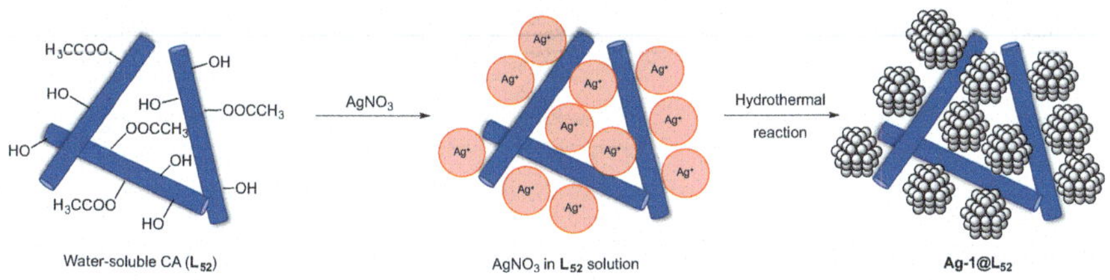

Scheme 38 Synthesis of AgNPs stabilized by cellulose acetate (**L_{52}**) (Adapted with permission from Ref. [49] Copyright 2016 Elsevier)

conventional solid form but rather as hybrid films, very different from typical synthesis methods. Characterization revealed nanoparticle sizes ranging from 1 to 10 nm, with the expected *fcc* crystalline structure characteristic of metallic silver. XPS analysis corroborated the metallic nature of the Ag NPs, detecting no presence of silver oxides. Furthermore, these Ag NPs demonstrated remarkable efficacy in the reduction of 4-nitrophenol to 4-aminophenol in the presence of $NaBH_4$, achieving conversion rates of up to 97.8% within 3 min. Impressively, they exhibited potent antibacterial activity against both *S. aureus* and *E. coli*, with a minimum inhibitory concentration (MIC) of 4.9 mg/L, surpassing the performance of conventional silver dispersions previously employed.

In a separate investigation, He et al. undertook the synthesis of Ag NPs using $AgNO_3$ as metal precursor and a bio-extract derived from *Cornus officinalis*, a plant renowned for its abundance of flavonoids and anthocyanins, as both stabilizer and reducing agent [50]. Characterization of the resulting nanoparticles revealed average sizes of approximately 11–12 nm and quasi-spherical morphologies, as determined by TEM and DLS analyses. The formation of Ag NPs was further confirmed via UV-Vis spectroscopy, where an increase in the intensity of the surface plasmon resonance (SPR) band was observed during nanoparticle synthesis. Furthermore, employing FTIR spectroscopy to examine the functional groups present, such as hydroxyls and carbonyls, the authors speculated on the involvement of flavonoids and anthocyanins in the stabilization and reduction processes. Notably, these Ag NPs exhibited promising anticancer activity against human prostate cancer and human liver cancer cell lines, hinting at their potential utility as novel anticancer agents.

Scratch has also emerged as a viable stabilizer for Ag NPs, as demonstrated by Khan, who utilized amylose (**L_{53}**), a polysaccharide abundant in scratch, for both stabilization and reduction of Ag^+ ions (Scheme 39) [51]. To explore the impact of amylose concentration on MNP synthesis, several concentrations were employed. Interestingly, an increase in amylose concentration, up to a limit of 6% w/w, correlated with enhanced Ag NP formation, evidenced by a sharper SPR peak indicative of increased reduction of Ag^+ ions. However, beyond this threshold, concentrations exceeding 6% w/w led to a decrease in SPR peak intensity. Moreover, the morphology of the Ag NPs was influenced by amylose concentration, transitioning from spherical to chain-like aggregates as concentration increased.

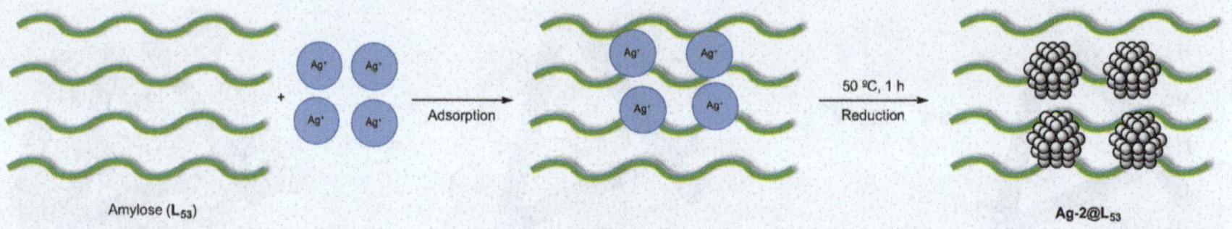

Scheme 39 Synthesis of AgNPs stabilized by amylose (L_{53}) (Adapted with permission from Ref. [51] Copyright 2017 Elsevier)

Notably, the synthesized Ag NPs exhibited a bimodal size distribution, with distinct populations having mean sizes around 10 nm and 50 nm, as determined by DLS. Additionally, the author observed that amylose may serve as a reducing agent owing to the presence of OH groups, which undergo oxidation concurrent with the reduction of Ag ions. Leveraging these properties, the Ag NPs were utilized as sensors for H_2O_2 detection and evaluated for antimicrobial activity against *S. aureus* and *E. coli*.

Zhou et al. introduced an innovative approach to the synthesis of MNPs by leveraging an external electric field, thereby addressing common challenges encountered in conventional MNP synthesis, such as sintering and limited control over nanoparticle size [52]. In their methodology, chitosan (**L$_{54}$**) served as a stabilizer, while formic acid was employed as a reducing agent. This choice was made due to the favorable stability of chitosan under acidic conditions (Scheme 40). The application of an external electric field enabled precise control over the orientation of the dipoles associated with the polar functionalities of the stabilizer. This controlled orientation effectively mitigated hydrogen bond formation, thereby enhancing the stabilization of Ag NPs and promoting stronger interactions between Ag atoms and the stabilizer. The resulting Ag NPs exhibited colloidal stability and were immobilized within a film matrix. The impact of varying voltages on the synthesis of Ag NPs was investigated, revealing intriguing findings. Initially, an increase in voltage from 0 kV to 1 kV led to a growth in MNP size attributed to increased hydrogen bond formation, thereby facilitating enhanced stabilization of Ag NPs. However, at voltages exceeding 1 kV, a decrease in MNP size was observed, attributable to the disruption of hydrogen bonds due to the reorientation of polar functionalities. This reorientation facilitated stabilization through the interaction of -NH_2 groups with Ag atoms. The synthesized Ag NPs demonstrated superior performance in biocofactor NADH regeneration compared to Pt NPs and Ir complexes, highlighting the potential of electric field-assisted Ag NPs in catalytic applications. Moreover, MNP-supported films exhibited notable bactericidal activity against *S. aureus*, underscoring their potential utility in antimicrobial applications.

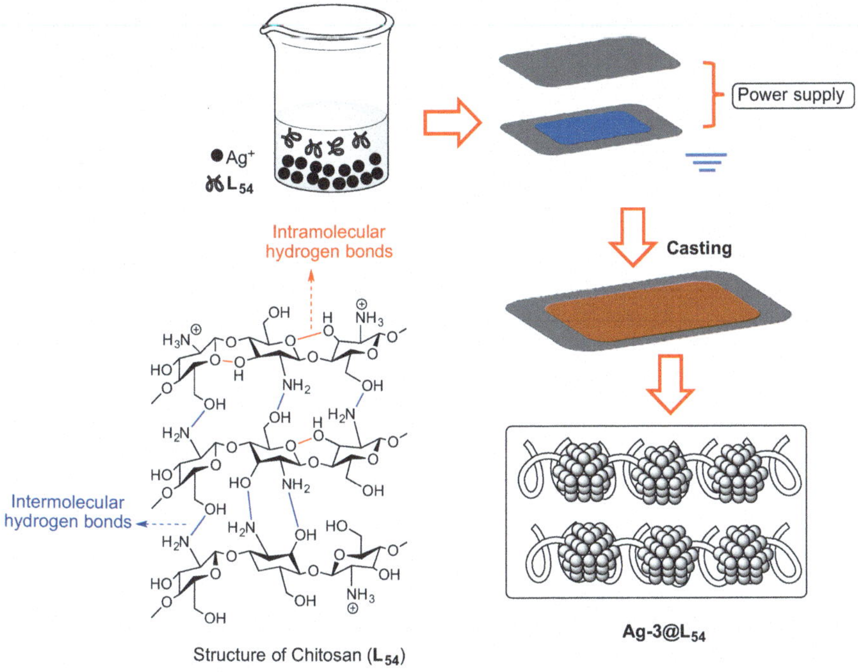

Scheme 40 Synthesis of Ag NPs stabilized by $\mathbf{L_{54}}$ assisted with an external electric field (Adapted with permission from Ref. [52] Copyright 2019 ACS Publications)

7 Rhodium-Based Nanoparticles

In the land of Rh NPs, diverse strategies have been explored, particularly focusing on the selection of surfactants as stabilizing agents, as highlighted by the work of Denicourt-Nowicki and Roucoux [53]. Herein, we aim to delve into pertinent investigations where the interplay between the nature of the surfactant counter anion and the potential role of optically active stabilizers was scrutinized. An illustrative example is provided by Bilé et al., who engineered various Rh NPs stabilized by surfactants with varying counter anions, encompassing fluoride (F^-), hydrogen carbonate (HCO_3^-), triflate ($CF_3SO_3^-$), tetrafluoroborate (BF_4^-), and chloride (Cl^-) (Scheme 41). This synthesis route employed $RhCl_3$ as metal precursor in the presence of $NaBH_4$ [54]. The resulting MNPs exhibited distinct morphologies contingent upon the counter anion: spherical and monodispersed MNPs were achieved with HCO_3^- and Cl^-, characterized by a mean size of approximately 3 nm. In contrast, BF_4^- and $CF_3SO_3^-$ yielded worm-like structures exhibiting high polydispersity, also with a mean size of approximately 3 nm. Notably, the F^- counter anion engendered both geometries, owing to its weak stabilizing influence, thus favoring MNP sintering and the formation of worm-like structures. This phenomenon possibly extends to other F-containing counter anions. Disparities in

Scheme 41 Different surfactants used as stabilizers of Rh NPs (Adapted with permission from Ref. [54] Copyright 2011 The Royal Society of Chemistry)

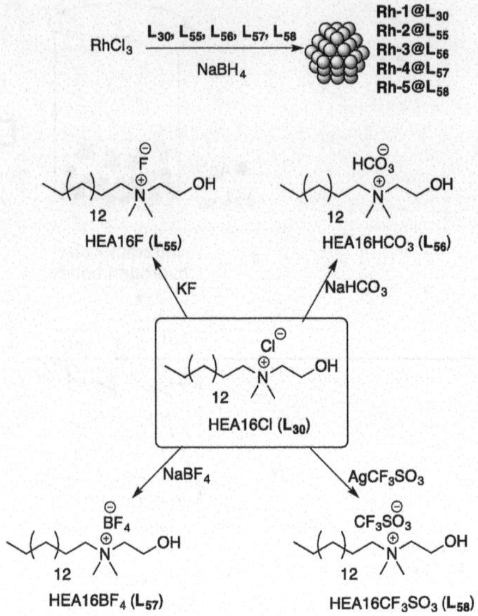

hydrodynamic diameter, as assessed via DLS, were also evident, with bulkier anions ($CF_3SO_3^-$ and BF_4^-) yielding larger systems. Turning to catalytic activity, the efficiency of these Rh NPs was evaluated in hydrogenation reactions involving various aromatic rings. Notably, benzene, toluene, and anisole were fully hydrogenated, with subtle variations observed among different counter anions. Particularly noteworthy was the superior performance of HCO_3^-, attributed to its promotion of smaller, more spherical MNPs, thereby exposing greater facets for hydrogenation. Conversely, F-containing anions demonstrated lower catalytic activity, attributable to their polydispersity and propensity to form worm-like morphologies.

In an alternate investigation, Bilé et al. explored the utilization of diverse chiral ammonium salts as stabilizers for Rh NPs in the enantiospecific hydrogenation of ethyl pyruvate and the stereospecific hydrogenation of *m*-methylanisole (Scheme 42) [55]. Employing the same methodology as previously outlined, Rh NPs were synthesized using $RhCl_3$ as metal precursor in the presence of $NaBH_4$. Two distinct counter anions, namely, bromide or lactate, were investigated. For the bromide surfactants, a range of MNP sizes was obtained, spanning from 0.79 to 2.6 nm, with larger MNPs observed for the less bulky ligands (**L_{59}-Br** and **L_{60}-Br**). Analogous trends in size variation were observed for the lactate surfactants, yielding MNPs ranging from 1.37 to 2.47 nm. Furthermore, DLS measurements revealed larger hydrodynamic diameters for the lactate-stabilized MNPs compared to those stabilized by bromide surfactants, attributable to differences in counter ion volume. In terms of catalytic performance, the Rh NPs were employed in the selective hydrogenation of ethyl pyruvate to either (*R*)- or (*S*)-ethyl lactate in water (Scheme 43). High conversion rates were achieved under 4 MPa of H_2, albeit with low enantiomeric excess (ee%) values toward the *R* configuration, peaking at 12% and 13% with either bromide or lactate surfactant stabilizers. Furthermore, altering the

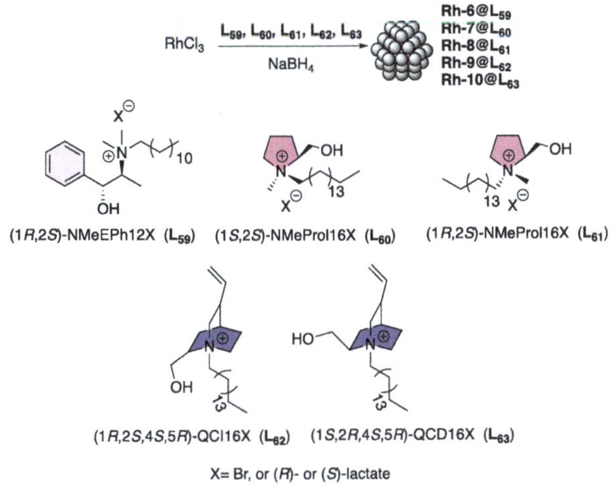

Scheme 42 Chiral surfactants used as stabilizers for Rh NPs (Adapted with permission from Ref. [55] Copyright 2012 Wiley)

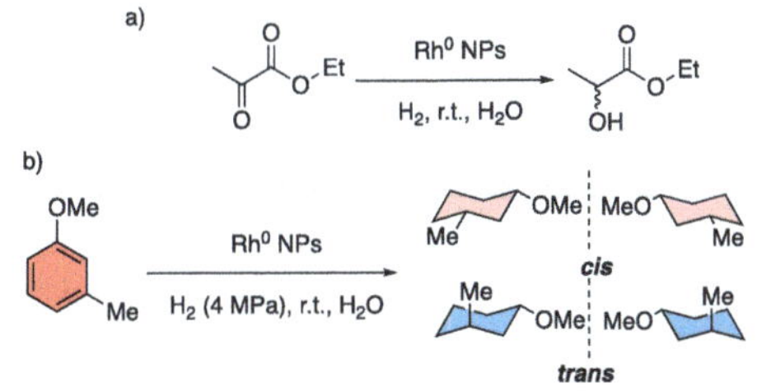

Scheme 43 Stereo- and enantioselective hydrogenation of (**a**) pyruvate and (**b**) *m*-methylanisole catalyzed by Rh NPs (Adapted with permission from Ref. [55] Copyright 2012 Wiley)

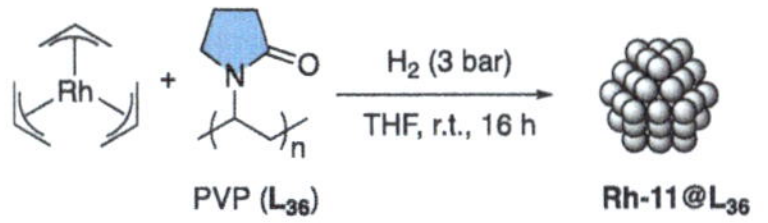

Scheme 44 Synthesis of **Rh-11@L$_{36}$** NPs (Adapted with permission from Ref. [56] Copyright 2016 The Royal Society of Chemistry)

polar head did not yield any enhancement in ee%. When *m*-methylanisole served as the substrate, the Rh NPs predominantly produced the *cis* isomer over the *trans* isomer, despite the thermodynamic preference for the latter. This outcome can be rationalized by the coordination mode of the substrate, favoring the addition of hydrogen atoms to only one face of the arene.

Polymers have also emerged as effective stabilizers for Rh NPs, akin to their utilization in stabilizing other metals. Ibrahim et al. synthesized 2.2 nm Rh NPs via the organometallic approach, employing Rh(η^3-C_3H_5)$_3$ as Rh precursor and L_{36} (PVP) as a stabilizer (Scheme 44) [56]. In terms of catalytic efficacy, these Rh NPs were evaluated in the hydrogenation of various arenes, with benzamide serving as the model substrate for reaction optimization. Full conversion and selectivity were achieved toward cyclohexane carbamide in aqueous media, whereas diminished catalytic activity was observed with other organic solvents. Subsequent examination of different substrates, including acetophenone, phenol, benzyl alcohol, pyridine, toluene, methylphenyl ether, and benzoic acid, revealed high conversion rates ranging from 74% to 100%, with exceptional selectivity (>97%) toward ring hydrogenation, except for acetophenone. Moreover, successful hydrogenation of levulinic acid and methyl levulinate to γ-valerolactone was achieved with full conversion and high selectivity, under milder conditions compared to conventional methods (120°C and 20 bar of H_2 vs 200°C and 30–40 bar H_2).

8 Iron-Based Nanoparticles

Iron oxide nanoparticles (FeO NPs) have garnered significant interest for their applications in biology and biomedicine, particularly in targeted delivery, localized therapy, magnetic hyperthermia, and serving as contrast agents for magnetic resonance imaging (MRI). One effective approach to obtaining water-soluble FeO NPs involves ligand exchange, wherein hydrophobic systems are initially employed as starting materials, and subsequently, water-soluble stabilizers are introduced to yield hydrophilic FeO NPs. A notable example of this strategy was demonstrated by Xu et al., who synthesized water-soluble FeO NPs using MNPs capped with oleic acid (OA) and trioctylphosphine oxide (TOPO) as precursors and stabilized them with poly(acrylic acid) (PAA, $\mathbf{L_{64}}$), polyethylenimine (PEI, $\mathbf{L_{65}}$), and glutathione (GSH, $\mathbf{L_{66}}$) (Scheme 45) [57]. Interestingly, the ligand exchange process did not affect the size and morphology of the nanoparticles, maintaining an average size of 12 nm with a spherical shape. DLS measurements revealed the coordination of the water-soluble ligands, resulting in hydrodynamic radii of approximately 20–30 nm, attributed to the chain length of the ligands. The successful ligand exchange was further confirmed by FTIR, which exhibited changes in the signals corresponding to the

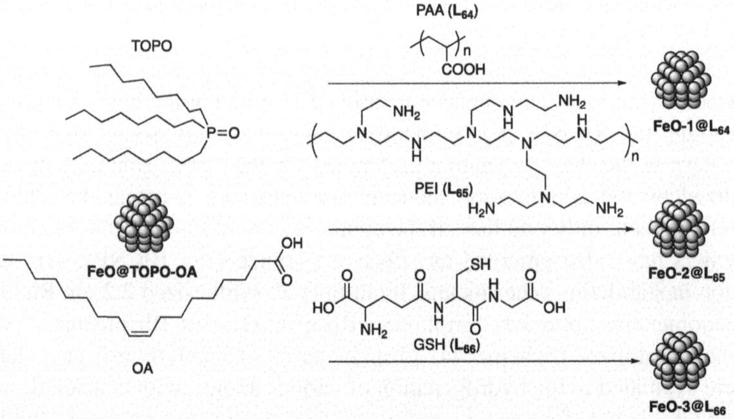

Scheme 45 Synthesis of FeO NPs stabilized by PAA ($\mathbf{L_{64}}$), PEI ($\mathbf{L_{65}}$), and GSH ($\mathbf{L_{66}}$) (Adapted with permission from Ref. [57] Copyright 2011 ACS Publications)

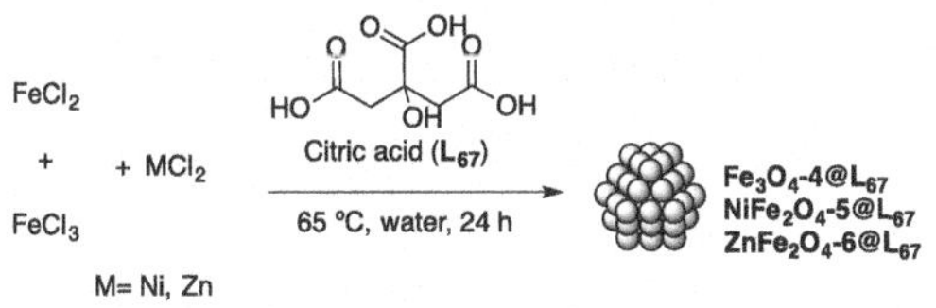

Scheme 46 Synthesis of MFe_2O_4 (M = Fe, Zn, Ni) NPs (Adapted with permission from Ref. [58] Copyright 2012 The Royal Society of Chemistry)

-COOH, $-NH_2$, and -SH groups. Moreover, the authors investigated the influence of the molecular weight of the stabilizer and the ligand-to-metal ratio on the properties of the FeO NPs. While minor differences were observed with varying molecular weights, significant variations in nanoparticle size were observed with changes in the ligand-to-metal ratio, with larger MNPs obtained at lower ratios. Finally, the stability of these MNPs was assessed in different buffers, revealing stability in 4-(2-hydroxyethyl)-1-piperazineethanesulfonic acid (HEPES) and 2-(N-morpholino)ethanesulfonic acid (MES), highlighting their potential for biological applications.

As previously mentioned, FeO NPs hold a promise for applications in MRI, as demonstrated by Zeng et al., who investigated FeO-based NPs along with other metals such as Zn and Ni for MRI purposes [58]. The synthesis of Fe_3O_4 NPs involved the use of a mixture of $FeCl_2$ and $FeCl_3$ in the presence of citric acid (L_{67}). For the bimetallic systems, $ZnCl_2$ and $NiCl_2$ replaced $FeCl_2$ to obtain $ZnFe_2O_4$ and $NiFe_2O_4$, respectively (Scheme 46). The FeO NPs exhibited sizes ranging from 3 nm to 5 nm in all cases, as determined by TEM, with a mean size of 4 nm for Fe_3O_4 and $ZnFe_2O_4$ and 5 nm for $NiFe_2O_4$. XPS analysis revealed the presence of Fe^{+2} and Fe^{+3} species in the monometallic system. In terms of MRI measurements, both the monometallic and bimetallic FeO-based systems displayed promising results as T_1-weighted contrast agents but were less effective as T_2-weighted agents. Importantly, these MNPs exhibited low cytotoxicity, underscoring their potential for biomedical applications.

Magnetic nanoparticles have garnered significant attention in biomedical applications, particularly for their potential in magnetic hyperthermia. Iron-based materials, in particular, are sought after for their water-soluble and non-toxic properties. For example, Duong et al. synthesized cobalt ferrite ($CoFe_3O_4$) NPs stabilized by citric acid and assessed their heating capabilities [59]. The synthesis involved a ligand exchange process with citric acid replacing oleic acid as the stabilizer for previously synthesized $CoFe_3O_4$ NPs (Scheme 47). Various concentrations of oleic acid were employed for MNP stabilization, resulting in $CoFe_3O_4$ nanoparticles with sizes ranging from 6 to 7 nm, as determined by TEM. Interestingly, contrary to previous reports, higher concentrations of oleic acid led to larger MNP sizes. The addition of citric acid did not alter the morphology or size of the MNPs. Hysteresis curves were constructed for both $CoFe_3O_4$ NPs stabilized with oleic acid and those stabilized with citric acid. Both exhibited a superparamagnetic nature, with a slightly higher magnetization saturation value observed for the MNPs stabilized by oleic acid

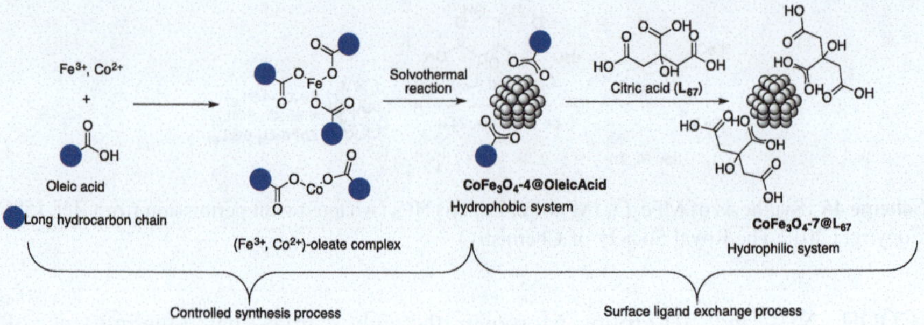

Scheme 47 Synthesis of $CoFe_3O_4$ NPs stabilized by citric acid (**L₆₇**) (Adapted with permission from Ref. [59] Copyright 2023 Elsevier)

compared to those stabilized by citric acid. However, the addition of citric acid did not induce significant changes in this property. The heating properties of the MNPs were also investigated, revealing superior results for nanoparticles with a mean size of 7 nm and a high ratio of oleic-to-citric acids (1:35). These MNPs exhibited a specific absorption rate (SAR) value of 152.7 W/g, demonstrating promising potential for application in magnetic hyperthermia.

In addition, iron-based magnetic nanoparticles have found applications beyond biomedical uses, extending into magnetically induced catalysis. An illustrative example is demonstrated by Raya-Barón and Mazarío et al., who utilized bimetallic FeNi NPs stabilized by L-lysine ($\mathbf{L_{68}}$) to achieve the selective hydrogenation of hydroxymethylfurfural (HMF) through magnetically induced heating, offering an alternative to conventional heating methods [60]. The magnetic NPs were synthesized via the ligand exchange method, wherein $\mathbf{L_{68}}$ was introduced to previously synthesized $FeNi_3$ and $FeNi_3$@Ni NPs stabilized by palmitic acid (Scheme 48). Characterization via TEM revealed MNP sizes in the range of 12–14 nm, while scanning TEM coupled with energy-dispersive X-ray spectroscopy (EDX) confirmed their bimetallic nature. XPS analysis confirmed the presence of $\mathbf{L_{68}}$ following the ligand exchange process. In terms of catalytic activity, **$\mathbf{FeNi_3}$-9@Ni@$\mathbf{L_{68}}$** exhibited superior performance in the reduction of HMF to bis(hydroxymethyl)furan (BHMF) under magnetic induction conditions in water over 6 h, achieving full conversion and 96% selectivity. This enhanced catalytic performance can be attributed to the magnetic-heating core of $FeNi_3$ and the catalytic activity of the Ni shell. Furthermore, the nanoparticles demonstrated satisfactory recyclability, retaining their catalytic activity after three cycles. To underscore the advantageous effects of magnetic heating, the authors compared the performance of **$\mathbf{FeNi_3}$-9@Ni@$\mathbf{L_{68}}$** NPs under conventional heating conditions. The reaction required a longer duration (15 h) to achieve similar selectivity but lower conversion, highlighting the efficiency of magnetically induced catalysis.

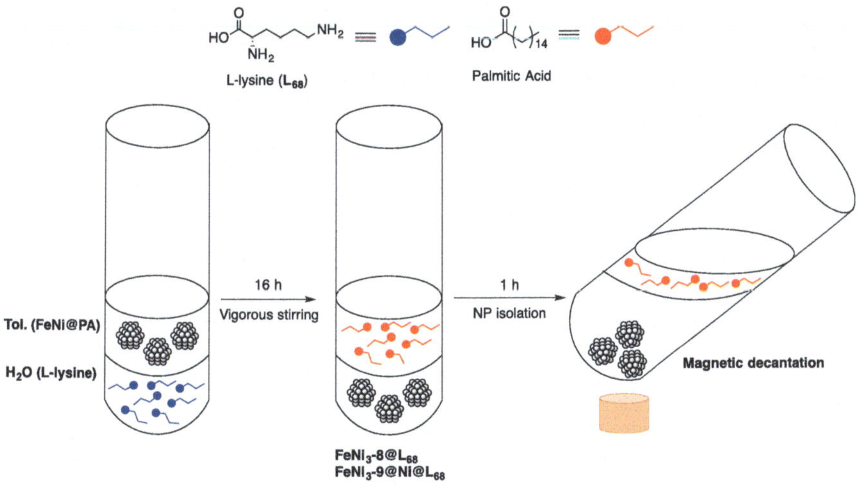

Scheme 48 Synthesis of FeNi NPs stabilized by L-lysine (**L_{68}**) (Adapted with permission from Ref. [60] Copyright 2023 Wiley)

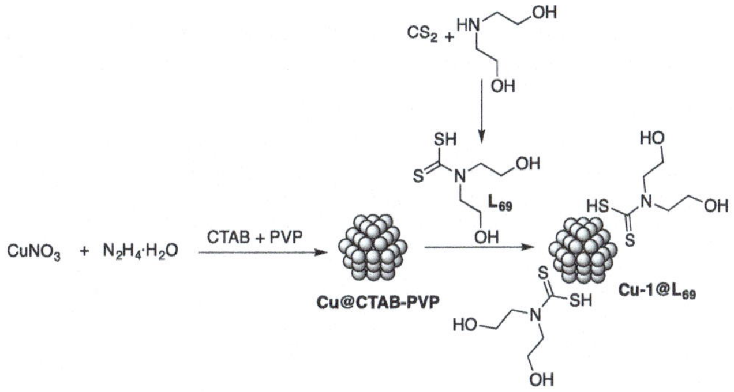

Scheme 49 Synthesis of CuNPs stabilized by **L_{69}** (Adapted with permission from Ref. [61] Copyright 2018 Springer)

9 Copper-Based Nanoparticles

Limited studies have explored the realm of water-soluble Cu NPs, with notable research conducted by Zhao et al., focusing on their application as additives in water-based lubricants as an eco-friendly alternative to oil lubricants, promoting energy efficiency and resource conservation (Scheme 49) [61]. The Cu NPs synthesis involved the utilization of $CuNO_3$ as copper precursor, alongside hydrazine hydrate as reducing agent, with CTBAB and PVP serving as initial stabilizers. Subsequently, a ligand exchange process was conducted with (2-hydroxyethyl)dithiocarbamic acid (**L_{69}**) to render the water-soluble Cu NPs. TEM analysis revealed Cu NPs with a spherical morphology and sizes around 3 nm. On the other hand, XPS analysis corroborated the presence of metallic Cu species, with no discernible signals indicative of Cu^+ or Cu^{2+} ions. FTIR spectroscopy provided evidence of ligand coordination through the S atom to the Cu surface. In practical application, the incorporation of these Cu NPs into distilled water resulted in improved heat transfer processes and reduced friction, highlighting their potential utility in lubrication systems.

10 Conclusions

Throughout this chapter we have presented several water-soluble MNP systems with different metals and stabilizing agents obtained by different synthetic routes and which present outstanding properties and diverse applications. Regarding their synthetic approach, several strategies have been developed to obtain these systems, among which we can mention the organometallic approach method by using organometallic precursors in the presence of the ligand or by using an organometallic complex which already contains the stabilizing agent; the ligand exchange methodology with previously synthesized MNPs; and the reduction of inorganic salts in which many of the cases the stabilizing agent also plays the role of reducing agent. These methods allowed modulating the size, shape, and properties of the MNPs by adding different types of ligands and changing their molar ratio with the metal.

As part of the study of these water-soluble systems, a wide variety of characterization techniques have been employed, such as TEM, UV-Vis, and DLS, mainly dedicated to determine the size of the MNPs. On the other hand, solid-state MAS-NMR, XPS, and FTIR were employed to study the surface and the ligand coordination on the surface of the MNPs and the atom, or atoms, involved in it, which is of main interest considering that the ligand is not only responsible for the hydrophilicity of the MNPs, but also it might be responsible for the selectivity and/or enhanced activity of the systems in catalytic applications.

Regarding the different applications mentioned throughout the chapter, these systems can be used in diverse fields, starting from catalytic reactions in which hydrogenation of various arenes and nitro compounds was found to be the main interest. In addition, oxidation, coupling, and hydrogen isotope exchange reactions were studied along the different reports, presenting water in all of them as the solvent and avoiding the use of toxic and hazardous solvents. Furthermore, and aside from the catalytic field, these systems have been used in biomedical applications, such as in magnetic hyperthermia and used as antimicrobial and anticancer agents, due to their high stability at different temperatures and media (various pH values, NaCl concentration, etc.).

In conclusion, several studies have been presented where different synthetic routes were employed to obtain water-soluble MNPs using a wide range of ligands with hydrosoluble moieties. These MNPs were completely characterized showing the versatility of the techniques to study and understand the basics of the hydrosoluble systems. Finally, different applications were found in which water-soluble MNPs are presented as a green alternative, showing promising results for the needs of using more biocompatible and environmentally friendly systems in the future.

Acknowledgments The authors extend their appreciation to the Vicerrectoría de Investigación of the Universidad Nacional de Colombia, for funding this work through the projects with HERMES codes 59659 and 56495.

References

1. Denicourt-Nowicki A, Mordvinova N, Roucoux A (2021) Metal nanoparticles in water: a relevant toolbox for green catalysis. In: Nanoparticles in catalysis. Wiley, pp 43–71. https://doi.org/10.1002/9783527821761.ch3
2. Kitanosono T, Masuda K, Xu P, Kobayashi S (2018) Catalytic organic reactions in water toward sustainable society. Chem Rev 118(2) American Chemical Society:679–746. https://doi.org/10.1021/acs.chemrev.7b00417
3. Polshettiwar V (2013) Nanomaterials in catalysis. Edited by Philippe Serp, Karine Philippot. Angew Chem Int Ed 52(43):11199. https://doi.org/10.1002/anie.201305828
4. Denicourt-Nowicki A, Roucoux A (2016) Odyssey in polyphasic catalysis by metal nanoparticles. Chem Rec:2127–2141. https://doi.org/10.1002/tcr.201600050
5. Alam AM, Shon YS (2021) Water-soluble Noble metal nanoparticle catalysts capped with small organic molecules for organic transformations in water. ACS Appl Nano Mater 4(4) American Chemical Society:3294–3318. https://doi.org/10.1021/acsanm.1c00335
6. Guo W, Pleixats R, Shafir A (2015) Water-soluble gold nanoparticles: from catalytic selective nitroarene reduction in water to refractive index sensing. Chem Asian J 10(11):2437–2443. https://doi.org/10.1002/asia.201500290
7. MacLeod MJ, Johnson JA (2015) PEGylated N-heterocyclic carbene anchors designed to stabilize gold nanoparticles in biologically relevant media. J Am Chem Soc 137(25):7974–7977. https://doi.org/10.1021/jacs.5b02452
8. MacLeod MJ, Goodman AJ, Ye HZ, Nguyen HVT, Van Voorhis T, Johnson JA (2019) Robust gold nanorods stabilized by bidentate N-heterocyclic-carbene–thiolate ligands. Nat Chem 11(1):57–63. https://doi.org/10.1038/s41557-018-0159-8
9. Salorinne K, Man RWY, Li C, Taki M, Nambo M, Crudden CM (2017) Water-soluble N-heterocyclic carbene-protected gold nanoparticles: size-controlled synthesis, stability, and optical properties. Angew Chem Int Ed 56(22):6198–6202. https://doi.org/10.1002/anie.201701605
10. Arabzadeh Nosratabad N, Jin Z, Du L, Mattoussi H (2023) N-heterocyclic carbene-based polymer coating of gold nanoparticles and luminescent quantum dots. MRS Adv 8(7):336–342. https://doi.org/10.1557/s43580-023-00555-9
11. Nguyen DTH, Bélanger-Bouliga M, Shultz LR, Maity A, Jurca T, Nazemi A (2021) Robust water-soluble gold nanoparticles via polymerized mesoionic N-heterocyclic carbene–gold (I) complexes. Chem Mater 33(24):9588–9600. https://doi.org/10.1021/acs.chemmater.1c02899
12. Thomas SR et al (2022) Bottom-up synthesis of water-soluble gold nanoparticles stabilized by N-heterocyclic carbenes: from structural characterization to applications. Chem A Eur J 28(56). https://doi.org/10.1002/chem.202201575
13. Monti GA, Fernández GA, Correa NM, Falcone RD, Moyano F, Silbestri GF (2017) Gold nanoparticles stabilized with sulphonated imidazolium salts in water and reverse micelles. R Soc Open Sci 4(7). https://doi.org/10.1098/rsos.170481

14. Monti GA, Correa NM, Falcone RD, Silbestri GF, Moyano F (2019) Gold nanoparticles stabilized by sulfonated-imidazolium salts as promising catalyst in water. ChemistrySelect 4(46):13496–13502. https://doi.org/10.1002/slct.201903396
15. Monti GA, Correa NM, Falcone RD, Silbestri GF, Moyano F (2020) Water-soluble gold nanoparticles: recyclable catalysts for the reduction of aromatic nitro compounds in water. RSC Adv 10(26):15065–15071. https://doi.org/10.1039/d0ra02131h
16. Enciso AE et al (2017) Facile synthesis of stable, water soluble, dendron-coated gold nanoparticles. Nanoscale 9(9):3128–3132. https://doi.org/10.1039/c6nr09679d
17. Ahmad AAL et al (2019) Synthesis of water-soluble gold-aryl nanoparticles with distinct catalytic performance in the reduction of the environmental pollutant 4-nitrophenol. Cat Sci Technol 9(21):6059–6071. https://doi.org/10.1039/c9cy01402k
18. Chen C, Ng DYW, Weil T (2020) Denatured proteins as a novel template for the synthesis of well-defined, ultra-stable and water-soluble metal nanostructures for catalytic applications. J Leather Sci Eng 2(1). https://doi.org/10.1186/s42825-020-00020-5
19. Pozdnyakov AS, Ivanova AA, Emel'yanov AI, Bolgova YI, Trofimova OM, Prozorova GF (2020) Water-soluble stable polymer nanocomposites with AuNPs based on the functional poly (1-vinyl-1,2,4-triazole-co-N-vinylpyrrolidone). J Organomet Chem 922. https://doi.org/10.1016/j.jorganchem.2020.121352
20. Traynor DJ et al (2022) Aqueous (co)polymer stabilisers for size-controlled 2-5 nm gold nanoparticle synthesis with tuneable catalytic activity. New J Chem 46(36):17282–17291. https://doi.org/10.1039/d2nj03257k
21. Pramanik S, Saha A, Devi PS (2015) Water soluble blue-emitting AuAg alloy nanoparticles and fluorescent solid platforms for removal of dyes from water. RSC Adv 5(43):33946–33954. https://doi.org/10.1039/c4ra14176h
22. Ledendecker M et al (2022) Engineering gold-platinum core-shell nanoparticles by self-limitation in solution. Commun Chem 5(1). https://doi.org/10.1038/s42004-022-00680-w
23. Venditti I et al (2017) Hydrophilic metal nanoparticles functionalized by 2-diethylaminoethanethiol: a close look at the metal-ligand interaction and interface chemical structure. J Phys Chem C 121(14):8002–8013. https://doi.org/10.1021/acs.jpcc.7b01424
24. Guerrero M et al (2012) Alkyl sulfonated diphosphines-stabilized ruthenium nanoparticles as efficient nanocatalysts in hydrogenation reactions in biphasic media. Catal Today 183(1): 34–41. https://doi.org/10.1016/j.cattod.2011.09.012
25. Debouttière PJ, Coppel Y, Denicourt-Nowicki A, Roucoux A, Chaudret B, Philippot K (2012) PTA-stabilized ruthenium and platinum nanoparticles: characterization and investigation in aqueous biphasic hydrogenation catalysis. Eur J Inorg Chem 8:1229–1236. https://doi.org/10.1002/ejic.201101159
26. Martínez-Prieto LM et al (2017) Monitoring of nanoparticle reactivity in solution: interaction of l-lysine and Ru nanoparticles probed by chemical shift perturbation parallels regioselective H/D exchange. Chem Commun 53(43):5850–5853. https://doi.org/10.1039/c7cc02445b
27. Pirali T, Serafini M, Cargnin S, Genazzani AA (2019) Applications of deuterium in medicinal chemistry. J Med Chem 62(11):5276–5297. https://doi.org/10.1021/acs.jmedchem.8b01808
28. Palazzolo A et al (2019) Efficient access to deuterated and tritiated nucleobase pharmaceuticals and oligonucleotides using hydrogen-isotope exchange. Angew Chem Int Ed 58(15):4891–4895. https://doi.org/10.1002/anie.201813946
29. Bouzouita D et al (2019) Tuning the catalytic activity and selectivity of water-soluble bimetallic RuPt nanoparticles by modifying their surface metal distribution. Nanoscale 11(35): 16544–16552. https://doi.org/10.1039/c9nr04149d
30. Zuluaga-Villamil A, Mencia G, Asensio JM, Fazzini PF, Baquero EA, Chaudret B (2022) N-heterocyclic carbene-based iridium and ruthenium/iridium nanoparticles for the hydrogen isotope exchange reaction through C-H bond activations. Organometallics 41(22):3313–3319. https://doi.org/10.1021/acs.organomet.2c00288

31. Rauchdi M, Ait Ali M, Roucoux A, Denicourt-Nowicki A (2018) Novel access to verbenone via ruthenium nanoparticles-catalyzed oxidation of A-pinene in neat water. Appl Catal A Gen 550: 266–273. https://doi.org/10.1016/j.apcata.2017.11.016
32. Bouriazos A, Sotiriou S, Stathis P, Papadogianakis G (2014) Superior aqueous-phase catalytic hydrogenation activity of palladium modified with nitrogen-containing ligands compared with the TPPTS benchmark modifier in micellar nanoreactors. Appl Catal B 150–151:345–353. https://doi.org/10.1016/j.apcatb.2013.12.032
33. Lee JY, Tzeng RJ, Wang MC, Lee HM (2017) Application of a zwitterionic palladium complex as a metal precursor of recyclable palladium nanoparticles for catalyzing Suzuki-Miyaura coupling reactions. Inorg Chim Acta 464:74–80. https://doi.org/10.1016/j.ica.2017.04.046
34. Yun G et al (2014) Highly stable, water-dispersible metal-nanoparticle-decorated polymer nanocapsules and their catalytic applications. Angew Chem Int Ed 53(25):6414–6418. https://doi.org/10.1002/anie.201403438
35. Zhu JF, Tao GH, Liu HY, He L, Sun QH, Liu HC (2014) Aqueous-phase selective hydrogenation of phenol to cyclohexanone over soluble Pd nanoparticles. Green Chem 16(5): 2664–2669. https://doi.org/10.1039/c3gc42408a
36. Iben Ayad A et al (2019) 'Water soluble' palladium nanoparticle engineering for C-C coupling, reduction and cyclization catalysis. Green Chem 21(24):6646–6657. https://doi.org/10.1039/c9gc02546d
37. Avila E, Nixarlidis C, Shon YS (2023) Water-soluble Pd nanoparticles for the anti-Markovnikov oxidation of allyl benzene in water. Nanomaterials 13(2). https://doi.org/10.3390/nano13020348
38. Ferry A et al (2015) Negatively charged N-heterocyclic carbene-stabilized Pd and au nanoparticles and efficient catalysis in water. ACS Catal 5(9):5414–5420. https://doi.org/10.1021/acscatal.5b01160
39. Asensio JM, Tricard S, Coppel Y, Andrés R, Chaudret B, de Jesús E (2017) Knight shift in 13 C NMR resonances confirms the coordination of N-heterocyclic carbene ligands to water-soluble palladium nanoparticles. Angew Chem Int Ed 56(3):865–869. https://doi.org/10.1002/anie.201610251
40. Asensio JM, Tricard S, Coppel Y, Andrés R, Chaudret B, de Jesús E (2017) Synthesis of water-soluble palladium nanoparticles stabilized by sulfonated N-heterocyclic carbenes. Chem A Eur J 23(54):13435–13444. https://doi.org/10.1002/chem.201702204
41. Suárez-Riaño O et al (2022) Water-soluble NHC Pd/Ni bimetallic nanoparticles for H/D exchange in aromatic amino-acids. Chem Commun 59(8):1062–1065. https://doi.org/10.1039/d2cc06019a
42. Li N et al (2023) Metal nanoparticles with a clean surface via clay-nanosheets as stabilizers and supports for catalytic hydrogenation reactions in aqueous-phase. Fuel 334. https://doi.org/10.1016/j.fuel.2022.126648
43. Corra S, Lewandowska U, Benetti EM, Wennemers H (2016) Size-controlled formation of Noble-metal nanoparticles in aqueous solution with a thiol-free tripeptide. Angew Chem Int Ed 55(30):8542–8545. https://doi.org/10.1002/anie.201510337
44. Baquero EA, Tricard S, Flores JC, De Jesffls E, Chaudret B (2014) Highly stable water-soluble platinum nanoparticles stabilized by hydrophilic N-heterocyclic carbenes. Angew Chem Int Ed 53(48):13220–13224. https://doi.org/10.1002/anie.201407758
45. Baquero EA, Tricard S, Coppel Y, Flores JC, Chaudret B, De Jesús E (2018) Water-soluble platinum nanoparticles stabilized by sulfonated N-heterocyclic carbenes: influence of the synthetic approach. Dalton Trans 47(12):4093–4104. https://doi.org/10.1039/c8dt00240a
46. Ruiz-Varilla AM, Baquero EA, Chaudret B, De Jesús E, Gonzalez-Arellano C, Flores JC (2020) Water-soluble NHC-stabilized platinum nanoparticles as recoverable catalysts for hydrogenation in water. Cat Sci Technol 10(9):2874–2881. https://doi.org/10.1039/d0cy00481b
47. Mao H, Liao Y, Ma J, Zhao SL, Huo FW (2016) Water-soluble metal nanoparticles stabilized by plant polyphenols for improving the catalytic properties in oxidation of alcohols. Nanoscale 8(2):1049–1054. https://doi.org/10.1039/c5nr07897k

48. Agasti N, Singh VK, Kaushik NK (2015) Synthesis of water soluble glycine capped silver nanoparticles and their surface selective interaction. Mater Res Bull 64:17–21. https://doi.org/10.1016/j.materresbull.2014.12.030
49. Cao J, Sun X, Zhang X, Lu C (2016) Homogeneous synthesis of Ag nanoparticles-doped water-soluble cellulose acetate for versatile applications. Int J Biol Macromol 92:167–173. https://doi.org/10.1016/j.ijbiomac.2016.06.092
50. He Y et al (2017) Synthesis, characterization and evaluation cytotoxic activity of silver nanoparticles synthesized by Chinese herbal Cornus officinalis via environment friendly approach. Environ Toxicol Pharmacol 56:56–60. https://doi.org/10.1016/j.etap.2017.08.035
51. Khan Z (2019) Encapsulation of silver nanoparticles into the helix of water soluble starch and their sensing properties. Int J Biol Macromol 136:165–176. https://doi.org/10.1016/j.ijbiomac.2019.05.012
52. Zhou Q et al (2019) Electric-field-assisted facile synthesis of metal nanoparticles. ACS Sustain Chem Eng 7(1):1271–1278. https://doi.org/10.1021/acssuschemeng.8b04927
53. Denicourt-Nowicki A, Roucoux A (2015) From hydroxycetylammonium salts to their chiral counterparts. A library of efficient stabilizers of Rh(0) nanoparticles for catalytic hydrogenation in water. Catal Today 247:90–95. https://doi.org/10.1016/j.cattod.2014.05.031
54. Guyonnet Bilé E, Sassine R, Denicourt-Nowicki A, Launay F, Roucoux A (2011) New ammonium surfactant-stabilized rhodium(0) colloidal suspensions: influence of novel counter-anions on physico-chemical and catalytic properties. Dalton Trans 40(24):6524–6531. https://doi.org/10.1039/c0dt01763a
55. Guyonnet Bilé E, Cortelazzo-Polisini E, Denicourt-Nowicki A, Sassine R, Launay F, Roucoux A (2012) Chiral ammonium-capped rhodium(0) nanocatalysts: synthesis, characterization, and advances in asymmetric hydrogenation in neat water. ChemSusChem 5(1):91–101. https://doi.org/10.1002/cssc.201100364
56. Ibrahim M, Poreddy R, Philippot K, Riisager A, Garcia-Suarez EJ (2016) Chemoselective hydrogenation of arenes by PVP supported Rh nanoparticles. Dalton Trans 45(48):19368–19373. https://doi.org/10.1039/C6DT03668F
57. Xu Y, Qin Y, Palchoudhury S, Bao Y (2011) Water-soluble iron oxide nanoparticles with high stability and selective surface functionality. Langmuir 27(14):8990–8997. https://doi.org/10.1021/la201652h
58. Zeng L, Ren W, Zheng J, Cui P, Wu A (2012) Ultrasmall water-soluble metal-iron oxide nanoparticles as T 1-weighted contrast agents for magnetic resonance imaging. Phys Chem Chem Phys 14(8):2631–2636. https://doi.org/10.1039/c2cp23196d
59. Duong HDT, Yoon SH, Nguyen DT, Kim KS (2023) Magnetic heating of water dispersible and size-controlled superparamagnetic cobalt iron oxide nanoparticles. Powder Technol 427. https://doi.org/10.1016/j.powtec.2023.118720
60. Raya-Barón Á, Mazarío J, Mencia G, Fazzini PF, Chaudret B (2023) l-lysine stabilized FeNi nanoparticles for the catalytic reduction of biomass-derived substrates in water using magnetic induction. ChemSusChem 16(12). https://doi.org/10.1002/cssc.202300009
61. Zhao J, Yang G, Zhang C, Zhang Y, Zhang S, Zhang P (2019) Synthesis of water-soluble Cu nanoparticles and evaluation of their tribological properties and thermal conductivity as a water-based additive. Friction 7(3):246–259. https://doi.org/10.1007/s40544-018-0209-7

Top Organomet Chem (2024) 75: 243–272
https://doi.org/10.1007/3418_2024_117
© The Author(s), under exclusive license to Springer Nature Switzerland AG 2024
Published online: 8 October 2024

Organometallic Nanoparticles for Magnetically Induced Catalysis in Solution

Jaime Mazarío, Víctor Varela-Izquierdo, and Bruno Chaudret

Contents

Abstract In the starting exploration of magnetically induced catalysis, organometallic nanoparticles (NPs) have proven themselves to serve as efficient and versatile heating agents, demonstrating remarkable potential for diverse liquid-phase reactions. The present chapter delves into the synthesis and application of various organometallic magnetic NPs (MagNPs), i.e., $Fe_{2.2}C$, FeCo, $FeNi_3$, and NiCo NPs. These magnetic cores, decorated with metals like Ru, Cu, or Ni, exhibit remarkable performance in hydrogenation and hydrodeoxygenation reactions, with promising results in the valorization of biomass derivatives under low hydrogen pressures. The superior catalytic activity compared to those achieved under similar apparent conditions through conventional heating, as well as some recent advances in stability, add to the well-proven advantages of this technology, such as rapid heating, improved energy efficiency, and adaptability to renewable energy sources. The present chapter also establishes the need for precise temperature control, optimized

J. Mazarío, V. Varela-Izquierdo, and B. Chaudret (✉)
Laboratoire de Physique et Chimie des Nano-Objets, Université de Toulouse, UMR 5215 INSA, CNRS, UPS, Toulouse, Cedex 4, France
e-mail: chaudret@insa-toulouse.fr

ligand strategies, and strategic stability enhancement in aqueous media, signaling avenues for future progress in magnetically induced catalysis.

Keywords Induction heating · Heterogeneous catalysis · Magnetic nanoparticles · Magnetically induced catalysis · Organometallic approach

Abbreviations

AC	Alternating current
AMF	Alternating magnetic field
bcc	Body-centered cubic
DPE	Diphenylether
FA	Furfuryl alcohol
FeC, $Fe_{2.2}C$	Iron–Carbide nanoparticles
$Fe_{2.2}C$@Ru	Iron–Carbide nanoparticles coated with ruthenium
$Fe_{2.2}C$@Cu	Iron–Carbide nanoparticles coated with copper
$Fe_{2.2}C/Cu_2Cr_2O_5$	Iron–Carbide nanoparticles ($Fe_{2.2}C$) supported on $Cu_2Cr_2O_5$
FeCo	Iron–Cobalt nanoparticles
$FeNi_3$	Iron–Nickel nanoparticles
$FeNi_3$-PA	Iron–Nickel nanoparticles stabilized with palmitic acid
$FeNi_3$-Lys	Iron–Nickel nanoparticles stabilized with L-Lysine
fcc	Face-centered cubic
GVL	γ-Valerolactone
hcp	Hexagonal closed packed
HDA	Hezadecylamine
HDO	Hydrodeoxygenation
HR-TEM	High-resolution transmission electron microscopy
LA	Levulinic acid
MagNPs	Magnetic nanoparticles
MFA	5-2-Hydroxymethyl-5-dimethylfuran
NiCo, Ni_7Co_3	Nickel–Cobalt nanoparticles
Ni_7Co_3@Ru	Nickel–Cobalt nanoparticles coated with Ru
NPs	Nanoparticles
PA	Palmitic acid
SAR	Specific absorption rate
T_c	Curie temperature
TEM	Transmission electron microscopy
XAS	X-ray absorption spectroscopy
XRD	X-ray diffraction
2,5-BHMF	2,5-(Bishydroxymethyl)furan
2,5-DMF	2,5-Dimethylfuran
2-MF	2-Methylfuran
5-HMF	5-Hydroxymethylfurfural

1 Introduction

In the first two decades of the twenty-first century, the field of heterogeneous catalysis has witnessed the emergence of a transformative domain, namely *magnetically induced catalysis*. This innovative field capitalizes on the synergies between magnetic susceptibility and efficient catalytic transformations. In essence, *magnetically induced catalysis* is a technique that uses alternating magnetic fields to thermally activate catalytic nanomaterials and improve their performance. This approach has emerged as a promising alternative to traditional heating methods in catalysis [1]. In fact, it has proven its potential to overcome limitations such as long heating and cooling times and replace undesired thermal gradients with highly localized hotspots in the active site or its vicinity. Therefore, it offers a pathway to energy-efficient, localized, and rapid heating of chemical processes [2–5]. Based on these unique features, recent studies have shown the potential of this technology in integrating chemical reactions within a future widespread intermittent energy production framework [6–9]. Hence, the implementation of magnetically induced catalysis is believed to hold the potential to participate in transforming the way energy is used by the chemical industry and pave the way for sustainable and cost-efficient chemical processes. Moreover, it is well known that magnetic catalysts possess a unique attribute of magnetic recovery, which facilitates their retrieval, purification, and reusability in multiple catalytic cycles [10]. Therefore, these catalysts are a promising option for various industrial applications from multiple viewpoints.

Furthermore, the rapidly growing rate of *magnetically induced catalysis* has demonstrated that its appeal extends beyond its initial advantages. In particular, the ability to integrate thermal and catalytic functions not only enhances overall energy efficiency but also enables precise control over reaction conditions. This, in turn, improves control over reaction rates, selectivity, and yields [11, 12]. Particularly, we have found that using induction heating in liquid solutions causes the temperature near magnetic nanoparticles (MagNPs) to rise significantly above the boiling point of the solvent (Fig. 1) [13]. This phenomenon has allowed for the accomplishment of several chemical transformations under H_2 pressure and bulk temperature conditions where catalytic performance would otherwise be inadequate [14–18]. These unique features have already shown remarkable potential in various catalytic applications, spanning from pharmaceuticals and fine chemicals to energy and environmental remediation.

The underlying principle behind magnetically induced catalysis is the generation of heat through the interaction of MagNPs with an alternating magnetic field (AMF), which can be modified by adjusting the material properties and the field characteristics (frequency and amplitude) [19]. The principal appeal of magnetically induced catalysis lies in the ability to control the heating properties of MagNPs. This control is governed by different heat-dissipation mechanisms, primarily for MagNPs under a high-frequency AMF, hysteresis losses [20–23]. Researchers can control magnetic properties such as coercivity and remanence by tuning particle size, composition,

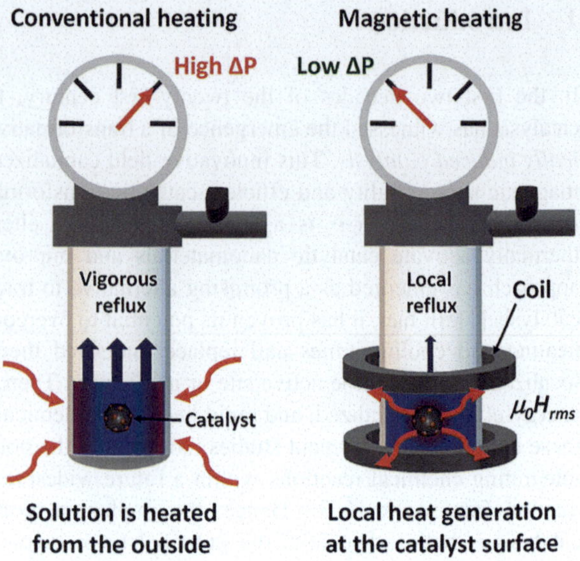

Fig. 1 Comparison between conventional and magnetic heating based on heat dissipation. Reproduced with permission from Ref. [14] Copyright 2019 Wiley-VCH

and magnetic anisotropy. This provides a means to achieve desired Specific Absorption Rate (SAR) values and control heat dissipation precisely. Equations governing hysteresis losses provide a systematic framework, elucidating the intricate relationship between material parameters and heating efficiency. For instance, Brownian and Néel relaxation mechanisms exhibit sensitivity to factors such as particle size, shape and the surrounding medium [20, 24]. Thus, for a comprehensive approach to optimizing magnetically induced catalysis, it is crucial to adjust parameters governing the hysteresis losses, thereby unlocking the full potential of MagNPs for controlled and efficient heating under an AMF. In that sense, organometallic nanoparticles (NPs), which are synthesized by the controlled decomposition of organometallic precursors [25], are significant elements in this brand-new field. By selectively choosing organometallic (or metal-organic) precursors containing magnetic elements such as iron, cobalt or nickel and meticulously tuning the reaction conditions (e.g., temperature, pressure, time), as well as designing strategic ligand stabilization during synthesis, researchers can lay the foundation for desired growth rates, control agglomeration, and in this way efficiently control nanoparticle properties such as compositions, shapes (spheres, cubes, rods, wires, and fractal structures), sizes, and surface properties. These factors, in turn, govern other properties pivotal for magnetic heating, such as anisotropy, magnetic saturation, coercivity, and remanence. Moreover, additional catalytic phases can be incorporated alongside the magnetic susceptor element. This integration can occur in the form of decorating NPs [26, 27] or well-formed shells [28, 29], thereby creating a catalytic system with a complex interplay of structures and functionalities.

In fact, the controlled decomposition inherent in the organometallic approach has already proven instrumental in *magnetically induced catalysis*, providing a pathway for synthesizing a variety of stable nanocatalysts capable of facilitating diverse temperature regimes and, therefore, supporting the execution of various chemical transformations in liquid media. In essence, this chapter highlights the vital role of organometallic NPs as catalysts meticulously crafted for the intricacies of *magnetically induced catalysis* within liquid solutions. Therefore, the subsequent sections will delve into the area of *magnetically induced catalysis*, providing a historical background and focusing on a detailed overview of several examples and applications that have led to significant advancements in liquid-phase catalytic processes using organometallic NPs. We believe that examining various instances where NPs approached through organometallic means have been utilized as catalysts in magnetic induction can enhance our comprehension and establish a practical path for effectively utilizing these NPs in *magnetically induced catalysis* in liquid-phase. As our exploration unfolds, we will also provide an overview of the measures taken to address issues related to deactivation and degradation, as well as those concerning stirring and temperature measurement. This will allow us to gain a deeper understanding of the challenges faced in these areas and the strategies that have been implemented to overcome them.

2 Induction-Assisted Chemistry

The history of induction heating spans two centuries, beginning with the groundbreaking experiments of Michael Faraday in 1831, who used two coils of wire wound around opposite sides of a ring of soft iron. In the experiment, one wire was connected to a galvanometer and observed while connecting the other to a battery. A transient current was detected when the wire was connected to the battery, and another when it was disconnected [30, 31].

As the nineteenth century ended, there was an increased understanding that certain materials could release heat when exposed to a time-varying electromagnetic field. Rather than trying to avoid heat losses, engineers began to explore the potential of exploiting these materials for practical purposes. Although debates lingered on the energy efficiency of induction heating, its high heating rates and undeniable advantages (rapidity, repeatability, reliability, and seamless automation) fueled its adoption. Initially harnessed for foundry work and cooking applications, induction heating soon found its way into biomedicine technologies for drug release and disease treatment through magnetic hyperthermia [30–34].

In the latter half of the twentieth century, engineers began investigating the use of induction heating in chemical reactions. During the 1960s, researchers at Shell Company secured a patent for applying metal induction heating to selectively heat specific components within a chemical reactor [35]. This is likely the first reported instance of electromagnetically induced heating being applied to a chemical process. They achieved this by heating a piece of rotating metal with induction and transferring the heat to a nearby fluid. This method had several advantages, including continuously reaching very high temperatures (up to 2,500°C) and a fast heating and cooling rate. This helped to avoid side reactions during the temperature change and increased the product yield by reducing temperature inhomogeneity.

This approach of chemical processes heated through the electromagnetic heating of macroscopic objects has been extensively studied, and notable instances emerged, such as Leclercq et al.'s innovative use of macroscale induction heating to swiftly address accidental plant contamination [36]. The method involved supplying heat to a steel tube with a catalyst film, showcasing the technology's potential in critical environmental applications. Also, Camacho et al. have demonstrated the feasibility of reforming dry biogas to stabilize the electricity grid while providing robust energy storage. The study employed a stainless-steel reactor with a perovskite $Na_{0.5}La_{0.5}Ni_{0.3}Al_{0.7}O_{2.5}$ as catalyst, and the process was facilitated by wind power and an induction heating system. By converting renewable biogas into hydrogen, the technology is claimed to have the potential to store and release energy as needed, thereby mitigating the intermittent nature of renewable energy sources [7]. Finally, steel balls have also been used as heating agents under electromagnetic induction in chemical reactors for many years, with examples extending to the present day [37, 38].

Selective heating also took a significant leap forward with the direct electromagnetic heating of the catalytic bed in 1975, when B.H. Bartley demonstrated dual functional catalysis within one reactor, thanks to electromagnetic heating of 20 μm iron spheres acting as catalysts [39]. This concept evolved over the years, enabling consecutive reactions within a reactor, as illustrated by Chatterjee et al.'s work with a quartz reactor with heating zones made of nickel ferrite NPs and catalysts separated by glass beads [40].

The heating of micro-nanoscale chemical agents inside a reactor, with metal wools [41–44] and MagNPs as the most notorious examples, has recently gained attention as a promising alternative to heating macroscopic reactor components, which current research groups have largely abandoned. Specifically, this method allows for more precise and selective heating and minimizes security risks. Additionally, using MagNPs as heating agents offers greater tunability and, therefore, enhanced efficiency and control over the reaction, leading to improved yields and reduced energy consumption. Kirschning's group pioneered magnetically induced heating of chemical reactions using MagNPs encapsulated in silica (Fe_2O_3/Fe_3O_4@SiO_2) in organic synthesis in both flow and batch reactors, achieving higher yields compared to traditional heating methods [45]. Subsequent works expanded applications to diverse reactions, including transfer hydrogenations or some pericyclic reactions [46].

Furthermore, the unique properties of metal NPs, such as their high surface area-to-volume ratio and tunability in composition and size, make them ideal candidates for catalytic activity. As a result, they have established themselves as dual agents for electromagnetic heating and catalysis. In that sense, Ovenston and Walls were the first to consider the possibility of electromagnetically heating catalytic nickel NPs for hydrocarbon reforming, modelling their heat dissipation behavior and demonstrating the potential of catalytic NP heating [47]. More recently, Kirschning successfully integrated heating and catalytic functions further combining their silica-coated MagNPs (Fe_2O_3/Fe_3O_4@SiO_2) with palladium and gold. The resulting catalysts demonstrated remarkable reusability and efficiently catalyzed Suzuki-Miyaura-Heck coupling reactions [45], as well as O_2-activated oxidation [48]. Rebrov and his team also developed a new core-shell magnetic catalyst called TiO_2@SiO_2@$NiFe_2O_4$ [49]. This catalyst was synthesized using the sol-gel method and has shown good catalytic activity in the direct amide synthesis reaction between aniline under radiofrequency induction.

In recent times, the combination of the organometallic NPs with induction heating introduced by our group not only propels chemical reactions in the liquid phase to new frontiers but also positions induction heating as a promising player in the valorization of alternative raw materials.

3 Organometallic Systems

Over the past two decades, the organometallic approach developed by our group in Toulouse has been applied to synthesize MagNPs, proving to be highly beneficial [50]. This approach has significantly contributed to the field by the creation of multiple magnetic nanostructures (i.e., Co rods, Co wires, Co urchins, Fe cubes, Ni, FeC, Fe spheres, among others) and offers a promising avenue for creating new systems. The methodology is characterized by its ability to synthesize small metal NPs with precisely controlled sizes, shapes, and compositions through the regulated decomposition of organometallic precursors under mild reaction conditions. This technique is notable for its capacity to produce clean NP surfaces and magnetic properties that are analogous to those achieved through ultra-high vacuum methods. In conjunction with accurate control over reaction conditions, the deliberate selection of organometallic precursors and ligands facilitates the growth and shaping of organometallic NPs and the formation of intricate bimetallic species, including alloys, core-shell structures and NPs with concentration gradients [25, 51, 52]. Furthermore, stabilizing ligands play a crucial role in tuning the properties of these organometallic NPs. They influence not only the structure of the NPs [53, 54], but also their stability, solubility, activity, and selectivity. For instance, modulating the structure of diphosphite ligands has been shown to enhance the stability and the catalytic activity of Ir NPs in the hydrogenation of m-methylanisole and o-methylanisole [55]. On the other hand, the number of equivalents and the structure of N-heterocyclic carbene ligands influence NP properties such as stability, solubility in organic or aqueous media and catalytic activity [56]. Similarly, the aminoacid l-lysine can significantly improve the solubility of FeNi-based NPs in aqueous media [57]. This ligand-based approach, although still predominantly utilized in homogeneous catalysis with inorganic and organometallic complexes, can be readily extrapolated to the synthesis and application of organometallic nanoparticles, offering a versatile platform for designing advanced catalytic systems. In essence, this methodology enables the production of uniform nano-objects with precise chemical arrangements, thereby facilitating the creation of materials possessing requisite characteristics not just in magnetic-based applications but also in the fields of catalysis and energy. Although organometallic precursors come at a high cost and require careful handling, the unique properties and versatility of the organometallic approach in NP synthesis have led several research groups worldwide to adopt them [58–61]. The following lines delve into the diverse magnetic cores prepared through this organometallic route, highlighting their applications so far reported in magnetically induced catalysis in liquid-phase solutions.

3.1 FeC

In one of the earliest literature reports on hybrid functional materials exhibiting high heating powers, Asensio et al. studied the viability of hydrodeoxygenation (HDO) of aliphatic and aromatic carbonylic compounds in liquid-phase solutions using magnetic heating [14]. The study utilized $Fe_{2.2}C$-NPs as magnetic cores, a previously synthesized material with superior heating properties to any prior material reported [62]. The synthesis involves first the preparation of bcc Fe(0) NPs through the reduction of $[Fe\{N(SiMe_3)_2\}_2]_2$ under H_2 in the presence of palmitic acid (PA) and hexadecylamine (HDA) [63]. Then, the $Fe_{2.2}C$ NPs were obtained through the carburization of the preformed Fe(0) NPs (mean size of 12 nm) as follows: Fe (0) NPs were dispersed in mesitylene, pressurized with a gas mixture of CO and H_2 and heated at 150°C. The resulting $Fe_{2.2}C$ particles exhibit a pseudohexagonal structure and display a mean size of 15 nm [62, 64].

The $Fe_{2.2}C$-NPs were decorated with Ru to enhance their catalytic activity ($Fe_{2.2}C$@Ru). This was done by decomposing Ru_3CO_{12} at 150°C. A high catalytic activity of $Fe_{2.2}C$@Ru in HDO reactions was observed (Scheme 2, top). The use of an AMF of 300 kHz and 80 mT (apparent temperature ≈150–160°C, measured with an IR camera) resulted in a complete conversion of acetophenone to ethylbenzene.

On the other hand, only 10% yield to the same product was obtained in an analogous experiment with conventional heating at 200°C. The study was extended to the HDO of p-substituted acetophenone derivatives (R(p) = hydrogen, methyl, isopropyl, methoxy, chloride, trifluoromethane). It was observed that electron-donating groups on the substrate led to high conversion rates (Table 1, entries 2–4). In contrast, electron-withdrawing groups (Table 1, entries 5 and 6) induced lower conversions than that obtained for unsubstituted acetophenone. The conversions of alkyl-substituents fell in the 88–100% range, while those of halogenated ones ranged from 5 to 54%. Total selectivity to the HDO product was observed in all the examples described above. However, the reaction with benzylideneacetone provided the hydrogenation of the double bond in 74% of the yield. Traces of 4-phenyl-2-butanol product (1%) were also observed (Table 1, entry 7). The researchers also extended the reaction to furfural and 5-hydroxymethylfurfural (5-HMF) at a low H_2 pressure of 3 bars (300 kHz, 58 mT). These compounds are platform molecules that are extracted from biomass. They are formed through the removal of water from C5 and C6 sugars, using an acidic catalyst [65]. Their HDO

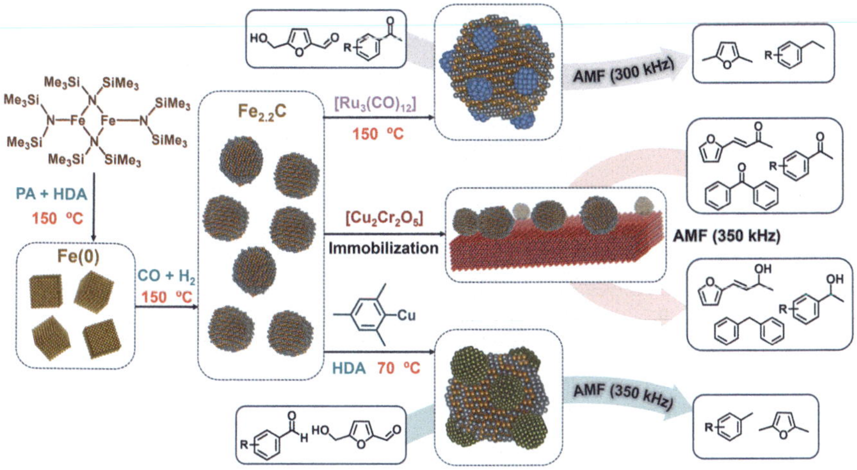

Scheme 2 Schematic overview of liquid-phase magnetically induced catalysis with $Fe_{2.2}C$ NPs as magnetic cores

products, 2-methylfurane (2-MF) and 2,5-dimethylfurane (2,5-DMF), are regarded as promising biofuel options [66]. With a catalyst loaded with a 1.0 mol% of Ru, 100% conversion of furfural was reached with 100% selectivity towards 2-MF (Table 1, entry 8). Similarly, 5-HMF could be selectively transformed into 2,5-DMF under the same reaction conditions (Table 1, entry 9). Notably, the catalytic HDO of carbonyl groups, which obtains the corresponding alkyl derivatives in a single step, is a significant reaction to produce platform molecules and customized fuels. This process is generally carried out industrially with high catalyst loadings, temperatures up to 280°C and high H_2 pressures (ranging from 10 to 50 bar) [67, 68]. Indeed, Asensio et al. obtained poor catalytic results using conventional heating (i.e., 200°C) with 3 H_2 bar. This suggests that magnetic catalysis could be a more efficient alternative to traditional methods. The authors credit the exceptional performance of the methodology to its capability of achieving higher temperatures at the catalyst surface. This amplifies the catalytic activity, forming "hot spots" in a relatively cold environment. This phenomenon enables local heating above the boiling point of the solvent by forming a vapor shell around the particle, which reduces heat transfer by the solvent.

Our team has also used $Fe_{2.2}C$ NPs to make noteworthy contributions to understanding the behavior of MagNPs in liquid phase media when subjected to an AMF. For instance, a model has been proposed to explain why the particles exhibit different heating powers despite having similar structures and static magnetic properties (SAR ranging from 0 to 2 kW g^{-1}) [69]. The team took into account the hysteresis losses in $Fe_{2.2}C$ nanoparticles, which are related to their capacity to form chains when exposed to magnetic excitation and the time required for this process. The formation of these chains is influenced by the presence or absence of strong

Table 1 Scope of substrates in the HDO of ketones catalyzed by $Fe_{2.2}C@Ru$ NPs in the presence of an AMF

Entry	Substrate	Product	Time (h)	Conv. (%)	Yield (mol %)
1	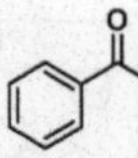	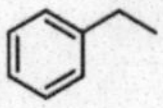	4	61	61
2	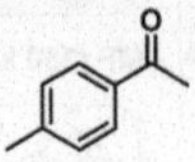	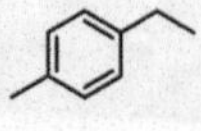	4	88	88
3	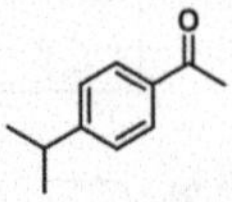	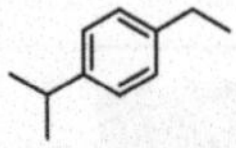	4	89	89
4	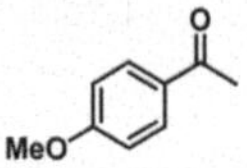	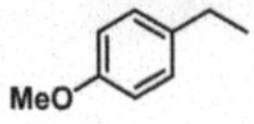	4	100	100
5	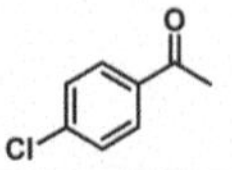	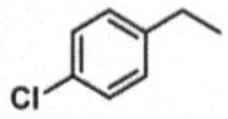	4	32	32
6	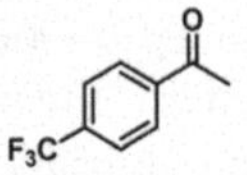	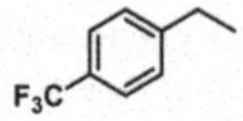	4	5	5
7	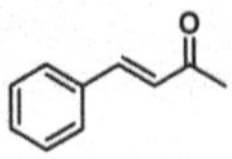	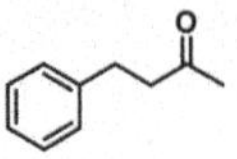	4	75	74
8	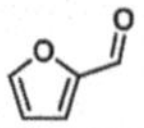	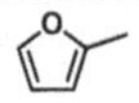	15	100	100
9	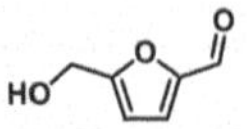	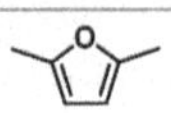	15	100	100

Reaction conditions: Substrate (1.36 mmol), $Fe_{2.2}C@Ru$ (5.0 mg, 0.25 mol% of Ru), mesitylene (5 mL), 3 H_2 bar, 58 mT (300 kHz)

dipolar couplings, which are caused by variations in the concentration of ligands on the surface of the particles. These differences lead to notable variations in heat production. In that sense, Scheme 1 depicts the syntheses carried out in that work. Fe(0) NPs were thoroughly washed to form Fe(0) NPs **Fe-1** with 73% Fe content. During the carburization of these NPs, the resulting FeC NPs (**FeC-1**) agglomerated, and the SAR of this **FeC-1** was 0 W g^{-1} (at 47 mT and 93 kHz). Even after the dispersion of the **FeC-1** NPs at 150°C in mesitylene in the presence of PA (**FeC-3** NPs), large agglomerates were still observable by TEM (Transmission Electron Microscopy), with a SAR value of ca. 20 W g^{-1}. On the other hand, six different carburization experiments (**FeC-4** to **FeC-9**) were carried out, starting from the **Fe-1**

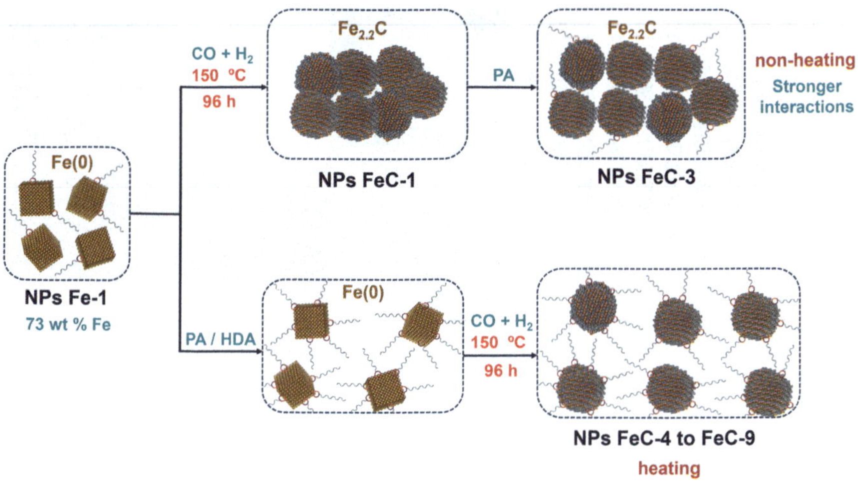

Scheme 1 Experimental diagram showing how the ligands affect the heating properties of $Fe_{2.2}C$ NPs

NPs (Scheme 1). In these experiments, various amounts of HDA and PA were introduced into the reactor before reacting with syngas, leading to heating NPs with SAR values ranging from 610 to 2,020 W g^{-1}. Therefore, the researchers concluded that it is crucial to meticulously control the concentration of surface ligands to produce nanoparticles with controllable and reproducible heating power. In another important contribution, a new method for determining the surface temperature of magnetically heated nanoparticles in solution was proposed in 2021. Diaz-Prieto et al. used the temperature dependence of the catalytic performance of iron carbide nanoparticles coated with ruthenium ($Fe_{2.2}C$@Ru) for acetophenone hydrogenation. They established an Arrhenius-based correlation between the acetophenone conversion, the surface temperature of the nanoparticle, and, therefore, the magnetic field. The study concluded that the surface temperatures in different solvents were very high, ranging from 260°C to above. These temperatures were found to be similar at a given magnetic field regardless of the solvent, and well above their different boiling points [13]. For example, when the boiling point of the solvent was reduced from 287°C (hexadecane) to 110°C (toluene), the surface temperatures of the nanoparticles remained within the same temperature range (230–260°C). This indicates that the reaction sites locally reached very high temperatures, much higher than the average temperature of the reaction medium. This is the first evidence of a nanoscale temperature gradient in a magnetically activated chemical reaction. The high temperatures generated are likely the reason behind the excellent catalytic performance of this system for the HDO of acetophenone and other aromatic ketones in toluene or mesitylene.

In 2021, Leitner's group accomplished a significant milestone by decorating a commercially available heterogeneous catalyst, copper chromite, with iron carbide

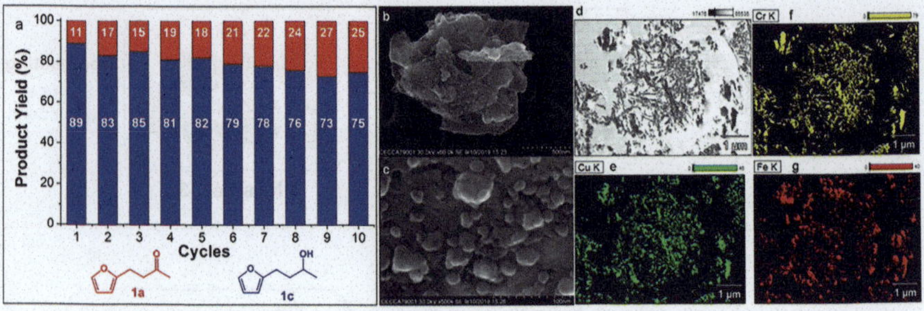

Fig. 2 Commercial $Cu_2Cr_2O_5$ decorated with Iron Carbide ($Fe_{2.2}C$) NPs as a multifunctional catalyst for magnetically induced hydrogenation catalysis. Reproduced with permission from Ref. [9] Copyright 2021 Wiley-VCH

NPs ($Fe_{2.2}C$ NPs) [9]. The resulting composite, $Fe_{2.2}C/Cu_2Cr_2O_5$, demonstrated efficient and selective hydrogenation of aromatic ketones to aromatic alcohols under magnetic induction and mild conditions (3–5 bar H_2, reactor temperature around 80°C; measured using an infrared camera). For example, >99% of conversion and selectivity were obtained for a range of phenyl ketones with different substituents (i.e., acetophenone, 4-methyl acetophenone, 4-methoxyacetophenone, benzophenone) to the corresponding alcohol, working at 64 mT in batch conditions for 4 h. Interestingly, methyl β-naphthyl ketone and 4-phenyl-3-buten-2-one not only formed the alcohol but also produced other products, such as the hydrogenation of the double bond or the HDO reaction, respectively.

The hydrogenation of 4-(2-furanyl)-2-butanone in batch system was carried out in a Fischer-Porter bottle with $Fe_{2.2}C/Cu_2Cr_2O_5$ and 0.1 mmol of the substrate in 0.5 mL heptane. This was done under 3 bar of H_2 for 2 h at 64 mT, without stirring. These conditions resulted in an 89% conversion of the substrate, with >99% selectivity to α-methyl-2-furanpropanol. The catalytic activity was observed to decrease only slightly after several cycles, with the last cycle leading to 75% conversion and full selectivity to α-methyl-2-furanpropanol (Fig. 2, left). The observed slight decline in selectivity across the cycles is attributed to the loss of minute amounts of catalyst during the inter-cycle washing process. These findings indicate that both the catalytically active and magnetically heating components exhibit excellent stability.

The same reaction was studied under flow conditions as follows: the inductor was fixed to μ_0Hmax = 42 mT, and the substrate, 4-(2-furanyl)-2-butanone (0.05 mol L^{-1} in heptane), was pumped with a flow of 0.1 g min^{-1} (0.15 mL min^{-1}, liquid residence time = 8.2 min), which is shorter than in batch. A gas flow of 0.001 g min^{-1} (2.47 mL min^{-1}) and a pressure of 5 bar, WHSV = 0.1 h^{-1} gave complete substrate conversion, giving α-methyl-2-furanpropanol in 88% of yield.

Most importantly, the authors have reported that they were able to promptly initiate and terminate the catalytic activity of their catalyst by applying and removing the alternating current (AC) magnetic field. They explain that the rapid heating and cooling of the catalyst through magnetic induction activation was responsible for this phenomenon. This feature allows the system to adjust to changes in electricity supply, which is especially important for alternative renewable energy sources. The catalyst was recycled and tested in both batch (10 cycles) and continuous flow conditions, proving its ability to be used for at least 17 h.

Remarkably, this research suggests incorporating magnetically induced heating can enhance traditional heterogeneous catalyst performance. This technology offers several advantages, including rapid and localized heating and improved energy efficiency, allowing current heterogeneous catalytic processes to adapt to intermittent power sources. This is also the first time a magnetically heated catalyst has been employed in a continuously operated hydrogenation reaction in the liquid phase. The study's results demonstrate that the $Fe_{2.2}C/Cu_2Cr_2O_5$ composite catalyst has enormous potential for various industrial applications, including producing high-value chemical products coupled with using renewable energy sources.

In their recent study, Lin et al. employed $Fe_{2.2}C$ NPs as magnetic heating agents under an AMF to create a multifunctional catalytic system for the selective hydrogenation of various aromatic aldehydes, including benzaldehyde derivatives, furfural, 5-HMF, vanillin, and syringaldehyde [26]. By decorating iron carbide NPs with copper, $Fe_{2.2}C$@Cu, the researchers were able to generate heat from the AC magnetic field (54–72 mT, 350 kHz) to rapidly heat the local area, thereby heating and activating the catalytically active Cu-containing NPs on the catalyst surface. The catalyst activity was shown to stop or restart concomitantly when the AC magnetic field was regularly stopped or restarted. Furthermore, no significant hydrogenation activity was observed using conventional heating methods, even at temperatures up to 200°C. In contrast, $Fe_{2.2}C$@Cu demonstrated high activity and selectivity for synthesizing alkyl derivatives from the starting aromatic aldehydes under mild conditions (3 bar of H_2 at an apparent reaction temperature of 105°C, measured with an infrared thermometer) in the presence of a magnetic field. A range of several *p*-substituted benzaldehyde derivatives (R = hydrogen, isopropyl, propyl, methoxy, chloride, trifluoromethane) were employed to study the HDO reaction producing the corresponding substituted alkanes at 72 mT, 350 kHz during 16 h. In that sense, more than 99% of conversion was achieved in all the examples. Nevertheless, electron-withdrawing substituents such as –Cl and $-CF_3$ led to a significantly lower selectivity, consistent with previous observations for the HDO of aromatic ketones. Additionally, non-benzylic aldehydes such as phenylacetaldehyde and 3-phenylpropionaldehyde underwent complete conversion. However, the selectivity was diminished due to the generation of dimers and decarboxylation by-products. Finally, the catalytic process was also successfully extended to the HDO reaction of biomass derivatives. In this sense, furfural and 5-HMF were converted in quantitative yields into 2-MF and 2,5-DMF, respectively (Table 2, entries 1 and 2). In the previous two instances, the amplitude of the magnetic field was reduced to 54 mT in order to prevent any unwanted side reactions. This is an excellent illustration of how adjusting the magnetic field amplitude can enhance the selectivity of specific chemical reactions. In a similar manner, vanillin and syringaldehyde were easily transformed into 2-methoxy-4-methylphenol and 2,6-dimethoxy-4-methyl-phenol, respectively, at 72 mT (Table 2, entries 3 and 4).

Table 2 Scope of substrates in the HDO of biomass derivatives by $Fe_{2.2}C$@Cu NPs in the presence of an AMF

Entry	Substrate	Product	Magnetic field (mT)	Solvent	Conv. (%)	Yield (mol %)
1	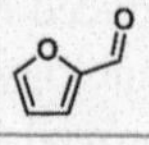	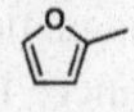	54	Mesitylene	>99	99
2	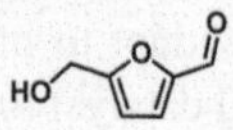	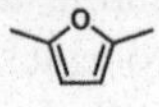	54	Mesitylene	>99	99
3	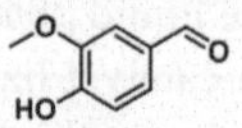	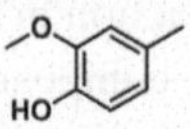	72	Mesitylene	>99	99 (91)
4	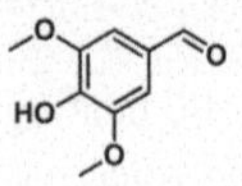	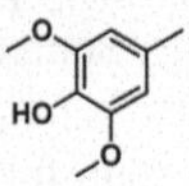	72	Dioxane	>99	99 (95)

Reaction conditions: Substrate (0.33 mmol), $Fe_{2.2}C$@Cu (10.0 mg, 0.023 mmol Cu, 7 mol %), solvent (0.5 mL), 3 H_2 bar, 300 kHz, 16 h. Isolated yields are given in parentheses

Remarkably, the pretreatment of the $Fe_{2.2}C$@Cu catalyst under magnetic induction heating (72 mT, 350 kHz) for 1 h, at 200°C, resulted in the production of toluene from benzaldehyde in a quantitative yield when using conventional heating (at 200° C). This discovery raises an intriguing question regarding the possibility of magnetic induction heating being capable of generating active species that would typically require substantially higher bulk liquid temperatures. Scheme 2 illustrates the different catalysts and catalytic reactions accomplished based on the use of $Fe_{2.2}C$ NPs.

3.2 *FeCo*

After the first reports describing the synthesis and applications of $Fe_{2.2}C$ NPs, the research in our group focused on synthesizing NPs that could enhance the heating of iron carbide NPs at high temperatures, i.e., with a higher Curie temperature (T_C). Eventually, these efforts crystallized in the optimization of monocrystalline FeCo NPs from precursors $\{Fe\{N[Si(CH_3)_3]_2\}_2\}$ and $\{Co\{N[Si(CH_3)_3]_2\}_2\}$, which were mixed in mesitylene in the presence of HDA and HDA·HCl. The next step involved pressurizing the solution with 3 H_2 bar and heating it at 150°C for 48 h to reduce the formed species. The resulting MagNPs display a mean diameter of 11 ± 1 nm, exhibiting a highly crystalline body-centered cubic (*bcc*) structure [70]. These $Fe_{0.5}Co_{0.5}$ NPs could then be used as universal heating agents, allowing gas-phase reactions up to ca. 700°C upon association with the appropriate active phases (i.e., Ni or Pt). Some examples of this reactivity include CO_2 methanation, dry reforming of propane and methane, and propane dehydrogenation; these reactions require temperatures of 350°C, 600°C, and 700°C, respectively

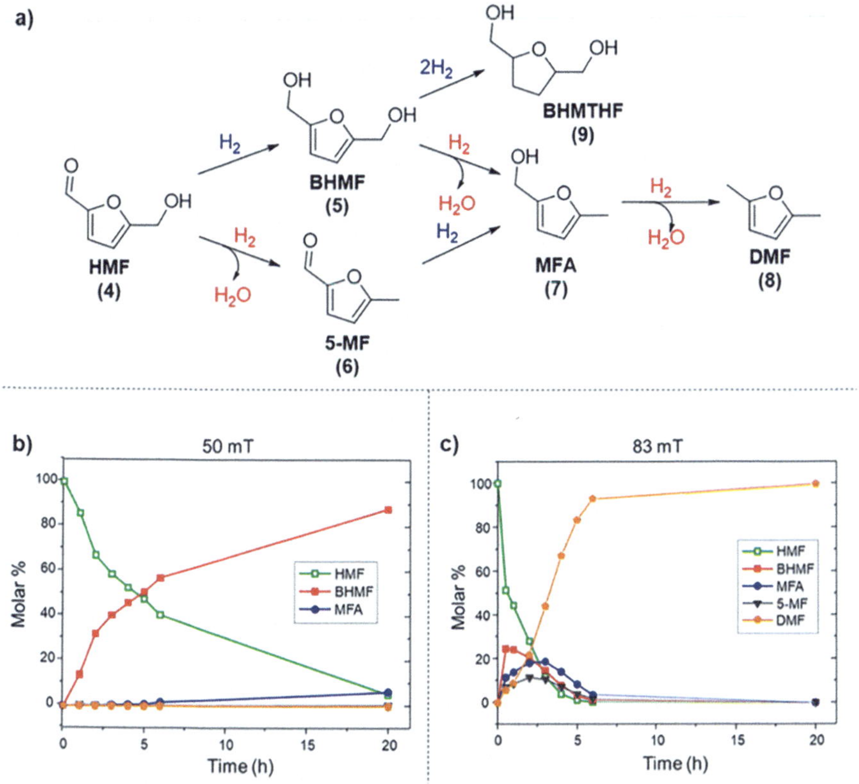

Fig. 3 (**a**) Catalytic transformation of 5-HMF using FeCo@Ni NPs at (**b**) 50 and (**c**) 83 mT. Reproduced with permission from Ref. [72] Copyright 2022 ACS

[71]. Furthermore, it was found that $Fe_{0.5}Co_{0.5}$NPs are less energy-demanding than $Fe_{2.2}C$, allowing CO_2 hydrogenation at lower magnetic fields and frequencies as low as 100 kHz. Following these discoveries and, as part of our group's collaborative efforts, Corma and Martínez-Prieto used this same magnetic core $Fe_{0.5}Co_{0.5}$ to report the first example of magnetically induced reduction of oxygenated compounds derived from biomass in aqueous media [72]. Initially, they decorated FeCo with Ni through the decomposition of [Ni(COD)$_2$] in the presence of $Fe_{0.5}Co_{0.5}$, forming core@shell FeCo@Ni NPs. The TEM analysis of FeCo@Ni NPs showed that the NPs were well-dispersed with a mean size of 13 ± 2 nm. The X-ray diffraction showed peaks corresponding to *bcc* for FeCo and face centered-cubic *fcc* for Ni, along with a small amount of nickel oxide, indicating the crystallinity of the NPs. These NPs were used in the reduction of 5-HMF in both aqueous and organic (dioxane) media. When tested in dioxane, the system completely converted to 2,5-DMF in just 20 h. Moreover, by using lower magnetic field (i.e., 58 mT), the hydrogenation product, namely 2,5-BHMF could be selectively obtained (Fig. 3b, c). Unfortunately, this system did not perform well in water. After being

subjected to 83 mT and 300 kHz for 20 h (with a temperature of 99°C measured by an optic-fiber temperature sensor), only 24% conversion was achieved with 100% selectivity to 2,5-bishydroxymethylfuran (2,5-BHMF). The low activity in water was attributed to multiple factors, such as surface oxidation and higher aggregation of NPs in water, even after being sonicated for an hour. On the contrary, when dispersed in dioxane and subjected to only 10 min of sonication, the FeCo@Ni NPs are uniformly dispersed. To solve this issue, a protective methodology was developed to create carbon-coated core-shell FeCo@Ni@C MagNPs that can act as both catalysts and heating agents in aqueous media. The encapsulation of the FeCo@Ni NPs was accomplished by adsorbing them on activated carbon (C Norit D10), and a thermal process, namely pyrolysis at a temperature of 300°C, was carefully designed to ensure accessibility of the substrates to the active metal surface. This way, the core-shell structure of the NPs was achieved while creating cracks in the carbon layer, which facilitated the necessary accessibility to the active sites. The size and shape of the FeCo@Ni@C nanoparticles were similar to those of the MagNPs before encapsulation, but their average diameter increased slightly to 14 ± 6 nm. The size distribution also became broader due to the formation of larger nanoparticles during the pyrolysis process. The heating power of FeCo, FeCo@Ni, and FeCo@Ni@C was measured at 47 mT. The values obtained were 1,600 $W{\cdot}g^{-1}$, 620 $W{\cdot}g^{-1}$, and 82 $W{\cdot}g^{-1}$, respectively. The results suggest that incorporating the nickel shell in these nanoparticles changes their magnetic anisotropy and, therefore, their heating capacity. The even lower values observed in FeCo@Ni@C can be attributed to the encapsulation of the nanoparticles into carbon.

Upon the encapsulation in carbon, 5-HMF underwent selective hydrogenation to form 5-BHMF with a high conversion rate (93.4%) after 20 h under a magnetic excitation of 100 mT. The only product observed was the diol. Nonetheless, the T_{bulk} of 95°C at 100 mT exhibited by the aqueous solution was far from that achieved in dioxane at the same field amplitude (i.e., $T_{bulk} = 143$°C at 100 mT), which also enabled to carry out HDO to 2,5-DMF. The discrepancy in bulk temperature observed between aqueous and dioxane media is likely attributable to the higher specific heat capacity of water ($Cp = 4.184\ J\ kg^{-1}\ K^{-1}$) compared to that of dioxane ($Cp = 1.702\ J\ kg^{-1}\ K^{-1}$). This highlights the necessity for high-heating-power agents when operating in aqueous environments, whereby it would be possible to widen the range of applications for this technology.

Remarkably, this study has also introduced a pioneering approach to magnetic-field-controlled selectivity between hydrogenation and HDO. For instance, during the magnetically induced hydrogenation of 5-HMF over FeCo@Ni NPs in dioxane, after 20 h at a magnetic field of 50 mT, which corresponds to a T_{bulk} of 75°C, the conversion rate reached 95%, with 92.8% selectivity towards 2,5-BHMF. Conversely, at a higher magnetic field amplitude of 83 mT ($T_{bulk} = 106$°C), total conversion to 2,5-DMF was observed (Fig. 3b, c). These results show that modulating the applied magnetic field can lead to a complete selectivity change between 2,5-BHMF and 2,5-DMF with FeCo@Ni NPs. Therefore, it was found that it is possible to control product distribution by adjusting the field amplitude during magnetically induced catalysis. In this sense, the catalytic application of

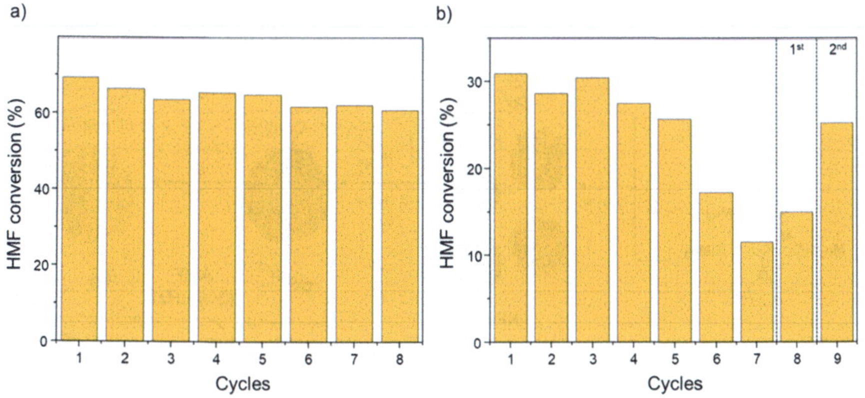

Fig. 4 Recycling experiments using FeCo@Ni@C in (**a**) dioxane and (**b**) water (regeneration treatments after the seventh run are separated by dashed lines). Reproduced with permission from Ref. [72] Copyright 2022 ACS

FeCo@Ni@C could be extended to other biomass-derived oxygenated compounds such as furfural, levoglucosenone, and vanillin, finding that exceptional selectivity for both hydrogenation and HDO can also be achieved in water using the same catalyst by simply changing the magnetic field applied amplitude.

Finally, encapsulating MagNPs in carbon has remarkably enhanced their stability. This helps prevent their agglomeration and allows them to maintain their activity and heating capacity. In that sense, FeCo@Ni@C catalysts were able to be reused up to eight times in dioxane. A similar study was conducted in aqueous media, where seven consecutive reactions were conducted using the same catalyst in water. Despite experiencing a decrease in activity after the fourth run, the catalyst was able to regain almost all its original activity following a successful thermal regeneration process. After the seventh run, the catalyst was heated at 110°C under a N_2 flow over 12 h. This method allowed for partial recovery of its catalytic activity (Fig. 4b, cycle 8). Subsequently, the catalyst was heated at 180°C under 3 bar H_2 over 12 h, and almost recovered its initial activity, suggesting that the Ni surface was also partially oxidized during the catalysis in water (Fig. 4b, cycle 9). This demonstrates the potential for continued use and optimization of the catalyst in future experiments (Scheme 3).

3.3 *FeNi$_3$*

In 2020, De Masi et al. synthesized bimetallic FeNi NPs with a SAR at 47 mT and 93 kHz of 600 W g^{-1} through the co-decomposition of $\{Fe[N(SiMe_3)_2]_2\}_2$ and Ni $[^iPrNC(CH_3)N^iPr]_2$ complexes in mesitylene under 3 bar of H_2. PA was used as stabilizer in a ratio of 1.8 eq. per Fe complex [73]. The resulting $FeNi_3$ NPs with a mean size of 17 ± 2 nm display an alloy *fcc* structure but with a surface enrichment

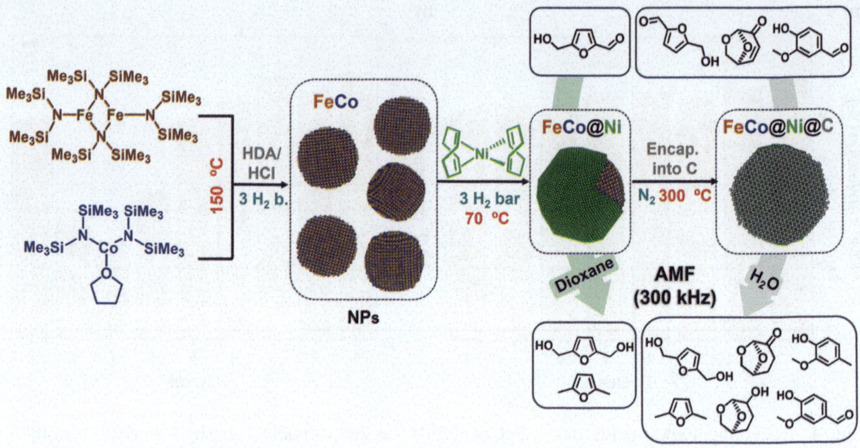

Scheme 3 Schematic overview of liquid-phase magnetically induced catalysis with FeCo NPs as magnetic cores

in Fe. This accounts for the relatively low reactivity of these particles for CO_2 hydrogenation. Nonetheless, full CO_2 conversion into CH_4 at a low magnetic field could be achieved by depositing an additional layer of Ni onto the nanoparticle surface. Following this discovery, Mustieles Marín et al. demonstrated the catalytic performance of these MagNPs after surface enrichment with nickel. These MagNPs, so-called $FeNi_3$@Ni, had a mean size of 19 ± 2 nm, were polycrystalline, and exhibited a *fcc* structure. As expected, the SAR of these NPs dropped from 600 to 350 W g^{-1} after Ni enrichment. $FeNi_3$@Ni were tested in valorizing lignocellulose-derived products in solution using magnetic induction [16]. The $FeNi_3$@Ni NPs, acting as both catalysts and magnetic heating agents, were subjected to a high-frequency magnetic field (300 kHz, 49 mT) to catalyze the total conversion of furfural and 5-hydroxymethyl furfural into 2-methyl furan and 2,5-DMF, respectively, under mild conditions (12 mol% catalyst, 3 H_2 bar). The results obtained for the two reactions in question are similar to those previously reported for $Fe_{2.2}C$@Ru NPs [14]. However, it is noteworthy that these findings were obtained at lower magnetic field amplitude (49 vs. 58 mT), resulting in lower energy consumption and without using a precious metal. Furthermore, furfural-derived cyclopentanone was subjected to oligomerization. After 8 h, the conversion rate was 100%, and a mixture of dimeric and trimeric products was obtained, with cyclopentanol only observed at shorter reaction times. The main product of the reaction was cyclopentylcyclopentane, which accounted for 79% of the final mixture. The remaining 21% of the product consisted of small amounts of dimers, including 2-cyclopentyl cyclopentanone (7%) and 1-cyclopentylcyclopentene (8%), as well as the trimer 1,3-dicyclopentylcyclopentane (6%). After 16 h, all the unsaturated dimers had been consumed, and the final product mix contained 72% of cyclopentylcyclopentane and 28% of 1,3-dicyclopentylcyclopentane. Finally,

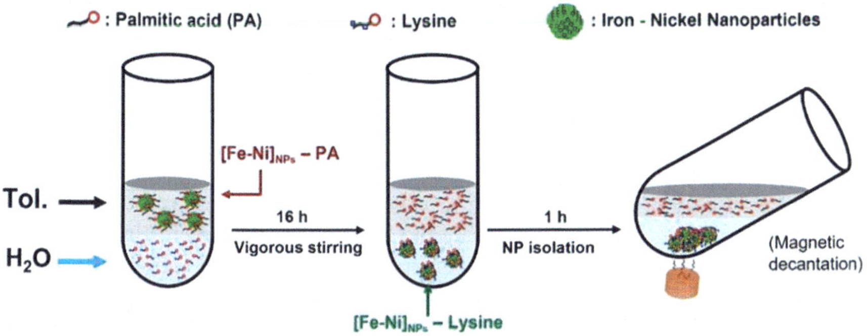

Fig. 5 Transfer of $FeNi_3$ and $FeNi_3$@Ni NPs from organic to aqueous dispersion via ligand exchange. Reproduced with permission from Ref. [57] Copyright 2023 Wiley

diphenyl and benzyl phenyl ether (DPE and BPE) were used as lignin linkage model molecules, observing a cleaving of the C–O of 64% and 100%, respectively, in 5 h at 49 mT to yield a mixture of hydrogenolysis and hydrogenation products. Furthermore, the catalytic performance of $FeNi_3$@Ni was evaluated under two heating conditions: conventional oil bath heating and magnetic heating. Thus, the results showed more than a three-fold increase in the cleavage of diphenylether (DPE) for magnetic induction heating compared to the classical conditions. Moreover, magnetic-induced heating led to a 37% increase in the conversion of furfural. Additionally, the selectivity to 2-MF is much lower when using conventional heating. The magnetic heating was carried out at 300 kHz and 33–49 mT, while the oil bath heating reactions were carried out at 170–180°C. The difference in performance between the two conditions was not that significant in the case of biphenyl phenyl ether (100 vs 94% conversion), which may be attributed to the lower C–O dissociation energy in this case. These findings suggest that magnetically induced catalysis is a promising strategy for achieving challenging transformations of biomass-derived molecules, offering the benefits of a simple experimental setup and short reaction times, as well as an added benefit in chemical transformations requiring higher surface temperatures.

After 2 years of research, our group produced a significant example of how surface functionalization can complement the organometallic approach. By tuning the hydrophilicity of the $FeNi_3$ and $FeNi_3$@Ni NP surfaces through a simple post-synthetic treatment involving a ligand exchange of palmitic acid by l-lysine, (Fig. 5) we succeeded in enhancing the catalytic stability and dispersibility of these NPs in water, enabling the hydrogenation of biomass derivatives under more realistic conditions [57]. The new lysine-stabilized NPs ($FeNi_3$-Lys) display a mean size of 12 ± 2 nm. The XRD patterns correspond to $FeNi_3$ or Ni *fcc* structure, showing that even when subjected to air and water during the ligand exchange reaction, there was no evidence of the formation of oxidized crystalline Fe or Ni phases that could be detected by XRD.

The lysine-coated NPs were applied in the hydrogenation of 5-HMF into 2,5-BHMF, generating a 90% yield after 15 h under AMF, outperforming conventional heating, which produced only 77% at the same apparent temperature. This remarkable result was made possible using only 3 bar of H_2, after applying a 65 mT magnetic field with a 300 kHz frequency. The temperature on the reactor wall registered with the IR camera was around 70–80°C. The catalytic performance of $FeNi_3$-Lys surpasses that of $FeNi_3$-PA, especially in terms of reusability, with 10 consecutive runs of catalysis compared to 8 for $FeNi_3$-PA. Interestingly, despite the presence of agglomerates of lower-density materials (namely, oxides or hydroxides) surrounding the magnetic core after the reaction, the addition of l-lysine aids in the preservation of the crystalline structure of the magnetic core. This, in turn, allows for the retention of its induction heating properties. Furthermore, it has been observed that $FeNi_3$-Lys showcases a substantially higher level of success regarding the expansion of catalytic scope in comparison to $FeNi_3$-PA in aqueous solutions. In that sense, the catalyst proved to be highly versatile, as it could perform the hydrogenation of levulinic acid (LA) to γ-valerolactone (GVL) and the 4′--hydroxyacetophenone HDO to 4-ethylphenol. The $FeNi_3$-Lys catalysts have yielded conversions up to 70%, with selectivities above 85% in both cases. Again, the catalytic efficiency in levulinic acid hydrogenation through magnetically induced catalysis was compared against that achieved under similar apparent conditions through conventional heating at 120°C. The results showed that the former was superior to the latter, earning only 38% conversion. In terms of selectivity, the primary product was GVL (82% selectivity), which is similar to the outcome obtained under the presence of magnetic induction. Table 3 summarizes the comparison between thermal and magnetic-induced catalysis with $FeNi_3$-based MagNPs.

After Martínez-Prieto's work on FeCo encapsulation, this is the second example showing that it is possible to transfer the magnetic heating technology, typically used in organic solutions, to water. The outcomes of the study affirm the possibility of using magnetically induced catalysis in upcoming industrial processes that involve renewable raw materials present in aqueous environments. Nonetheless, it also suggests that more research in this direction is needed to explore and expand its range of capabilities. Scheme 4 summarizes the synthesis, modification, and use in magnetically induced catalysis of $FeNi_3$ MagNPs.

3.4 *NiCo*

In the most recent example of MagNPs synthesized by the organometallic approach, NiCo NPs with several compositions, i.e., $Ni_{10-x}Co_x$ (x: 2, 3, 4, 5), were synthesized through the co-decomposition of $Ni[^iPrNC(CH_3)N^iPr]_2$ and $Co[N(SiMe_3)_2]_2(thf)$ complexes at 120°C in mesitylene under 3 bar of H_2 [74]. PA was chosen as stabilizer in a ratio of 1.0 eq. per metallic center. The resulting NiCo NPs exhibited mean sizes between 13 and 16 nm with a predominant *fcc* structure. However, deeper structural analyses of the MagNPs combining High-Resolution Transmission

Table 3 Comparison between Thermal and Magnetic Induced Catalysis based on MagNPs catalysts [16, 63]

Heating method	Catalyst	Reaction	App. T (°C)[a]	Conv. (%)	Yield (mol%) (Product)
Conv. heating	$FeNi_3$@Ni	DPE hydrogenation[b]	170	18	14 (C–O cleavage)
AMF (49 mT, 300 kHz)			150–160	64	53 (C–O cleavage)
Conv. heating	$FeNi_3$@Ni	5-HMF hydrogenation[c]	180	60	53 (FA) and 8 (2-MF)
AMF (49 mT, 300 kHz)			150–160	97	61 (FA) and 36 (2-MF)
Conv. heating	$FeNi_3$-PA	5-HMF hydrogenation[d]	120	44	44 (2,5-BHMF)
	$FeNi_3$-Lys			77	77 (2,5-BHMF)
AMF (65 mT, 300 kHz)	$FeNi_3$-PA		80–90	98	98 (2,5-BHMF)
AMF (65 mT, 300 kHz)	$FeNi_3$-Lys		70–80	90	90 (2,5-BHMF)
Conv. heating	$FeNi_3$-PA	LA hydrogenation[d]	120	21	10 (GVL)
	$FeNi_3$-Lys			38	32 (GVL)
AMF (65 mT, 300 kHz)	$FeNi_3$-PA		80–90	20	18 (GVL)
AMF (65 mT, 300 kHz)	$FeNi_3$-Lys		70–80	73	65 (GVL)

[a] In reactions under AMF, temperature was measured with an IR camera at the external reactor wall
[b] Reaction conditions: 1 mmol of substrate, 12 mol% of catalyst, 5 mL of decane, 3 H_2 bar and 5 h
[c] Reaction conditions: 1 mmol of substrate, 12 mol% of catalyst, 5 mL of mesitylene, 3 H_2 bar and 3 h
[d] Reaction conditions: 1 mmol substrate, 35 mol% of catalyst, 5 mL water, 3 H_2 bar and 15 h

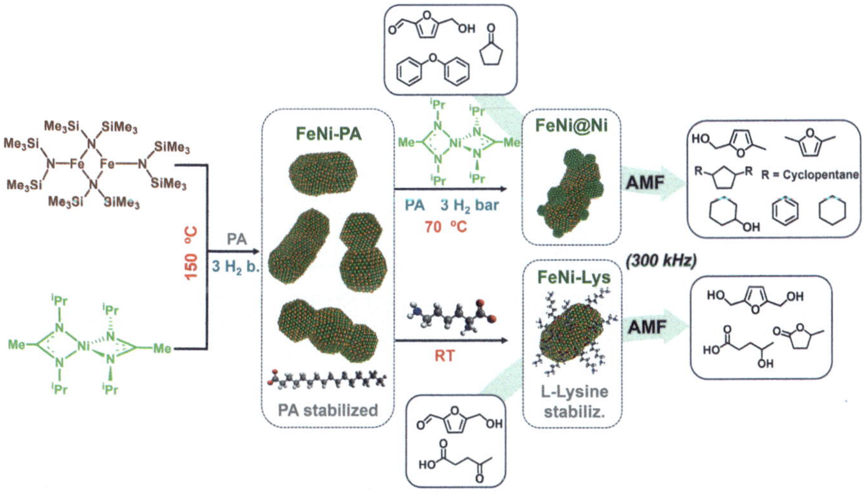

Scheme 4 Schematic overview of liquid-phase magnetically induced catalysis using $FeNi_3$ NPs as magnetic cores

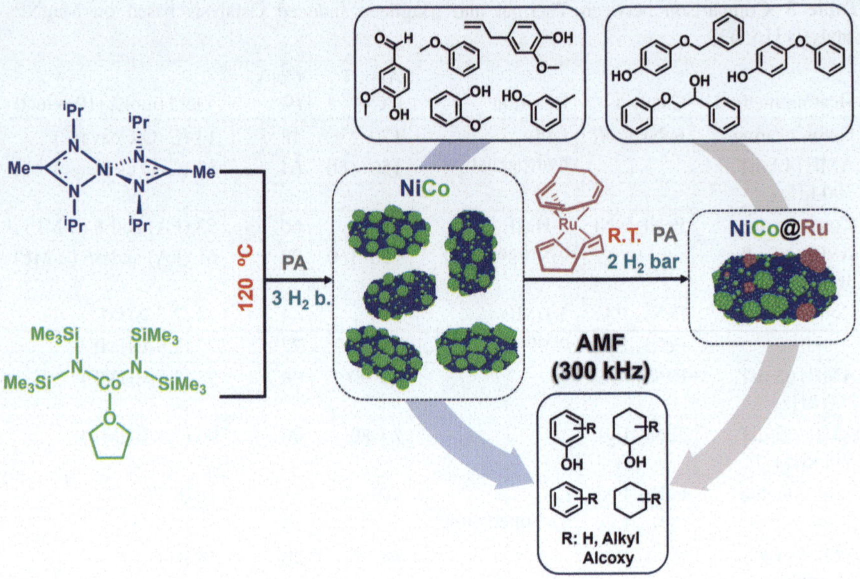

Scheme 5 Schematic overview of liquid-phase magnetically induced catalysis using NiCo NPs as magnetic cores.

Table 4 Magnetic properties and SAR values of $Ni_{10-x}Co_x$-NPs

Material	M_S ($A\,m^2\,kg^{-1}$)[a]	M_R ($A\,m^2\,kg^{-1}$)[a]	H_C (mT)[a]	SAR (at 93 kHz, 49 mT) ($W\,g^{-1}$)[b]
Ni_5Co_5	115	14	18	513
Ni_6Co_4	99	3	3	492
Ni_7Co_3	91	3	4	310
Ni_8Co_2	77	2	5	126

[a] Measured at the 300 K
[b] SAR values are normalized with respect to the metal content in the sample

Electron Microscopy (HR-TEM), X-ray absorption spectroscopy (XAS) and XRD revealed a bimetallic structure composed of an *fcc* Ni-rich core (≈10 nm) decorated with ultra-small (≤2 nm) hexagonal closed packed (*hcp*) Co NPs. The NiCo synthesis and their use as catalysts and/or heating agents in several reactions is summarized in Scheme 5.

As expected, not only the catalytic properties but also the hyperthermia properties of these MagNPs are highly dependent on their composition, particularly the magnetic hardness and softness provided by the elements cobalt (Co) and nickel (Ni), respectively. As such, the SAR of these MagNPs can be fine-tuned based on their composition (Table 4). In that sense, at a magnetic field of 47 mT and a frequency of 93 kHz, the SAR of the MagNPs was found to vary from 513 $W\cdot g^{-1}$ for Ni_5Co_5 to

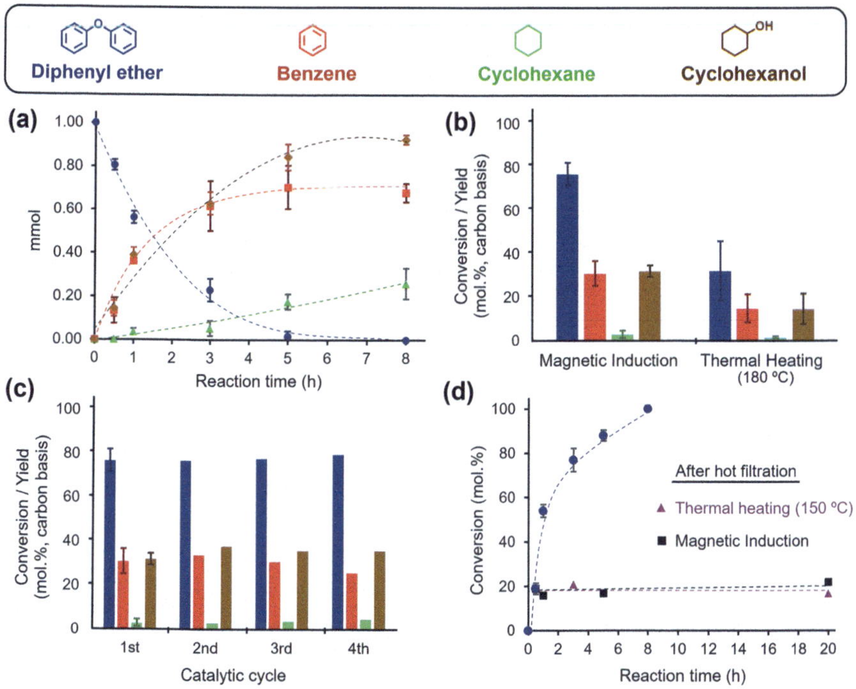

Fig. 6 Catalytic performance of Ni_7Co_3-NPs in DPE hydrogenation under magnetic induction. (**a**) Temporal evolution of product distribution; (**b**) Comparison against conventional heating with analogous reaction conditions; (**c**) Catalyst reusability after washing with toluene; (**d**) Hot filtration tests. Reaction conditions: Batch-type reactor, 1.0 mmol DPE, 0.7 mmol dodecane, 5 mL decane, 8 mg cat (≈0.1 metallic mmol), AMF 300 kHz, 39 mT, 3 H_2 bar. In (**b**) and (**c**), reaction time has been adjusted to 3 h to keep conversion below 80%. Notes: (i) Dashed lines intend to serve only as visual guidelines. (ii) Error bars correspond to the standard deviation of at least 2 repetitions of the same catalytic test. Adapted with permission from Ref. [74] Copyright 2024 ACS

126 $W\cdot g^{-1}$ for Ni_8Co_2. Moreover, at the same AMF frequency of 93 kHz, the MagNPs with a higher proportion of Ni (Ni_8Co_2) were found to be activated at low magnetic fields (10 mT), whereas the activation of the Ni_5Co_5 NPs occurred at higher magnetic fields (around 30 mT). Specifically, the Ni_7Co_3 compound demonstrated remarkable catalytic activity, heating power, and stability. This was observed in the hydrogenolysis of DPE under an AMF (300 kHz, 39 mT) at low H_2 pressure (3 bar). The catalyst composition was able to achieve 100% C–O cleavage after 8 h and showed a considerable tendency towards aromatic ring hydrogenation (Fig. 6).

After the promising results attained with Ni_7Co_3-NPs as a catalyst for DPE hydrogenolysis, a comprehensive evaluation of the potential of this system for lignin depolymerization was conducted at higher magnetic field amplitudes, utilizing the maximum coil power (12.1 kW) which, at 300 kHz, resulted in a magnetic field amplitude of 65 mT. The evaluation was performed on various hydroxylated lignin linkage model compounds, including β-O-4 (2-phenoxy-1-phenylethanol), α-O-4

(4-benzyloxyphenol), and 4-O-5 (4-phenoxyphenol). However, it was observed that the conversion of substrates with a 100% C–O bond cleavage was difficult even after long reaction times (≥10 h). For instance, in the case of 2-phenoxy-1-phenylethanol, the cleavage of the C–O bond stops at approximately 60 mol%. Similarly, conversions above 50% are difficult to achieve in 4-benzyloxyphenol and 4-phenoxyphenol. These results suggest that Ni_7Co_3-NPs will face challenges completing lignin depolymerization. In view of this, C–O bond cleavage in these hydroxylated lignin linkage models was significantly enhanced by adding Ruthenium (Ru). In fact, a minimum amount of Ru (ca. 1 wt.%) was enough to improve catalyst prospects for lignin depolymerization by enhancing C–O cleavage under the same reaction conditions. The extent of improvement varies depending on the type of lignin linkage, with an increment of 10 to 20 points in the case of the β-O-4 (2-phenoxy-1-phenylethanol) and 4-O-5 (4-phenoxyphenol), but almost twofold (45 vs 88 mol%) in the case of the α-O-4 model compound (4-benzyloxyphenol). Although the addition of Ru could significantly enhance the catalytic efficiency, when using a more realistic β-O-4 Type C model, i.e., 1-(4-hydroxy-3,5-dimethoxyphenyl)-2-(2-methoxyphenoxy)propane-1,3-diol [75], it can be seen that only 30% of C–O cleavage was attained after 60 h of reaction at 3 H_2 bar under an AMF of 65 mT and a frequency of 300 kHz. The main product obtained after this transformation was 4-(3-hydroxy-2-(2-methoxyphenoxy)propyl)-2,6-dimethoxyphenol (74 mol% yield), which corresponds to the loss of the –OH group in α. Minor amounts of the products from the C–O cleavage reaction (i.e., guaiacol, dihydrosynapyl alcohol, 4-propylsyringol) were detected. As a result, it has become apparent that further enhancements need to be made to the catalyst to render it appropriate for actual lignin depolymerization. To this end, the authors have suggested that future research should focus on optimizing the Ru loading and speciation or incorporating supports to regulate substrate adsorption.

Furthermore, anisole, guaiacol, vanillin, cresol, and eugenol were used as typical lignin-oil model compounds to probe catalyst performance for lignin derivatives upgrading to arenes, cyclohexanes and cyclohexanols (300 kHz, 39 to 65 mT). Vanillin could be selectively transformed into 2-methoxy-4-methylphenol; removing the methoxy group is challenging under those reaction conditions. Anisole, on the other hand, exhibits a high degree of demethoxylation, which proceeds primarily via direct deoxygenation to benzene, as revealed by the reaction kinetics. In the case of m-cresol, only the aromatic ring could be reduced. As for eugenol, the demethoxylation and hydrogenation were observed, while its deoxygenation (i.e., –OH group removal) was found to be minimal. This reaction mechanism is also evident in guaiacol, as evidenced by the absence of 2-methoxycyclohexanol in the final reaction crude. Additionally, a reaction mixture comprising the five aforementioned molecules was utilized to evaluate the performance of the Ni_7Co_3@Ru system. The results indicated that these MagNPs also exhibited a significant selectivity to cyclohexanols, indicating good activity for aromatic ring hydrogenation and demethoxylation of ortho-substituted monomethoxyphenols. These promising findings suggest that the Ni_7Co_3 and Ni_7Co_3@Ru systems may effectively upgrade lignin oils through HDO. On the contrary, the catalyst cannot perform more complex deoxygenations, such as –OH removal from phenol and cyclohexanol dehydrations. In that sense, the authors again acknowledged room for improvement in HDO due to the absence of distinct oxygen adsorption sites, such as oxygen vacancies, acid sites, or metal/oxide interfaces.

4 Conclusions

This chapter intends to introduce the application of organometallic MagNPs in magnetically induced catalysis in liquid-phase solutions, aiming to identify critical challenges and contribute to expanding the scope of magnetically induced catalysis in solution. Overall, the advancements discussed in the text highlight the potential of magnetically induced catalysis through organometallic magnetic cores for various liquid-phase reactions. These advancements also showcase previously postulated benefits, such as rapid and localized heating, improved energy efficiency, and adaptability to renewable energy sources.

Implementing the organometallic approach in synthesizing magnetic cores to be utilized as magnetic susceptors in liquid-phase reactions for magnetically-induced catalysis has resulted in diverse structures with varying heating and catalytic characteristics. By and large, utilizing these organometallic MagNPs in magnetically induced catalysis has significantly enhanced catalytic activity. These NPs, when exposed to an AMF, have demonstrated the ability to achieve quantitative transformations of various substrates of interest, including acetophenone, furfural, and 5-HMF, at lower temperatures than conventional heating methods. Specifically, magnetically induced catalysis has shown great promise in selective HDO reactions, particularly for biomass utilization. This technology enables the selective conversion of compounds derived from biomass into valuable biofuel candidates like 2-MF and 2,5-DMF, under lower hydrogen pressures than traditional industrial processes. Achieving localized high temperatures at the catalyst surface, creating "hot spots" in a relatively cold environment, has been crucial in attaining enhanced catalytic activity and facilitating selective reactions at the nanoscale.

Studies have addressed the reusability and stability of the magnetic catalysts. For instance, both $Fe_{2.2}C/Cu_2Cr_2O_5$ and FeCo@Ni demonstrated efficient and selective hydrogenation over multiple cycles in organic solvents, highlighting the stability of the magnetically heated catalyst components. Additionally, further to the organometallic approach, additional post-synthetic modifications involving surface modifications such as encapsulation or ligand exchange, have shown significant promise in protecting the magnetic core from aqueous media. This, in turn, has widened the scope of application media for these catalysts. This strategic stability opens avenues for broader applications in catalysis within aqueous environments, laying the groundwork for a focused examination of magnetically induced catalysis in the specific context of biomass valorization. More recently, the use of organometallic-derived nanostructures for magnetically heated catalysis in continuously operated reactions in the liquid phase is also a noteworthy advancement towards the scalability of this technology.

While the promise of magnetically induced catalysis is immense, challenges persist in the application to liquid-phase reactions. For instance, it would be essential to have precise control and optimization of temperatures at the catalyst surface under AMF to ensure efficient catalysis. In that sense, accurate techniques to monitor the surface temperatures of magnetically heated MagNPs need to be established. To date, two direct measurement methods have been used, but both target the bulk liquid temperature. In the first one, a fiber-optic temperature sensor is placed inside the reaction mixture. In the second method, the temperature is measured on the external wall of the reactor using an IR camera. One attractive alternative is using Arrhenius-derived methods to determine the surface temperature for a given coil condition [13, 72]. These approaches have indeed confirmed the presence of high local temperatures within the solution, which exceed the mean temperature of the bulk and could be higher than the boiling point of the solvent. However, the practical applications of these methodologies to nanoparticle hyperthermia may be hindered when entering into diffusional regimes or suffering from specific surface interactions that determine NP behavior in a given chemical reaction. Moreover, controlling surface ligand concentration is crucial for reproducible heating power by controlling the agglomeration of MagNPs during synthesis and catalytic processes, as agglomerates may reduce heating power and catalytic efficiency. Therefore, it is essential to put the spotlight on the influence of stabilizing ligand concentration on the catalytic efficacy and heating potential of MagNPs in future research endeavors involving the development of novel magnetic cores via the organometallic approach.

On the other hand, it would be important to keep on developing effective strategies to enhance the stability of MagNPs when exposed to severe reaction conditions, especially during the transition from organic solvents to aqueous environments. The stability of MagNPs is crucial as it affects their heating and catalytic performance. Although some strategies have been mentioned before, there is still room for improvement to make them even more stable for future industrial applications. It is also advisable to carefully tailor the reaction conditions when using magnetic catalysts to achieve more efficient and selective reactions. This can be done by adjusting parameters such as magnetic field amplitude, pellet preparation, reactor design, coil geometry, and frequency optimization. Working with higher amounts of reactants and catalyst loadings is also essential to consider the scale-up potential of magnetically induced catalysis for practical industrial applications.

As the field continues to evolve, the synergy between organometallic chemistry and magnetically induced catalysis holds promise for creating new nanostructures with diverse heating capabilities to accomplish a broader range of catalytic transformations, specifically within liquid environments. Achieving more intricate transformations with a high selectivity of biomass-derived molecules that contain diverse functional groups presents a significant challenge. To overcome these challenges and broaden the versatility of magnetically induced catalysis, it is necessary to combine MagNPs with oxide supports to enhance their acid-base properties and provide oxygen adsorption sites. However, the expected heating power loss when MagNPs are combined with non-susceptors and used in solvents with high specific heat capacity, such as water, makes exploring the search for nanoparticle systems with even higher heating power imperative. At the same time, designing novel reactors that can accommodate these magnetic catalysts while withstanding higher temperatures and pressures is also crucial to advancing induction-heated catalysis. Although quartz reactors are commonly used, they have obvious limitations that require exploring alternative materials such as PEEK or advanced ceramics. Integrating organometallic-approached magnetic cores with the aforementioned chemical and physical coadjutants is vital to fully harness the potential of induction-heated catalytic systems and enable operation under higher temperatures and pressures.

References

1. Wang W, Tuci G, Duong-Viet C, Liu Y, Rossin A, Luconi L, Nhut J-M, Nguyen-Dinh L, Pham-Huu C, Giambastiani G (2019) ACS Catal 9:7921–7935
2. Houlding TK, Rebrov EV (2012) Green Process Synth 1:19–31
3. Niether C, Faure S, Bordet A, Deseure J, Chatenet M, Carrey J, Chaudret B, Rouet A (2018) Nat Energy 3:476–483
4. Muley PD, Henkel C, Abdollahi KK, Boldor D (2015) Energy Fuel 29(11):7375–7385
5. Bursavich J, Abu-Laban M, Muley PD, Boldor D, Hayes DJ (2019) Energy Conv Manag 183: 689–697
6. Marbaix J, Kerroux P, Montastruc L, Soulantica K, Chaudret B (2020) Int J LCA 25:733–743
7. Pérez-Camacho MN, Abu-Dahrieh J, Rooney D, Sun K (2015) Catal Today 242:129–138
8. Marbaix J, Mille N, Carrey J, Soulantica K, Chaudret B (2021) Philippot K, Roucoux A (eds) Nanoparticles in catalysis: advances in synthesis and applications, vol 1. Wiley, New York, pp 307–329
9. Kreissl H, Jin J, Lin S-H, Schüette D, Störtte S, Levin N, Chaudret B, Vorholt AJ, Bordet A, Leitner W (2021) Angew Chem Int Ed 60:26639–26646
10. Wang D, Astruc D (2014) Chem Rev 114(14):6949–6985
11. Mehdizadeh M, Koc TA, Krause KR, Sengupta SK, Blackwell BE (2001) US patent 6315972
12. Nielsen PEH, McKenna BT, Hansen JB, Nielsen RM (2017) US patent 11319284
13. Díaz-Puerto ZJ, Raya-Barón Á, van Leeuwen PWNM, Asensio JM, Chaudret B (2021) Nanoscale 13:12438–12442
14. Asensio JM, Miguel AB, Fazzini P-F, van Leeuwen PWNM, Chaudret B (2019) Angew Chem Int Ed 58:11306–11310
15. Gyergyek S, Kocjan A, Grilc M, Likozar B, Hočevarc B, Makovec D (2020) Green Chem 22: 5978–5983
16. Mustieles Marin I, De Masi D, Lacroix L-M, Fazzini P-F, van Leeuwen PWNM, Asensio JM, Chaudret B (2021) Green Chem 23:2025–2036
17. Gyergyek S, Grilc M, Likozar B, Makovec D (2022) Green Chem 24:2788–2794
18. Huang C, Wang Y, Zhon R, Sun Z, Deng Y, Duan L (2023) Chin Chem Lett 34:108101
19. Chikazumi S (1997) Physics of ferromagnetism.2nd edn. Oxford University Press, Oxford
20. Deatsch AE, Evans BA (2014) J Magn Magn Mater 354:163–172
21. Ruta S, Chantrell R, Hovorka O (2015) Sci Rep 5:9090
22. Hergt R, Dutz S, Röder M (2008) J Phys Condens Matter 20:385214
23. Carrey J, Mehdaoui B, Respaud MJ (2011) J Appl Phys 109:083921
24. Zhang X, Chen S, Wang H-M, Hsieh S-L, Wu C-H, Chou H-H, Hsieh S (2010) Biomed Eng 22: 393–399

25. Amiens C, Chaudret B, Ciuculescu-Pradines D, Colliere V, Fajerwerg K, Fau P, Kahn M, Maisonnat A, Soulantica K, Philippot K (2013) New J Chem 37:3374–3401
26. Lin S-H, Hetaba W, Chaudret B, Leitner W, Bordet A (2022) Adv Energy Mater 12:2201783
27. Kanwal Z, Raza MA, Riaz S, Manzoor S, Tayyeb A, Sajid I, Naseem S (2019) R Soc Open Sci 6:182135
28. Raya-Barón Á, Ghosh S, Mazarío J, Varela-Izquierdo V, Fazzini P-F, Tricard S, Esvan J, Chaudret B (2023) Mater Horiz 10:4952–4959
29. Shviro M, Polani S, Dunin-Borkowski RE, Zitoun D (2018) Adv Mater Interfaces 5(9):1701666
30. Rudnev V (2003) Handbook of induction heating. CRC Press
31. Lucía O, Maussion P, Dede EJ, Burdío JM (2014) Induction heating technology and its applications: past developments, current technology, and future challenges. IEEE Trans Ind Electron 61(5):2509–2520
32. Matesa JM, Won KJ, Demarest HM Jr (1986) U.S. Patent 4610711
33. Forschungsverbund Berlin e. V. (1997) Patent DE 196 38 966 A 1
34. Sunicon AG (2011) DE 10 2009 035 041
35. Schaafsma Y (1963) GB925025A, 1 May
36. Leclercq J, Giraud F, Bianchi D (2014) Appl Catal B 146:131–137
37. Wu L, Ma H, Yan Z, Xu Q, Li Z (2022) Energ Conver Manage 270:116242
38. Li N, Pan Y, Yan Z, Liu Q, Yan Y, Liu Z (2023) Fuel 336:127124
39. Bartley BH (1975) J Catal 40:86–93
40. Chatterjee S, Degirmenci V, Rebrov EV (2015) Chem Eng J 281:884–891
41. Lach D, Tomiczek B, Siudyga T, Kapkowski M, Sitko R, Klimontko J, Golba S, Dercz G, Matus K, Borek W, Polanski J (2023) Int J Mol Sci 24:4729
42. Ghosh S, Ourlin T, Mazarío J, Cayez S, Daccache S, Carrey J, Chaudret B (2023) Chem Mater 35:7542–7553
43. Chen J, Su S, Wang C, Li Q, Wang H, Xu W, Li X, Jia H (2023) J Chem Eng 476:146712
44. Siudyga T, Kapkowski M, Lach D, Bartczak P, Kocot K, Jendrzejewska I, Dercz G, Zubko M, Balin K, Golba S, Tomiczek B, Pacultová K, Polański J (2023) Chem Eng Res Des 199:102–114
45. Ceylan S, Friese C, Lammel C, Mazac K, Kirschning A (2008) Angew Chem Int Ed 47:8950–8953
46. Ceylan S, Coutable L, Wegner J, Kirschning A (2011) Chem A Eur J 17:1884–1893
47. Ovenston A, Walls JR (1983) J Chem Soc Faraday Trans 79:1073–1084
48. Chaudhuri SR, Hartwig J, Kupracz L, Kodanek T, Wegner J, Kirschning A (2014) Adv Synth Catal 356:3530–3538
49. Liu Y, Cherkasov N, Gao P, Fernández J, Rebrov EV (2017) J Catal 355:120–130
50. Estrader M, Soulantica K, Chaudret B (2022) Angew Chem Int Ed 61:e202207301
51. Martínez-Prieto LM, Chaudret B (2018) Acc Chem Res 51:376–384
52. Cormary B, Dumestre F, Liakakos N, Soulantica K, Chaudret B (2013) Dalton Trans 42:12546–12553
53. Rodriguez A, Amiens C, Chaudret B, Casanove M-J, Lecante P, Bradley JS (1996) Chem Mater 8:1978–1986
54. Cordente N, Respaud M, Senocq F, Casanove M-J, Amiens C, Chaudret B (2001) Nano Lett 1(10):565–568
55. Gual A, Godard C, Philippot K, Chaudret B, Denicourt-Nowicki A, Roucoux A, Castillón S, Claver C (2009) ChemSusChem 2:769–779
56. Cerezo-Navarrete C, Lara P, Martínez-Prieto LM (2020) Catalysts 10(10):1144
57. Raya-Barón Á, Mazarío J, Mencia G, Fazzini P-F, Chaudret B (2023) ChemSusChem 16: e2023000
58. Ramírez-Meneses E, Betancourt I, Morales F, Montiel-Palma V, Villanueva-Alvarado CC, Hernández-Rojas ME (2011) J Nanopart Res 13:365–374
59. Herman DAJ, Cheong-Tilley S, McGrath AJ, McVey BFP, Leina M, Tilley RD (2015) Nanoscale 7:5951–5954

60. Reina A, Favier I, Pradel C, Gómez M (2018) Adv Synth Catal 360(18):3544–3522
61. Braga AH, Costa NJS, Philippot K, Gonçalves RV, Szanyi J, Rossi LM (2020) ChemCatChem 12(11):2967–2976
62. Bordet A, Lacroix L-M, Fazzini P-F, Carrey J, Soulantica K, Chaudret B (2016) Angew Chem Int Ed 55:15894–15898
63. Dumestre F, Chaudret B, Amiens C, Renaud P, Fejes P (2004) Science 303:821–823
64. Bordet A, Lacroix L-M, Soulantica K, Chaudret B (2016) ChemCatChem 8:1727–1731
65. Mamman AS, Lee J-M, Kim Y-C, Hwang IT, Park N-J, Hwang YK, Chang J-S, Hwang J-S (2008) Biofuels Bioprod Biorefin 2:438–454
66. Leitner W, Klankermayer J, Pischinger S, Pitsch H, Kohse-Hoeinghaus K (2017) Angew Chem Int Ed 56:5412–5452
67. Hu L, Lin L, Liu S (2014) Ind Eng Chem Res 53:9969–9978
68. Huang Y-B, Chen M-Y, Yan L, Guo Q-X, Fu Y (2014) ChemSusChem 7:1068–1072
69. Asensio JM, Marbaix J, Mille N, Lacroix LM, Soulantica K, Fazzini P-F, Carrey J, Chaudret B (2019) Nanoscale 11:5402–5411
70. Garnero C, Lepesant M, Garcia-Marcelot C, Shin Y, Meny C, Farger P, Warot-Fonrose B, Arenal R, Viau G, Soulantica K, Fau P, Poveda P, Lacroix L-M, Chaudret B (2019) Nano Lett 19(2):1379–1386
71. Marbaix J, Mille N, Lacroix L-M, Asensio JM, Fazzini P-F, Soulantica K, Carrey J, Chaudret B (2020) ACS Appl Nano Mater 3(4):3767–3778
72. Cerezo-Navarrete C, Mustieles Marin I, García-Miquel H, Corma A, Chaudret B, Martínez-Prieto LM (2022) ACS Catal 12:8462–8475
73. De Masi D, Asensio JM, Fazzini P-F, Lacroix L-M, Chaudret B (2020) Angew Chem Int Ed 6: 6187–6191
74. Mazarío J, Mustieles Marin I, Mencia G, Lopes CW, Varela-Izquierdo V, Agostini G, Fazzini P-F, Ratel-Ramond N, Chaudret B (2024) ACS Appl Nanomater 7(8):9412–9427
75. Lahive CW, Kamer PC, Lancefield CS, Deuss PJ (2020) ChemSusChem 13(17):4238–4265

Top Organomet Chem (2024) 75: 273–332
https://doi.org/10.1007/3418_2024_122
© The Author(s), under exclusive license to Springer Nature Switzerland AG 2024
Published online: 16 October 2024

Ligand-Capped Heterogeneous Catalysts from Groups 8 to 10

María Dolores Fernández-Martínez and Cyril Godard

Contents

M. D. Fernández-Martínez and C. Godard (✉)
Departament de Química Física i Inorgánica, Universitat Rovira i Virgili, Tarragona, Spain
e-mail: cyril.godard@urv.cat

Abstract This chapter focuses on the functionalization of heterogeneous catalysts of group 8–10 by capping ligands. The synthetic approaches used for the introduction of ligands were particularly looked at, and according to the type of ligands, three main approaches were used: the colloidal formation of metal nanoparticles (MNPs) stabilized by ligands and their subsequent deposition over a solid support, the direct synthesis by colloidal formation of the ligand-capped active phase in the presence of the support, and the ligand functionalization of a previously synthesized heterogeneous catalyst. Depending on the aim of the study and the catalytic application, one of the synthetic strategies was selected. The ligands were shown to have a range of effect on the catalytic performance of the catalyst by tuning the electronic properties of the metal(s), the steric hindrance at the metal surface, or blocking some of the catalyst active sites. These modifications can therefore influence the activity, the selectivity, and/or the stability of the materials, and as such, ligand functionalization appears as a powerful method to fine-tune the properties of heterogeneous catalysts.

Keywords Catalytic performance · Heterogeneous catalysts · Ligands · Nanoparticles · N-Heterocyclic carbenes · Nitrogen · Phosphorus · S-donor

1 Introduction

Over the last few decades, the preparation of metal nanoparticles (MNPs) by colloidal strategies has become an efficient tool for the controlled synthesis of heterogeneous catalysts [1–3]. In this approach, the binding of organic ligands (also called capping agents) at the metallic surface enables the precise control of MNP size, shape, and composition (i.e., surface structure). However, the presence of such ligands was traditionally believed to induce the blockage of the catalytic active sites and therefore to negatively affect the catalyst activity and various strategies were reported to remove these ligands without altering the surface properties of the MNPs.

More recently, the beneficial roles of surface ligands in terms of activity or selectivity of nanocatalysts have been demonstrated [4]. Indeed, the presence of ligands on metal surfaces can induce steric effect between ligands and reactants, electron transfer through the metal–organic interface, etc. [5] These capping ligands can be introduced in different manners depending on the nature of the support and future application of the nanocatalysts. Indeed, when the fine control of the MNP size and composition is aimed for, the ligands can be used as unique stabilizers in colloidal synthesis prior to their deposition onto a solid support. However, when the metal–support interactions are strong enough, the decomposition of the metal precursor can be performed in the presence of both the support and the capping agents. Moreover, the effect of the capping agents can also be directly probed by functionalization of a heterogeneous catalyst previously synthesized under classical conditions (Fig. 1).

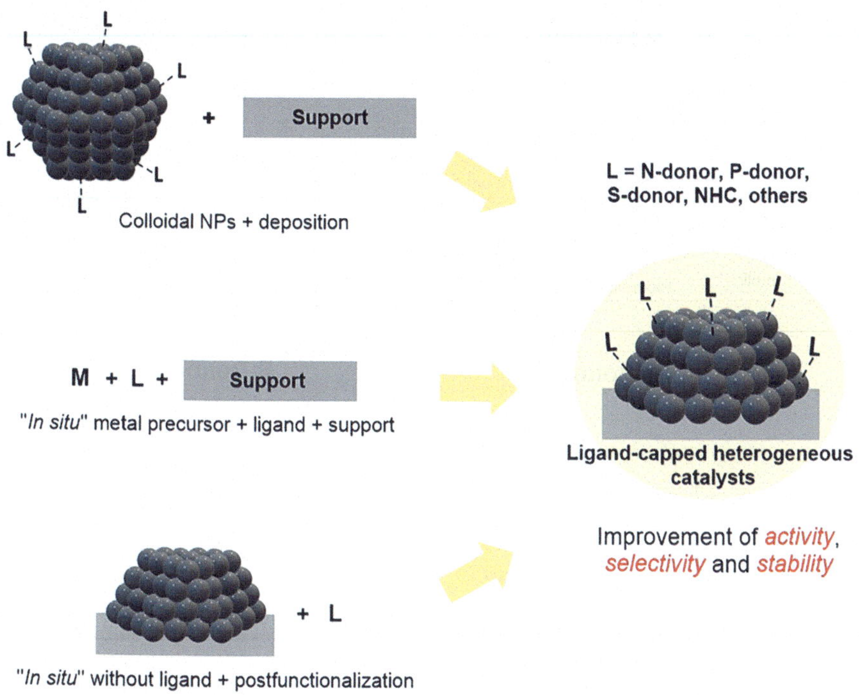

Fig. 1 Synthetic strategies for the formation of ligand-capped heterogeneous catalysts

In this chapter, the effects of ligands on heterogeneous nanocatalysts are presented through literature examples and classified by the donor atoms of the ligands (Fig. 2) and the synthetic methodology utilized (Fig. 1). In view of the somewhat unclear definition of ligands in heterogeneous catalysis, the effects of capping agents described here will be limited to N-, P-, and S-donor molecules and NHC ligands. Recent results demonstrated that appropriate modifications of metallic nanoparticle-based catalysts by ligands can dramatically improve the activity/selectivity of specific catalytic reactions, mainly due to steric and electronic modifications of the heterogeneous nanocatalysts. The stability of ligand-promoted nanocatalysts before and after catalytic reactions is also discussed.

2 N-Donor Ligand-Capped Heterogeneous Catalysts

Generally, N-donor ligands were mainly utilized to finely tune the physical properties of the heterogeneous catalyst active phases through colloidal synthesis. Using this approach, the metallic active phases were formed in solution using N-based capping agents followed by their deposition over a solid support. This approach presents the advantage of precisely controlling the structure and composition of

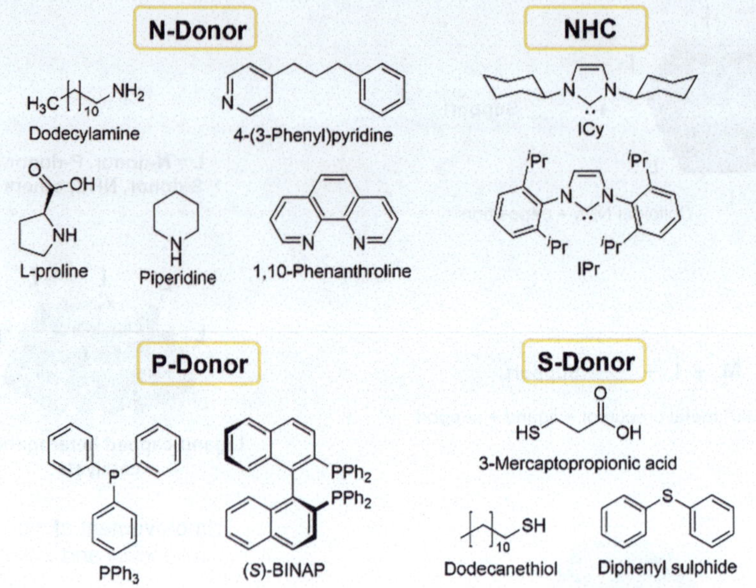

Fig. 2 Examples of ligands used to enhance the performances of heterogeneous catalysts

MNPs and to reach very high particle loadings. In colloidal chemistry, N-donor molecules were among the first ligands that were reported due to their ability to stabilize MNPs [6]. Since these early reports, a variety of N-donor ligands were used to improve the catalytic behavior of heterogeneous catalysts during the synthesis of colloidal nanoparticles, during the formation of the nanoparticles in the presence of the support, or by functionalization of the previously synthesized catalysts by addition of ligands.

The introduction of N-functions at the surface of the support during the catalyst synthesis has also attracted much interest [7]. In these cases, the N-based capping agents can be designed to either retain their structural properties or to be integrated within the solid support structure during a calcination step. Depending on the strategy to be used, either amines containing aliphatic chains or pyridine derivatives were reported as capping agents.

For instance, Guerrero, Rossi, Philippot, and co-workers reported the preparation of palladium NPs supported over a magnetite-based support decorated by terpyridine ligands acting as stabilizers (Fig. 3) [8, 9]. The role of the terpyridine ligand was to control the size and stability of the palladium nanoparticles. Moreover, in hydrogenation reactions, the activity and selectivity of this nanocatalyst were enhanced compared to those provided by palladium nanoparticles supported on an amino-modified magnetic support.

Using a distinct approach, Chen and co-workers used ethylenediamine (EDA) as a nitrogen precursor for the synthesis of a Fe/Co-N_x composite onto nanoporous carbon black [10]. In this case, the N-donor ligands were used to stabilize the metal

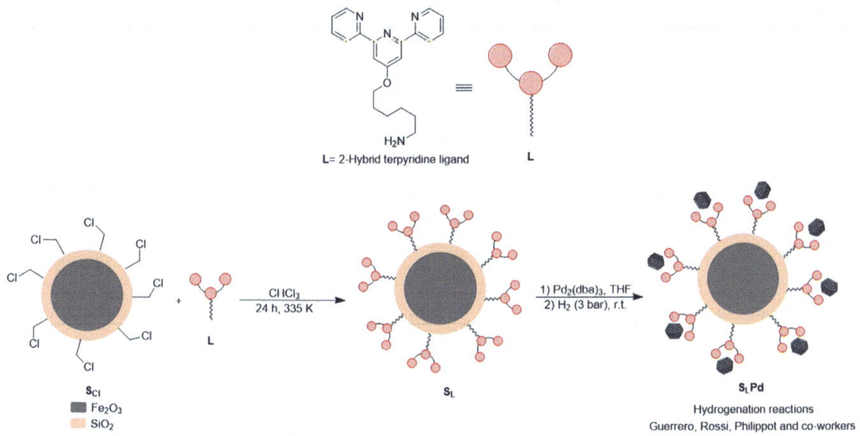

Fig. 3 Ligand modification of a support reported by Guerrero, Rossi, Philippot, and co-workers

precursor prior to their decomposition upon heating the mixture. The resulting pyridinic-type nitrogen introduced in the catalysts was revealed to be crucial for ORR catalytic activity and selectivity.

In the following sections, representative examples of heterogeneous catalysts containing N-donor ligands will be described and classified according to the synthetic approach used for the introduction of the N-donor ligands and its nature (i.e., aliphatic vs aromatic).

2.1 N-Capped Heterogeneous Catalysts Formed by Colloidal Methods Prior to Their Deposition Over a Solid Support

2.1.1 Aliphatic N-Donor Ligands

Long-chain aliphatic amines have been successfully used for the stabilization of well-defined transition MNPs and their application in a range of applications.

For instance, in 2007, Nikles and co-workers prepared well-dispersed Pt NPs by polyalcohol reduction using oleylamine as capping agent that were subsequently loaded onto a carbon support for application in the electrocatalytic oxidation of methanol [11]. The ligand to metal ratio used in the synthesis affected the particle size, and variations from 3.5 nm to 11.5 nm were observed when decreasing the amount of oleylamine. The as-prepared particles were subsequently deposited onto a carbon support but showed no electrocatalytic activity. An activation step involving the removal of the ligand from the surface of the Pt NPs was carried out by heating the particles in air at 185°C. Cyclic voltammetry showed that the ligand-free particles were electrocatalytically active for methanol oxidation. Interestingly, the smallest Pt NPs (3.5–4.0 nm) demonstrated a higher intrinsic activity for methanol

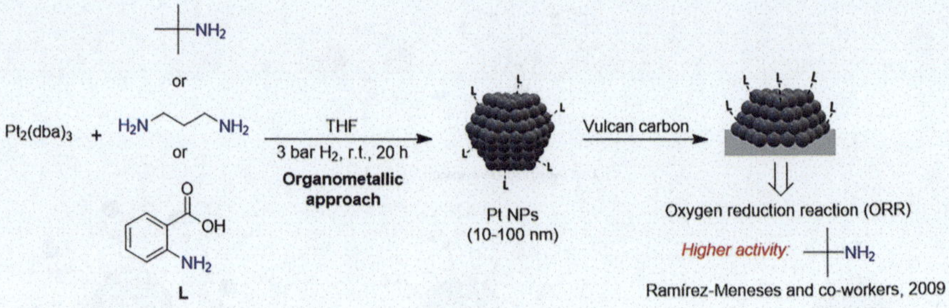

Fig. 4 Synthesis of N-capped Pt nanocatalysts reported by Ramírez-Meneses and co-workers

oxidation, but a lower tolerance to CO poisoning, compared with the larger particles (6.0–11.5 nm).

In 2009, Peng and Yang described a new approach to address synergetic effects on both the activity and stability issues of electrocatalysts using palladium MNPs as support for Pt catalysts [12]. The Pt-on-Pd nanoparticles were prepared using a sequential synthetic method. The Pd NPs were formed by thermal decomposition of palladium acetylacetonate in the presence of oleylamine and trioctylphosphine and exhibited an average diameter of 5.3 ± 0.6 nm [13, 14]. The Pt NPs were formed by subsequent decomposition of Pt acetyl acetonate in the presence of the Pd NPs. The Pt NPs exhibited an average diameter of $\sim$3 nm and were distributed evenly on the surface of Pd NPs. Similarly to the previous examples, activation was necessary and these bimetallic catalysts exhibited much higher activity and stability than the corresponding pure Pt catalyst, which was attributed to the favored interfacial structures between Pt and the Pd support.

Ramírez-Meneses and co-workers compared the performance of Pt NPs stabilized by different primary amines (i.e., tert-butylamine, 1,3-diaminopropane, and anthranilic acid) in the oxygen reduction reaction (ORR) (Fig. 4). The MNPs were synthesized through the organometallic approach and subsequently supported over Vulcan carbon [15]. The resulting particle sizes ranged from 10 to 100 nm depending on the stabilizer. The catalytic results revealed an effect of the amine stabilizer and higher activities were obtained when the Pt NPs were capped by tert-butylamine.

More recently, Miyabayashi, Nishihara, and Miyake reported the preparation of Pt NPs (2.4–2.8 nm) protected with octylamine (OA) and partly substituted by alkylamine with a pyrene group (PA) and supported over carbon black (i.e., Vulcan XC72) [16]. The synthesis of the ligand-stabilized MNPs was performed by chemical reduction with $NaBH_4$ [17]. The MNPs without any pretreatment to remove the introduced organic-protected agents were applied as catalysts for oxygen reduction reaction (ORR). The modification of the Pt NP surface with OA and PA considerably enhanced the electrocatalytic activity of the catalysts. Indeed, the area and mass-specific activities increased at high PA/OA ratios. The potential cycle test demonstrates that the modification of Pt NPs by PA improves the stability of the catalyst and maintains high area and mass-specific activities of ORR.

In 2017, Miyabayashi and Miyake reported the preparation of Pt NPs (2.6 nm) using undecafluorohexylamine (UFHA) and octylamine as stabilizing ligands and their subsequent deposition over acid-treated carbon black [10, 18]. The synthesis of Pt NPs using both ligands was performed following the modified two-phase liquid reduction method using $NaBH_4$ as reductant [16]. The carbon black supported systems were applied as electrocatalysts for the ORR. The affinity between the perfluorinated alkyl chains of UFHA and the Nafion® used as binder between the support and the catalyst improved the activity of the latter by generating a proton conduction path. Moreover, modification by UFHA enhanced robustness by avoiding Pt dissolution and carbon corrosion due to limited water accessibility.

In 2019, Ung and Cossairt prepared bare CoP nanocrystals that were decorated with carboxylates (i.e., oleate, octanoate, acetate) and amines (i.e., oleylamine, octylamine, dioctylamine, trioctylamine, butylamine) and deposited over a carbon fiber electrode support with the aim of studying the effect of the ligand in catalysis [19]. Carboxylates and amines were selected as common stabilizers in colloidal syntheses and to evaluate the impact of their steric properties and hydrophobicity on catalysis. These systems were tested in the hydrogen evolution reaction (HER). Long-chain carboxylates caused a larger increase in overpotential than their equivalent amines (e.g., oleate vs oleylamine), due to higher ligand density and stronger coordination of carboxylates. Moreover, carboxylates bearing longer carbon chain resulted in higher overpotential whereas no such effect was observed with primary amines. These results suggested that the surface ligands hinder catalysis on CoP for HER mainly by preventing substrate access to surface active sites.

Wang, Lou, and co-workers prepared novel 3D Pt nano assemblies consisting of 1D single-crystal Pt nanowires (diameter and length were in the range of 5–10 nm and 100–200 nm, respectively) by a one-pot method using oleylamine and cetyltrimethylammonium bromide (CTAB) as stabilizers [20]. In this case, oleylamine acted as solvent, reducing agent, and stabilizer for the formation of metal nanostructures [21, 22]. Cetyltrimethylammonium bromide (CTAB) plays an important role as the structure-directing agent in the formation of multi-wire structures since only irregular multipods were obtained in the absence of CTAB [23]. When evaluated as electrocatalysts for low-temperature fuel cells, these interconnected 3D Pt nanostructures manifested high electrochemical surface area and significant long-term stability.

OA ligands were also used for thermocatalytic processes. For instance, Son and co-workers reported the synthesis of tetrahedral Rh nanoparticles (4.9 ± 0.4 nm) using oleylamine as stabilizing ligand that were subsequently loaded on charcoal [24]. Spherical Rh nanoparticles were also synthesized by hot injection method for comparison. Both catalysts were tested in the hydrogenation of arenes. The tetrahedral Rh NPs on charcoal provided much higher activity than the spherical nanoparticles and commercial Rh/C in the hydrogenation of anthracene. Moreover, the tetrahedral nanoparticles showed outstanding activity for a range of arenes.

Borchert and co-workers reported the preparation of Pt NPs and nanowires using dodecylamine (DDA) and/or hexanethiol as stabilizers for CO oxidation catalysis. These nano-objects were subsequently deposited over γ-alumina or silica [25]. CO

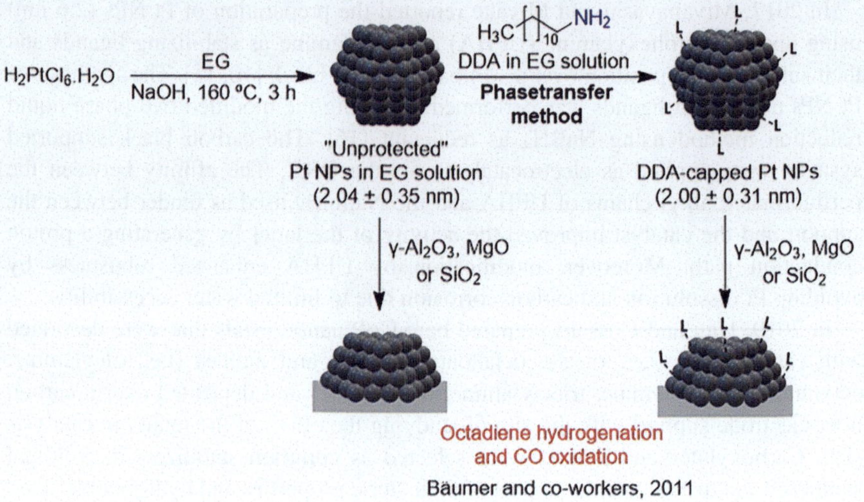

Fig. 5 Preparation of Pt catalysts by Bäumer and co-workers

adsorption experiments were performed for these systems and revealed that small molecules can pass through the ligand shell and adsorb on free areas of the Pt surface. These systems were active in the catalytic CO oxidation reaction. The differences in activity observed were ascribed to the morphology of the catalysts, nature of the support, or differences in surface coverage by the ligand molecules. Interestingly, the Pt nanowires were restructured upon catalytic conditions and the shape of the Pt species in the spent catalyst was quasi-spherical.

Bäumer and co-workers reported a distinct approach based on the initial synthesis of colloidal ligand-free Pt NPs (2.2 ± 0.3 nm) by metal salt reduction in ethylene glycol (EG) and the subsequent functionalization with DDA (2.2 ± 0.3 nm) [26]. Then, the coating of three different monolithic supports (i.e., cordierite honeycomb, Al_2O_3 foam, and Nickel fleece) was performed with preformed ligand-free and ligand-stabilized Pt NPs. Interestingly, prior to the deposition of the MNPs the supports were functionalized with the ligands. The formation of this ligand double layer was shown to avoid the formation of an alloy between the MNP and the support, and as such, high activity for CO oxidation was maintained. In contrast, when no ligand was used, the catalytic activity decreased over time. Similarly, the same authors later reported the deposition of ligand-free (2.3 ± 0.2 nm), dodecylamine-capped (2.3 ± 0.2 nm), or dodecanethiol (DDT) Pt NPs supported over Fe_3O_4 and showed that the presence of ligands at the Pt surface avoids unfavorable strong metal–support interactions (SMSI) and, as such, enhances the activity of the catalysts in CO oxidation [27]. They also reported the deposition of ligand-free and dodecylamine-capped Pt NPs (1.9–2.2 nm) over three different support materials with varying acid/base properties, namely γ-Al_2O_3, SiO_2, and MgO [28]. The nanoparticles were prepared in ethylene glycol (EG) and subsequent functionalization with DDA was performed using a phase-transfer method [29, 30] (Fig. 5).

This constituted the first systematic study on the influence of ligands in heterogeneous gas-phase catalysis using identical nanoparticles synthesized by colloidal methods. The catalytic properties of these catalysts were investigated in two reactions. First, octadiene hydrogenation was used as a test reaction to probe the influence of the ligands under reducing conditions temperatures ($T < 100°C$). The ligands strongly modified the metal–support interactions and exerted a protecting effect with respect to support-induced oxidation during the deposition step. When SiO_2 was the support, where surface oxidation is more prominent, the ligand-capped catalyst was significantly more active than the ligand-free analog. When the catalysts were tested in CO oxidation at high temperatures ($T \approx 200°C$), decomposition, desorption, and spillover of ligands on the support became important. Depending on the supports, the ligand coverage varied and spillover was most evident on Lewis acidic γ-Al_2O_3 compared to the Brønsted acidic SiO_2.

In 2013, Altmann, Bäumer, and co-workers reported the preparation of colloidal ligand-capped (i.e., DDA) Pt (2.2 ± 0.3 nm) and bimetallic PtSn (2.3–2.4 nm) nanoparticles that were subsequently deposited on SiO_2 for the selective semi-hydrogenation of acetylene in ethylene-rich streams [31]. To assess the effect of ligands on catalysis, monometallic and bimetallic DDA-capped MNPs were compared with ligand-free catalysts. Both surface ligands and Sn enhance the selectivity and the most selective catalyst was the DDA-capped PtSn catalyst (metal atom ratio of 1:1).

Philippot, Rossi, and co-workers reported the preparation of bimetallic NiPd nanoparticles (5–6 nm) via the organometallic approach using hexadecylamine (HDA) as stabilizing ligand and supported over magnetic silica [32]. Nickel–palladium nanoparticles were supported onto a magnetic silica using various Ni/Pd ratios. The supported catalysts were evaluated in the hydrogenation of cyclohexene and their performance compared to the corresponding monometallic catalysts. The catalysts prepared using a 1:9 Ni/Pd molar ratio provided the highest initial activity (i.e., $>50{,}000\ h^{-1}$), greater than the monometallic Pd catalyst. Moreover, excellent separation from the products could be performed without metal leaching, leading to an efficient recycling.

More recently, Domingos, Philippot, and co-workers reported the preparation of a series of Ni(0) nanocatalysts prepared through the organometallic approach in the presence of hexadecylamine (HAD), polyvinylpyrrolidone (PVP), PVP/triphenylphosphine, octanoic acid (OA), and stearic acid. The resulting MNPs exhibited diameters of ca. 5–6 ± 2 nm and were evaluated in the selective hydrogenation of α,β-unsaturated carbonyl compounds by H_2 under mild reaction conditions [33]. The system in which OA was used as ligand was also deposited over SiO_2. All catalysts were active in the chemoselective hydrogenation of the C=C bond, which was attributed to the reduced nature of the Ni NPs surface. Moreover, the hydrogenation reaction rate was revealed to be sensitive to the stabilizer type, with the carboxylic acid-stabilized systems providing the best performances.

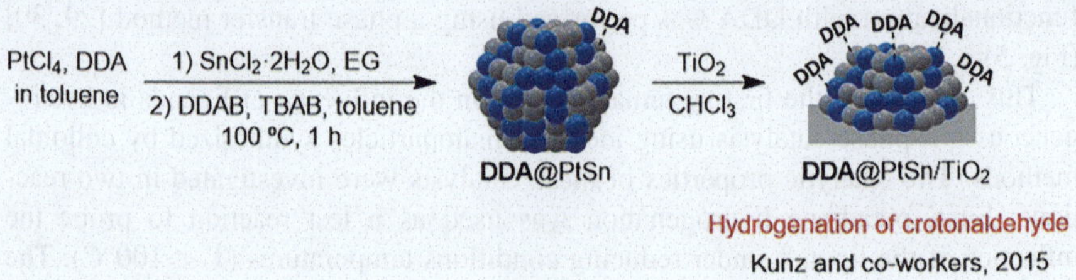

Fig. 6 Preparation of well-defined amine-capped bimetallic PtSn nanoparticles and their subsequent deposition on TiO_2

Moreover, the supported catalyst could be reused up to 10 times without significant loss of activity.

Pérez-Yuste and co-workers presented a novel strategy based on the immobilization of oleylamine-stabilized Pd NPs (3.0 ± 0.4 nm) on filter paper for the development of a catalytic system with high efficiency and recyclability [34, 35]. The final material was prepared by a simple dip-coating process using highly concentrated MNP dispersions in volatile organic solvents and tested in several model reactions such as the oxidative homocoupling of arylboronic acids, the Suzuki cross-coupling reaction, and nitro to amine reduction. These N-capped heterogeneous catalysts displayed efficient catalytic activity and excellent recyclability and reusability.

The same year, Kunz and co-workers reported the preparation of well-defined DDA-capped bimetallic PtSn nanoparticles (ca. 2 nm) by colloidal methods that were subsequently deposited on TiO_2 (Fig. 6) [36]. The influence of the Pt/Sn ratio on the catalytic properties for the hydrogenation of crotonaldehyde was looked at. The activity of Pt increased significantly by adding small amounts of Sn (10%), which was attributed to an electronic effect. However, a gradual decrease was observed when the Sn content was further increased, which attributed to the loss of Pt sites needed for the activation of H_2. The lower activity was accompanied by a constant selectivity increase toward crotylalcohol. In this study, activity and selectivity followed clear trends related to the Sn content, which emphasizes the potential of colloidal chemistry for model studies in heterogeneous catalysis.

More recently, Fan and co-workers used a slightly distinct approach for the synthesis of rhodium nanoparticles bearing L-proline as capping agent [37]. The preparation was performed by chemical reduction of $RhCl_3$ with $NaBH_4$ in the presence of L-proline, and without isolation, the MNPs were then deposited on carbon (Fig. 7). In this synthesis, L-proline was employed as a size controller for the synthesis of small (~2 nm) and uniform Rh NPs for the first time.

The Rh catalysts showed high activity for ammonia borane (AB) methanolysis with a total turnover frequency (TOF) up to 1,035 mol_{H2} $(mol_{Rh}min)^{-1}$ under basic conditions. This excellent catalytic performance was attributed to the well-dispersed Rh NPs and the functionalized metal surface, which provided more active sites for the reaction. Other ligands such as ethylenediamine, triethylamine, ethylene glycol, polyvinylpyrrolidone, and citric acid were used for the preparation of analogous

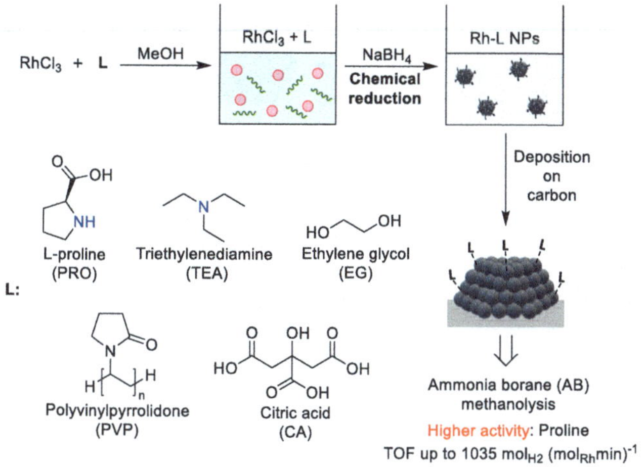

Fig. 7 Ligand-stabilized Rh NPs supported on carbon reported by Fan and co-workers

catalysts; however, the best results were obtained with the L-proline-functionalized Rh catalyst.

2.1.2 Aromatic N-Donor Ligands

Aromatic N-based donor molecules were also used for the preparation of N-ligand-capped heterogeneous catalysts.

For instance, in 1993, Schmid and co-workers reported the deposition of Pd clusters (between 31.5 and 36.0 Å) stabilized with phenanthroline over TiO_2 [38]. The resulting material was used as catalyst in the semi-hydrogenation of alkynes with selectivity to the corresponding alkene up to 98%. In 1996, Schmid and co workers reported the use of ligand-stabilized palladium clusters (3–4 nm) on various supports such as TiO_2, Al_2O_3, and $CaCO_3$ for applications in gas-phase and liquid-phase hydrogenation and semi-hydrogenation reactions [39, 40]. For example, N-donor ligands such as di-2,9-(2-methyl-butyl)-1,10-phenanthroline and (−)-cinchonidine were employed in the catalytic semi-hydrogenation of 2-hexyne. The phenanthroline stabilizer improved the selectivity of the reaction by suppressing isomerization and over-hydrogenation reactions. In the hydrogenation of acetophenone, the stabilizers mainly affected the activity of the catalysts.

In 1997, Schmid and co-workers reported the preparation of $p\text{-}H_2NC_6H_4SO_3Na$ (sodium sulfanilate)-stabilized Pt shell on Au or Pd cores supported onto alumina and TiO_2 for their application in hydrosilylation reactions [41–46]. The aim of these works was to study the influence of the electronegativity of the colloidal core metals on the activity and selectivity of the Pt surface in hydrosilylation reactions.

Gómez and co-workers reported the synthesis of small and well-dispersed Ru NPs stabilized by 4-(3-phenylpropyl)pyridine as ligand and supported over different

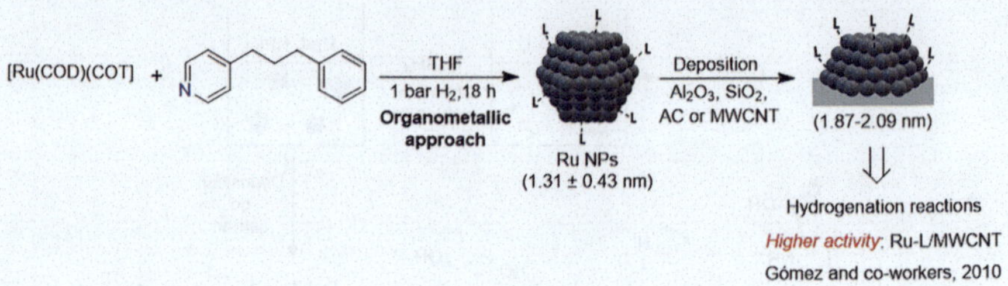

Fig. 8 Ru NPs stabilized by 4-(3-phenylpropyl)pyridine and supported over different supports, including MWCNT

supports, such as Al_2O_3, SiO_2, activated carbon (AC), and multiwalled carbon nanotubes (MWCNT) [47]. The synthesis of the catalyst was carried out in two steps: first, the preparation of colloidal MNPs (1.3 ± 0.4 nm, Fig. 8) [48] and their subsequent immobilization over the corresponding support.

These supported catalysts were used in the hydrogenation of several unsaturated substrates. The effect of the supports was particularly looked at and the highest activity, chemoselectivity, and recyclability were observed for the Ru-L system immobilized over MWCNT.

In 2011, the same group reported the preparation of well-dispersed Pd NPs in ionic liquids in the presence of 4-(3-phenylpropyl)pyridine and their subsequent immobilization onto multiwalled carbon nanotubes (MWCNTs) modified with imidazolium moieties [49]. When the synthesis of the MNPs was performed in the presence of the ligand in [BMI][PF_6] as solvent, the obtained MNPs exhibited a size of 4.5 ± 0.8 nm. In the absence of the ligand, the particle size was similar (4.5 ± 1.0 nm) but a more heterogeneous dispersion was observed. Once dispersed over the support, this catalyst was tested in hydrogenation and Heck coupling reactions and could be recycled up to four times without significant activity loss.

Recently, García-Antón, Sala, and co-workers reported the synthesis of a hybrid material (Ru@RuO_2PPTiO_2-RuP) based on 4-phenylpyridine-capped Ru nanoparticles (RuPP), TiO_2 nanocrystals, and [Ru(bpy)$_2$(4,4′-(PO_3H_2)$_2$(bpy))]Cl_2 (RuP) as catalyst for the hydrogen evolution reaction (HER) [50]. The preparation of this catalyst was performed in three steps. First, the Ru NPs (1.4 ± 0.3 nm) capped with 4-phenylpyridine were prepared though the organometallic approach [51] followed by their deposition onto TiO_2 by impregnation (1.7 ± 0.4 nm) and the decoration of the latter with a visible-light photoabsorber (RuP). TiO_2 acted as an efficient dispersing agent and stabilizer for the catalyst (Ru@RuO_2PP) and electron relay between Ru@RuO_2PP and the photoabsorber (RuP) under irradiation. Thus, this double role of TiO_2 as support and electron mediator enabled the first efficient visible-light-driven HER photocatalysis based on a Ru-based photocatalyst and a molecular photoabsorber.

In the area of asymmetric catalysis, Bönnemann and Braun reported in 1997 the first example of MNPs stabilized by chiral molecules as a new type of enantioselective heterogeneous catalysts, combining good selectivity control and high activity in asymmetric hydrogenation reactions [52]. In their work, colloidal Pt NPs were stabilized by the alkaloid dihydrocinchonidine and subsequently supported over silica. Both colloidal and supported systems exhibited the same MNP size. These systems were tested in catalysis and compared with a commercial heterogeneous catalyst. The colloidal systems provided higher conversions and enantiomeric excess (*ee*) values than the two supported heterogeneous catalysts.

2.1.3 Mixture of Amine and Carboxylic Acid as Capping Agents

In 2010, Yang and co-workers reported the synthesis of Pt_3Ni catalysts with using long-chain amines as capping agent and Vulcan carbon as support [53]. In this system, adamantanecarboxylic acid (ACA) or adamantaneacetic acid (AAA) was also employed [54, 55]. At a later stage, butylamine was used to replace the capping agents used during the synthesis of Pt_3Ni nanocrystals and to generate carbon-supported active electrocatalysts that exhibited enhanced activity compared to commercial catalysts.

In 2012, Yang and co-workers described the synthesis of PtM (M = Au, Ni, Pd) (PtAu particle size: 11 ± 0.2 nm; Pt_3Ni particle size: 13 ± 0.3 nm) icosahedral nanocrystals [56]. The synthesis was performed using the gas reducing agent in liquid solution (GRAILS) method [57, 58], which requires the use of carbon monoxide (CO) [59–62]. In this study, oleylamine (OAm) and oleic acid (OA) were used in the synthetic mixture and the resulting MNPs were supported over carbon. Both CO gas and organic surface-capping agents play critical roles in stabilizing the icosahedral shape with {111} surfaces. Among the PtM alloy nanocrystals, Pt_3Ni was revealed to be very active with an impressive ORR-specific activity of 1.83 mA/cm^2 Pt and 0.62 A/mg$_{Pt}$.

In 2012, Strasser and co-workers reported the preparation of a series of dealloyed Pt_xNi_{1-x} core–shell nanoparticles [63] in the presence of oleic acid (OAc), oleylamine (OAm), and 1,2-tetradecandiol (TDD), which were subsequently supported over carbon. The particle size of the PtNi, $PtNi_3$, and $PtNi_5$ NPs was 5.1 ± 0.6, 8.3 ± 0.8, and 12.1 ± 1.1 nm, respectively. The radial location and the composition of the Ni-enriched inner shells were dependent on the initial alloy compositions. These self-organized Ni-enriched inner shells were revealed to play a key role in maintaining surface lattice strain and thus control the surface catalytic activity in the oxygen reduction reaction (ORR).

Bäumer and co-workers also prepared colloidal PtNi nanoparticles (~4 nm) using OAm and OA as organic stabilizers [64] and their deposition onto γ-Al_2O_3 and MgO powders [65]. These materials were tested in the CO oxidation reaction and no pretreatment was needed to obtain immediate activity. The results prove that well-defined catalysts with exceptional activities can be obtained with this approach and the authors concluded that the residual ligands could have a protecting effect.

In 2014, Yin, Ma, and co-workers reported for the first time the preparation of monodispersed bimetallic PdAg NPs using OAm and OA as capping agents and carbon black as support [66]. The bimetallic PdAg NPs were synthesized via the thermal decomposition of palladium acetate and reduction of silver acetate in the emulsion-assisted ethylene glycol ternary system [67]. Various compositions of bimetallic PdAg nanoparticles, $Pd_{80}Ag_{20}$ (3.8 nm), $Pd_{65}Ag_{35}$ (3.6 nm), and $Pd_{46}Ag_{54}$ (3.3 nm), were obtained by modifying the reaction parameters. The PdAg systems were supported on carbon black without any post-treatment and provided high electro-oxidation activity toward methanol oxidation in alkaline media. The carbon-supported $Pd_{80}Ag_{20}$ nanoparticles revealed superior activity than that of the commercial Pt/C electrocatalyst. It was concluded that the enhanced activity was dependent on the unique twinning structure of the catalyst.

In 2009, Rinaldi, Zanchet, and co-workers reported the preparation of colloidal Ni NPs (5.3 nm) stabilized by trioctylamine and OA and their deposition over silica and activated carbon. Subsequently, trioctylphosphine was added [68–70]. These heterogeneous catalysts were applied in the hydrogenation of cyclohexene and the steam reforming of ethanol [71]. The impregnation onto silica resulted in high metal dispersion. However, no or low catalytic activity was observed. The detailed surface analysis of the Ni/SiO_2 sample showed that the capping ligands are not easily removed from the Ni surface in the posterior steps of the catalyst preparation, inhibiting the surface activation of the reactants.

In 2022, Serp, Soulantica, and co-workers reported the synthesis of freestanding and few-layer graphene (FLG)-supported ε-Co nanoparticles and hexagonal close-packed hcp-Co nanorods stabilized by lauric acid (LA) and hexadecylamine (HDA) [72]. Cobalt nanostructures were synthesized by the reduction of $[Co\{N(SiMe_3)_2\}_2(thf)]$ by H_2 in the presence of LA and HDA [73, 74]. The authors showed that while hcp-Co nanorods exposing a majority of $\{11\bar{2}0\}$-type facets are the most active in the selective hydrogenation of cinnamaldehyde to cinnamyl alcohol, the supported spherical ε-Co nanoparticles combined high activity and excellent selectivity. The concentration-dependent ligand conformation on the surface of the nanostructures also influences their catalytic performances, as higher concentrations favored both activity and selectivity to cinnamyl alcohol.

In 2010, Sun and co-workers described the preparation of monodisperse Ni NPs by $Ni(acac)_2$ reduction using borane tributylamine (BTB) in the presence of OAm and OA. The Ni NPs dispersed on Ketjen carbon were highly active for hydrolytic dehydrogenation of the ammonia borane (AB) complex at room temperature [75]. In this process, OAm acted as both solvent and surfactant, BTB was the reducing agent, and OA was added as cosurfactant for Ni NP stabilization.

2.2 N-Capped Heterogeneous Catalysts Formed by Colloidal Methods in the Presence of the Solid Support

Another approach for the synthesis of N-donor ligand-capped heterogeneous catalysis is the direct reduction of the metal precursor in the presence of both the support and the stabilizing ligands. In this approach, both the support and the ligands participate in the stabilization of the MNPs.

2.2.1 Aliphatic N-Donor Ligands

Jang and co-workers reported the synthesis of Pt NPs supported over carbon using OAm as ligand for their application as electrocatalysts in proton exchange fuel cells [76]. For the preparation of this material, a mixture of the metal precursor, OAm, carbon black, and $NaBH_4$ as reducing agent was used to produce MNPs with diameters between 2 and 3 nm. The performance of this catalyst was compared with that of a commercial Pt/C modified by OAm. The electrocatalyst prepared by the authors showed higher ORR activity and kinetic current density. They concluded that the surface ligands selectively blocked large anion adsorption and as such enhanced the catalytic activity since small oxygen molecules can still adsorb and react.

In 2020, Serp, Soulantica, and co-workers reported the preparation of octadecylamine (ODA)-stabilized platinum concave nanocubes supported on graphene exposing {110} facets using an in situ process and their subsequent use as catalysts in the selective hydrogenation of cinnamaldehyde [77]. The nanocubes were directly grown onto the FLG support in the presence of ODA as stabilizing agent. Two sets of unsupported nanocubes stabilized by ODA were also prepared following literature procedures for comparison purposes [78]. Immobilization improved both activity and selectivity toward the desired unsaturated alcohol. The FLG-supported concave nanocubes presented outstanding activity and selectivity to cinnamyl alcohol higher than 80%. The presence of the organic stabilizer ensured the stability of the nanoparticle morphology and influenced the catalyst performance. While higher activity was obtained using small nanoparticles and low ligand contents, the selectivity to the unsaturated alcohol was favored by high ligand contents, probably due to steric effects favoring the substrate adsorption through the C=O moiety. Apart from stabilizing the nanocubes, the authors reported the presence of interactions between the support and the long-chain amine ligand and as such behaved as a ligand reservoir.

2.2.2 Aromatic N-Donor Ligands

In 2010, Esteruelas and co-workers reported the synthesis of Rh NPs (1.5 ± 0.2 nm) by decomposition of the pre-synthesized chloro[2,6-bis{1-(phenyl)iminoethyl}pyridine]rhodium(I) complexes [79]. The obtained Rh NPs were stabilized by the partially hydrogenated ligand. The nanoparticles were also generated in the presence of alumina as support. Under a constant atmospheric pressure of hydrogen, the nanoparticles catalyzed the dehalogenation and the hydrogenation of different substrates. However, Hg(0) poisoning test revealed that both homogeneous and heterogeneous catalysts coexist during the dehalogenation reactions, whereas the hydrogenation processes were only catalyzed by heterogeneous catalysts.

In 2017, Hayashi and co-workers reported the preparation of nickel nanoparticles (ca. 4 nm) stabilized with pyridine ligands and supported over ZSM-5 zeolite [80]. The synthesis of the catalyst was carried out by $NaBH_4$ chemical reduction of the metal precursor in the presence of pyridine and the ZSM-5 zeolite support. Pyridine played a crucial role in the selective formation of small and well-dispersed Ni NPs and increased their stability by enhancing the interaction between the Ni and the support. This catalyst provided the first complete hydrogenation of phenolic compounds as lignin monomers with higher activity than that of conventional Ni catalysts. Upon recycling, no relevant increase in size was observed after each catalytic cycle when pyridine was used in the synthesis of the MNPs, whereas a large increase in diameter was evidenced for those synthesized in the absence of pyridine.

In 2009, Serp and co-workers reported the preparation of PtRu NPs (ca. 2 nm) stabilized by 4-(3-phenylpropyl) pyridine ligand containing a pyridine unit that has an affinity for the surface of the MNPs and a phenyl group able to interact with the carbon nanotube (CNT) graphene layers (π–π interaction) [81]. The authors also functionalized the CNT external surface with carboxylic acid and amide groups bearing a long alkyl chain to induce weak interaction or repulsion between the MNPs and the CNT external surface. These nanocatalysts displayed excellent catalytic performance in the selective hydrogenation of cinnamaldehyde owing to confinement of both the active phase and the reactants in the inner cavity of the CNTs.

2.2.3 Mixture of N-Donor and Carboxylic Acid

Serp, Soulantica, and co-workers reported the preparation of non-promoted Co/Al_2O_3–SiO_2 catalyst for Fischer–Tropsch synthesis (FTS) using hexadecylamine (HDA) and lauric acid (LA) ligands as MNP stabilizers [82]. Cobalt nanorods were prepared by growing anisotropic hcp-cobalt nanostructures on conventional Co/Al_2O_3–SiO_2 catalyst through the so-called seeded growth method [83]. The as-obtained catalyst showed a better stability than a reference catalyst for FTS. The removal of most of the ligands from the nanorod further enhanced the catalyst activity while maintaining its stability [84]. Previously, the same group reported monolithic *hcp*-cobalt nanowires bearing HDA and LA as ligands and supported over copper and nickel foams through a wet chemistry approach [73, 74, 85–87]. The Co/Cu_{foam} catalyst was tested in the Fischer–Tropsch synthesis in a fixed-bed reactor, providing superior activity and selectivity toward C_{5+} than a Co/SiO_2-Al_2O_3 reference catalyst. In this work, a negative effect of the ligands was described [82].

2.3 N-Capped Heterogeneous Catalysts Formed by Classical Methods Followed by Ligand Functionalization

Other N-capped heterogeneous catalysts only involve the functionalization of the metal surface by the N-ligand in a final step of the synthesis. Usually, the capping of heterogeneous catalyst by N-donor molecules was performed to tune the selectivity of the catalytic reactions.

In an early example, Vasudevan and co-workers reported the effect of piperidine addition over Pd NPs supported on Al_2O_3 prepared by wet impregnation for the hydrogenation of unsaturated hydrocarbons [88–90]. The addition of piperidine during catalysis improved the hydrogenation rate and the selectivity, which was attributed to the electron-donating properties of the piperidine ligands that decreased the bonding energy of the hydrocarbons.

In 2013, Sung and co-workers reported the modification of Pt/C (~5 nm) with oleylamine to prepare a carbon-supported platinum electrocatalyst for oxygen reduction reaction (ORR) [91]. The addition of a small amount of OAm on Pt NPs resulted in a significant enhancement of the catalyst activity, which was attributed to both the downshift of the frontier d-band structure of platinum and the hindering of anionic spectator species that leads to an enhancement of ORR kinetics. These results suggest that the use of capping organic molecules could be a strategy to design advanced electrocatalysts via the alteration of both structural and electronic properties.

More recently, Chen and Xu immobilized some ligands containing secondary or tertiary amine groups and hydrophobic chains (Fig. 9) onto TiO_2-supported Ru

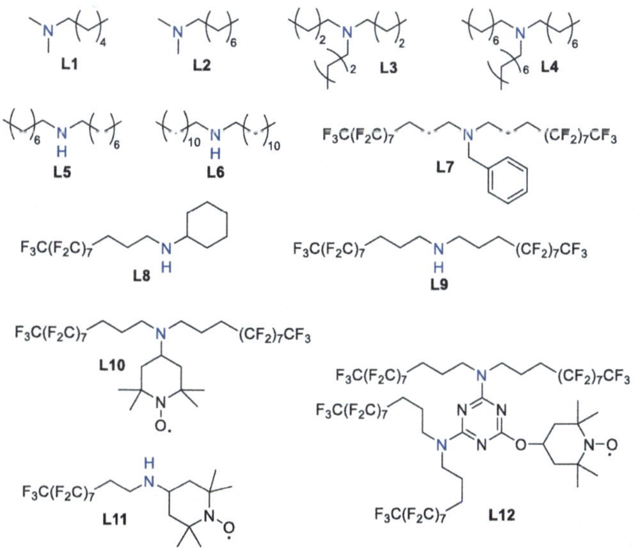

Fig. 9 Amine ligands used by Chen and Xu to functionalize TiO_2-supported Ru NPs

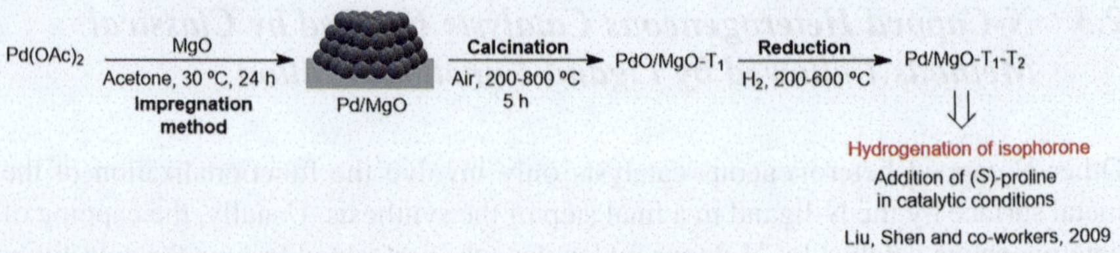

Fig. 10 Preparation of Pd/MgO catalysts modified by (*S*)-proline for the hydrogenation of isophorone

nanoparticles [29]. The immobilization of the ligands onto Ru/TiO_2 was based on a simple two-phase centrifugation technique and driven by the hydrophobic effect. Additionally, the synthesis of "unprotected" Ru NPs (2.5 nm) was carried out according to previously reported methods [92] and subsequently immobilized onto TiO_2 by sol immobilization (2.0 nm) [93].

These systems were tested in the aerobic oxidation of alcohols and ligands containing 2,2,6,6-tetramethyl-1-piperidine-Noxyl (TEMPO) moieties (L10, L11, and L12) significantly improved the selectivity to the aldehyde at high conversions.

Later, the same authors reported the modification of Pt/SiAl catalyst with ethylenediamine (EDA) by co-impregnation method and its application in the selective catalytic oxidation of ammonia [94]. TEM measurements demonstrated that the average particle sizes of Pt species decreased from 2.8 to 2.0 nm upon diamine addition. The results based on precise characterization indicated that the addition of EDA could adjust the states and particle sizes of Pt and the reduction property of Pt/SiAl catalyst.

The functionalization of heterogeneous catalysts by chiral N-donor amines was also reported. Indeed, the first examples of asymmetric hydrogenations catalyzed by metal surface modified by the presence of (*S*)-proline (and other chiral molecules) as additive under catalytic conditions were first reported in the 1980s by Petró and co-workers [95–97]. This strategy was used to induce enantioselectivity in the asymmetric hydrogenation of α,β-unsaturated ketones.

More recently, Liu, Shen, and co-workers reported the use of a series of Pd/MgO catalysts modified by (*S*)-proline for the hydrogenation of isophorone (Fig. 10) [98]. The Pd/MgO catalysts were synthesized by the impregnation method and variations in MNP size were observed depending on the temperatures employed during calcination and hydrogen reduction. Particles sizes between 1 and 25 nm were obtained. In this study, it was demonstrated that Pd surfaces not only supply chemisorbed hydrogen but also participate in the enantio-differentiating step. Indeed, the configuration of the chiral product could be tuned by varying the size of Pd NP. In this process, two competitive reactions were identified: the formation of excess (*R*)-TMCH enantiomer was reached through the enantioselective hydrogenation of intermediates formed between isophorone and (*S*)-proline on Pd particles with a mean diameter >10 nm, while the direct hydrogenation of isophorone yielded racemic TMCH over Pd particles smaller than 4 nm, followed by kinetic resolution that yielded the (*S*)-TMCH enantiomer in excess.

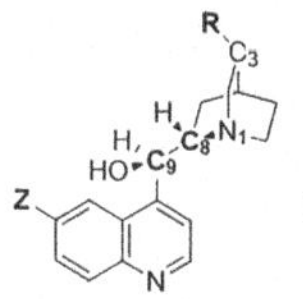

Cinchona alkaloids

8-(*R*), 9-(*S*)	R	Z	8-(*S*), 9-(*R*)
Cinchonidine (Cd)	Vin	H	Cinchonine (Cn)
10,11-Dihydrocinchonidine (HCd)	Et	H	10,11-Dihydrocinchonine (HCn)
Quinine (Qn)	Vin	OMe	Quinidine (Qd)
10,11-Dihydroquinine (HQn)	Et	OMe	10,11-Dihydroquinidine (HQd)

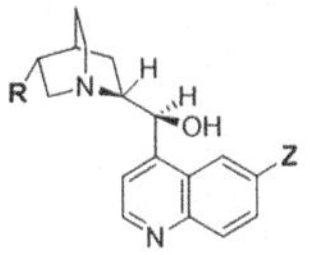

Fig. 11 Chiral modifiers derived from cinchona alkaloids used for the modification of Pt/Al_2O_3 catalysts for application in asymmetric hydrogenation of ethyl pyruvate

In 2010, Győffry and co-workers also reported the use of Pd NPs supported over various supports for the kinetic resolution of 3,5,5-trimethyl cyclohexanone (TMCH) and asymmetric hydrogenation of isophorone (3,5,5-trimethyl cyclohex-2-enone) in the presence of (*S*)-proline [99].

Aromatic N-donor ligands were also used as metal surface-capping agents. In an early example, Baiker and co-workers reported the modification of commercial Pt/Al_2O_3 with cinchonidine as additive for the asymmetric hydrogenation of activated ketones such as ketopantolactone and ethyl pyruvate [100]. The catalytic reactions were carried out in a fixed-bed reactor with continuous feeding of the chiral modifier into the reactant stream. Total chemoselectivity was obtained with enantiomeric excesses of 83.4% for ketopantolactone and 89.9% *ee* for ethyl pyruvate.

Blaser and co-workers also reported the use of commercial Pt/Al_2O_3 for the asymmetric hydrogenation of ethyl pyruvate using chiral modifiers derived from natural cinchona alkaloids [101]. Effects related to the structure of the chiral modifiers were studied. The strongest effects on *ee* were observed for changes in the O-C_9-C_8-N part of the cinchona alkaloid and for partial or total hydrogenation of the quinoline rings (Fig. 11).

The solvent also demonstrated to have a significant effect on enantioselectivity and reaction rate, and it was concluded that an efficient modifier for the hydrogenation of R-keto esters must contain a basic nitrogen center close to stereogenic center (s) and connected to an aromatic system.

Similarly, the modification of Pt NPs (4.4 nm) immobilized over silica-coated Fe_3O_4 by cinchonidine was reported for the enantioselective hydrogenation of activated ketones (Fig. 12) [102]. The catalyst could be recycled up to eight times with little variation in activity although the enantioselectivity slightly decreased from 57% to 52% *ee* after the eighth cycle.

In 2010, Sugimura and Ogawa reported the use of cinchonidine (CD) and cinchonine (CN) as chiral modifiers for Pd/C in enantioselective hydrogenation reactions (Fig. 13) [103]. The catalytic hydrogenation with CD always provided higher *ee* and faster reaction rate than that with CN and it was concluded that the difference between CD and CN was not in their adsorption properties, but in the enantio-induction ability of these modifiers.

Baiker and co-workers used commercial Pt/Al_2O_3 in the presence of different modifiers such as NMe_3, NEt_3, quinuclidine, quinoline, (*R*,*R*)-PNEA, CD, and *O*-

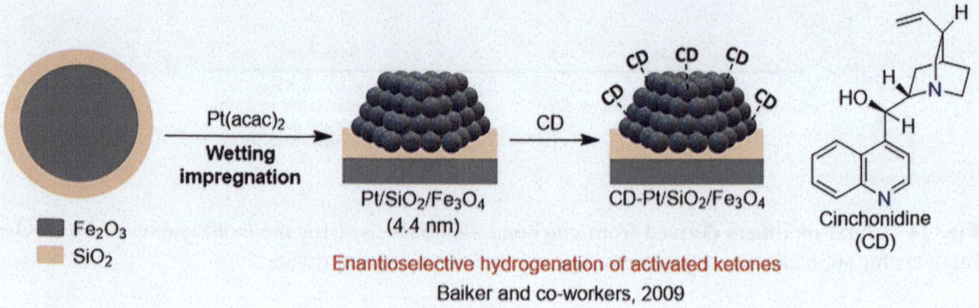

Fig. 12 Modification of Pt catalysts on silica-coated Fe_3O_4 by cinchonidine

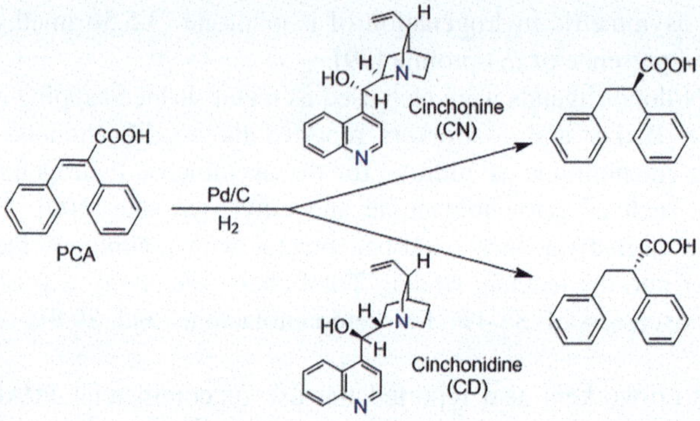

Fig. 13 Cinchonidine (CD) and cinchonine (CN) as chiral modifiers for Pd/C in enantioselective hydrogenation

methyl-CD as additives in the hydrogenation of different substrates [104]. The addition of trace amounts of the chiral modifiers enhanced the chemoselectivity up to 100% in the hydrogenation of activated carbonyl groups with up to 93% *ee*. This study showed a higher adsorption for the chiral ligands than for simple amines and the high chemo- and enantioselectivities were attributed to the formation of an ion pair involving the protonated amine function of the chiral modifier and the enolate form of the substrate.

3 Phosphorus Donor Ligand-Capped Heterogeneous Catalysts

Another common type of capping agents for heterogeneous catalysts is P-donor ligands. Similarly to N-donor ligands, depending on the focus of the study and the catalytic application, several synthetic strategies were reported to modify the MNP surface with these ligands.

3.1 Colloidal Synthesis of P-Capped MNPs Followed by Deposition Over a Solid Support

In order to form well-defined MNPs with small size and narrow distribution, one of the most efficient synthetic methods consist in the colloidal formation of the MNPs followed by their deposition onto a solid support. The drawback of this strategy is, in some cases, the weak interactions between the MNPs and the support that can result in the release of the MNPs under catalytic conditions.

For instance, Philippot and co-workers synthesized a series of ligand-stabilized Pd NPs and their subsequent deposition over mesoporous carbon (Fig. 14) [105]. The Pd NPs were prepared via the organometallic approach [81] and stabilized with triphenylphosphine or trioctylphosphine [106, 107]. Direct impregnation of the colloids onto the solid support provided the final catalysts which were tested in aerobic oxidation of benzyl alcohol to benzaldehyde, providing excellent conversions and high selectivity to the aldehyde.

Later, the same group reported a series of carbon-supported Pd NPs for the dual selective aqueous oxidation of benzyl alcohol and the hydrogenation of furfural in water under microwave irradiation [108]. Two synthetic routes for the catalysts were compared: the impregnation of the carbon support with the previously synthesized Pd nanoparticles stabilized with trioctylphosphine and triphenylphosphine as ligands and the impregnation of the support with the palladium metal precursor (in the absence of phosphine ligand) followed by hydrogenation. The nanoparticles stabilized by trioctylphosphine followed by their anchoring onto carbon support provided the highest activity. When trioctylphosphine was ligand, the steric hindrance induced by the aliphatic chains limited the interaction between the Pd NPs and the carbon surface oxygen groups, thus preventing their agglomeration that was observed when triphenylphosphine was used. The presence of oxygen groups on the surface of the carbon support improved the Pd NPs immobilization and the water affinity and consequently enhanced the catalytic performances of the system.

In 2016, Lee and co-workers reported the preparation of Ni NPs by a colloidal method using oleylamine (OAm) and trioctylphosphine (TOP) as stabilizers and their subsequent deposition over a carbon support. These systems were used as catalysts for the selective hydrogenation of unsaturated furanic aldehydes to their

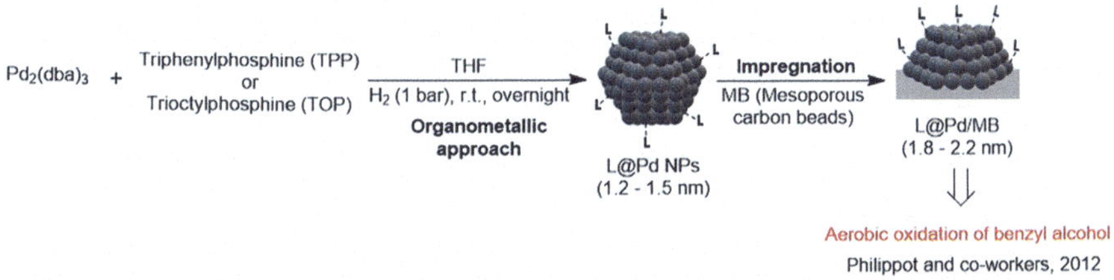

Fig. 14 Phosphine-stabilized Ru NPs supported on carbon beads for their application in the aerobic oxidation of benzyl alcohols

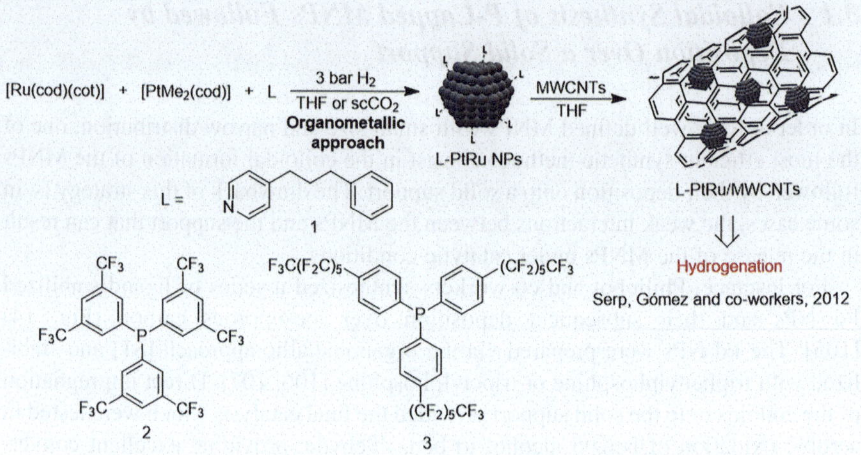

Fig. 15 Preparation of ligand-capped bimetallic platinum–ruthenium hydrogenation catalysts supported over functionalized multiwalled carbon nanotubes

corresponding alcohols [109]. In a previous study, it was reported that OAm was the reductant that controlled the nucleation rate while TOP provides a tunable surface stabilization through its coordination to the Ni surface [110]. The capped nanoparticles provided high selectivities (>90%) in the hydrogenation of unsaturated furanic aldehydes. In comparison, the uncapped calcined Ni/SiO_2 catalysts presented selectivity to the over-hydrogenation product tetrahydrofurfuryl alcohol. DRIFT results confirmed that the approach of the furanic ring to the Ni surface was limited at the capped nanoparticles due to the steric hindrance induced by the organic surface layer.

In 2012, Serp, Gómez, and co-workers synthesized bimetallic platinum–ruthenium nanoparticles stabilized by pyridine- and monophosphine-based ligands and supported over functionalized multiwalled carbon nanotubes in supercritical CO_2 or in THF (Fig. 15) [111]. The bimetallic nanoparticles were confined into functionalized multiwalled carbon nanotubes (MWCNT), which led to a higher degree of agglomeration for the PtRu nanoparticles stabilized by phosphine ligands than for those stabilized by the pyridine ligands. These catalysts were applied to cinnamaldehyde hydrogenation and confined PtRu nanoparticles provided higher activity and selectivity than the unsupported nanoparticles.

Ruiz and co-workers synthesized colloidal Rh NPs bearing a chiral ligand ((−)-DIOP) by chemical reduction of the Rh precursor with $NaBH_4$ (Fig. 16 (a)) prior to their deposition over SiO_2 [112]. The range of NP sizes was 3.9–5.8 nm depending on the metallic loading and the ligand concentration. When the catalyst prepared in the absence of ligand was used, the diameter of the corresponding MNPs was ca. 14 nm, which highlights the stabilization role of the ligand during the synthesis. These MNPs were tested in the asymmetric hydrogenation of acetophenone, 1-phenyl-1,2-propanedione, 3,4-hexanedione, 2,3-butanedione, and ethyl pyruvate with *ee's* up to 53%.

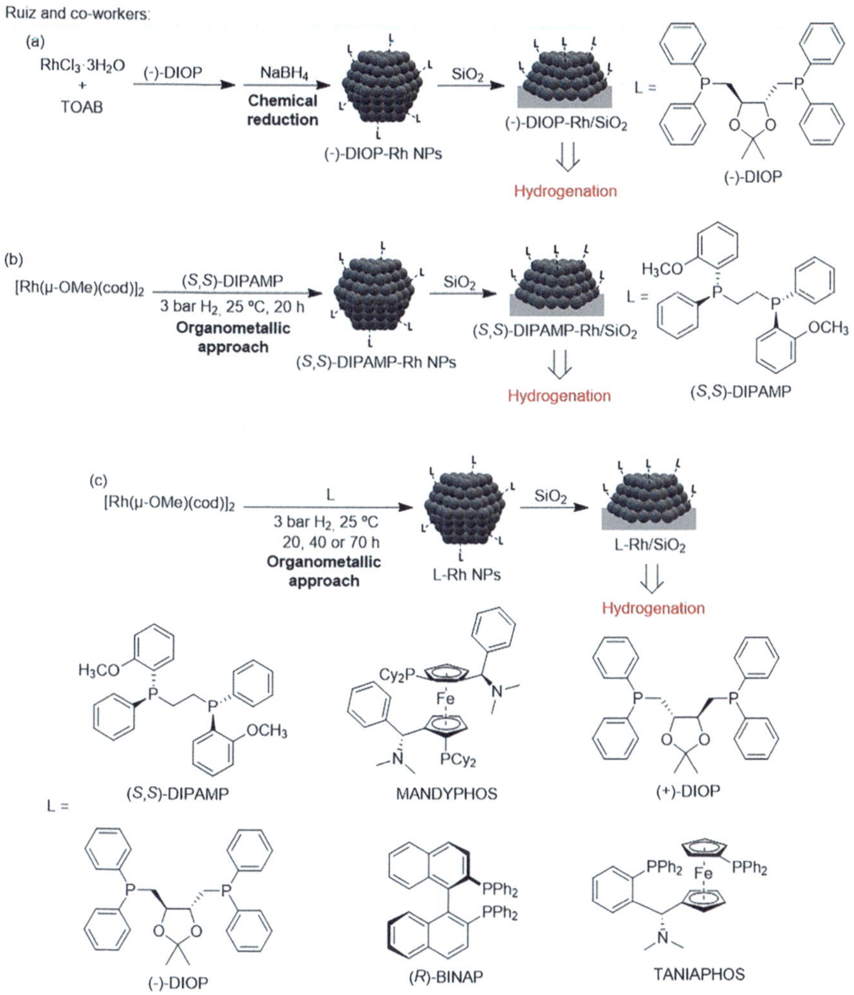

Fig. 16 Asymmetric hydrogenation Rh catalysts stabilized by chiral P-ligands and supported over silica reported by Ruiz and co-workers

The same authors reported a series of Rh NPs stabilized with (*S,S*)-1,2-ethanediylbis[(2-ethoxyphenyl)phenylphosphine] ((*S,S*)-DIPAMP) (Fig. 16 (b)) subsequently deposited over SiO_2 [113]. Catalysts formed using various Rh/L ratios were tested in the asymmetric hydrogenation of hexanodione, ethyl pyruvate, ketopantolactone, and acetophenone. Good conversion (99%) and enantiomeric excess (*ee* up to 54%) were obtained.

In 2019, they also reported the preparation of Rh NPs (<3 nm) stabilized with a series of P-chiral ligands supported over SiO_2 (Fig. 16 (c)) [114].

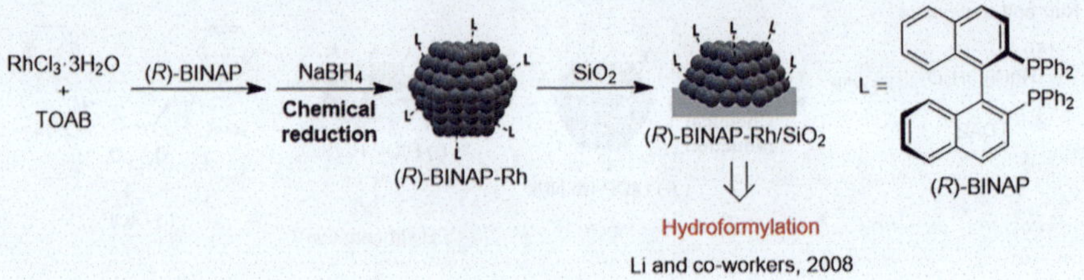

Fig. 17 Preparation of heterogeneous asymmetric hydroformylation catalysts reported by Li and co-workers

These systems were tested in 1-phenyl-1,2-propanedione hydrogenation and obtained up to 98% conversion and 67% enantiomeric excess for (*R*)-1-hydroxy-1-phenyl-propan-2-one and 59% for (*R*)-2-hydroxy-1-phenylpropan-1-one.

In 2016, Rossi, Philippot, and co-workers synthesized a series of Rh NPs stabilized with PVP and mono- and diphosphines (i.e., PPh_3 and dppb) using the organometallic approach [115]. The resulting MNPs presented diameters of 2.2, 1.3, and 1.7 nm, respectively, and were subsequently supported over an amino-functionalized silica-coated magnetite.

The colloidal and supported systems were tested in the hydrogenation of cyclohexene. Lower activity was observed for the phosphine-stabilized catalysts due to the blocking of the active sites by the ligands. It is noteworthy that the presence of the phosphine ligands was confirmed even after several recycling cycles of the supported nanocatalysts.

Li and co-workers reported the synthesis of colloidal Rh NPs stabilized by (*R*)-BINAP by chemical reduction with $NaBH_4$ (Fig. 17) [116]. The size of these MNPs was between 1.5 and 2.0 nm. The stabilized MNPs were later immobilized by impregnation over silica. The coordination of the ligand was demonstrated by ^{31}P MAS NMR and IR. These systems were tested in the asymmetric hydroformylation of olefins with total regioselectivity and *ee* up to 59%. The supported MNPs exhibited higher activity than the colloidal systems. The system was also successfully recycled and reused upon addition of (*R*)-BINAP ligand after a few runs.

3.2 *Colloidal Synthesis of P-Capped MNPs in the Presence of a Solid Support*

Chen and co-workers reported the preparation of Ir NPs by reduction of a metal salt in the presence of PPh_3 and SiO_2 modified with a chiral amine derived from natural cinchona alkaloids (Fig. 18) [117]. This catalyst was employed in the hydrogenation of aromatic ketones and provided high activity and high enantioselectivity. The authors later expanded their study with the synthesis of Ir NPs stabilized with a series of phosphines and supported over SiO_2 modified with cinchonine, cinchonidine, quinone, and derivatives (9-amino(9-deoxy)epicinchonine, 9-amino(9-deoxy)

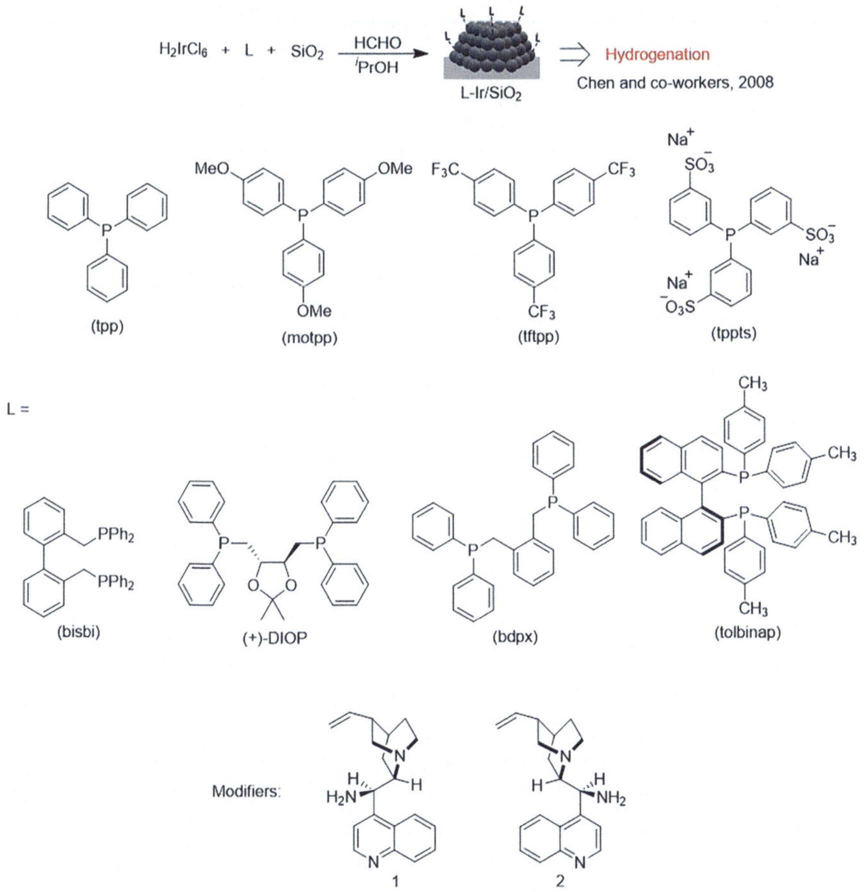

Fig. 18 Preparation of Ir NPs by reduction of a metal salt in the presence of phosphines and SiO_2 modified with a chiral amine

epicinchonidine (1), and 9-amino(9-deoxy)epiquinine) (2) [118]. When cinchonine, cinchonidine, and quinine were used, low catalytic activity and high chemoselectivity to saturated ketone were observed. The best results were obtained when PPh_3 was the ligand used for the stabilization of the catalyst but rather low enantioselectivity was obtained in all cases (up to 46%).

In 2017, Godard, Taoufik, and co-workers prepared by surface organometallic chemistry (SOMC) methodology a series of Rh and Pt NPs ligated by phosphine ligands and supported on phosphine-functionalized silica [119]. The presence of the P-donor ligand both in solution and at the support surface was revealed to be important. These systems were employed in the hydrogenation of *p*-xylene, 3-hexyne, 4-phenyl-2-butanone, benzaldehyde, and furfural and demonstrated high activities. Moreover, excellent results were obtained for the selective hydrogenation of furfural to furfuryl alcohol, with high activities in the case of the Pt catalyst. It was

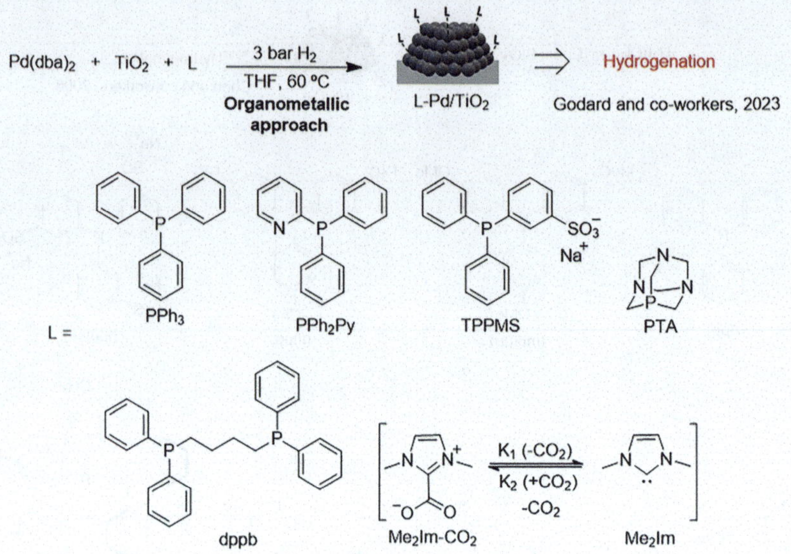

Fig. 19 Preparation of phosphine-stabilized heterogeneous Pd catalysts for CO_2 hydrogenation into formate

concluded that the phosphine-stabilized moiety played an important role in the selectivity of this reaction by blocking some surface sites.

Later, the same authors reported the synthesis of monometallic Pd and Cu and bimetallic PdCu NPs stabilized by triphenylphosphine supported over carbon nanotubes (CNTs) and phosphorus-functionalized silica (P-SiO_2) [120]. When the support employed was the P-SiO_2, MNP sizes were between 1 and 2.4 nm. However, when CNTs were used, the nanoparticle size was larger, in the range of 2.4–2.6 nm. These catalysts were tested in the semi-hydrogenation of alkynes. PdCu/CNTs was the most active system when the substrates were alkyl-substituted alkynes and also showed good recyclability during several cycles.

Godard and co-workers also used this approach to synthesize a Pd-based catalyst supported on filter paper [121]. In this work, a Pd precursor was decomposed under H_2 atmosphere in the presence of PCy_3 and CNTs. The resulting material was then used to make an ink via dispersion in water in the presence of a surfactant. Painting of the ink on the filter paper support followed by drying provided a low-cost and easy handling catalyst that was applied in the semi-hydrogenation of alkynes and alkynols. Recycling of this catalyst could be performed at least 5 times without loss of activity or selectivity.

In 2023, the same group reported a series of Pd NPs stabilized with ligands of different nature (mono- and bidentate phosphines organic and water-soluble and NHCs) (Fig. 19) via the organometallic approach in the presence of different supports (metal oxides and carbon-based) [122]. The Pd NPs supported over TiO_2 presented sizes between 1.92 (for PTA) and 2.83 nm (for NHC). When the synthesis of Pd/TiO_2 was performed in the absence of ligand, the particle size was 5.36 nm, demonstrating the beneficial effect of the ligands on the MNP stabilization. All these catalysts were tested in the hydrogenation of CO_2 in basic media to produce formate. The system containing PPh_3 provided the highest catalytic activity (TON = 1,032).

In 2008, Xiong and co-workers reported the synthesis of the Ru-phosphine/γ-Al_2O_3 catalyst (ca. 5 nm) by reduction of $RhCl_3$ in the presence of the support and the phosphine ligand. For application in catalysis, modification of this material by addition of (1*S*,2*S*)-DPEN was performed. This catalyst was used in the asymmetric hydrogenation of acetophenone and derivatives [123]. The effect of various phosphine stabilizers was examined in catalysis. When the Ru/γ-Al_2O_3 catalyst was stabilized by phosphine ligands, the catalytic activity and enantioselectivity increased. The best results were obtained when PPh_3 was used. Under optimum conditions, total conversion of acetophenone was reached with enantioselectivity up to 78% *ee*. The catalyst could be reused several times and although *ee* was maintained for several cycles, the catalytic activity decreased due to the loss of Ru-PPh_3/γ-Al_2O_3 catalyst during the separation process and the partial oxidation of the PPh_3 stabilizer.

3.3 *Functionalization of Previously Synthesized Heterogeneous Catalysts Using P-Donor Ligands*

In 2000, Anderson and co-workers reported the use of $Rh/SiO_2 \cdot Al_2O_3$ in the presence of the chiral diphosphines [i.e., (−)-Chiraphos and (−)-DIOP] in the hydroformylation of styrene [124]. The catalytic activity was significantly decreased after functionalization with the diphosphines although high chemo- and regioselectivity toward the chiral aldehyde were obtained. Optical yields were moderate (up to 9% *ee*) and dropped to zero when the catalysts were reused. The interactions of these diphosphines with the supported rhodium catalysts were studied after exposing the catalyst to a solution of the P-additive in toluene. Mechanistic investigations revealed the formation of a rhodium gem-dicarbonyl complex, Rh(I)(CO)(P–P) attached to the solid. The modifiers were adsorbed on the support and to a lesser extent on the metal surface. Controlling the water content of the solvent and the degree of hydration of the catalyst surface reduced the Rh leaching.

In 2005, Ding and co-workers prepared a novel heterogeneous Rh/SiO_2 catalyst by impregnation of $RhCl_3$ over commercial SiO_2 through the incipient wetness method followed by calcination. The resulting material was then modified with PPh_3 [125]. This catalyst employed in the hydroformylation of propylene provided higher catalytic performance than the unmodified heterogeneous Rh/SiO_2 catalyst. Spectroscopic techniques confirmed that PPh_3 molecules were chemically adsorbed at the surface of the modified heterogeneous Rh/SiO_2. The same authors later reported a novel catalyst based on Rh NPs supported over SiO_2 [126] modified with triphenylphosphite ($P(OPh)_3$). This system was employed in the hydroformylation of internal olefins to linear aldehydes and reached high activity and regioselectivity.

In 2009, Yamashita and co-workers synthesized bimetallic FePd NPs by thermal decomposition of iron carbonyl ($Fe(CO)_5$), followed by reduction of palladium acetylacetonate ($Pd(acac)_2$) and using oleic acid and oleylamine as stabilizers [127]. These MNPs contained a Fe_xO_y-rich core and a Pd-rich shell which were functionalized by ligand exchange with the chiral ligands, (*S*)-BINAP and (*R*)-BINAP. These MNPs exhibited sizes of ca. 5.6 nm. These materials were the first example of asymmetric catalytic application using chiral magnetic FePd NPs. They were used in asymmetric Suzuki–Miyaura coupling reaction giving >99% yield in both cases and *ee* values of 48% when (*S*)-BINAP was used and 46% in the case of (*R*)-BINAP.

Felpin and Landais reported the first Pd/C-mediated allylic substitution in water using PPh_3 as additive under catalytic conditions [128]. It was concluded that the phosphine probably acts as a stabilizing agent for Pd, generating in situ a more reactive Pd(0)-PPh_3 complex. Finally, an asymmetric version of this new catalytic system was performed using chiral diphosphine ligands. (*R*)-BINAP afforded diethyl 2-[(*E*)-1,3-diphenyl-2-propenyl]malonate in 80% *ee*, albeit in low yield.

In this field, Baiker and co-workers reported the use of commercial Pd/Al_2O_3 (ca. 3 nm) modified by chiral ligands (Fig. 20 (a)) for their application in asymmetric allylic substitution of (*E*)-1,3-diphenylallyl acetate with dimethyl malonate [129]. The enantioselectivity of the Pd/Al_2O_3-BINAP system [58–60% *ee* to (*S*)-2] was independent of the reaction temperature (60 and 120°C), and very low amounts of BINAP were required. Surprisingly, the addition of BINAP induced significant rate acceleration (by a factor of almost 7 at 60°C) and improved the chemoselectivity of Pd/Al_2O_3. Among the diphosphine ligands tested, (*R*)-Solphos provided the highest enantioselectivity (67% *ee*).

In 2008, the same group reported the use of commercial Pd/Al_2O_3 in the presence of optically active ferrocenyl phosphines (Fig. 20 (b)) as chiral modifiers for the enantioselective allylic substitution of (*E*)-1,3-diphenylallyl acetate with dimethyl malonate [130]. These catalytic systems exhibit good activity along with good enantioselectivity (up to 88% *ee*). Addition of the modifiers to Pd led to significant rate acceleration and, in some cases, kinetic resolution of the racemic allylic acetate substrate.

More recently, the same group reported the functionalization of commercial Pd/Al_2O_3 by (*R*)-BINAP for the enantioselective allylic alkylation of (*E*)-1,3-diphenylallyl acetate with enantio-induction in the range 20–30% *ee* (Fig. 20) [131].

Using a different approach, Park and Chung prepared colloidal CoRh heterobimetallic nanoparticles (ca. 2 nm) by thermal decomposition of cobalt-rhodium carbonyl clusters [i.e., $Co_2Rh_2(CO)_{12}$ and $Co_3Rh(CO)_{12}$] in the presence of oleic acid and trioctylphosphine oxide, prior to their immobilization on charcoal [132]. These catalysts were applied in the Pauson–Khand-type reaction under 1 atm of CO. The catalytic activity depended on the Co:Rh ratio of the bimetallic nanoparticles and the highest activity for intra- and intermolecular Pauson–Khand-type reactions was observed when the ratio was 1. When the catalyst was used in the

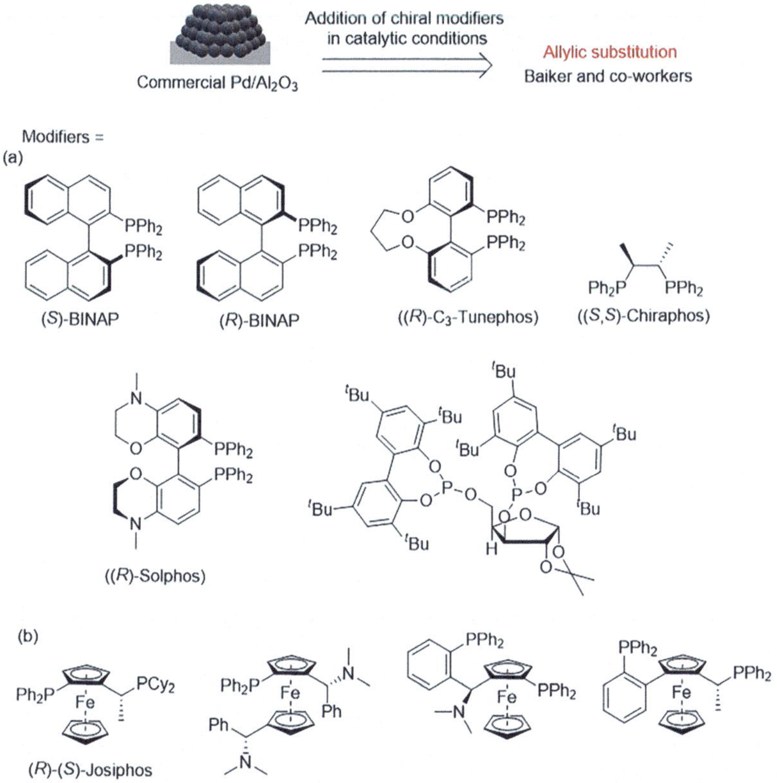

Fig. 20 Modification of commercial Pd/Al_2O_3 for application in asymmetric allylic substitution of (*E*)-1,3-diphenylallyl acetate with dimethyl malonate

presence of an aldehyde instead of carbon monoxide, high efficiency was reported. In addition, when the reaction was carried out in the presence of chiral diphosphines, up to 87% *ee* was achieved. The catalytic system could be reused at least five times in the presence of chiral diphosphines (added in catalytic conditions as modifiers) without loss of catalytic activity and enantioselectivity.

In 2007, Zhao and co-workers reported the asymmetric hydrogenation of acetophenone catalyzed by Ru/γ-Al_2O_3 in the presence of PPh_3 and the chiral amine (*R*,*R*)-DPEN, obtaining up to 60% *ee* [133]. The results demonstrated that the phosphine ligand and amine were crucial for this reaction. The catalyst could be reused four times without loss of activity nor enantioselectivity although the addition of the modifier and the ligand together with the substrate was required for the recycling.

Recently, Cui and co-workers employed the combination of the non-noble metal heterogeneous catalyst Ni/Al_2O_3 (ca. 5 nm) with the Xantphos ligand as catalytic system in the regioselective hydrosilylation of a wide range of aromatic and aliphatic terminal alkynes with $PhSiH_3$, affording the corresponding (*E*)-vinylsilanes in good

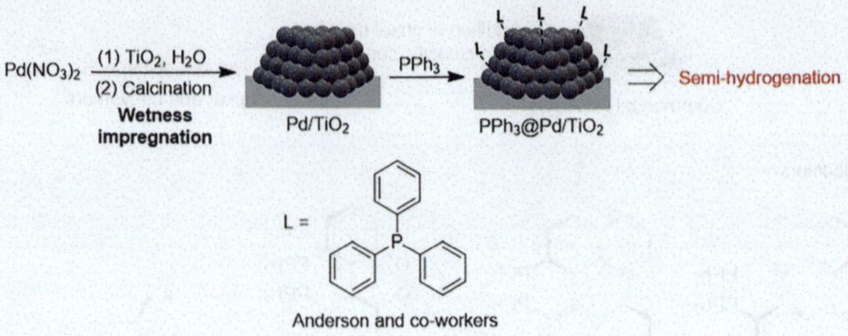

Fig. 21 Preparation and modification of a Pd/TiO_2 catalyst for application in the semi-hydrogenation of acetylene

to excellent yields with high regioselectivity [134]. Several ligands were employed but the best results were obtained with Xantphos, which was attributed to the efficient modulation of the steric and electronic environment of the supported Ni/Al_2O_3 catalyst by this ligand.

Anderson and co-workers previously reported the preparation of Pd/TiO_2 catalyst by impregnation and subsequently modified via the adsorption of triphenylphosphine for acetylene semi-hydrogenation (Fig. 21) [135, 136]. Spectroscopic and theoretical calculations provided evidence for charge transfer from PPh_3 to the metal centers [137, 138], and it has been postulated that the impact of this ligand on a metal surface is to polarize the metal–hydrogen bond and subsequently enhance selectivity in the semi-reduction of alkynes [139].

Very recently, Ballesteros-Soberanas and Leyva-Pérez reported the use of Pd/C catalyst modified by electron-poor phosphines in the semi-hydrogenation of alkynes [140]. The presence of the ligands increased the catalyst activity by ca. 7-folds while the selectivities were comparable to that of the industrial benchmark Lindlar catalyst. In terms of recyclability, when PPh_3 was employed as modifier, the catalyst could be reused up to four cycles without loss of activity or selectivity.

Li and co-workers prepared Rh NPs supported over silica modified with chiral phosphorus ligands (Fig. 22) by impregnation methods [141]. Different particle sizes were obtained depending on the metal loading (2.0, 3.0–4.0, and 5.0–6.0 nm for loadings of 1.0, 3.0, and 5.0 wt%, respectively). The coordination of the chiral phosphine to the Rh was demonstrated by ^{31}P MAS NMR and IR spectra of adsorbed CO. These materials were tested as catalysts in hydroformylation reactions. 100% selectivity of branched aldehyde and up to 72% *ee* were obtained in the hydroformylation of vinyl acetate using the (*R*)-BINAP–Rh/SiO_2 catalyst. It is noteworthy that the modification of Rh/SiO_2 with (*S*,*S*)-DIOP resulted in an increased activity for the hydroformylation of vinyl acetate with a TOF of 128 h^{-1}.

In 2014, Niemczyk, Polanski, and co-workers reported the performance of a bimetallic copper-supported nanopalladium catalyst Pd/PdO/Cu in the Sonogashira coupling reaction using PPh_3 as additive [142]. The presence of the phosphine additive improved the reactivity of the catalyst.

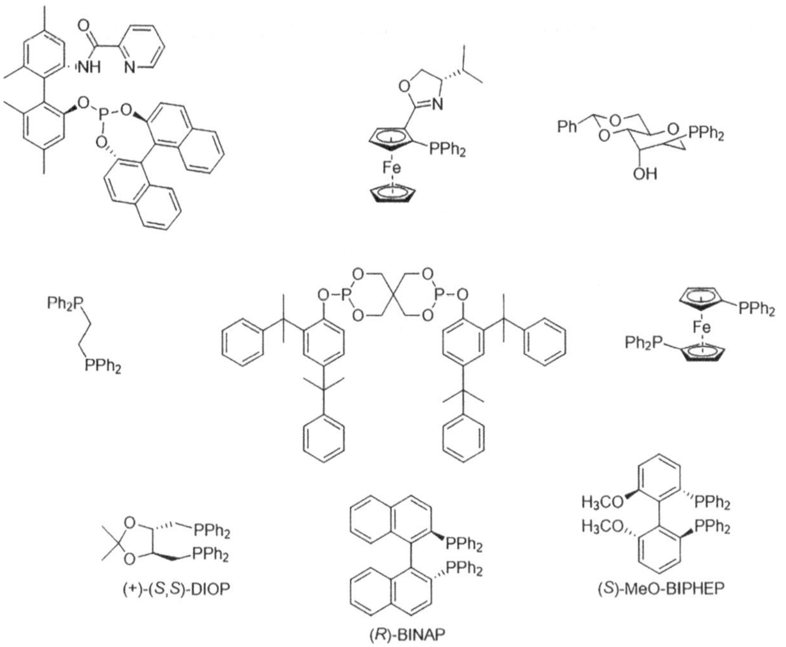

Fig. 22 Chiral phosphorus ligands used by Li and co-workers for the modification of Rh NPs supported over silica

Mika and co-workers reported the use of 5% Ru/C, 10% Pd/C, and RANEY Ni CatCart® in the presence of butyl-bis(*m*-sulfonatophenyl)phosphine for the continuous synthesis of γ-valerolactone (GVL) using a continuous-flow reactor [143]. The conversions exceeded 82.9% with 5% Ru/C catalyst, which was further increased to 98.5% by the addition of ligand.

Chatterjee and Jensen reported the use of the heterogeneous commercial Pd/C (ca. 20 nm) catalyst in the presence of phosphine ligands for α-olefin formation in deoxygenation of fatty acids [144]. Several phosphines were employed, but best results were obtained using DPEphos as modifiers. The catalyst was recycled at least twice without loss of activity or selectivity. However, after the third cycle, a decrease in activity was observed. This could be attributed to an increase of NP size (ca. 70 nm after the third cycle) due to sintering.

In 2021, Hou and co-workers synthesized Pd NPs over mesocellular silica foams by reduction with $NaBH_4$ and subsequent modification with a chiral phosphine [145]. Sizes (from 3.9 to 6.3 nm) could be regulated by increasing $NaBH_4$ reduction times and enlarging support pores. Resulting catalysts were tested in the catalytic intramolecular carboiodination. This was the first report on the asymmetric synthesis of iodomethyl dihydrobenzofuran over supported MNPs. The presence of capping agents positively affected the catalysis outcome by modulation of the size and electronic properties of catalyst surface.

3.4 *Functionalization of the Support/Solvent Using P-Donor Ligands*

In 2018, Guo et al. reported the preparation of Pd NPs over a support modified (FDU-12) with PPh_3 by impregnation-reduction method [146]. PPh_3 was cross-linked in the nanopores of a mesoporous silica. Very small Pd NPs were obtained (1.1 nm). Interaction between the phosphine and Pd NPs was demonstrated by XPS, in situ FTIR adsorption of CO, and D/H exchange reactions. The catalytic performance of this system in hydrogenation reactions was compared with an analogous unmodified material. Interestingly, the activity of Pd NPs was enhanced by the presence of PPh_3 in the hydrogenation of electrophilic nitro-compounds.

In 2010, Wang and co-workers synthesized very small Rh NPs (1.8 nm) stabilized by PEG-substituted triphenyl-phopshine [147]. The enhanced stabilization of these MNPs was due to the presence of the PPh_3 groups, and it was evidenced by comparison with MNPs stabilized by PEG. This system was tested in the biphasic hydrogenation of benzene with high activity (TOF = 3,333 h^{-1}) and stability. In contrast, the catalyst only stabilized by PEG showed intense aggregation and resulted in very low catalytic activity.

In 2017, Yang and co-workers synthesized a series of Pt NPs over silica modified with triphenylphosphine through an impregnation-reduction method [148]. Very small MNPs were obtained (<1 nm) and in situ CO adsorption FTIR spectroscopy demonstrated the existence of strong interactions between PPh_3 and Pt NPs. The catalyst in which silica was modified with PPh_3 showed the highest chemoselectivity in the hydrogenation of various substrates compared with other catalysts in which silica was not modified, demonstrating the benefit in terms of stabilization and catalytic performance.

More recently, Yu, Yang, and co-workers reported the synthesis of Ru NPs incorporated into silica nanoreactors modified with phosphines and amines [149]. The MNPs were synthesized by chemical reduction with $NaBH_4$ and the particle sizes were ca. 2.5 nm.

These systems were tested in the hydrogenation of benzoic acid. In the absence of phosphine, the adsorption of benzoic acid was disfavored and DFT calculations showed that phosphine ligands could modulate the adsorption energy of the benzoic acid over the MNPs tuning their surface.

In 2020, Lei and co-workers reported the synthesis of Pd NPs stabilized with phosphine-functionalized porous inorganic polymer through a two-step approach of anion exchange followed by chemical reduction with $NaBH_4$ [150]. These MNPs were tested in the hydrogenation of nitroarenes with high yield (99.7%) and could be at least 5 times recycled with no significant loss of activity.

In 2021, Philippot and Poli synthesized Rh NPs inside triphenylphosphine-functionalized core-crosslinked micelle latexes [151]. The synthesis was performed by reduction of the Rh precursor with H_2. Rh NPs stabilized with PPh_3 and PEOMA were also synthesized. These systems were tested in several hydrogenation reactions under biphasic conditions. Using this system, the efficient hydrogenation of the vinyl group of styrene was observed. For acetophenone, reduction of both benzene ring and carbonyl reduction was detected. The use of neat 1-octene was also successful, although an increase in activity was obtained when the substrate was diluted in 1-nonanol. For 1-octene hydrogenation, the molecular Rh catalyst was more active than the Rh NPs while when styrene was the substrate, the opposite trend was observed.

In 2022, Doherty, Knight, Chamberlain, and co-workers synthesized Ru NPs through impregnation and reduction of the metal precursor over a phosphine and phosphine oxide-decorated styrene-based polymer-immobilized ionic liquid [152]. This was the first report of RuNPs stabilized by a phosphine oxide-modified support. Very small MNPs were obtained and the catalysts were employed in the hydrogenation of ketones, aldehydes, and other substrates such as levulinic acid and ethyl levulinate, with high yields and selectivities. The catalyst in which phosphine oxide is present could be recovered 10 times without relevant loss in activity (from 85% to 84%) or selectivity.

In 2010, Rossi and co-workers synthesized Pd NPs over a phosphine-functionalized support (Fe_3O_4@SiO_2–iminophosphine–Pd catalyst) [153]. After impregnation of the Pd complex over the previously synthesized support, refluxing conditions led to the formation of Pd NPs [154]. In Suzuki cross-coupling reactions, the catalysts were revealed to be active without the addition of other sources of phosphine ligands. This catalytic system could be easily recovered with the use of an external magnetic field. However, deactivation of the catalyst by etching and MNP aggregation was observed. The addition of the base to the arylboronic acid solution before the introduction of the catalyst solved this issue.

Lu, Chen, and co-workers also reported a system based on supported Ru NPs over phosphorus-rich polymer [155]. This system was tested in the reduction of nitro-compounds providing high yields using $NaBH_4$ as reductant.

4 NHC Ligand-Capped Heterogeneous Catalysts

4.1 NHC-Capped Heterogeneous Catalysts Formed by Colloidal Synthesis Followed by Deposition onto a Solid Support

In 2018, Ravoo and co-workers reported the first preparation of water-soluble NHC-stabilized Au-Pd alloy NPs by ligand exchange (Fig. 23) [156]. First, the synthesis of the AuPd NPs was carried out in the presence of oleylamine (OAm) as ligand [157–159]. In the second step, the replacement of the hydrophobic OAm by NHCs was performed by ligand exchange, which led to an increase in Au content of the bimetallic MNPs, probably due to Au atom migration within the MNPs. Subsequently, the deposition over TiO_2 of the NHC-stabilized AuPd NPs was carried out. The size of the obtained MNPs varied from 4.1 to 4.6 nm. These systems were

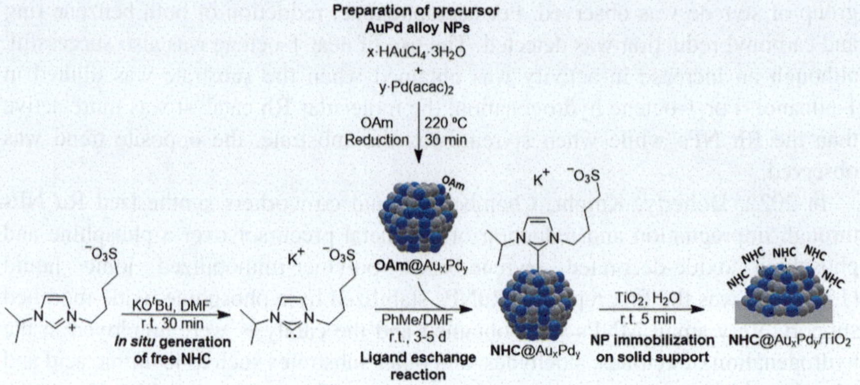

Fig. 23 First preparation of water-soluble NHC-stabilized Au-Pd alloy NPs by ligand exchange

applied in semi-hydrogenation reactions with good conversions and selectivities. They were also successfully tested in the hydrogenation of nitroarenes.

Ravoo, Evangelisti, and co-workers also reported a new strategy including the NHC functionalization of ligand-free Pd NPs previously obtained by metal vapor synthesis and the subsequent dispersion of these NHC-Pd NPs onto a carbon support [160]. This NHC-modified system was tested in the hydrogenation of limonene and higher regioselectivity was obtained than when the bare Pd NPs supported over carbon were used.

4.2 *NHC-Capped Heterogeneous Catalysts Formed by Colloidal Synthesis in the Presence of a Solid Support*

In 2017, Godard and co-workers reported the synthesis of Ni NPs stabilized with NHCs that were obtained by a new methodology involving the decarboxylation of zwitterionic CO_2 adducts and using multiwalled carbon nanotubes (MWCNT) as support (Fig. 24) [161]. Immobilization of these Ni NPs in the inner cavity of carbon nanotubes was carried out by hydrogenation of the Ni precursor in the presence of both the NHC precursor and the support. This system was tested in the selective hydrogenation of alkynes into (*Z*)-alkenes, showing excellent activities and selectivities. Moreover, no over-hydrogenation of the Z-alkene products was observed under prolonged reaction times. Later, the same group extended the study to the synthesis of NHC-stabilized CNT-supported monometallic (Cu and Pd) and bimetallic (PdCu and NiCu) MNPs via decarboxylation of imidazolium carboxylate adducts (Fig. 24) [162]. In terms of size (2–7 nm), the trend was that the supported MNPs were slightly smaller than the colloidal ones in all cases, except the NiCu NPs. These systems were tested in the semi-hydrogenation of alkynes and alkynols. PdCu/CNTs were revealed to be highly selective in the hydrogenation of terminal

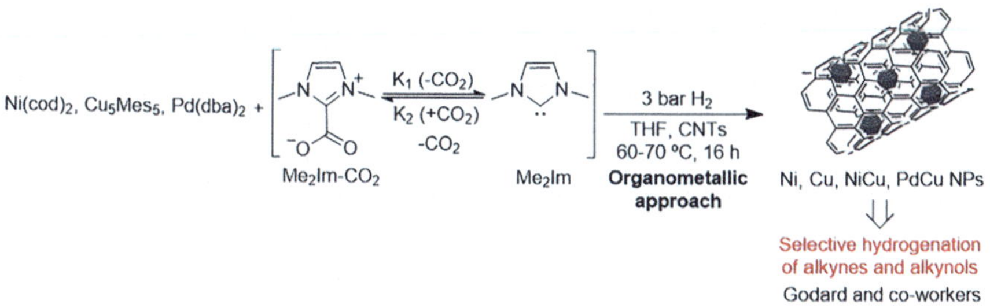

Fig. 24 Preparation of NHC-stabilized mono- and bimetallic MNPs supported on MWCNT by decarboxylation of imidazolium carboxylate

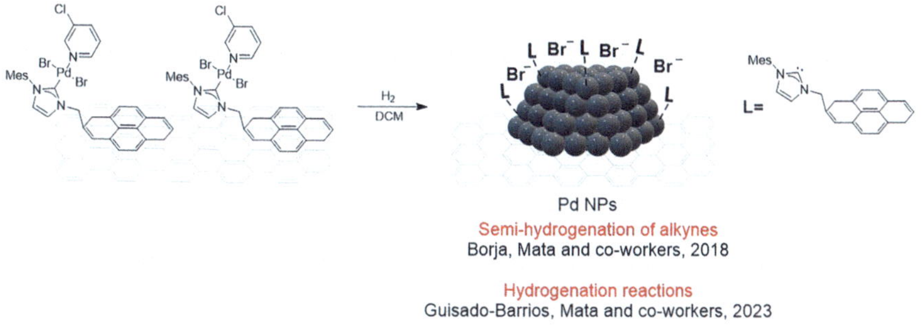

Fig. 25 Synthesis of NHC-capped heterogeneous Pd catalysts by decomposition of anchored palladium NHC complexes at the surface of reduced graphene oxide (rGO)

and internal alkynes and showed high stability in the semi-hydrogenation of acetylene in acetylene/ethylene-rich feeds.

In 2018, Borja, Mata, and co-workers synthesized Pd NPs by decomposition of palladium complexes stabilized by NHC ligands, which were previously anchored onto the surface of reduced graphene oxide (rGO) (Fig. 25) [163]. These systems were successfully employed in the semi-hydrogenation of alkynes and could be recycled up to 10 times with no relevant loss in activity. Very recently, Guisado-Barrios, Mata, and co-workers synthesized a similar catalyst (Fig. 25) [164]. This system was tested in the hydrogenation and dehydrogenation of N-heterocycles without additives. In this case, the ligand did not only act as stabilizer of the Pd NPs, but also played a role in the π-interaction with the support, increasing the catalyst stability. Moreover, the support (rGO) also acted as carbocatalyst in the acceptorless dehydrogenation of N-heterocycles. This catalyst demonstrated high stability and could be recycled eight times without loss of activity.

In 2018, Oberhauser and co-workers prepared Ir NPs stabilized by a bulky NHC [165]. However, when the catalyst was produced in the presence of carbon support (C^K), the NHC stabilizer could not be detected. These systems were tested in the oxidation of glycerol into lactic acid in the presence of NaOH. The colloidal system gave high selectivity to the lactic acid (93%) and good recyclability even under air.

In 2019, Fadakar and co-workers synthesized a new hybrid nanoporous material based on Pd NPs (50–200 nm determined by SEM) stabilized by an NHC ligand and supported on a zeolitic imidazole framework [166]. The procedure for the synthesis of these new materials was based on the impregnation of the NHC-palladium precursor over the support, followed by a reduction using $NaBH_4$. These materials were tested in the Suzuki–Miyaura cross-coupling reaction and provided high activity and stability.

4.3 *Functionalization of Previously Synthesized Heterogeneous Catalysts Using NHC Ligands*

Recently, Liu, He, and co-workers modified previously synthesized supported Au and Pd NPs by addition of polymers containing NHC ligands (mono- and polydentate) [167] and compared their performance in the CO_2 electroreduction with analogous catalysts modified by oleic amine and thiols (Fig. 26). The Pd catalysts containing OA and DT were less selective for CO_2 reduction. The highest activity was obtained for the system based on Au NPs, although, in both cases, the presence of the polymeric NHCs resulted in higher activities compared with the unmodified Au/C and Pd/C catalysts. The authors indicated that the polymeric NHCs formed a layer that does not only stabilize the MNPs against sintering, but also promote CO_2 reduction by limiting proton diffusion and binding to MNPs. As a result, the selectivity for CO_2 reduction was relevantly improved due to the hydrophobicity of the polymer ligands and the higher surface electron density of MNPs through σ-donation from NHCs.

In 2020, Pieters and co-workers reported new air-stable catalysts based on the commercial Ru/C system modified by NHCs [168]. The coordination of the carbenes was confirmed via the use of ^{13}C-labeled NHC ligands and subsequent analysis by solid-state ^{13}C MAS NMR. TEM analysis demonstrated that the morphology of the Ru NPs was not affected by the NHC modification. These catalysts were employed

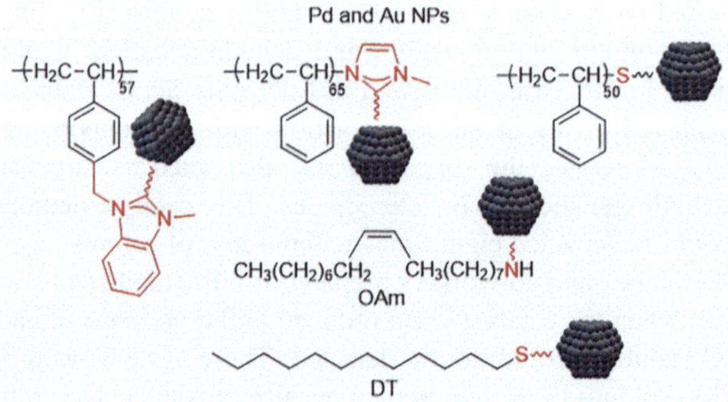

Fig. 26 Au and Pd catalysts modified by addition of polymers containing NHC ligands

in C-H deuteration reactions. Depending on the Ru/NHC ratios, the reactivity of these systems could be tuned. The NHC modification increased the catalyst chemoselectivity and regioselectivity in the deuteration of alcohols and allowed the synthesis of pharmaceutically relevant deuterated heterocycles (i.e., azine, purine) that are completely reduced using unmodified commercial catalysts. Moreover, a novel reactivity was revealed for such heterogeneous Ru catalysts, namely the selective C-1 deuteration of aldehydes.

In 2022, Martínez-Prieto and co-workers reported the synthesis of graphene-supported Ru NPs (ca. 1.5 NPs) that were tuned by the modification of their metal surface with pyrene-tagged NHCs ligands [169]. XPS analysis showed that NHC ligands were present as protonated carbenes, coordinated to the Ru surface and interacting with the graphene support. Moreover, colloidal Ru NPs were also synthesized to study the coordination of the ligand [170]. These systems were tested in the hydrogenation of various molecules of interest. At high surface coverage, the activity of the catalyst was reduced, but the selectivity was enhanced.

In 2016, Muratsugu, Glorius, and co-workers reported for the first time the extensive characterization of an NHC-modified supported heterogeneous catalyst (Fig. 27 (a)) [171]. The modification was carried out over the previously synthesized Ru/K-Al_2O_3 catalyst [172]. The modification with NHCs did not induce structural changes in the catalyst, even upon variations of the Ru/NHC ratios. The bond between the carbene ligands and Ru was probed by solid-state ^{13}C MAS NMR. The resulting catalysts were tested in hydrogenation of acetophenone, *trans*-stilbene, and phenylacetylene. In the hydrogenation of *trans*-stilbene, a decrease in the conversion was observed at high NHC to surface Ru ratios, although the extent of this variation was dependent on the nature of the NHC ligand. In contrast, in the hydrogenation of acetophenone, a decrease in activity was already observed at lower NHC to surface Ru ratios. The reduced activity observed upon NHC addition was explained by the blockage of active sites of the catalyst by the NHC.

In 2010, Glorius and co-workers synthesized Pd NPs by chemical reduction of a palladium salt with $NaBH_4$ over magnetite (Fe_3O_4) [173]. They subsequently carried out the modification of this material using enantiomerically pure NHCs (Fig. 27 (b)) [157]. These functionalized MNPs were tested in asymmetric α-arylation reactions reaching *ee's* up to 85%. Moreover, these catalysts could be separated by magnetic removal from the reaction media and no loss of activity nor selectivity was observed after several cycles.

Very recently, the same group reported the modification of a ceria-based catalyst incorporating Cr and Rh with NHCs (Fig. 27 (c)) [174]. The synthesis of the $Cr_{0.19}Rh_{0.06}CeO_z$ was performed by hydrothermal method [175, 176] and followed by the modification with NHCs onto H_2-reduced $Cr_{0.19}Rh_{0.06}CeO_z$. These systems were tested in the 1,4-arylation of cyclohexanone with phenylboronic acid. When the NHC employed was 1,3-dicyclohexylimidazol-2-ylidene (ICy), high activity was obtained. However, in the absence of NHC ligand, the catalyst was inactive. DFT calculations were performed and suggested that the ICy controlled the adsorption sites of the phenyl group on the Rh nanocluster to promote the C–C bond formation of the phenyl group and cyclohexanone.

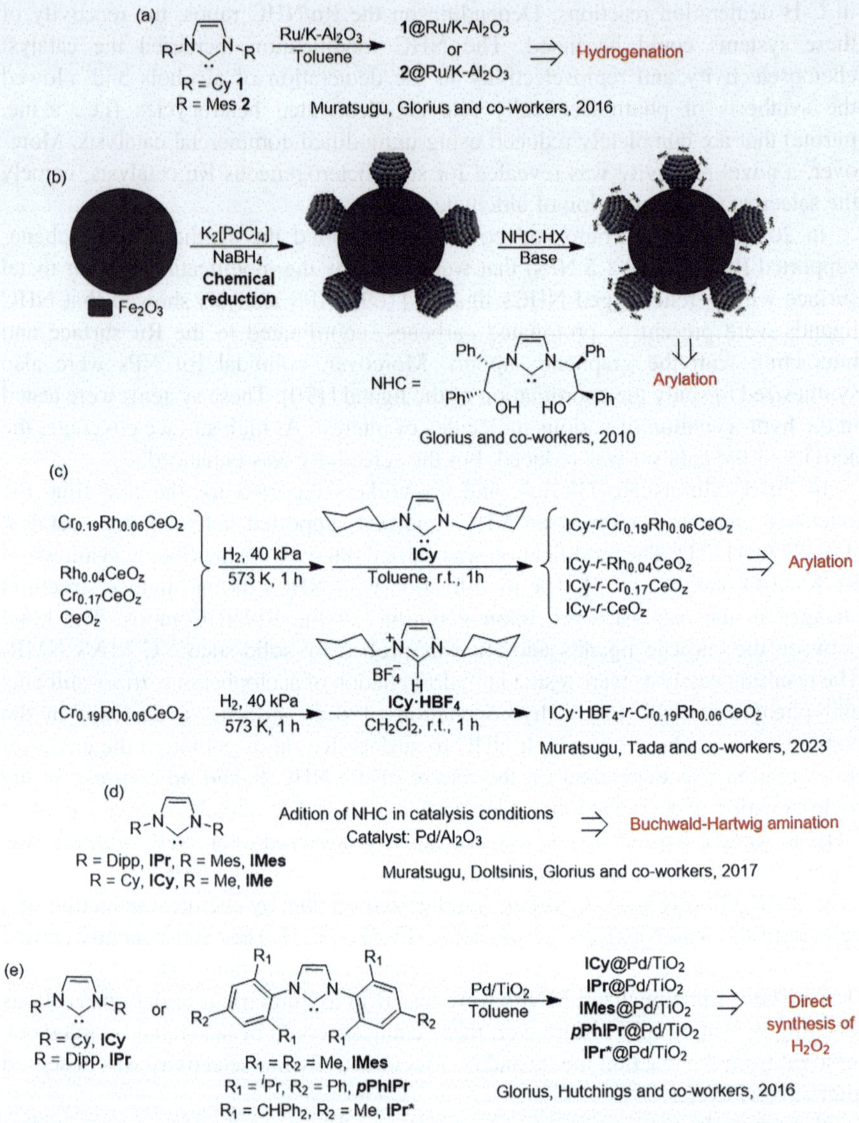

Fig. 27 Synthetic approaches used by Glorius and co-workers for the preparation of NHC-capped catalysts

Glorius and co-workers also reported the modification of Pd/Al_2O_3 NPs with NHCs [177]. The NHC was directly added in the catalytic Buchwald–Hartwig amination of aryl halides (Fig. 27 (d)). Additionally, further insights into the electronic effect of the coordination with the NHC were obtained by precise characterization of the catalyst surface and ab initio calculations.

In 2022, Glorius, Hutchings, and co-workers reported a heterogeneous Pd/TiO_2 catalyst modified by NHCs (2.4 nm) (Fig. 27 (e)) [178]. The Pd/TiO_2 (2.0 nm) system was first prepared by wet-impregnation method prior to the NHC modification [179]. The systems were tested in the direct synthesis of hydrogen peroxide (H_2O_2) in the absence of promoters. Depending on the NHC substituent, activity could be improved compared with the unmodified catalyst due to the ability of the NHC to electronically modify Pd NPs. Moreover, the modification had a positive effect on the recyclability of the catalysts.

Levratovsky and Gross reported the functionalization of previously synthesized Pt NPs (100 ± 30 nm) supported over SiO_2 with NHCs [180]. Pt NPs were synthesized by the deposition of a Pt film over a Si (110) surface which was coated with a layer of SiO_2. Subsequent modification with NHCs was carried out. The formation of a carbene self-assembled layer onto the surface of the Pt nanoparticles was confirmed and its stability under oxidizing conditions studied by infrared measurements.

In 2018, Wen, Chang, and co-workers reported the modification of Pd electrodes with NHCs for the electrochemical reduction of CO_2 [181]. When the tris-N-heterocyclic carbene ligand-modified Pd electrode was employed, a 32-fold increase in activity was measured in the electrochemical reduction of CO_2 to C1 products, observing a high Faradaic efficiency compared to the unmodified sample.

4.4 *Functionalization of the Support by NHC Ligands Prior to MNP Formation*

In 2009, Jia and co-workers immobilized Pd NPs onto NHC-modified mesoporous SBA-15 by a post-grafting route [182]. First, a Pd bromide complex coordinated to the supported NHC ligand was formed, which, upon reduction under H_2 at 200°C, yielded Pd NPs with a diameter of ca. 3.5 nm bearing NHC ligands. Both the initial complex and the resulting MNPs were tested in the aerobic oxidation of benzyl alcohol. The MNP-based catalyst showed much higher activity and could be recycled up to 8 times despite a slight loss in activity.

In 2011, Glorius and co-workers reported the immobilization of an enantiomerically pure imidazolium salt on the surface of magnetite (Fe_3O_4) and the formation of three catalysts of different nature: an organocatalyst, a molecular Pd complex, and Pd NPs [183]. The Pd NPs were obtained by chemical reduction in water using hydrazine as reductant. All of them were tested in the allylation of 4-nitrobenzaldehyde, resulting in enantioselectivities between 43 and 74% *ee*'s.

In 2016, Ervithayasuporn and co-workers described a new synthetic methodology for the fabrication of polyhedral oligomeric silsesquioxane-supported NHC/imidazolium salt palladium(II) complexes [184]. Upon addition of phenylboronic acid as reductant, the resulting Pd NPs exhibited a diameter of 3.0 ± 1.5 nm. This material was employed in the Suzuki–Miyaura cross-coupling reactions with high efficiency and could be reused up to 10 times without losing activity.

In 2013, Kobayashi and co-workers synthesized a series of Ni NPs over polymer supports which incorporated NHC ligands [185]. The synthesis of the Ni NPs was carried out by the reduction with triethylborohydride ($LiEt_3BH$) and their sizes were between 1 and 4 nm. The presence of the embedded NHCs was checked by NMR analysis and these systems were tested in Corriu–Kumada–Tamao reactions with good results for a wide range of substrates. Moreover, the catalyst could be recycled several times without loss of activity.

5 S-Donor Ligand-Capped Heterogeneous Catalysts

In view of the ability of S-donor molecules to stabilize group 11 metals and their strong coordination on metal surfaces, S-based ligands were also reported to enhance the catalytic performance of heterogeneous catalysts.

5.1 S-Ligand-Capped Heterogeneous Catalysts Formed by Colloidal Synthesis Followed by Deposition onto a Solid Support

The first example using this synthetic methodology was reported by Obare and co-workers who described the synthesis of colloidal Pd NPs by thermal decomposition of $Pd_3(OAc)_6$ in the presence of thioethers as stabilizing ligands (sizes: 1.7–3.5 nm) [186]. The Pd NPs were subsequently immobilized over commercial SiO_2 and the resulting catalysts evaluated in the hydrogenation of styrene, cyclooctadiene, and 6-bromo-1-hexene (Fig. 28). The results revealed that the MNP size was a key parameter for catalytic performance as the smallest nanoparticles provided the highest activity. Moreover, these heterogeneous systems were robust since they could be recycled up to 8 times without loss of activity or selectivity.

In 2012, Khlobystov reported the use of Rh and RhPt NPs ($\approx$3 nm) supported over hollow graphitized carbon nanofibers (GNF) and stabilized with sodium S-dodecylthiosulfate (SSDS) in the hydrosilylation of terminal alkynes [187]. Colloidal MNPs were first prepared prior to their deposition over the support (Fig. 29) [17]. Variations in the conditions for the deposition of the MNPs yielded two distinct catalysts containing the active phase either at the surface of the nanofibers or confined within the walls of the support. The regioselectivity was affected by the confinement of the MNPs, and depending on the nature of the substituents at the substrates, three main trends were observed while using the confined catalysts: (1) when the reagents did not contain aromatic moieties, no confinement effect

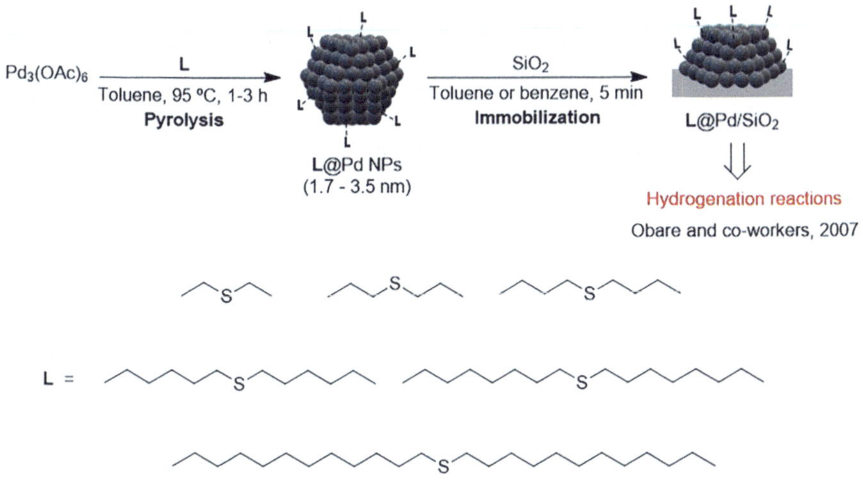

Fig. 28 Colloidal formation of S-donor ligand-capped Pd hydrogenation catalysts followed by deposition over silica

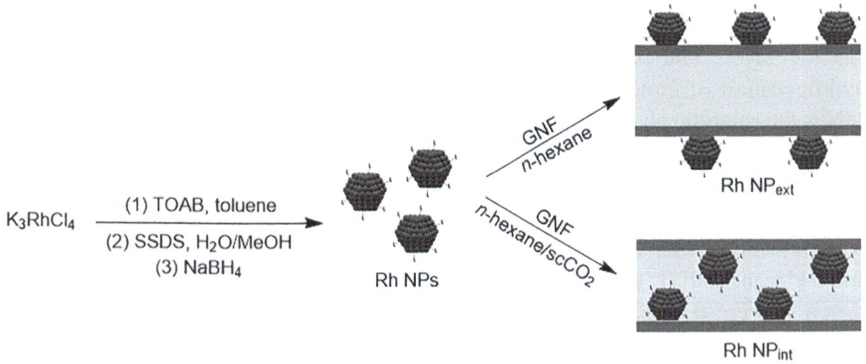

Fig. 29 Preparation of Rh alkyne hydrosilylalation catalysts supported either at the surface or within the cavity of hollow carbon nanofibers

was observed; (2) when only the alkyne contained an aromatic substituent, a higher selectivity to dehydrogenative silylation products was observed together with a decrease in the β-(*Z*): β-(*E*) isomers ratio; and (3) when both reagents possess aromatic groups, the resulting interactions favored the formation of the thermodynamically less stable β-(*Z*) regioisomer.

In 2013, Lopez-Sanchez and co-workers reported the synthesis of colloidal Rh NPs (3–5 nm) stabilized by 1-octadecanethiol (ODT), polyvinyl alcohol (PVA), and tetraoctylammonium bromide (TOAB), and their subsequent immobilization for application in hydrogenation reactions [188]. These nanoparticles were immobilized over silica-coated magnetic MNPs modified with 3-aminopropyltriethoxysilane, 3-mercaptopropyltriethoxysilane, and thiol [8]. Both colloidal and supported

MNPs were tested in the hydrogenation of cyclohexene. Distinct results were obtained according to the stabilizing agents since Rh–ODT NPs were inactive, Rh–PVA NPs exhibited distinct activities in solution (aqueous biphasic catalysis) and as a supported catalyst, and colloidal Rh–TOAB NPs exhibited similar activities with their supported counterpart. Additionally, the stability of the nanoparticles depended on the nature of the ligand modifier and on the functionalization of the support surface before immobilization. The highest turnover frequencies (up to 700,000 h^{-1}) were observed for the colloidal catalyst Rh–TOAB that was stable and recyclable.

In 2014, Kunz and co-workers studied the influence of organic ligands on the geometric and electronic surface properties of colloidally prepared Pt nanoparticles (Pt NPs) (1.8 ± 0.3 nm) stabilized with dodecylamine (DDA) and dodecylthiol (DDT) [189]. These MNPs were subsequently supported over TiO_2 and a surface study was carried out by means of diffuse reflectance infrared spectroscopy (DRIFTS) using CO as probe molecule. Marked differences were evidenced according to the stabilizing agent used. The amine capping agent provided an electronic donor influence and a geometric effect, while the thiol was mainly geometric in nature. Moreover, DDA was only weakly adsorbed and could therefore be displaced by strongly binding adsorbates (e.g., CO), while a strong coordination of DDT at the particle surface was observed, forming a uniformly distributed capping layer. The influence of both ligands on the catalyst selectivity in the hydrogenation of crotonaldehyde was also investigated. For amine-functionalized MNPs the catalytic results were similar to those of "unprotected" Pt NPs. In contrast, an enhancement of the selectivity was evidenced for thiol-functionalized particles.

More recently, Chen, Lian, and co-workers modified Pt NPs prepared in basic solutions of glycol using a series of stabilizers such as triphenylphosphine (PPh_3), octadecylamine (ODA), poly(vinylpyrrolidone) (PVP), poly(vinyl alcohol) (PVA), and dodecanethiol (DT), prior to their deposition over activated carbon [190]. These materials were tested in the hydrogenation of *para*-chloronitrobenzene (*p*-CNB). Direct or indirect interactions between nanocrystals and stabilizers affected their catalytic performance. The electron-donating ability of the capping agents directly altered the surface electronic state of Pt nanocrystals and consequently affected the adsorption and reactivity of reactants, intermediates, and products. Interestingly, using DT as stabilizer, the coordinated Pt atoms were oxidized to generate cationic Pt species at the surface of the nanocrystals, which improved both the hydrogenation rate and selectivity to *para*-chloroaniline.

5.2 Functionalization of Previously Synthesized Heterogeneous Catalysts Using S-Donor Ligands

The strong coordination of S-donor ligands was mainly utilized to tune the selectivity of heterogeneous catalysts. However, in some cases, the activity of these catalysts was affected.

For instance, Rusling, Suib, He, and co-workers reported the synthesis of Au, Pt, and Pd catalysts supported on N-doped carbon by chemical reduction of metallic precursors with $NaBH_4$.[191] These catalysts were subsequently modified by 3-mercaptopropionic acid (MPA) or dodecanethiol (DT). However, after comparison of their catalytic performance in electrocatalytic oxidation of methanol with the unmodified catalysts, the authors concluded that the presence of the modifiers had a negative effect in terms of activity.

In contrast, McKenna and Anderson reported the modification of a Pd/TiO_2 catalyst with diphenyl sulfide for testing in acetylene hydrogenation (Fig. 30) [192]. The Pd/TiO_2 catalyst was prepared by incipient wet impregnation, and after calcination of the material, the modification was carried out by addition of a hexane solution of the diphenyl sulfide. The modification of the catalyst resulted in an enhancement of the selectivity to ethylene by inhibition of the over-hydrogenation process. Interestingly, even after thermal treatments which led to decomposition of the adsorbed phenyl sulfide, the remaining sulfur could enhance the selectivity in mixed acetylene/ethylene feed streams.

In a similar study, Anderson and co-workers reported the use of triphenylphosphine and diphenyl sulfide as modifiers of Pd/TiO_2 catalysts in the selective hydrogenation of acetylene from ethylene-rich feed streams at high pressures and in the presence of CO (Fig. 30) [193]. Pd/TiO_2 was synthesized by incipient wet impregnation prior to deposition of the modifiers. Initial ethylene selectivities of 85–92% were obtained with acetylene conversions up to 100%. Given the differences in the steric and electronic properties of the surface modifiers, the origin of the enhanced selectivity was attributed to the creation of ensembles of Pd atoms able to readily accommodate acetylene but limit the access of ethylene at the metal surface. The use of diphenyl sulfide enhanced ethylene selectivity but also increased the conversion of acetylene. However, the by-product formation was appreciably higher than with the unmodified or triphenylphosphine-modified catalysts.

Sajiki and co-workers also reported the modification of commercial Pd/C by diphenyl sulfide and its application in chemoselective hydrogenation reactions [194]. Using this system, olefin, acetylene, nitro, and azide functionalities were efficiently hydrogenated. In contrast, aromatic carbonyl and halide, benzyl ester,

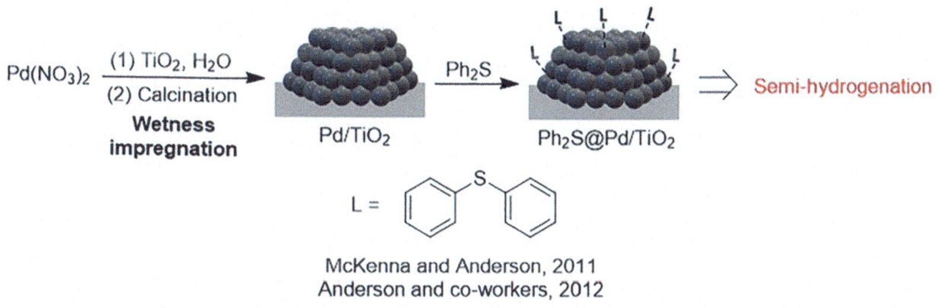

Fig. 30 Preparation of diphenyl sulfide-modified Pd/TiO_2 catalysts for the semi-hydrogenation of acetylene

and N-Cbz functionalities were not. The enhanced chemoselectivity observed was explained by the partial interaction between the active sites of the Pd surface with Ph_2S (i.e., Pd-S bonding). The catalyst could be recovered and reused at least five times without any significant loss of the reactivity.

In 2012, Bokhoven and co-workers reported the modification of Pt/TiO_2 (unmodified: 1 nm and modified: 1.4 nm) with organic thiols for the hydrogenation of 4-nitrostyrene [195]. The Pt/TiO_2 catalyst was synthesized by wet incipient impregnation and functionalized by thioglycerol, 1,6-dithiolhexane, α-lipoic acid, and 1-dodecanethiol. The use of a modifier containing a polar group adjacent to the chain of the organic thiol provided higher selectivities and the best results in terms of activity and selectivity were obtained using 1,6-dithiolhexane as modifier.

The group Medlin and co-workers reported the modification of several heterogeneous catalysts with S-donor ligands for various catalytic applications (Fig. 31). In 2012, they described the coating of Pd/Al_2O_3 and Pt/Al_2O_3 catalysts with

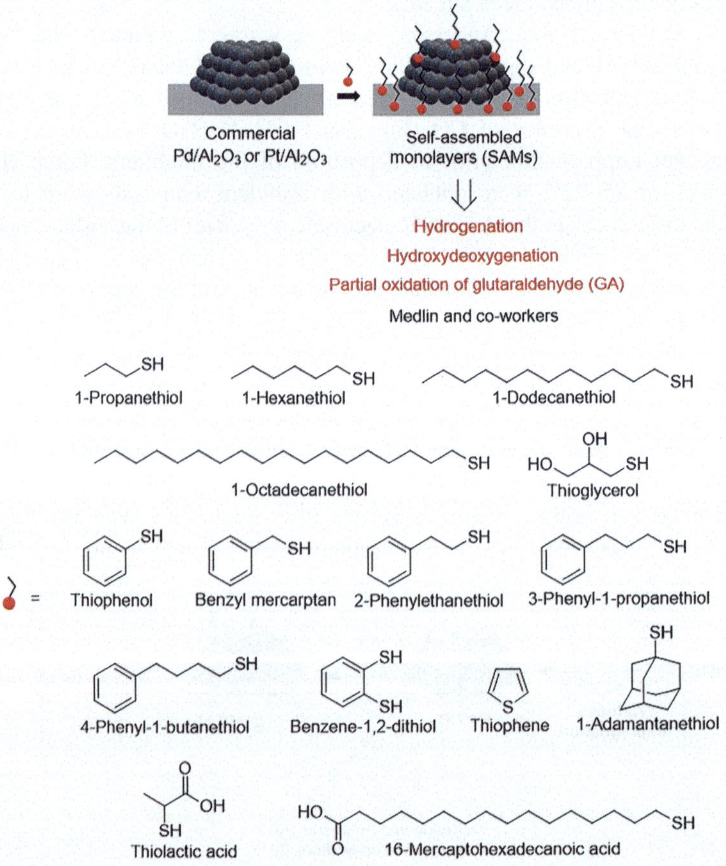

Fig. 31 Modification of heterogeneous catalysts with S-donor ligands reported by Medlin and co-workers

thioglycerol, octadecanethiol, dodecanethiol, hexanethiol, and propanethiol for the hydrogenation of epoxybutene [196]. Depending on the nature of the thiol used, positive or negative effects on activity and selectivity were observed. Coatings with 1-octadecanethiol resulted in marked improvements in selectivity in the liquid-phase hydrogenation of epoxybutene. Interactions of the solvent with the self-assembled monolayer (SAM) were also important in controlling selectivity.

Later, the same authors used thiolate (1-octadecanethiol and 1-adamantanethiol) self-assembled monolayers (SAMs) as modifiers on commercial Pd/Al_2O_3 (6.7 nm) catalysts (Fig. 31) for their use in hydrogenation reactions [197]. In acetylene hydrogenation, the modification by thiols did not provide improvements. In contrast, for reactions where coking is less problematic, the reactivity of the catalysts could be tuned using SAMs, and for instance, ethylene hydrogenation was much faster on adamantanethiol-coated surfaces than on C18-coated surfaces, which was attributed to the impediment of active sites using C18-modified surfaces.

The modification of commercial Pd/Al_2O_3 catalyst with alkanethiol SAMs (i.e., 1-propanethiol, 1-hexanethiol, 1-dodecanethiol, 1-octadecanethiol, and thioglycerol) was also applied in the partial hydrogenation of 18-carbon polyunsaturated fatty acids (Fig. 31) [198]. While the use of the bare Pd/Al_2O_3 catalyst led to the fast formation of saturated fatty acid, the addition of alkanethiol SAMs increased the reaction selectivity to the monounsaturated product oleic acid up to 80–90%, even at high conversions. This effect was attributed to steric effects between the SAMs and reactants.

Similarly, Pd/Al_2O_3 catalysts coated with 1-octadecanethiol (C18) and benzene-1,2-dithiol (BDT) SAMs were applied in the hydrogenation of furfural [199]. Addition of these surface modifiers systematically increased the hydrogenation selectivity of furfural to furfuryl alcohol and methylfuran.

In 2014, the same authors reported the modification of commercial Pt/Al_2O_3 (Pt particle size: 3.9 ± 1.1 nm for 5 wt % Pt/Al_2O_3 and 0.8 ± 0.1 nm for 0.5 wt % Pt/Al_2O_3) with thiolate self-assembled monolayers (SAMs) (Fig. 31) [200]. The selective hydrogenation of both prenal and cinnamaldehyde substrates to the corresponding unsaturated alcohols was enhanced using thiol-modified catalysts. However, it is noteworthy that the selective hydrogenation of cinnamaldehyde was further controlled through interactions of the substrate phenyl ring with the aromatic ligands of the SAM layer.

Later, they reported the use of 3-phenyl-1-propanethiol, thiophenol, and 1-octadecanethiol for covering the surface of commercial Pt/Al_2O_3 (Fig. 31) [201]. These systems were tested in the selective hydrogenation of cinnamaldehyde. Once again, the presence of modifiers enhanced the selectivity of the reaction, but upon recycling, redeposition of the thiol modifier was necessary to maintain the efficiency of the process.

In 2016, Lien and Medlin reported the use of a mixture of modifiers (1-adamantanethiol (AT) and 1-octadecanethiol (C18)) (Fig. 31) on commercial Pd/Al_2O_3 for the benzyl alcohol hydrodeoxygenation [202]. The authors varied the relative amounts of modifiers and looked at the effects on catalyst performance. Compared with the performance of the unmodified catalyst, the use of AT as sole

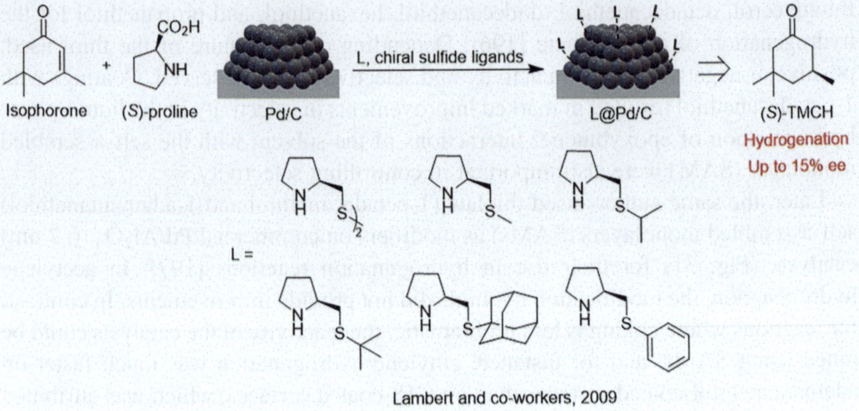

Fig. 32 Chiral sulfide ligands used as modifiers of commercial Pd/C for applications in the enantioselective catalytic hydrogenation of isophorone

modifier improved both the selectivity to toluene and the activity. As the fractional C18 coverage increased, the selectivity to toluene increased and the selectivity to benzene decreased at the cost of activity. The authors showed that at higher C18 coverage, the availability of terrace sites was reduced due to selective poisoning, inhibiting the decarbonylation to benzene.

In a similar study, the same authors deposited thiol self-assembled monolayers (SAMs) onto Pt/Al_2O_3 catalyst for benzyl alcohol hydroxydeoxygenation (HDO) [203]. These SAMs consisted in the use of 16-mercaptohexadecanoic acid (MHDA) and tetraethylammonium (TEA) (Fig. 31). The density of the SAM affected the catalyst activity and the hydrodeoxygenation (HDO) selectivity was improved by suppressing decarbonylation and ring hydrogenation pathways, which was attributed to the ability of the ligands to control the binding orientation of benzyl alcohol.

The same group reported the impact of pretreatment and use of hydrophilic and hydrophobic thiolate SAMs modifiers (1-thioglycerol (TG), thiolactic acid (TLA), adamantanethiol (AT), and 1-octadecanethiol (C18)) (Fig. 31) over Pd/Al_2O_3 for the partial oxidation of glutaraldehyde (GA) to glutaric semialdehyde (GSA) [204]. When using catalysts coated with hydrophilic thiols such as TG and TLA, an enhancement in selectivity was observed, although conversion was affected.

S-donor ligands were also used as chiral modifiers for application in asymmetric catalysis. For instance, Lambert and co-workers reported the use of chiral sulfide ligands (Fig. 32) as modifiers of commercial Pd/C for applications in the enantioselective catalytic hydrogenation of isophorone [205]. Using chiral sulfide ligands that strongly bind to Pd NPs and resist displacement, they reported the first heterogeneous enantioselective catalytic hydrogenation of isophorone with ee's up to 15%. The variation in asymmetric induction across the series of ligands was correlated with the size of the alkyl group in the sulfide.

Another example was reported by Weng and Zaera who modified a previously synthesized Pt/Al_2O_3 catalyst with thiols (Fig. 33). The Pd catalyst was prepared via

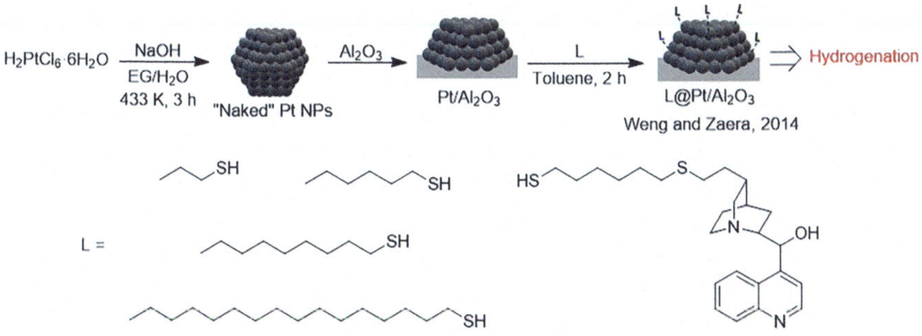

Fig. 33 Preparation of thiol-modified Pt catalysts supported onto Al_2O_3

the formation of colloidal Pt NPs and their subsequent immobilization onto Al_2O_3. The modifiers used were 1-propanethiol, 1-hexaethiol, 1-nonanethiol, 1-hexadecanethiol, and a cinchonidine-derivatized thiol [206]. First, colloidal naked MNPs were synthesized according to the literature [92], and subsequently, they were coated using the modifiers. For the preparation of the supported ones, the previously naked colloidal MNPs were dispersed in ethanol in the presence of the support. They were also covered with the previously mentioned modifiers. The modification of this catalyst by thiols resulted in enhancement of both activity and selectivity in the hydrogenation of ethyl pyruvate. These heterogeneous catalysts can promote the α-keto ester hydrogenation reaction with similar enantioselectivity than the best naked supported Pt catalysts [207, 208] without requiring the addition of cinchonidine to the reaction mixture.

6 Conclusions and Perspectives

This chapter highlights how the functionalization of heterogeneous catalysts by capping ligands constitutes an efficient method for the tuning of the structure and catalytic performance of supported metal nanoparticles.

The capping ligands can be introduced through several synthetic methodologies according to the objectives to be reached. Introduction of these capping agents via colloidal methods, either by formation of colloids followed by deposition onto the support or by direct synthesis of the supported MNPs, can influence the size, shape, and distribution of the MNPs at the support surface. Functionalization of pre-synthesized heterogeneous catalysts usually causes minor structural modifications of the nanoparticles but can enhance the catalytic performance through various effects such as alterations of the electronic properties of the surface atoms, poisoning of active sites, and increase of the steric hindrance at the metal surface.

The design of functionalized heterogeneous catalysts has been largely influenced by the knowledge acquired in homogeneous catalysis where precise mechanistic insights were acquired by spectroscopic investigations as well as theoretical calculations. The effects caused by the capping ligands are mainly influenced by their nature and properties such as electron-donating/withdrawing abilities, steric hindrance, and binding strength.

Consequently, through the selection of the appropriate synthetic methodology and ligand, the activity of the catalysts can be enhanced by either modifying the electronic properties of the active phase or increasing the relative surface area (size dependent) and/or active sites (shape dependent). The selectivity is usually influenced by the steric hindrance of the capping ligands and their ability to strongly bind the metal surface and, as such, limit their mobility over the catalyst. In the case of asymmetric catalysis, this latter parameter is particularly important to condition the approach of the substrate to the active sites of the catalysts and thus provide enantio-induction. To improve the catalyst stability and allow its recycling, the synthetic strategy is key and the addition of extra ligands between each cycle is sometimes necessary to maintain the activity/selectivity of the catalysts.

The large numbers of ligands that were originally designed for homogeneous catalysis still provide a wide range of possibilities for the efficient functionalization of the metal surface of heterogeneous catalysts. More recently, some of these ligands were also adapted to be introduced on the support surface and offer additional possibilities.

The specific design of ligands for heterogeneous catalysts to favor interactions with the metal surface, the support, and/or the substrates to be transformed during catalysis remains at an early stage and will probably be a focus of interest in the future.

References

1. Sau TK, Rogach AL (2010) Nonspherical Noble metal nanoparticles: colloid-chemical synthesis and morphology control. Adv Mater 22:1781–1804
2. You H, Yang S, Ding B, Yang H (2013) Synthesis of colloidal metal and metal alloy nanoparticles for electrochemical energy applications. Chem Soc Rev 42:2880–2904
3. Cheong S, Watt JD, Tilley RD (2010) Shape control of platinum and palladium nanoparticles for catalysis. Nanoscale 2:2045–2053
4. See for instance Delgado JA, Benkirane O, Claver C, Curulla-Ferré D, Godard C (2017) Advances in the preparation of highly selective nanocatalysts for the semi-hydrogenation of alkynes using colloidal approaches. Dalton Trans 46:2381–12403
5. Lu L, Zou S, Fang B (2021) The critical impacts of ligands on heterogeneous nanocatalysis: a review. ACS Catal 11:6020–6058
6. Heuer-Jungemann A, Feliu N, Bakaimi I, Hamaly M, Alkilany A, Chakraborty I, Masod A, Casula MF, Kostopoulou A, Oh E, Susumu K, Stewart MH, Medintz IL, Stratakis E, Parak WJ, Kanaras AG (2019) The role of ligands in the chemical synthesis and applications of inorganic nanoparticles. Chem Rev 119:4819–4880
7. Pujari S, Scheres L, Marcelis AT, Zuihof H (2014) Covalent surface modification of oxide surfaces. Angew Chem Int Ed 53:6322–6356
8. Jacinto MJ, Kiyohara PK, Masunaga SH, Jardim RF, Rossi LM (2008) Recoverable rhodium nanoparticles: synthesis, characterization and catalytic performance in hydrogenation reactions. Appl Catal A 338:52–57

9. Guerrero M, Costa NJS, Vono LLR, Rossi LM, Gusevskaya EV, Philippot K (2013) Taking advantage of a terpyridine ligand for the deposition of Pd nanoparticles onto a magnetic material for selective hydrogenation reactions. J Mater Chem A 1:1441–1449
10. Choi J-Y, Hsu RS, Chen Z (2010) Highly active porous carbon-supported nonprecious metal–n electrocatalyst for oxygen reduction reaction in PEM fuel cells. J Phys Chem C 114:8048–8053
11. Liu Z, Shamsuzzoha M, Ada ET, Reichert WM, Nikles DE (2007) Synthesis and activation of Pt nanoparticles with controlled size for fuel cell electrocatalysts. J Power Sources 164:472–480
12. Peng Z, Yang H (2009) Synthesis and oxygen reduction electrocatalytic property of Pt-on-Pd bimetallic heteronanostructures. J Am Chem Soc 131:7542–7543
13. Kim SW, Park J, Jang Y, Chung Y, Hwang S, Hyeon T, Kim YW (2003) Synthesis of monodisperse palladium nanoparticles. Nano Lett 3:1289–1291
14. Liu QS, Bauer JC, Schaak RE, Lunsford JH (2008) Supported palladium nanoparticles: an efficient catalyst for the direct formation of H_2O_2 from H_2 and O_2. Angew Chem Int Ed 47: 6221–6224
15. Ramírez-Meneses E, Domínguez-Crespo MA, Montiel-Palma V, Chávez-Herrera VH, Gómez E, Hernández-Tapia G (2009) Electrochemical characterization of platinum nanoparticles stabilized by amines. J Alloys Compd 483:573–577
16. Miyabayashi K, Nishihara H, Miyake M (2014) Platinum nanoparticles modified with alkylamine derivatives as an active and stable catalyst for oxygen reduction reaction. Langmuir 30:2936–2942
17. Brust M, Walker M, Bethell D, Schiffrin D, Whyman R (1994) Synthesis of thiol derivatised gold nanoparticles in a two-phase liquid-liquid system. J Chem Soc Chem Commun:801–802
18. Miyabayashi K, Miyake M (2017) Platinum nanoparticles modified with perfluorinated alkylamines as a model cathode catalyst for fuel cells. Electroanalysis 29:898–906
19. Ung D, Cossairt BM (2019) Effect of surface ligands on CoP for the hydrogen evolution reaction. ACS Appl Energy Mater 2:1642–1645
20. Xia BY, Ng WT, Wu HB, Wang X, Lou XW (2012) Self-supported interconnected Pt nanoassemblies as highly stable electrocatalysts for low-temperature fuel cells. Angew Chem 124:7325–7328
21. Xu Z, Shen C, Hou Y, Gao H, Sun S (2009) Oleylamine as both reducing agent and stabilizer in a facile synthesis of magnetite nanoparticles. Chem Mater 21:1778–1780
22. Tzitzios V, Niarchos D, Gjoka M, Boukos N, Petridis D (2005) Synthesis and characterization of 3D CoPt nanostructures. J Am Chem Soc 127:13756–13757
23. Lim B, Xia Y (2011) Metal nanocrystals with highly branched morphologies. Angew Chem Int Ed 50:76–85
24. Park KH, Jang K, Kim HJ, Son SU (2007) Near-monodisperse tetrahedral rhodium nanoparticles on charcoal: the shape-dependent catalytic hydrogenation of arenes. Angew Chem Int Ed 46:1152–1155
25. Borchert H, Fenske D, Kolny-Olesiak J, Parisi J, Al-Shamery K, Bäumer M (2007) Ligand-capped Pt nanocrystals as oxide-supported catalysts: FTIR spectroscopic investigations of the adsorption and oxidation of CO. Angew Chem Int Ed 46:2923–2926
26. Sonström P, Halbach B, Djakpou ST, Ritz B, Ahrenstorf K, Grathwohl G, Weller H, Bäumer M (2011) Foam, fleece and honeycomb: catalytically active coatings from colloidally prepared nanoparticles. Cat Sci Technol 1:830–838
27. Sonström P, Arndt D, Wang X, Zielasek V, Bäumer M (2011) Ligand capping of colloidally synthesized nanoparticles – a way to tune metal–support interactions in heterogeneous gas-phase catalysis. Angew Chem Int Ed 50:3888–3891
28. Wang X, Sonström P, Arndt D, Stöver J, Zielasek V, Borchert H, Thiel K, Al-Shamery K, Bäumer M (2011) Heterogeneous catalysis with supported platinum colloids: a systematic study of the interplay between support and functional ligands. J Catal 278:143–152

29. Chen T, Xu Z (2018) Immobilization of small-molecule ligands containing secondary or tertiary amine groups onto TiO_2-supported Ru catalysts driven by the hydrophobic effect. ChemistrySelect 3:6421–6425
30. Jana NR, Peng X (2003) Single-phase and gram-scale routes toward nearly monodisperse Au and other noble metal nanocrystals. J Am Chem Soc 125:14280–14281
31. Altmann L, Wang X, Stçver J, Klink M, Zielasek V, Thiel K, Kolny-Olesiak J, Al-Shamery K, Borchert H, Parisi J, Bäumer M (2013) Impact of organic ligands on the structure and hydrogenation performance of colloidally prepared bimetallic PtSn nanoparticles. ChemCatChem 5:1803–1810
32. Costa NJS, Guerrero M, Collière V, Teixeira-Neto E, Landers R, Philippot K, Rossi LR (2014) Organometallic preparation of Ni, Pd, and NiPd nanoparticles for the design of supported nanocatalysts. ACS Catal 4:1735–1742
33. Zaramello L, Albuquerque BL, Domingos JB, Philippot K (2017) Kinetic investigation into the chemoselective hydrogenation of α,β-unsaturated carbonyl compounds catalyzed by Ni (0) nanoparticles. Dalton Trans 46:5082–5090
34. Zheng G, Kaefer K, Mourdikoudis S, Polavarapu L, Vaz B, Cartmell SE, Bouleghlimat A, Buurma NJ, Yate L, de Lera AR, Liz-Marzán LM, Pastoriza-Santos I, Pérez-Juste J (2015) Palladium nanoparticle-loaded cellulose paper: a highly efficient, robust, and recyclable self-assembled composite catalytic system. J Phys Chem Lett 6:230–238
35. Sato R, Kanehara M, Teranishi T (2011) Homoepitaxial size control and large-scale synthesis of highly monodisperse amine-protected palladium nanoparticles. Small 7:469–473
36. Altmann L, Wang X, Borchert H, Kolny-Olesiak J, Zielasek V, Parisi J, Kunz S, Bäumer M (2015) Influence of Sn content on the hydrogenation of crotonaldehyde catalysed by colloidally prepared PtSn nanoparticles. Phys Chem Chem Phys 17:28186–28192
37. Luo W, Cheng W, Hu M, Wang Q, Cheng X, Zhang Y, Wang Y, Gao D, Bi J, Fan G (2019) Ultrahigh catalytic activity of L-proline-functionalized Rh nanoparticles for methanolysis of ammonia borane. ChemSusChem 12:535–541
38. Schmid G, Harms M, Malm J-O, Bovin J-O, Ruitenbeck JV, Zandbergen HW, Fu WT (1993) Ligand-stabilized giant palladium clusters: promising candidates in heterogeneous catalysis. J Am Chem Soc 115:2046–2048
39. Schmid G, Emde S, Maihack V, Meyer-Zaika W, Peschel S (1996) Synthesis and catalytic properties of large ligand stabilized palladium clusters. J Mol Catal A Chem 107:95–104
40. Schmid G, Maihack V, Lantermann F, Peschel S (1996) Ligand-stabilized metal clusters and colloids: properties and applications. J Chem Soc Dalton Trans:589–595
41. Schmid G, West H, Mehles H, Lehnert A (1997) Hydrosilation reactions catalyzed by supported bimetallic colloids. Inorg Chem 36:891–895
42. Turkevich J, Stevenson PC, Hillier J (1951) A study of the nucleation and growth processes in the synthesis of colloidal gold. Discuss Faraday Soc 11:55–75
43. Turkevich J, Kim G (1970) Palladium: preparation and catalytic properties of particles of uniform size. Science 169:873–879
44. Schmid G, Lehnert A (1989) The complexation of gold colloids. Angew Chem Int Ed Engl 28: 780–781
45. Schmid G, Lehnerf A, Malm J-O, Bovin J-O (1991) Ligand-stabilized bimetallic colloids identified by HRTEM and EDX. Angew Chem Int Ed Engl 30:874–876
46. Schmid G, West H, Malm J-O, Bovin J-O, Grenthe C (1996) Catalytic properties of layered gold–palladium colloids. Chem A Eur J 2:1099–1103
47. Jahjah M, Kihn Y, Teuma E, Gómez M (2010) Ruthenium nanoparticles supported on multi-walled carbon nanotubes: highly effective catalytic system for hydrogenation processes. J Mol Catal A Chem 332:106–112
48. Favier I, Massou S, Teuma E, Philippot K, Chaudret B, Gómez M (2008) A new and specific mode of stabilization of metallic nanoparticles. Chem Commun:3296–3298

49. Rodríguez-Pérez L, Pradel C, Serp P, Gómez M, Teuma E (2011) Supported ionic liquid phase containing palladium nanoparticles on functionalized multiwalled carbon nanotubes: catalytic materials for sequential heck coupling/hydrogenation process. ChemCatChem 3:749–754
50. Romero N, Guerra RB, Gil L, Drouet S, Salmeron-Sánchez I, Illa O, Philippot K, Natali M, García-Antón J, Sala X (2020) TiO_2-mediated visible-light-driven hydrogen evolution by ligand-capped Ru nanoparticles. Sustain Energy Fuels 4:4170–4178
51. Creus J, Drouet S, Suriñach LP, Collière V, Poteau R, Philippot K, García-Antón J, Sala X (2018) Ligand-capped Ru nanoparticles as efficient electrocatalyst for the hydrogen evolution reaction. ACS Catal 8:11094–11102
52. Bönnemann H, Braun GA (1997) Enantioselectivity control with metal colloids as catalysts. Chem A Eur J 3:1200–1202
53. Wu J, Zhang J, Peng Z, Yang S, Wagner FT, Yang H (2010) Truncated octahedral Pt_3Ni oxygen reduction reaction electrocatalysts. J Am Chem Soc 132:4984–4985
54. Mullen TJ, Dameron AA, Saavedra HM, Williams ME, Weiss PS (2007) Dynamics of solution displacement in 1-adamantanethiolate self-assembled monolayers. J Phys Chem C 111:6740–6746
55. Maksimuk S, Teng X, Yang H (2007) Roles of twin defects in the formation of platinum multipod nanocrystals. J Phys Chem C 111:14312–14319
56. Wu J, Qi L, You H, Gross A, Li J, Yang H (2012) Icosahedral platinum alloy nanocrystals with enhanced electrocatalytic activities. J Am Chem Soc 134:11880–11883
57. Wu JB, Gross A, Yang H (2011) Shape and composition-controlled platinum alloy nanocrystals using carbon monoxide as reducing agent. Nano Lett 11:798–802
58. Kang Y, Ye X, Murray CB (2010) Size- and shape-selective synthesis of metal nanocrystals and nanowires using CO as a reducing agent. Angew Chem Int Ed 49:6156–6159
59. Wu BH, Zheng NF, Fu G (2011) Small molecules control the formation of Pt nanocrystals: a key role of carbon monoxide in the synthesis of Pt nanocubes. Chem Commun 47:1039–1041
60. Schauermann S, Hoffmann J, Johanek V, Hartmann J, Libuda J, Freund HJ (2002) Catalytic activity and poisoning of specific sites on supported metal nanoparticles. Angew Chem Int Ed 41:2532–2535
61. Eichler A (2002) CO oxidation on transition metal surfaces: reaction rates from first principles. Surf Sci 498:314–320
62. Somorjai GA, Li Y (2010) Introduction to surface chemistry and catalysis.2nd edn. Wiley, Hoboken
63. Gan L, Heggen M, Rudi S, Strasser P (2012) Core−shell compositional fine structures of dealloyed Pt_xNi_{1-x} nanoparticles and their impact on oxygen reduction catalysis. Nano Lett 12:5423–5430
64. Ahrenstorf K, Albrecht O, Heller H, Kornowski A, Görlitz D, Weller H (2007) Colloidal synthesis of Ni_xPt_{1-x} nanoparticles with tuneable composition and size. Small 3:271–274
65. Jürgens B, Borchert H, Ahrenstorf K, Sonström P, Pretorius A, Schowalter M, Gries K, Zielasek V, Rosenauer A, Weller H, Bäumer M (2008) Colloidally prepared nanoparticles for the synthesis of structurally well-defined and highly active heterogeneous catalysts. Angew Chem Int Ed 47:8946–8949
66. Yin Z, Zhang Y, Chen K, Li J, Li W, Tang P, Zhao H, Zhu Q, Bao X, Ma D (2014) Monodispersed bimetallic PdAg nanoparticles with twinned structures: formation and enhancement for the methanol oxidation. Sci Rep 4:4288
67. Yin Z, Ma D, Bao XH (2010) Emulsion-assisted synthesis of monodisperse binary metal nanoparticles. Chem Commun 46:1344–1346
68. Winnischofer H, Rocha TC, Nunes WC, Socolovsky LM, Knobel M, Zanchet D (2008) Chemical synthesis and structural characterization of highly disordered Ni colloidal nanoparticles. ACS Nano 2:1313–1319
69. Yang J, Deivaraj TC, Too HP, Lee JY (2004) Acetate stabilization of metal nanoparticles and its role in the preparation of metal nanoparticles in ethylene glycol. Langmuir 20:4241–4245

70. Li D, Komarneni S (2006) Microwave-assisted polyol process for synthesis of Ni nanoparticles. J Am Ceram Soc 89:1510–1517
71. Rinaldi R, de Melo PA, Rocha TCR, Cassinelli WHC, Ribeiro RU, Bueno JMC, Zanchet D (2009) Construction of heterogeneous Ni catalysts from supports and colloidal nanoparticles – a challenging puzzle. J Mol Catal A Chem 301:11–17
72. Yi D, Min Y, Muzzi B, Marty A, Romana I, Fazzini P-F, Blon T, Viau G, Serp P, Soulantica K (2022) Epsilon cobalt nanoparticles as highly performant catalysts in cinnamaldehyde selective hydrogenation. ACS Appl Nano Mater 5:5498–5507
73. Liakakos N, Cormary B, Li X, Lecante P, Respaud M, Maron L, Falqui A, Genovese A, Vendier L, Koïnis S, Chaudret B, Soulantica K (2012) The big impact of a small detail: cobalt nanocrystal polymorphism as a result of precursor addition rate during stock solution preparation. J Am Chem Soc 134:17922–17931
74. Cormary B, Li T, Liakakos N, Peres L, Fazzini P-F, Blon T, Respaud M, Kropf AJ, Chaudret B, Miller JT, Mader EA, Soulantica K (2016) Concerted growth and ordering of cobalt nanorod arrays as revealed by tandem in situ SAXS-XAS studies. J Am Chem Soc 138: 8422–8431
75. Metin Ö, Mazumder V, Özkar S, Sun S (2010) Monodisperse nickel nanoparticles and their catalysis in hydrolytic dehydrogenation of ammonia borane. J Am Chem Soc 132:1468–1469
76. Chung Y-H, Kim SJ, Chung DY, Park HY, Sung Y-E, Yoo SJ, Jang JH (2015) Third-body effects of native surfactants on Pt nanoparticle electrocatalysts in proton exchange fuel cells. Chem Commun 51:2968–2971
77. Peres L, Axet MR, Yi D, Serp P, Soulantica K (2020) Selective hydrogenation of cinnamaldehyde by unsupported and few layer graphene supported platinum concave nanocubes exposing {110} facets stabilized by a long-chain amine. Catal Today 357:166–175
78. Peres L, Yi D, Bustos-Rodriguez S, Marcelot C, Pierrot A, Fazzini PF, Florea I, Arenal R, Lacroix LM, Warot-Fonrose B, Blon T, Soulantica K (2018) Shape selection through epitaxy of supported platinum nanocrystals. Nanoscale 10:22730–22736
79. Buil ML, Esteruelas MA, Niembro S, Oliván M, Orzechowski L, Pelayo C, Vallribera A (2010) Organometallics 29:4375–4383
80. Qi S-C, Zhang L, Einaga H, Kudo S, Norinaga K, Hayashi J-i (2017) Nano-sized nickel catalyst for deep hydrogenation of lignin monomers and first-principles insight into the catalyst preparation. J Mater Chem A 5:3948–3965
81. Castillejos E, Debouttière P-J, Roiban L, Solhy A, Martinez V, Kihn Y, Ersen O, Philippot K, Chaudret B, Serp P (2009) An efficient strategy to drive nanoparticles into carbon nanotubes and the remarkable effect of confinement on their catalytic performance. Angew Chem Int Ed 48:2529–2533
82. Harmel J, Berliet A, Dembélé K, Marcelot C, Gay A-S, Ersen O, Maury S, Fécant A, Chaudret B, Serp P, Soulantica K (2018) A seed-mediated approach for the preparation of modified heterogeneous catalysts. ChemCatChem 10:1614–1619
83. Xia Y, Gilroy KD, Peng H-C, Xia X (2017) Seed-mediated growth of colloidal metal nanocrystals. Angew Chem Int Ed 56:60–95
84. Campisi S, Schiavoni M, Chan-Thaw CE, Villa A (2016) Untangling the role of the capping agent in nanocatalysis: recent advances and perspectives. Catalysts 6:185
85. Harmel J, Peres L, Estrader M, Berliet A, Maury S, Fécant A, Chaudret B, Serp P, Soulantica K (2018) *hcp*-Co nanowires grown on metallic foams as catalysts for Fischer–Tropsch synthesis. Angew Chem Int Ed 57:10579–10583
86. Cormary B, Dumestre F, Liakakos N, Soulantica K, Chaudret B (2013) Organometallic precursors of nano-objects, a critical view. Dalton Trans 42:12546–12553
87. Dumestre F, Chaudret B, Amiens C, Fromen M-C, Casanove M-J, Renaud P, Zurcher P (2002) Shape control of thermodynamically stable cobalt nanorods through organometallic chemistry. Angew Chem Int Ed 41:4286–4289

88. Boitiaux JP, Cosyns J, Vasudevan S (1985) Hydrogenation of highly unsaturated hydrocarbons over highly dispersed Pd catalyst. Part II: ligand effect of piperidine. Appl Catal 15:317–326
89. Boitiaux JP, Cosyns J, Vasudevan S (1983) Hydrogenation of highly unsaturated hydrocarbons over highly dispersed palladium catalyst. Part I: behaviour of small metal particles. Appl Catal 6:41–51
90. Boitiaux JP, Cosyns J, Vasudevan S (1982) Proceedings of the 3rd international symposium on the scientific bases for the preparation of heterogeneous catalysts. Elsevier, Belgium
91. Chung Y-H, Chung DY, Jung N, Sung Y-E (2013) Tailoring the electronic structure of nanoelectrocatalysts induced by a surface-capping organic molecule for the oxygen reduction reaction. J Phys Chem Lett 4:1304–1309
92. Wang Y, Ren J, Deng K, Gui L, Tang Y (2000) Preparation of tractable platinum, rhodium, and ruthenium nanoclusters with small particle size in organic media. Chem Mater 12:1622–1627
93. Munnik P, de Jongh PE, de Jong KP (2015) Recent developments in the synthesis of supported catalysts. Chem Rev 115:6687–6718
94. Liu J, Sun M, Lin Q, Liu S, Xu H, Chen Y (2019) Promotional effects of ethylenediamine on the low-temperature catalytic activity of selective catalytic oxidation of ammonia over Pt/$SiAlO_x$: states and particle sizes of Pt. Appl Surf Sci 481:1344–1351
95. Tungler A, Máthé T, Petró J (1985) In: Proceedings of the second Czechoslovakian conference on Het cat, Bechyne
96. Tungler A, Kajtár M, Máthé T, Petró J (1987) In: Proceedings of the fourth Belgian-Hungarian Conference on catalysis, Herbeaumont
97. Tungler A, Kajtár J, Máthé T, Tóth T, Fogassy E, Petró J (1989) Enantioselective hydrogenation of α-β-unsaturated ketones. Catal Today 5:159–171
98. Li S, Chen C, Zhan E, Liu S-B, Shen W (2009) Chirality inversion in enantioselective hydrogenation of isophorone over Pd/MgO catalysts in the presence of (*S*)-proline: effect of Pd particle size. J Mol Catal A Chem 304:88–94
99. Győrffy N, Tungler A, Fodor M (2010) Stereodifferentiation in heterogeneous catalytic hydrogenation. Kinetic resolution and asymmetric hydrogenation in the presence of (*S*)-proline: catalyst-dependent processes. J Catal 270:2–8
100. Künzle N, Hess R, Mallat T, Baiker A (1999) Continuous enantioselective hydrogenation of activated ketones. J Catal 186:239–241
101. Blaser HU, Jalett HP, Lottenbach W, Studer M (2000) Heterogeneous enantioselective hydrogenation of ethyl pyruvate catalyzed by cinchona-modified Pt catalysts: effect of modifier structure. J Am Chem Soc 122:12675–12682
102. Panella B, Vargas A, Baiker A (2009) Magnetically separable Pt catalyst for asymmetric hydrogenation. J Catal 261:88–93
103. Sugimura T, Ogawa H (2010) Acceleration of enantioselective hydrogenation of olefins over Pd/C by cinchonidine as a chiral codifier. Comparison with cinchonine, pseudoenantiomer. Chem Lett 39:232–233
104. Diezi S, Ferri D, Vargas A, Mallat T, Baiker A (2006) The origin of chemo- and enantioselectivity in the hydrogenation of diketones on platinum. J Am Chem Soc 128:4048–4057
105. García-Suárez EJ, Tristany M, García AB, Collière V, Philippot K (2012) Carbon-supported Ru and Pd nanoparticles: efficient and recyclable catalysts for the aerobic oxidation of benzyl alcohol in water. Microporous Mesoporous Mater 153:155–162
106. Favier I, Massou S, Teuma E, Philippot K, Chaudret B, Gómez M (2008) A new and specific mode of stabilization of metallic nanoparticlesw. Chem Commun 28:3296–3298
107. Ramirez E, Jansat S, Philippot K, Lecante P, Gómez M, Masdeu A, Chaudret B (2004) Influence of organic ligands on the stabilization of palladium nanoparticles. J Organomet Chem 689:4601–4610

108. García-Suárez EJ, Balu AM, Tristany M, García AB, Philippot K, Luque R (2012) Versatile dual hydrogenation–oxidation nanocatalysts for the aqueous transformation of biomass-derived platform molecules. Green Chem 14:1434–1439
109. Jeong H, Kim C, Yang S, Lee H (2016) Selective hydrogenation of furanic aldehydes using Ni nanoparticle catalysts capped with organic molecules. J Catal 344:609–615
110. Carenco S, Boissiere C, Nicole L, Sanchez C, Le Floch P, Mezailles N (2010) Controlled design of size-tunable monodisperse nickel nanoparticles. Chem Mater 22:1340–1349
111. Castillejos E, Jahjah M, Favier I, Orejón A, Pradel C, Teuma E, Masdeu-Bultó AM, Serp P, Gómez M (2012) Synthesis of platinum–ruthenium nanoparticles under supercritical CO_2 and their confinement in carbon nanotubes: hydrogenation applications. ChemCatChem 4:118–122
112. Ruiz D, Mella C, Fierro JLG, Reyes P (2012) Silica supported rhodium metal nanoparticles stabilized with (−)-DIOP. Effect of ligand concentration and metal loading on the enantioselective hydrogentation of ketones. J Chil Chem Soc 57:1394–1399
113. Mella C, Ávila M, Sánchez A, Marzialetti T, Reyes P, Ruiz D (2013) Chiral Rh/SiO_2 catalysts for enantioselective hydrogenation reactions. The role of (*S*,*S*)-DIPAMP as chiral modifier and stabilizer on metallic nanoparticles synthesis. J Chil Chem Soc 58:2125–2130
114. Ruiz D, Mäki-Arvela P, Aho A, Chimentão R, Claver C, Godard C, Fierro JLG, Murzin DY (2019) Immobilized chiral rhodium nanoparticles stabilized by chiral P-ligands as efficient catalysts for the enantioselective hydrogenation of 1-phenyl-1,2-propanedione. Mol Catal 477:110551
115. Ibrahim M, Garcia MAS, Vono LLR, Guerrero M, Lecante P, Rossi LM, Philippot K (2016) Polymer versus phosphine stabilized Rh nanoparticles as components of supported catalysts: implication in the hydrogenation of cyclohexane model molecule. Dalton Trans 45:17782–17791
116. Han D, Li X, Zhang H, Liu Z, Hu G, Li C (2008) Asymmetric hydroformylation of olefins catalyzed by rhodium nanoparticles chirally stabilized with (*R*)-BINAP ligand. J Mol Catal A Chem 283:15–22
117. Jiang H-y, Yang C-f, Li C, Fu H-y, Chen H, Li R-x, Li X-j (2008) Heterogeneous enantioselective hydrogenation of aromatic ketones catalyzed by cinchona- and phosphine-modified iridium catalysts. Angew Chem Int Ed 47:9240–9244
118. Jiang H-y, Sun B, Zheng X-x, Chen H (2012) Heterogeneous selective hydrogenation of *trans*-4-phenyl-3-butene-2-one to allylic alcohol over modified Ir/SiO_2 catalyst. App Catal A Gen 421–422:86–90
119. Castelbou JL, Szeto KC, Barakat W, Merle N, Godard C, Taoufik M, Claver C (2017) Chem Commun 53:3261–3264
120. Sánchez-Resa D, Delgado JA, Fernández-Martínez MD, Didelot C, De Mallmann A, Szeto KC, Taoufik M, Claver C, Godard C (2021) Pd, Cu and bimetallic PdCu NPs supported on CNTs and phosphine-functionalized silica: one-pot preparation, characterization and testing in the semi-hydrogenation of alkynes. Eur J Inorg Chem:4970–4978
121. Montiel L, Delgado JA, Novell M, Andrade FJ, Claver C, Blondeau P, Godard C (2016) A simple and versatile approach for the fabrication of paper-based nanocatalysts: low cost, easy handling, and catalyst recovery. ChemCatChem:3041–3044
122. Fernández-Martínez MD, Godard C (2023) Hydrogenation of CO_2 into Formates by ligand-capped palladium heterogeneous catalysts. ChemCatChem:e202201408
123. Tang B, Xiong W, Liu D-R, Jia Y, Wang J-B, Chen H, Li X-J (2008) Asymmetric hydrogenation of aromatic ketones catalyzed by (1*S*,2*S*)-DPEN-modified $Ru\text{-}PPh_3/\gamma\text{-}Al_2O_3$ catalyst. Tetrahedron Asymmetry 19:1397–1401
124. Coronado JM, Coloma F, Anderson JA (2000) Styrene hydroformylation over modified $Rh/SiO_2{\cdot}Al_2O_3$ catalysts. J Mol Catal A Chem 154:143–154
125. Yan L, Ding YJ, Zhu HJ, Xiong JM, Wang T, Pan ZD, Lin LW (2005) Ligand modified real heterogeneous catalysts for fixed-bed hydroformylation of propylene. J Mol Catal A Chem 234:1–7

126. Li X, Ding Y, Jiao G, Li J, Yan L, Zhu H (2008) Hydroformylation of methyl-3-pentenoate over a phosphite ligand modified Rh/SiO_2 catalyst. J Nat Gas Chem 17:351–354
127. Mori K, Kondo Y, Yamashita H (2009) Synthesis and characterization of FePd magnetic nanoparticles modified with chiral BINAP ligand as a recoverable catalyst vehicle for the asymmetric coupling reaction. Phys Chem Chem Phys 11:8949–8954
128. Felpin F-X, Landais Y (2005) Practical Pd/C-mediated allylic substitution in water. J Org Chem 70:6441–6446
129. Reimann S, Mallat T, Baiker A (2007) Enantioselective allylic substitution on Pd/Al_2O_3 modified by chiral diphosphines. J Catal 252:30–38
130. Reimann S, Mallat T, Baiker A (2008) A new, efficient heterogeneous Pd catalyst for enantioselective allylic substitution. J Catal 254:79–83
131. Reimann S, Grunwaldt J-D, Mallat T, Baiker A (2010) Asymmetric C-C bond-formation reaction with Pd: how to favor heterogeneous or homogeneous catalysis? Chem A Eur J 16: 9658–9668
132. Park KH, Chung YK (2005) Immobilized Co/Rh heterobimetallic nanoparticle-catalyzed Pauson–Khand-type reaction. Adv Synth Catal 347:854–866
133. Cheng H, Hao J, Wang H, Xi C, Meng X, Cai S, Zhao F (2007) (*R*,*R*)-DPEN-modified $Ru/\gamma\text{-}Al_2O_3$ – an efficient heterogeneous catalyst for enantioselective hydrogenation of acetophenone. J Mol Catal A Chem 278:6–11
134. Wang H, Huang Y, Wang X, Cui X, Shi F (2020) Supported Ni nanoparticles with a phosphine ligand as an efficient heterogeneous non-noble metal catalytic system for regioselective hydrosilylation of alkynes. Org Biomol Chem 18:7554–7558
135. McKenna F-M, Wells RPK, Anderson JA (2011) Enhanced selectivity in acetylene hydrogenation by ligand modified Pd/TiO_2 catalysts. Chem Commun 47:2351–2353
136. McCue AJ, McKenna F-M, Anderson JA (2015) Triphenylphosphine: a ligand for heterogeneous catalysis too? Selectivity enhancement in acetylene hydrogenation over modified Pd/TiO_2 catalyst. Cat Sci Technol 5:2449–2459
137. Hu G, Feng Z, Han D, Li J, Jia G, Shi J, Li C (2007) Charge transfer between triphenyl phosphine and colloidal silver: a SERS study combined with DFT calculations. J Phys Chem C 111:8632–8637
138. Sizova OV, Varshavskii YS, Skripnikov LV (2007) Quantum-chemical study of donor-acceptor interactions in rhodium(I) carbonyl carboxylate complexes with phosphine ligands. Russ J Coord Chem 33:313–322
139. Strozier RW, Caramella P, Houk KN (1979) Influence of molecular distortions upon reactivity and stereochemistry in nucleophilic additions to acetylenes. J Am Chem Soc 101:1340–1343
140. Ballesteros-Soberanas J, Leyva-Pérez A (2023) Electron-poor phosphines enable the selective semihydrogenation reaction of alkynes with Pd on carbon catalysts. J Phys Chem Lett 14:965–970
141. Han D, Li X, Zhang H, Liu Z, Li J, Li C (2006) Heterogeneous asymmetric hydroformylation of olefins on chirally modified Rh/SiO_2 catalysts. J Catal 243:318–328
142. Korzec M, Bartczak P, Niemczyk A, Szade J, Kapkowski M, Zenderowska P, Balin K, Lelątko J, Polanski J (2014) Bimetallic nano-Pd/PdO/Cu system as a highly effective catalyst for the Sonogashira reaction. J Catal 313:1–8
143. Tukacs JM, Jones RV, Darvas F, Dibó G, Lezsák G, Mika LT (2013) Synthesis of γ-valerolactone using a continuous-flow reactor. RSC Adv 3:16283–16287
144. Chatterjee A, Jensen VR (2017) A heterogeneous catalyst for the transformation of fatty acids to α-olefins. ACS Catal 7:2543–2547
145. Cui H-Z, Zhang Z-M, Zhang H, Reheman A, Hong X, Zhan B, Zhang J, Hou X-F (2021) Mesocellular silica foams supported size-controlled Pd nanoparticles for racemic and asymmetric iodomethyl dihydrobenzofuran synthesis. Microporous Mesoporous Mater 322:111157
146. Guo M, Li H, Ren Y, Ren X, Yang Q, Li C (2018) Improving catalytic hydrogenation performance of Pd nanoparticles by electronic modulation using phosphine ligands. ACS Catal 8:6476–6485

147. Lu YD, Wang YH, Jin ZL (2010) Rh nanoparticles stabilized by PEG-substituted triphenylphosphine: a highly active and recyclable catalyst for aqueous biphasic hydrogenation of benzene. Chin Chem Lett 21:1067–1070
148. Jayakumar S, Modak A, Guo M, Li H, Hu X, Yang Q (2017) Ultrasmall platinum stabilized on triphenylphosphine-modified silica for chemoselective hydrogenation. Chem A Eur J 23: 7791–7797
149. Ren X, Guo M, Li H, Li C, Yu L, Liu J, Yang Q (2019) Microenvironment engineering of ruthenium nanoparticles incorporated into silica nanoreactors for enhanced hydrogenations. Angew Chem Int Ed 58:14483–14488
150. Lei Y, Chen Z, Lan G, Wang R, Zhou X-Y (2020) Pd nanoparticles stabilized with phosphine functionalized porous ionic polymer for efficient catalytic hydrogenation of nitroarenes in water. New J Chem 44:3681–3689
151. Wang H, Fiore AM, Fliedel C, Manoury E, Philippot K, Dell'Anna MM, Mastrorilli P, Poli R (2021) Rhodium nanoparticles inside well-defined unimolecular amphiphilic polymeric nanoreactors: synthesis and biphasic hydrogenation catalysis. Nanoscale Adv 3:2554–2566
152. Doherty S, Knight JG, Backhouse T, Tran TST, Paterson R, Stahl F, Alharbi HY, Chamberlain TW, Bourne RA, Stones R, Griffiths A, White JP, Aslam Z, Hardare C, Daly H, Hart J, Temperton RH, O'Shead JN, Rees NH (2022) Highly efficient and selective aqueous phase hydrogenation of aryl ketones, aldehydes, furfural and levulinic acid and its ethyl ester catalyzed by phosphine oxide-decorated polymer immobilized ionic liquid-stabilized ruthenium nanoparticles. Cat Sci Technol 12:3549–3567
153. Costa NJS, Kiyohara PK, Monteiro AL, Coppel Y, Philippot K, Rossi LM (2010) A single-step procedure for the preparation of palladium nanoparticles and a phosphine-functionalized support as catalyst for Suzuki cross-coupling reactions. J Catal 276:382–389
154. Rossi LM, Nangoi IM, Costa NJS (2009) Ligand-assisted preparation of palladium supported nanoparticles: a step toward size control. Inorg Chem 48:4640
155. Cai X, Nie J, Yang G, Wang F, Ma C, Lu C, Chen Z (2019) Phosphorus-rich network polymer supported ruthenium nanoparticles for nitroarene reduction. Mater Lett 240:80–83
156. Tegeder P, Freitag M, Chepiga KM, Muratsugu S, Möller N, Lamping S, Tada M, Glorius F, Ravoo BJ (2018) N-heterocyclic carbene-modified Au–Pd alloy nanoparticles and their application as biomimetic and heterogeneous catalysts. Chem A Eur J 24:18682–18688
157. Ranganath KVS, Kloesges J, Schäfer AH, Glorius F (2010) Asymmetric nanocatalysis: N-heterocyclic carbenes as chiral modifiers of Fe_3O_4/Pd nanoparticles. Angew Chem Int Ed 49:7786–7789
158. Richter C, Schaepe K, Glorius F, Ravoo BJ (2014) Tailor-made N-heterocyclic carbenes for nanoparticle stabilization. Chem Commun 50:3204–3207
159. Ferry A, Schaepe K, Tegeder P, Richter C, Chepiga KM, Ravoo BJ, Glorius F (2015) Negatively charged N-heterocyclic Carbene-stabilized Pd and Au nanoparticles and efficient catalysis in water. ACS Catal 5:5414–5420
160. Tegeder P, Marelli M, Freitag M, Polito L, Lamping S, Psaro R, Glorius F, Ravoo BJ, Evangelisti C (2018) Metal vapor synthesis of ultrasmall Pd nanoparticles functionalized with N-heterocyclic carbenes. Dalton Trans 47:12647–11265
161. Díaz de los Bernardos M, Pérez-Rodríguez S, Gual A, Claver C, Godard C (2017) Facile synthesis of NHC-stabilized Ni nanoparticles and their catalytic application in the Z-selective hydrogenation of alkynes. Chem Commun 53:7894–7897
162. Lomelí-Rosales DA, Delgado JA, Díaz de los Bernardos M, Pérez-Rodríguez S, Gual A, Claver C, Godard C (2019) A general one-pot methodology for the preparation of mono- and bimetallic nanoparticles supported on carbon nanotubes: application in the semi-hydrogenation of alkynes and acetylene. Chem A Eur J 25:8321–8331
163. Mollar-Cuni A, Ventura-Espinosa D, Martín S, Mayoral Á, Borja P, Mata JA (2018) Stabilization of nanoparticles produced by hydrogenation of palladium−N-heterocyclic carbene complexes on the surface of graphene and implications in catalysis. ACS Omega 3:15217–15228

164. Mollar-Cuni A, Martín S, Guisado-Barrios G, Mata JA (2023) Dual role of graphene as support of ligand-stabilized palladium nanoparticles and carbocatalyst for (de)hydrogenation of N-heterocycles. Carbon 206:314–324
165. Oberhauser W, Evangelisti C, Liscio A, Kovtun A, Cao Y, Vizza F (2018) Glycerol to lactic acid conversion by NHC-stabilized iridium nanoparticles. J Catal 368:298–305
166. Azad M, Rostamizadeh S, Nouri F, Estiri H, Fadakar Y (2019) Pd nanoparticles at N-heterocyclic Carbene at ZIF-8 as an ultrafine, robust and sustainable heterogeneous system for Suzuki-Miyaura cross coupling processes. Mater Lett 236:757–760
167. Zhang L, Wei Z, Thanneeru S, Meng M, Kruzyk M, Ung G, Liu B, He J (2019) Polymer solution to prevent nanoclustering and improve the selectivity of metal nanoparticles for electrocatalytic CO_2 reduction. Angew Chem Int Ed 58:15834–15840
168. Palazzolo A, Naret T, Daniel-Bertrand M, Buisson D-A, Tricard S, Lesot P, Coppel Y, Chaudret B, Feuillastre S, Pieters G (2020) Tuning the reactivity of a heterogeneous catalyst using N-heterocyclic Carbene ligands for C-H activation reactions. Angew Chem Int Ed 59: 20879–20884
169. García-Zaragoza A, Cerezo-Navarrete C, Mollar-Cuni A, Oña-Burgos P, Mata JA, Corma A, Martínez-Prieto LM (2022) Tailoring graphene-supported Ru nanoparticles by functionalization with pyrene-tagged N-heterocyclic carbenes. Cat Sci Technol 12:1257–1270
170. Martínez-Prieto LM, Ferry A, Lara P, Richter C, Philippot K, Glorius F, Chaudret B (2015) New route to stabilize ruthenium nanoparticles with non-isolable chiral N-heterocyclic carbenes. Chem A Eur J 21:17495–17502
171. Ernst JB, Muratsugu S, Wang F, Tada M, Glorius F (2016) Tunable heterogeneous catalysis: N-heterocyclic carbenes as ligands for supported heterogeneous Ru/K-Al_2O_3 catalysts to tune reactivity and selectivity. J Am Chem Soc 138:10718–10721
172. Muratsugu S, Kityakarn S, Wang F, Ishiguro N, Kamachi T, Yoshizawa K, Sekizawa O, Urugagh T, Tada M (2015) Formation and nitrile hydrogenation performance of Ru nanoparticles on a K-doped Al_2O_3 surface. Phys Chem Chem Phys 17:24791–24802
173. Liu J, Peng X, Sun W, Zhao Y, Xia C (2008) Magnetically separable Pd catalyst for carbonylative Sonogashira coupling reactions for the synthesis of α,β-alkynyl ketones. Org Lett 10:3933–3936
174. Ikemoto S, Muratsugu S, Koitaya T, Tsuji Y, Das M, Yoshizawa K, Glorius F, Tada M (2023) Coordination-induced trigger for activity: N-heterocyclic carbene-decorated ceria catalysts incorporating Cr and Rh with activity induction by surface adsorption site control. J Am Chem Soc 145:1497–1504
175. Ikemoto S, Huang X, Muratsugu S, Nagase S, Koitaya T, Matsui H, Yokota G, Sudoh T, Hashimoto A, Tan Y, Yamamoto S, Tang J, Matsuda I, Yoshinobu J, Yokoyama T, Kusaka S, Matsuda R, Tada M (2019) Reversible low-temperature redox activity and selective oxidation catalysis derived from the concerted activation of multiple metal species on Cr and Rh-incorporated ceria catalysts. Phys Chem Chem Phys 21:20868–20877
176. Ikemoto S, Muratsugu S, Koitaya T, Tada M (2022) Chromium oxides as structural modulators of rhodium dispersion on ceria to generate active sites for NO reduction. ACS Catal 12: 431–441
177. Ernst JB, Schwermann C, Yokota G-i, Tada M, Muratsugu S, Doltsinis NL, Glorius F (2017) Molecular adsorbates switch on heterogeneous catalysis: induction of reactivity by N-heterocyclic carbenes. J Am Chem Soc 139:9144–9147
178. Lewis RJ, Koy M, Macino M, Das M, Carter JH, Morgan DJ, Davies TE, Ernst JB, Freakley SJ, Glorius F, Hutchings GJ (2022) N-heterocyclic Carbene modified palladium catalysts for the direct synthesis of hydrogen peroxide. J Am Chem Soc 144:15431–15436
179. Sankar M, He Q, Morad M, Pritchard J, Freakley SJ, Edwards JK, Taylor SH, Morgan DJ, Carley AF, Knight DW, Kiely CJ, Hutchings GJ (2012) Synthesis of stable ligand-free gold-palladium nanoparticles using a simple excess anion method. ACS Nano 6:6600–6613
180. Levratovsky Y, Gross E (2016) High spatial resolution mapping of chemically-active self-assembled N-heterocyclic carbenes on Pt nanoparticles. Faraday Discuss 188:345–353

181. Cao Z, Derrick JS, Xu J, Gao R, Gong M, Nichols EM, Smith PT, Liu X, Wen X, Copéret C, Chang CJ (2018) Chelating N-heterocyclic carbene ligands enable tuning of electrocatalytic CO_2 reduction to formate and carbon monoxide: surface organometallic chemistry. Angew Chem Int Ed 57:4981–4985
182. Hou Y, Ji X, Liu G, Tang J, Zheng J, Liu Y, Zhang W, Jia M (2009) Immobilization of palladium in N-heterocyclic carbene functionalized SBA-15 for the catalytic application in aerobic oxidation of benzyl alcohol. Catal Commun 10:1459–1462
183. Ranganath KVS, Schäfer AH, Glorius F (2011) Comparison of superparamagnetic Fe_3O_4-supported N-heterocyclic carbene-based catalysts for enantioselective allylation. ChemCatChem 3:1889–1891
184. Mohapatra S, Chaiprasert T, Sodkhomkhum R, Kunthom R, Hanprasit S, Sangtrirutnugul P, Ervithayasuporn V (2016) Solid–state synthesis of polyhedral oligomeric silsesquioxane-supported N-heterocyclic carbenes/imidazolium salts on palladium nanoparticles: highly active and recyclable catalyst. ChemistrySelect 1:5353–5357
185. Soulé J-F, Miyamura H, Kobayashi S (2013) Copolymer-incarcerated nickel nanoparticles with N-heterocyclic Carbene precursors as active cross-linking agents for Corriu–Kumada–Tamao reaction. J Am Chem Soc 135:10602–10605
186. Ganesan M, Freemantle RG, Obare SO (2007) Monodisperse thioether-stabilized palladium nanoparticles: synthesis, characterization, and reactivity. Chem Mater 19:3464–3471
187. Solomonsz WA, Rance GA, Suyetin M, La Torre A, Bichoutskaia E, Khlobystov AN (2012) Controlling the regioselectivity of the hydrosilylation reaction in carbon nanoreactors. Chem A Eur J 18:13180–13187
188. Rossi LM, Vono LLR, Garcia MAS, Faria TLT, Lopez-Sanchez JA (2013) Screening of soluble rhodium nanoparticles as precursor for highly active hydrogenation catalysts: the effect of the stabilizing agents. Top Catal 56:1228–1238
189. Altmann L, Kunz S, Bäumer M (2014) Influence of organic amino and thiol ligands on the geometric and electronic surface properties of colloidally prepared platinum nanoparticles. J Phys Chem C 118:8925–8932
190. Yan M, Wu T, Chen L, Yu Y, Liu B, Wang Y, Chen W, Liu Y, Lian C, Li Y (2018) Effect of protective agents upon the catalytic property of platinum nanocrystals. ChemCatChem 10: 2433–2441
191. Liu B, Yao H, Song W, Jin MIM, Rusling JF, Suib SL, He J (2016) Ligand-free noble metal nanocluster catalysts on carbon supports via "soft" nitriding. J Am Chem Soc 138:4718–4721
192. McKenna F-M, Anderson JA (2011) Selectivity enhancement in acetylene hydrogenation over diphenyl sulphide-modified Pd/TiO_2 catalysts. J Catal 281:231–240
193. McKenna FM, Mantarosie L, Wells RPK, Hardacre C, Anderson JA (2012) Selective hydrogenation of acetylene in ethylene rich feed streams at high pressure over ligand modified Pd/TiO_2. Cat Sci Technol 2:632–638
194. Mori A, Mizusaki T, Kawase M, Maegawa T, Monguchi Y, Takao S, Takagi Y, Sajiki H (2008) Novel palladium-on-carbon/diphenyl sulfide complex for chemoselective hydrogenation: preparation, characterization, and application. Adv Synth Catal 350:406–410
195. Makosch M, Lin W-I, Bumbálek V, Sá J, Medlin JW, Hungerbühler K, van Bokhoven JA (2012) Organic thiol modified Pt/TiO_2 catalysts to control chemoselective hydrogenation of substituted nitroarenes. ACS Catal 2:2079–2081
196. Kahsar KR, Schwartz DK, Medlin JW (2012) Liquid- and vapor-phase hydrogenation of 1-epoxy-3-butene using self-assembled monolayer coated palladium and platinum catalysts. Appl Catal A Gen 445–446:102–106
197. Schoenbaum CA, Schwartz DK, Medlin JK (2013) Controlling surface crowding on a Pd catalyst with thiolate self-assembled monolayers. J Catal 303:92–99
198. Kahsar KR, Schwartz DK, Medlin JW (2013) Selective hydrogenation of polyunsaturated fatty acids using alkanethiol self-assembled monolayer-cated Pd/Al_2O_3 catalysts. ACS Catal 3: 2041–2044

199. Pang SH, Schoenbaum CA, Schwartz DK, Medlin JW (2014) Effects of thiol modifiers on the kinetics of furfural hydrogenation over Pd catalysts. ACS Catal 4:3123–3131
200. Kahsar KR, Schwartz DK, Medlin JW (2013) Control of metal catalyst selectivity through specific noncovalent molecular interactions. J Am Chem Soc 136:520–526
201. Kahsar KR, Schwartz DK, Medlin JW (2015) Stability of self-assembled monolayer coated Pt/Al_2O_3 catalysts for liquid phase hydrogenation. J Mol Catal A Chem 396:188–195
202. Lien C-H, Medlin JW (2016) Control of Pd catalyst selectivity with mixed thiolate monolayers. J Catal 339:38–46
203. Blanchette Z, Schwartz DK, Medlin JW (2023) Directing reaction pathways on supported metal catalysts with low-density self-assembled monolayers. ACS Appl Nano Mater 6:9059–9069
204. Al Khulaifi FM, Alsunni YA, Musgrave CB, Holewinski A, Medlin JW (2023) Impact of pretreatment and thiol modifiers on the partial oxidation of glutaraldehyde using Pd/Al_2O_3. App Catal A Gen 661:119229
205. Watson DJ, Jesudason RJBJ, Beaumont SK, Kyriakou G, Burton JW, Lambert RM (2009) Heterogeneously catalyzed asymmetric hydrogenation of C=C bonds directed by surface-tethered chiral modifiers. J Am Chem Soc 131:14584–14589
206. Weng Z, Zaera F (2014) Increase in activity and selectivity in catalysis via surface modification with self-assembled monolayers. J Phys Chem C 118:3672–3679
207. Azmat MU, Guo Y, Guo Y, Wang Y, Lu G (2011) An easy and effective approach towards heterogeneous Pt/SiO2-cinchonidine catalyst system for enantioselective hydrogenation of ethyl pyruvate. J Mol Catal A Chem 336:42–50
208. Huang Y, Xu S, Lin VSY (2011) New strategy for enantioselective heterogeneous catalysis: immobilization of both metal nanoparticles and chiral modifiers on mesoporous silica nanoparticles. ChemCatChem 3:690–694

SPRINGER NATURE

GPSR Compliance

The European Union's (EU) General Product Safety Regulation (GPSR) is a set of rules that requires consumer products to be safe and our obligations to ensure this.

If you have any concerns about our products, you can contact us on ProductSafety@springernature.com

In case Publisher is established outside the EU, the EU authorized representative is:

Springer Nature Customer Service Center GmbH
Europaplatz 3
69115 Heidelberg, Germany

The manufacturer's authorised representative in the EU is Springer Nature Customer Service Centre GmbH, Europaplatz 3, 69115 Heidelberg, Germany. If you have any concerns regarding our products, please contact ProductSafety@springernature.com

Printed and bound by CPI Group (UK) Ltd, Croydon, CR0 4YY
01/10/2026
02225898-0019